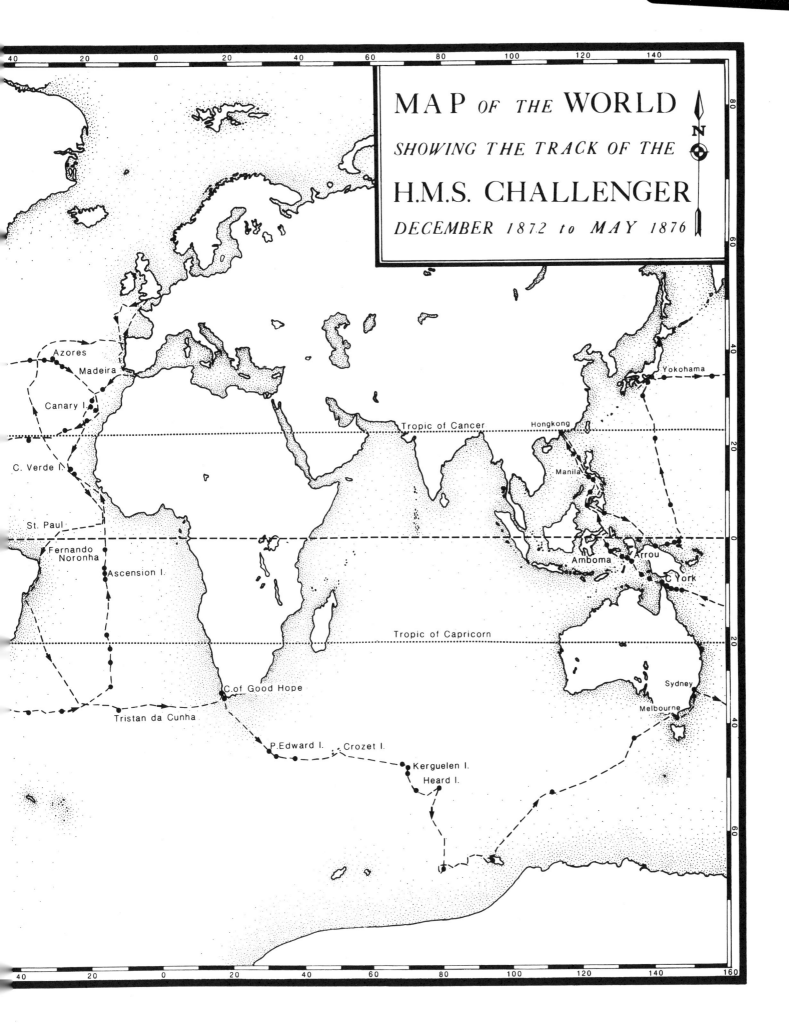

MAP OF THE WORLD

SHOWING THE TRACK OF THE

H.M.S. CHALLENGER

DECEMBER 1872 to MAY 1876

N

Azores
Madeira
Canary I.
C. Verde I.
St. Paul
Fernando Noronha
Ascension I.
Tropic of Cancer
Hongkong
Manila
Amboma
Arrou
C. York
Yokohama
Tropic of Capricorn
C. of Good Hope
Tristan da Cunha
Sydney
Melbourne
P. Edward I.
Crozet I.
Kerguelen I.
Heard I.

The *Challenger* Foraminifera

This report, lengthy as it appears, has no pretence to completeness, indeed a lifetime might be spent upon the collections of which it treats without exhausting their points of interest. But in a publication of this sort there are certain limits as to time which must be respected; and the present volume is offered as an instalment in which the details of the subject have been pursued as far as has been practicable . . .

The method of treatment has been made as comprehensive as possible, in the hope that the results may be of service as a starting point for further research.

H. B. Brady, *Challenger* Report (1884)

The Natural History Museum, London

The
Challenger Foraminifera

ROBERT WYNN JONES

Oxford   New York   Tokyo
OXFORD UNIVERSITY PRESS
1994

Oxford University Press, Walton Street, Oxford OX2 6DP

Oxford New York Toronto
Delhi Bombay Calcutta Madras Karachi
Kuala Lumpur Singapore Hong Kong Tokyo
Nairobi Dar es Salaam Cape Town
Melbourne Auckland Madrid
and associated companies in
Berlin Ibadan

Oxford is a trade mark of Oxford University Press

Published in the United States
by Oxford University Press Inc., New York

A catalogue record for this book is available from the British Library

Library of Congress Cataloging in Publication Data
Jones, Robert Wynn.
The Challenger foraminifera/Robert Wynn Jones.
At head of title: The Natural History Museum.
Includes bibliographical references and index.
1. Foraminifera.   2. Foraminifera, Fossil.   3. Challenger
Expedition, 1872–1876.   I. Brady, Henry Bowman, 1835–1891. *Report*
on the Foraminifera dredged by H.M.S. Challenger during the years
1873–1876.   II. British Museum (Natural History).   III. Title.
QL368.F6J64 1994   593.1'2–dc20   94–6288
ISBN 0 19 854096 5

Typeset by Cotswold Typesetting Ltd, Gloucester

Printed in Hong Kong

Preface

The 'Report on the Foraminifera dredged by HMS *Challenger* . . .' (hereafter referred to as the '*Challenger* Report') by Henry Bowman Brady (1884) dates from what might best be termed the systematic or descriptive phase of deep-sea research. It was conceived in the pervasive intellectual atmosphere of the 'English School' and in consequence embodies a broader ('lumping') species concept than is acceptable today (Cifelli and Richardson 1990; Haynes 1990). Because of its outdated taxonomy, it cannot be used without qualification in the type of analytical or synoptic work currently being undertaken.

Previous revisions to the taxonomy of the *Challenger* foraminifera

Previous revisions to the taxonomy of the *Challenger* foraminifera include those of H. E. Thalmann (1932, 1933, 1937, 1942) and R. Wright Barker (1960).

Barker's work was particularly important and useful in that it synthesized all of the great variety of names assigned to Brady's figured species in the literature up to 1960 (including those of Thalmann, op. cit.). However, Barker did not subjectively assess which names were valid and which not, referring simply to the '. . . return in . . . recent times to the principles of Reuss . . .', the '. . . splitting of a large number of Brady's "species" . . .' and '. . . their allocation to new genera . . .'. He proposed new names for four species whose earlier names were preoccupied, but did not formally describe the thirty-one species and three varieties he regarded as new, stating that '. . . a number of the forms figured by Brady, and as yet undescribed, have been indicated but have not been named, for the writer has not had the opportunity to study Brady's specimens in order to prepare adequate descriptions and to designate types . . .'. Moreover, in many cases he left problems of generic assignment unresolved, commenting in the case of the nodosariids, for instance, that '. . . there is still confusion in the treatment of this group, . . . and no attempt has been made to clarify the question here, the generic names in all cases being quoted unchanged . . .'. His work was essentially, as he himself stated, '. . . of a bibliographic nature and in no sense a critical revision of the '*Challenger* Report'.

Present revision

Since Barker's time, a number of additional references have appeared which have a significant bearing on the taxonomy of the *Challenger* foraminifera. Some of the more general of these (to the end of 1993) are those of Leroy (1964), Shchedrina (1964, 1969), Belford (1966), Saidova (1970, 1975), Haynes (1973*a*), Hofker (1976), McCulloch (1977, 1981), Lewis (1979), Whittaker and Hodgkinson (1979), Zheng (1979, 1988), Wang, Zhang and Min (in Wang *et al.* 1985), van Morkhoven *et al.* (1986), Hermelin (1989), and van Marle (1991) (monographs), Srinivasan and Sharma (1969, 1980), Hornibrook (1971), Rögl and Hansen (1984), Papp and Schmid (1985) (revisions to existing monographs), and Loeblich and Tappan (1964, 1974, 1981, 1987), Haynes (1981), and Saidova (1981) (suprageneric classification schemes).

Some more specific references dealing with particularly problematic groups are those of Loeblich and Tappan (1961), Macfadyen (1962), Banner (1966), Shchedrina (1969), Hofker (1972), Tendal and Hessler (1977), Brönnimann and Whittaker (1980*a,b*, 1983, 1984*a,b*,

1987, 1988), Banner and Pereira (1981), Gooday and Nott (1982), Brönnimann *et al.* (1983), Gooday (1983, 1986, 1988), Loeblich and Tappan (1984*a,b*, 1985*a,b*, 1986), Schröder (1986), Bender and Hemleben (1988), Cartwright *et al.* (1989), Charnock and Jones (1990), Hottinger *et al.* (1990), and Jones *et al.* (in press) (Astrorhizida and Lituolida), Bogdanovich (1969), Łuczkowska (1971, 1972, 1974), Ponder (1972, 1974*a,b*), Cherif and Flick (1974), Seiglie *et al.* (1977), Poignant (1981), and Haig (1988) (Miliolida), Schnitker (1970), Moncharmont-Zei and Sgarrella (1977, 1978, 1980), Popescu (1983), Jones (1984*a,b*), Poignant (1984), Knight (1986), Patterson (1986), Patterson and Pettis (1986), and Patterson and Richardson (1987, 1988) (unilocular Nodosariida and Buliminida), Hayward and Brazier (1980), Boersma (1984), Lamb and Miller (1984), van der Zwaan *et al.* (1986), Revets (1989, 1990*a,b*, 1991, 1992, 1993*a,b*), and Revets and Whittaker (1991) (multilocular Buliminida), Huang (1964), Shchedrina (1964), O'Herne (1974), Larsen (1976, 1977), Hottinger (1977), Banner and Culver (1978), Levy *et al.* (1979, 1982, 1986), Billman *et al.* (1980), Banner *et al.* (1985), Haynes and Whittaker (1990), Hottinger *et al.* (1990), Hottinger *et al.* (1991*a,b*), and Hansen and Revets (1992) (Rotaliida), and Banner and Blow (1960*a,b*, 1967), Blow (1969, 1979), Stainforth *et al.* (1975), Saito *et al.* (1981), Kennett and Srinivasan (1983), and Cifelli and Scott (1986) (Globigerinida).

In my opinion, sufficient data are now available in connection with the taxonomy of the *Challenger* foraminifera to enable a fresh revision, while the growing interest in Cenozoic cosmopolitan deep-water benthonic foraminifera and their palaeoecological and stratigraphic potential necessitates it.

The revision has particular application in increasing geological understanding and in reducing technical risk (and, ultimately, cost) in petroleum exploration in the Pleistogene (Quaternary) to Neogene of the Gulf of Mexico and offshore Niger Delta, the Neogene of the Gulf of Suez and South-East Asia, the Neogene to Palaeogene of California and offshore Angola, and the Palaeogene to Senonian of the North Sea.

The latter and greater part of this book is dedicated to this revision, which incidentally involves changes to over half (nearly 500) of Barker's identifications. The earlier part (based on a previous publication (Jones 1990)), concerns matters of essentially historical, biographical, and curatorial interest, but also includes some new comments on stratigraphic and palaeoecological applications of the taxonomic revision.

Uxbridge
May 1993   R.W.J.

Acknowledgements

First and foremost I wish to express my indebtedness to Fred Banner of the Palaeontology Department of the British Museum (Natural History) (BMNH) (now known as the Natural History Museum, London) and the Postgraduate Unit of Micropalaeontology, University College London (formerly of the BP Research Centre, Sunbury-on-Thames), to John Saunders of the Naturhistorisches Museum, Basel, Switzerland, and to John Whittaker of the Palaeontology Department of the BMNH for many hours of fruitful discussion in the BMNH (which I have now visited about fifty times in the last five years) and elsewhere (most memorably in the 'Kwality Tandoori'), for constructive criticism of various versions of the typescript and not least for their constant and unflagging support and encouragement; to Clive Jones and John Whittaker of the Palaeontology Department of the BMNH for their skill in capturing the images used on the supplementary plates; and to Myra Givans and Jenny Simcock of the Publications Department of the BMNH, and the staff of Oxford University Press, for helping me to see the project through to the publication stage.

Dr C. G. Adams (recently retired) and Mr R. L. Hodgkinson of the Palaeontology Department of the BMNH, Mrs Eileen Brunton of the Palaeontology Library of the BMNH, Mr H. A. Buckley of the Mineralogy Department of the BMNH, the staff of the General Library of the BMNH, Mr P. S. Davis of the Hancock Museum, Newcastle upon Tyne, Mr J. V. Howard and the staff of the Special Collections Department, Edinburgh University Library, Mr F. Manders and the staff of the Local Studies Department of the Central Library, Newcastle upon Tyne, Mr K. Moore and Ms Pippa Senior of the Royal Society, London, Dr A. L. Rice of the Institute of Oceanographic Sciences Deacon Laboratory, Wormley, Surrey, Dr F. Rögl of the Palaeontology Department of the Naturhistorisches Museum, Vienna, Austria, Mrs Ann Shirley of the Ships Department of the National Maritime Museum, Greenwich, London, Mr R. Symonds of the University Museum of Zoology, Cambridge, and Mr F. Woodward of the Natural History Department of the Museum and Art Gallery, Kelvingrove, Glasgow are thanked for their help with matters of historical or curatorial interest in the introductory chapters.

Extracts from papers in the Archives of the Royal Society are reproduced by kind permission of the President and Council of that Institution. The National Maritime Museum gave permission to reproduce the photographs of HMS *Challenger*. Edinburgh University Library gave permission to reproduce the photograph of the *Challenger* crew, Gulland's pen-and-wash sketch of Wyville Thomson, and Buchanan and Wild's original drawing of *Globigerina bulloides* d'Orbigny. Newcastle upon Tyne City Libraries and Arts gave permission to reproduce the photographic portraits from the Brady family album. The photographs of the *Challenger* slides appear courtesy of the Palaeontology Department of the BMNH.

My father helped with some of the biographical work. My wife Heather and small sons Wynn and Gethin are thanked for their forbearance.

Permission to publish was granted by the British Petroleum Company PLC.

Contents

The *Challenger* expedition, Henry Bowman Brady, and the *Challenger* foraminifera

Introduction

It is now over a hundred years since the cruise of HMS *Challenger* (1872–76) and the subsequent publication of the report on the *Challenger* foraminifera by Henry Bowman Brady (1884); indeed, 1991 marked the centenary of Brady's death. So why the interest? The answer is two-fold but simple.

Firstly, because the *Challenger* expedition fundamentally and significantly advanced numerous fields in the earth, life, and ocean sciences, and indeed has been described as representing 'the greatest advance in the knowledge of our planet since the celebrated geographical discoveries of the fifteenth and sixteenth centuries' (Linklater 1972). It also heralded the dawn of a new, modern, multi-disciplinary era of oceanography, such that its significance may be justly compared with that of the voyage of the *Beagle* (made famous by Charles Darwin) in other areas of scientific interest and endeavour.

Secondly, because Brady's work on the *Challenger* foraminifera detailed for the first time the dominant faunal element in the largest biotope on the face of the earth—the abyssal plains of the deep sea—until shortly beforehand thought to be incapable of sustaining life. The '*Challenger* Report' (Brady 1884), as herein revised, features 875 species (approximately 15 per cent of the estimated total number of extant species (6000)), belonging to 362 genera (45 per cent of the total number (800)), including the type-species of 238 (30 per cent)). The number of species excludes fossil species (either reworked or figured for comparison) and non-foraminiferal species (Allogromiida, Pogonophora, Xenophyophorea). The number of genera excludes those regarded as synonyms.

The comprehensiveness and attention to detail exhibited in this work is such that it remains the most famous of the classic monographs of foraminiferal studies and the most often cited. Moreover, of all the foraminiferal collections in the British Museum (Natural History), none is more important or more frequently consulted than the *Challenger* collection.

The succeeding short sections are concerned with historical notes on the *Challenger* expedition, biographical notes on H.B. Brady and curatorial notes on the *Challenger* foraminifera and are based on a previous publication (Jones 1990). The remainder and bulk of this book is taken up with a revision to the taxonomy of the *Challenger* foraminifera.

The *Challenger* expedition (1872–76)

Prelude to the expedition

The reasoning behind the commissioning of the *Challenger* expedition is best understood in its historical context. Britain in the latter half of the last century was at her mightiest as an imperial power. She was above all a maritime power, proud of her naval achievements at war and in peace-time.

When the news of impending American and German voyages of scientific exploration broke in London in 1871, William Benjamin Carpenter (1813–85) urged prompt action to maintain Britain's leading position in marine science. In an address to the Royal Institution and later in a letter to G.G. Stokes (Secretary of the Royal Society, of which Carpenter was President), Carpenter suggested the draughting and submission to parliament of a joint plan for a circumnavigation of the globe which would take the concepts and techniques developed and pioneered on the North-East Atlantic voyages of the *Lightning* (1868) and *Porcupine* (1869, 1870), and put them to work on the oceans of the world. In further correspondence with George Goschen (First Lord of the Admiralty), Carpenter received assurances that the government would give favourable consideration to such a proposal.

Carpenter was a remarkable man who held a number of honorary positions and titles and was at various times Lecturer on Medical Jurisprudence at the Bristol Medical School, Lecturer on Physiology at the London Hospital, Fullerian Professor of Physiology at the Royal Institution, Professor of Forensic Medicine at University College, London, and Registrar of the University of London. In the words of the *Dictionary of national biography* (Lee 1898), he was 'a man of no ordinary mental grasp and range of study' and 'one of the last examples of an almost universal naturalist'. He published variously on physiology, behavioural psychology, evolutionary theory (a deeply religious man who played the organ at the Unitarian Church in Hampstead, he was a rather reluctant ally of Darwin), and microscopy, always in a lucid and often ratiocinative style. He was a particular authority on oceanography, marine biology, and foraminiferology (see, for instance, Carpenter 1862, 1883; Murray 1971*a*, 1981, 1989; Murray and Taplin 1984*a,b*).

Once Carpenter's circumnavigation plan had been tacitly approved by the Admiralty, and the approval communicated to the Royal Society, a committee was set up under the auspices of that august body which comprised Carpenter himself, J.D. Hooker, T.H. Huxley, J.G. Jeffreys (see Plate A), the Navy Hydrographer Captain G.H. Richards, Professor Charles Wyville Thomson, and Sir William Thomson (later Lord Kelvin). On the recommendation of these eminent personages, the Council of the Royal Society made a formal request to the government to send out an expedition to undertake a scientific study of the oceans. The request was granted in April 1872, and preparations for departure began immediately. The organization of the voyage (Burstyn 1968, 1972) proceeded remarkably swiftly and efficiently, due doubtless in part to previous experience in similar ventures but probably also in no small measure to political factors (it is noteworthy that Carpenter had social contacts with the then Prime Minister Gladstone).

A suitable vessel, HMS *Challenger* (Rice 1986; Plate B herein) was

Dr. Carpenter F.R.S.

Dr. J. D. Hooker F.R.S.

Professor Huxley F.R.S.

J. Gwyn Jeffreys F.R.S.

Plate A   Top left: W.B. Carpenter. **Top right:** J.D. Hooker. **Bottom left:** T.H. Huxley. **Bottom right:** J.G. Jeffreys. (From the Brady family photograph album, reproduced by kind permission of Newcastle upon Tyne City Libraries and Arts.)

chosen in the summer of 1872, and schedule for her voyage drawn up for which Richards was largely responsible. *Challenger* was a 226′, 1462-ton (builder's measurement) or 2306-ton (displacement) 'Pearl' class steam-assisted screw corvette. Her engines were nominally of 400 hp but capable of 1234 hp, and she also carried 16 000 square feet of sail. She had been built at Woolwich in 1858 and had seen naval service off the Americas and later in Australasia before returning to England in 1871. Preparatory to what was to become her most famous voyage, she was fitted out at Sheerness, where all but two of her twenty-two cannon were removed to make way for a dredging platform over the upper deck forward of the main mast, extra laboratories, work-rooms, and storage space.

The Aberdonian Captain (later Rear-Admiral Sir) George Strong Nares (1829–1915) was chosen by the Admiralty to take the helm. Under him were some score of officers and crew (see Plate C), including Paymaster R.R.A. Richards, Lieutenant P. Aldrich, Lieutenant A. Balfour, Sub-Lieutenant Campbell, and Navigating Sub-Lieutenant Swire, all of whom kept journals (Swire's being memorable for some less than reverential references to Wyville Thomson's appearance and manner!). Nares left *Challenger* in 1874 to lead the *Alert* and *Discovery* Arctic expedition, and was replaced by Captain Frank Tourle Thomson. Aldrich left with him and was replaced by a Lieutenant A. Carpenter.

Wyville Thomson (see Plate C) was appointed by the Royal Society as head of the civilian scientific team, Carpenter at 59 having decided he was too old to put to sea again. He was to be assisted on board by secretary and ship's artist John James Wild, chemist John Young Buchanan (see Plate C) and three naturalists, Henry Nottidge Moseley, Rudolph von Willemöes-Suhm (whose promising career was cut short when he died on board *Challenger*), and John Murray.

Wyville Thomson was born in 1830 at Linlithgow in Scotland and received his early education at Merchiston Castle School. He matriculated at the age of sixteen as a student of medicine at Edinburgh University, where he seems largely to have pleased himself in the choice of lectures he attended, taking in such subjects as zoology, botany, and geology; he was also active in his role as Secretary of the Royal Physical Society. He gave up his medical studies in 1850 on the grounds of ill health and embarked instead on a career in natural science, succeeding to the posts of Lecturer in Botany at King's College, Aberdeen in 1850, Professor of Botany at Marischal College, Aberdeen in 1851, Professor of Natural History at Queen's College, Cork in 1853, Professor of Mineralogy and Geology at Queen's College, Belfast in 1854, Professor of Zoology and Botany at Belfast in 1860, and Allman Professor of Natural History at Edinburgh University in 1870. He was reputedly a delightful and instructive lecturer on a variety of scientific subjects, speaking without notes but with constant reference to the profusive array of specimens on his table.

Wyville Thomson was, with Carpenter, instrumental in setting up the *Lightning* and *Porcupine* expeditions, the results of which were published in his book *The depths of the sea* (1873). He was widely recognized, on account of this work, as an active instigator and leading spirit of new and successful investigations. It was natural, therefore, that he should be appointed as chief naturalist on the *Challenger* expedition. It was unfortunate that his health broke down in the wake of the expedition and that he did not have the freedom to finish his original research on the *Challenger* crinoids and sponges, having to concentrate instead on his administrative responsibilities.

Nonetheless he received international recognition for his career on the *Challenger*, which he wrote up in the form of the scholarly and erudite book *The voyage of the Challenger* (1877). He received a Royal Society Gold Medal in 1876 (having already acceded to the fellowship of that institution in 1869) and was made a Knight of the British Empire in 1876, and a Knight of the Polar Star (an honorary title bestowed at the University of Uppsala on the occasion of its quatercentenary celebrations) in 1877. Wyville Thomson died in 1882, whereupon a memorial window was installed in the Linlithgow Cathedral and a bust in Edinburgh University. Further details of his life are given by Herdman (1923).

Moseley was a no less able or enthusiastic fellow, of whom it was once said that 'you had only to put him down on a hillside with a piece of string and an old nail, and in an hour or two he would have discovered some natural object of surpassing interest' (Herdman 1923). He wrote up his experiences on board *Challenger* in the lively and enjoyable book *Notes by a naturalist* (1879), the enduring popularity of which is evident from the fact that it was reprinted as recently as 1944 in a series entitled *Live books resurrected*! He went on from his *Challenger* exploits to become a Fellow of the Royal Society and Linacre Professor of Human and Comparative Anatomy at Oxford.

Murray (see, for instance, Herdman 1923 and Boog Watson 1967) was Canadian-born but of Scots ancestry. He registered as a medical student at Edinburgh University, where one of his fellow students, Robert Louis Stevenson, came to criticize him for failing to pursue his studies in 'orderly, purposive and profitable fashion'. Rather like Wyville Thomson before him, he attended lectures not strictly connected with his course-work, for instance on chemistry, natural history, literature, law, and theology. He spent his summer vacations indulging his interests in marine biology and oceanography on dredging trips off the Scottish coasts (on one such occasion meeting Sir William Thomson on Skye), and had enterprisingly enrolled as a surgeon aboard the whaler *Jan Mayen* in 1868, spending 7 months in the Arctic. Enough was known of the rare talent of this remarkable man by 1872, when he was only 31, that he was appointed to the prestigious post of *Challenger* naturalist (albeit as a replacement for William Stirling, who had resigned his appointment) on the recommendation of the eminent Edinburgh University physicist Professor Peter Guthrie Tait. He is acknowledged as the 'father of modern oceanography' on account of his many achievements in the wake of the *Challenger* cruise. He went on to be made a Fellow of the Royal Society in 1896 and was knighted in 1898.

By December, 1872, the scientists and crew were assembled and all the necessary equipment was on board. Most of the equipment had been tried and tested on earlier voyages. Nets, trawls, and dredges were to be put out on hempen lines. Temperature measurements were to be taken with Miller–Casella thermometers (appropriately compensated for pressure), though some Siemen's and Johnson's instruments were also taken (as was Siemen's photometric apparatus). Intermediate depth water samples were to be collected by a stopcock water bottle designed by Buchanan and deep water samples by a Slip water bottle as used by the German North Sea expedition. Bottom sediment samples were to be brought up from the sea floor using a Hydra sounding tube, modified in 1873 by Lieutenant C.W. Baillie. Only Sir William Thomson's sounding device, comprising piano-wire wound around a drum, was an unknown quantity. This had proved successful in trials aboard the *Lalla Rookh* in comparatively shallow

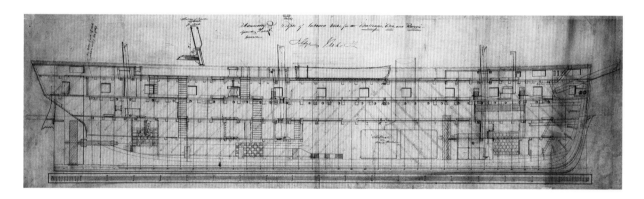

water, but when it was tried aboard *Challenger* its drum collapsed; hence, depth measurements aboard *Challenger* came to be taken with hempen lines and are inaccurate. Many of the instruments used aboard *Challenger* are figured by McConnell (1981).

Challenger finally set sail from Portsmouth on 21 December 1872. Her three-and-a-half year long voyage covered some 68 890 nautical miles and involved sampling at 362 stations (see map following p. 150). In addition, there were numerous coaling stops and more protracted periods ashore at a great many ports-of-call (at many of which her crew attended or organized lavish social functions). *Challenger* returned to Spithead on 24 May 1876, and was broken up at Chatham in 1921. Her figurehead survives to this day, outside the Institute of Oceanographic Sciences in Wormley, Surrey. 'Several pieces of rope' and 'a piece of wood from one of the masts with an enscribed plaque stating where it came from' survive in the '*Challenger* Relics' collection in the British Museum (Natural History) (H. Buckley, personal communication).

Aims of the expedition

Challenger was confidently expected to bring back the answers to all the questions posed by earlier studies of the North-East Atlantic. These were concerned largely with the nature and distribution of bottom sediments, with problems of oceanic circulation, and with the very existence of life itself in the deepest parts of the oceans.

Deep-sea sediments

With regard to the nature of bottom sediments, important new discoveries were made almost from the outset of the voyage. The first samples recovered from the sea bed proved, as expected, to be of pale grey '*Globigerina* ooze'. It had been predicted that this would cover the entire sea floor. However, as the cruise progressed westward on the leg from Tenerife to the West Indies and into ever deeper waters, the nature of the recovered sediments was observed gradually to change. They became darker and darker in colour, and upon microscopical examination proved to comprise fewer and fewer foraminiferal remains.

On 18 February 1873, deep-water samples from Station 3 (24° 45′N; 20° 14′W) were observed to contain 'a number of very peculiar black oval bodies about an inch long'. These were first thought to be fossils or lumps of pitch, but Buchanan's chemical analyses showed them to be composed of almost pure manganese peroxide. They were the first recorded examples of what we now term manganese nodules.

On 26 February 1873, a sample brought up from 3150fm. proved to comprise, in Wyville Thomson's words 'a perfectly smooth red clay, containing scarcely a trace of organic matter'. His initial surmise was that this would turn out to be a local phenomenon. This seemed to be borne out when, from the shallower Dolphin Rise, more '*Globigerina* ooze' was recovered. However, the passage into the deeper waters of the Western Atlantic again saw clay return as the dominant substrate.

The widespread distribution of clay on the sea floor was confirmed

on the leg between the West Indies by way of the Puerto Rico Trench to Bermuda. This necessitated a reconsideration by Wyville Thomson of its significance. He came to regard it as typical of deep areas, supposing these areas to be too deep to sustain the '*Globigerinae*' that made up shallower sediments. As Murray had by this time shown that '*Globigerinae*' were planktonic organisms ubiquitous in surface waters and raining down upon the whole sea floor upon death, a problem arose. What could account for their absence in deep areas? Close inspection of samples from progressively deeper waters showed a progressive disappearance not only of foraminifera but also of all other calcareous organisms. Wyville Thomson came to the conclusion that a chemical reaction was removing calcium carbonate at depth and that the product of this reaction was clay. Buchanan invoked carbonic acid as responsible for the carbonate dissolution, and modelled the reaction in his laboratory. The 'snow line' below which calcite passes into solution was first described by Murray and the Belgian Abbé Renard. It was with this that the modern idea of the calcite compensation depth was conceived, though it was not until the latter half of this century that it was named as such and fully quantified.

In 1875 *Challenger* set out into the Pacific for the first time, following the American USS *Tuscarora* and the German SS *Gazelle*. The combined efforts of all three vessels showed this ocean to differ markedly from all others, not only in terms of its greater areal extent but also in terms of its greater depth. By virtue of the latter feature, it was found also to comprise proportionately larger areas underlain by clay. Here, though, the clay was associated with larger mineral particles. This led Murray to question Wyville Thomson's theory that the clay represented the residue derived from dissolution of '*Globigerina* ooze'. Murray came to favour a volcanic origin for the clay, citing as evidence the proximity of numerous volcanic centres (the 'Pacific ring of fire'), and the relative ease with which their products could be transported into deep water. Militating somewhat against this was the associated occurrence of manganese nodules of apparently concretionary habit. Murray ascribed the origin of these nodules to volcanism also, while Buchanan had by this time come to favour mineralization of organic remains as the most likely mechanism for their formation. Later, the French geologist Dieulafait hypothesized that they originated by precipitation from sea water at the surface following a reaction between salt water and the atmosphere. Even to this day, their precise origin remains somewhat problematic.

Oceanic circulation

In 1873, on the leg between Bermuda and Halifax, Nova Scotia, a detour was made to enable *Challenger* to study the Gulf Stream. This had first been described by the sixteenth-century explorer Ponce de Leon, following a voyage from Puerto Rico to Teneriffe which crossed what is now termed the Florida Current in the vicinity of Cape Canaveral. Conjectures about its nature were published by Peter Marytr of Anghrera, also in the sixteenth century, and the line of demarcation between warm and cold water masses was first recorded by Lescarbot in the seventeenth. Cold water eddies were known by

---

Plate B **Top:** HMS *Challenger* in starboard view with funnel lowered. **Centre:** plans for her refit. **Bottom:** port view with funnel raised, Sydney approaches, June 1874. (Reproduced by kind permission of the Trustees of the National Maritime Museum, Greenwich.)

Plate C   Top left: the crew of HMS *Challenger*. Standing (left to right): Balfour, Buchanan, Willemöes-Suhm, Aldrich, Ass. Engineer W.A. Howlett, Wild, Swire, Ass. Paymaster J. Hynes, Moseley, Sub-Lt. A. Channer, Richards, Sub-Lt. H.E. Harston, Murray, Lt. A.C. Bromley. Seated (left to right): Cdr. J.F.L.P. Maclear, Surgeon G. Maclean, Nav. Sub-Lt. A. Havergal, Wyville Thomson, Engineer W.J.J. Spry, Nares, Staff Surgeon A. Crosby, Lt. G.R. Bethel, unknown. For additional portraits of senior crew members and of the non-shipboard scientists see Crane (1897). **Bottom left:** pen and wash sketch by Elizabeth Gulland showing Wyville Thomson (left) and Buchanan (right) at work on board *Challenger*. **Right:** original drawing made by Wild on board *Challenger* of the planktonic foraminifer *Globigerina bulloides* d'Orbigny, subsequently used in the '*Challenger* Report' (Brady 1884; Pl. 77), signed bottom left. (Reproduced by kind permission of Edinburgh University Library.)

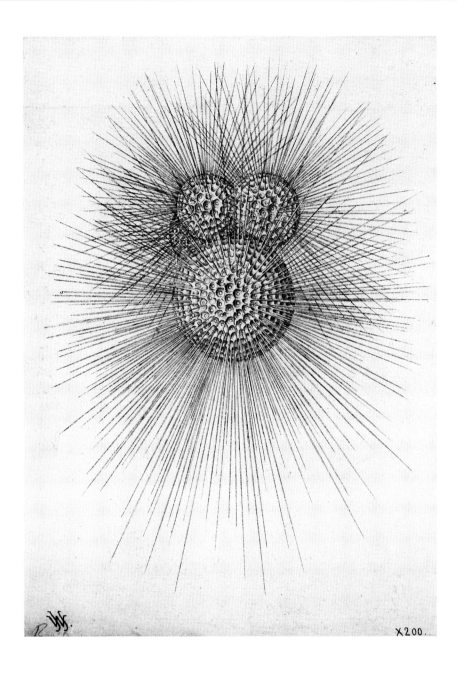

X200.

1810, and there had been a great deal of subsequent research done by the US Coast Survey: a synthesis of data had been published in 1868.

Serial temperature measurements made by the *Challenger* crew essentially confirmed what had been observed earlier, namely that there was a relatively shallow body of warm water forming the current, and a deeper body of cool water underlying it and rising to the surface at its western flank.

Velocity measurements were made, both at the surface and at depth, on the whole of the ensuing equatorial Atlantic leg in order to ascertain whether there was any subsurface movement of water in a direction counter to that at the surface. A drogue of similar design to that used by Nares and his crew aboard the *Shearwater* was used to track the undercurrents. At Station 106, it was shown using this apparatus that the strength of the surface current diminished with increasing depth and ceased to have any measurable effect at 75fm. Unfortunately, the shipboard scientists were content to note at this juncture 'how very superficial the Equatorial Current is'. Had they made additional measurements at greater depth, they would undoubtedly have discovered for the first time the existence of the

Equatorial Counter Current. As it transpired, this was discovered by Buchanan during a subsequent cruise on the cable vessel *Buccaneer*.

As *Challenger* sailed from the Cape Verde Islands to Brazil, the temperature readings that she was taking were observed to form a distinct pattern. Nares speculated that the 'cold stream' to the west was separated from warmer waters to the east by a north-south trending shoal system (subsequently confirmed by sounding) that acted as a barrier to mixing. We now recognize this as the Mid-Atlantic Ridge, a centre of sea-floor spreading, and the Walvis Ridge (discovered when *Challenger* re-entered the South Atlantic in 1876) running roughly at right angles to it as a transform fault system. The numerous 'sills' recorded in the East Indian archipelago are also now interpreted in terms of modern plate tectonic theory, as parts of an island arc system.

While the *Challenger* crew and scientists relaxed at one of their many ports-of-call (Sydney, Australia) in March 1874, controversy was raging back in England over the causes of ocean currents. It had been contended by Carpenter that the motive force lay in the superior weight of one column of water over another. This thesis was held to be untenable by James Croll, in that it presupposed the existence of a significant difference in level between one part of the ocean and another. Observations from the *Challenger* cruise did seem to indicate that a dynamic circulatory system could be generated and maintained by temperature gradients, as then advocated by Carpenter. But Croll remained unconvinced, citing wind-stress as an equally likely alternative explanatory mechanism. He attacked Carpenter from many fronts, having at his disposal a greater knowledge of physical processes and a commendable tenacity. But he was unable to win the war of words, in which Carpenter's literary grace was a great advantage, or indeed to shift Carpenter from his entrenched position. Later, Wyville Thomson was to argue that there was not the slightest ground for supposing that such a thing existed as 'a general vertical circulation of the water of the oceans depending upon differences of specific gravity'.

The debate ended somewhat acrimoniously and far from satisfactorily. The modern view (for example Sverdrup *et al.* 1942) is that no one simple model in isolation can explain the dynamics of ocean circulation. Rather, it is seen as due to the dynamic interaction of a large number of variable forces and as resulting in a complex series of inter-related movements. We can thus perhaps excuse Carpenter and Croll their simplistic models, which at least accounted for some of the phenomena observed by their time.

Life in the deep sea

As recently as the 1840s Edward Forbes, a lecturer in natural sciences, had stated quite categorically, following the voyages of the *Beacon* in 1841 and 1842, that conditions below 300fm. were incapable of sustaining life. He saw this as due to the lack of light penetration at those depths, which meant that there could be no photosynthesis. This in turn meant that there could be no plant life and consequently neither the oxygen nor the primary food source necessary for the sustenance of animal life.

Wyville Thomson, though, had cause to doubt Forbes's 'Azoic Theory'. He was aware of the findings of lowly life-forms at great depths during the voyage of the *Bulldog* in 1860 (Wallich 1862), and had first-hand reports of the forms attached to the Bona submarine cable off Sardinia at a depth of 1000fm. From his Edinburgh colleague Fleeming Jenkin. He had therefore set sail in 1868 aboard the *Lightning* and in 1869 aboard the *Porcupine* to investigate for himself the possibility of life in the deep-sea, finding abundant evidence of it to depths of 600fm. off the Shetlands and Faeroes on the former cruise, and to 2000fm. off Ushant on the latter. It was with the publication of the results of the voyages of the *Lightning* and *Porcupine* (Wyville Thomson 1873) that Forbes's 'Azoic Theory' finally came to be discredited (Rice *et al.* 1976; Mills 1978, 1984; Rehbock 1979; Rice 1986).

Nonetheless, it still came as something of a revelation when *Challenger* discovered worm tubes from a depth of about 3000fm. in March 1873. These were identified by Willemöes-Suhm as Annelida. This, and subsequent similar discoveries, prompted Wyville Thomson to note that 'conditions of the bottom are not only such as to admit the existence of life, but are such as to allow of the unlimited extension of distribution of animals high in the zoological series, and closely in relation to the faunae of shallower zones'. It was inevitable, in the light of this finding, that there would come to be developed a theory of abyssal circulation and regeneration of bottom waters.

Meanwhile, important observations were also being made on the surface-dwelling and intermediate faunae recovered by plankton tows. Muray was able to document for the first time the diurnal vertical migration of many species. He was also able to confirm that the planktonic foraminifera in surface waters (see Plate C) were of the same species as those comprising the bulk of the underlying sediments. Up until this time, it was only surmised and sometimes openly contested that the '*Globigerina* ooze' consisted essentially of planktonic rather than benthonic species.

Also of note was the finding that the abundance and diversity of planktonic foraminifera varied greatly with latitude. The diversity tended to be much lower in higher latitudes, and in the highest latitudes few if any specimens were found: here, the '*Globigerina* ooze' passed into diatom ooze.

Particularly noteworthy was the following observation, probably the first on any living planktonic foraminifer, by Wyville Thomson (1877):

On one occasion in the Pacific, when Mr Murray was out in a boat in a dead calm collecting surface creatures, he took gently up in a spoon a little globular gelatinous mass with a red centre and transferred it to a tube. The globule gave us our first and last chance of seeing what a pelagic foraminifer really is when in its full beauty. When placed under the microscope it proved to be a *Hastigerina* in a condition wholly different from anything we had yet seen. The spines, which were mostly unbroken, owing to its mode of capture, were enormously long, about fifteen times the diameter of the shell in length; the sarcode, loaded with yellow oil-cells, was almost all outside the shell, and beyond the fringe of yellow sarcode the space between the spines to a distance of about twice the diameter of the shell all round was completely filled up with delicate bullae . . . as if the most perfectly transparent portion of the sarcode has been blown out into a delicate froth of bubbles of uniform size. Along the spines fine double threads of transparent sarcode, loaded with white granules, coursed up one side and down the other, while between the spines independent thread-like pseudopodia ran out, some of them perfectly free and others anastomosing . . ., but all showing the characteristic fluid movement of living protoplasm. The [accompanying] woodcut [based on drawings by Wild, the originals of which are now among the *Challenger* MSS. in the Special Collections Department of Edinburgh University Library], excellent though it is, gives only a most imperfect idea of the complexity and the beauty of the organism with its swimming or floating machinery in its expanded condition.

In the deeper parts of the Pacific (below about 4000fm.) in 1875 a new type of sediment was discovered. As this was a type of red clay but contained a high proportion of siliceous radiolarian sclerocoma, it came to be termed radiolarian ooze. Work on this material led incidentally to the solution of the mystery surrounding *Bathybius*. *Bathybius haeckelii* had originally been described by Huxley (1868*a,b*) from samples collected by the *Cyclops* and preserved in spirit. It was regarded as a form of protoplasm and at the time was central to the debate on abiogenesis or spontaneous generation of life from mud. Wyville Thomson, with Buchanan's help, demonstrated apparently conclusively that it was of mineral origin and represented an artefact of preparation technique (a precipitate in spirit of the calcium sulphate in sea water). Latterly, Rice (1983) has put forward a possible alternative hypothesis, envisaging *Bathybius* as an aggregated mass of phytoplankton bound by an amorphous matrix.

Achievements of the expedition

From the point of view of increasing knowledge of the nature and distribution of deep-sea sediments, the voyage of the *Challenger* was hugely successful. Pioneer work by Buchanan and especially Murray opened up an almost entirely new field which Murray was to make peculiarly his own.

Despite primitive sampling methods, the biological aims of the cruise were also amply fulfilled. A tremendous amount of material was collected from trawling stations from all over the world and from dredging stations to the greatest depths ever sounded. This material had been sorted, where necessary preserved, and classified into the main systematic groups for further analysis, identification, and description. For this further study, Wyville Thomson carefully recruited seventy-six acknowledged international specialists, who for their services received a small honorarium to cover their expenses, a commemorative *Challenger* medal, and eventually a personal copy of the report in which their results were published: many also appear to have acquired, or sent to their colleagues, duplicate sets of specimens!

About half of the species described (including about a quarter of the foraminifera) proved to be entirely new to science. Not only was this taxonomic synthesis invaluable in its own right, it permitted ecological analysis on an ocean– or world-wide basis for the first time. The deep-sea fauna proved to be remarkably cosmopolitan. Wyville Thomson was able to note that it bore close affinities to the shallower water fauna of high latitudes, 'no doubt because the conditions of temperature, on which the distribution of animals mainly depends, are nearly similar [in the two types of environment]'. Also that its relations with Tertiary and newer Mesozoic faunae were much closer than those of the faunae of shallow water, though not as close as he had been led to expect.

Only in the area of physical oceanography did the voyage fail to achieve wholly satisfactory results, though here it fuelled debate and encouraged further study. Of all the criticisms that have been levelled at it, the most telling was that it did not incorporate a physicist. As it was, it was left to the chemist Buchanan to make whatever measurements he could of density, salinity, and dissolved oxygen. Instruments such as the current drogue and reversing thermometers were certainly not used to their full potential. It was Buchanan, too, who was able to demonstrate, by obtaining from his piezometers indirect but independent temperature values, that the Miller–Casella readings were inaccurate. In view of this finding, and of the crucial importance of temperature in the debate on oceanic circulation, all earlier data had to be reassessed. It was therefore not until 1877, when Wild's book *Thalassa* was published, that the physical results of the *Challenger* cruise were adequately synthesized.

Contemporary opinions on the value of the *Challenger* expedition were divided. One rather bitter contemporary British commentator wrote, in an article reproduced in the journal *Hydrospace* in December 1971:

The first volume recording the adventures of the *Challenger* yachting trip is now out, and the other fifty-nine will be ready in less than a century. Everyone knows that Mr. Lowe sent a man-of-war laden with Professors, and that these learned individuals amused themselves for four years. They played with thermometers, they fished to all depths from two feet to three miles, they brought up bucketfuls of stuff from the deep-sea bottom, and they plottered about and imagined they were furthering the Grand Cause of Science. Then the tons of rubbish were brought home, and the genius who bossed the expedition proceeded to employ a swarm of foreigners to write monographs on the specimens. There were plenty of good scientific men in England, but the true philosopher is nothing if not cosmopolitan; so the taxpayers' money was employed in feeding a mob of Germans and other aliens. The whole

business has cost two hundred thousand pounds; and in return for this sum we have got one lumbering volume of statistics, and a complete set of squabbles which are going on briskly wherever two or three philosophers are gathered together. I believe the expedition discovered one new species of shrimp, but I am not quite sure.

In contrast, a letter (not among the *Challenger* MSS. in the Special Collections Department of Edinburgh University Library) signed by the Austrian scientist Suess and several of his colleagues, dated 12 June 1876 and addressed to 'The Editors of the Periodical *Nature*, London', reads:

Gentlemen, After having followed the reports of the Naturalists of HM Ship *Challenger* with the utmost interest we beg leave to ask you kindly to transmit this simple but sincere expression of a hearty welcome and of thankful admiration to those distinguished gentlemen, as well as to the officers and crew of this gallant ship which has been called to render so prominent services to science.

The consensus of current opinion (as apparent from retrospective reviews written in commemoration of the *Challenger* centenary (Deacon (1971, 1972), Linklater (1972), Yonge (1972), and Charnock (1973)) is that the *Challenger* expedition was one of the most significant in the history of science.

Reports of the scientific results of the voyage of HMS Challenger

The publication of the 'Reports of the scientific results of the voyage of HMS *Challenger*' was the joint responsibility of the series editor and the Treasury. Wyville Thomson was the editor until his resignation on the grounds of ill health at the end of 1881, whereupon Murray was appointed as successor. Treasury parsimony then led to Murray drawing on his own personal fortune to finance the publication of the fifty lavishly-illustrated royal quarto *Challenger* volumes to appear between 1880 and 1895. These included *A general account of the scientific results of the expedition* by Murray, Buchanan, Moseley, and Staff-Commander Tizard, and six other volumes to which Murray contributed personally.

Incidentally, the cost of the whole enterprise (estimated at one hundred and seventy thousand pounds sterling) was more than offset by the profits brought about by the exploitation of the phosphate deposits of Christmas Island in the Indian Ocean. These had been discovered quite by chance in 1886 by the *Flying Fish*, on a cruise which had been organized by Murray for the purpose of increasing knowledge of coral reefs. When Murray head about the discovery from the Commanding Officer of the *Flying Fish*, who had served with him on the *Challenger*, he was quick to turn unexpected good fortune to greater advantage. He promptly hired the geologist H.P. Guppy to conduct a detailed survey, and, when he found out that the phosphate deposits were of commercial importance, urged annexation of the island and its valuable resources to Great Britain. On 4 April 1888, he wrote a letter (which survives to this day in the Palaeontology Library of the British Museum (Natural History) and makes interesting reading) to Sir Henry Thurston Holland, Secretary of State for the Colonies, on this subject, and in June of that year, the *Imperieuse* landed on Christmas Island and hoisted the Union Flag there.

Murray, ever astute in matters of business, then helped to found the Christmas Island Phosphate Company, which, as the Admiralty Hydrographer later wrote 'provided . . . more than the whole cost of the *Challenger* expedition'. After the death of co-founder Irvine, some of the company's profits were used to set up the Irvine Chair of Bacteriology at Edinburgh University which is still in place to this day.

This illuminating episode in Murray's life is chronicled in an article that appeared in the *Scotsman* of 8 March 1914 entitled 'Christmas Island: Fortune from a No-Man's Land'.

Henry Bowman Brady (1835–91)

Henry Bowman Brady (see Plate D) was born on 22 February 1835 in Gateshead in the north-east of England, where his parents had settled in 1829. His father, also called Henry (1805–83) was a respected medical practitioner and surgeon: he was also, as his biographers (his son-in-law and daughter) T.C. and H.B. Watson in Steel (1899) put it, 'emphatically a Christian physician' who had 'yielded to a call to the ministry' in 1861 and was a member of the Religious Society of Friends (Quakers), offering prayer with 'marked reverence and fervency' and inculcating in his children moral values which remained with them throughout their lives. His mother, Hannah (1802–72), daughter of Ebenezer and Ann Bowman of One Ash Grange, Derbyshire (where the Brady family spent its summer holidays) was an 'active worker in many of the charitable agencies of Gateshead'. His elder brother, George Stewardson (1832–1921) went on to occupy the Chair of Natural History at the University of Durham and to achieve international recognition for his work on the Ostracoda (see, for instance, Davis and Horne 1985). He had seven other brothers and sisters, including Thomas, who was born in 1837 and whose descendants survive to this day (he himself never married).

The young H.B. Brady no doubt had an interest in natural history fostered by his father (a keen amateur naturalist), by Tuffen West (an apprentice of his father's, who was involved in dredging expeditions supported by the British Association in the 1860s and later achieved fame as an illustrator of zoological monographs), by the teachers in the two Quaker schools at which he was educated (Ackworth and Tulketh Hall near Preston), by John Storey (a teacher, botanist, and Secretary to the Tyneside Naturalists' Field Club from 1849–57), and various other members of the colony of naturalists which had its headquarters in the north-east of England for many generations (Albany and John Hancock, Bewick and Joshua Alder among them).

It was evidently a lifelong interest: one of Brady's many letters to Dr A. Gunther, Keeper of Zoology in the British Museum (Natural History), (now in the archives of the General Library of that institution), dated 19 November 1878, concerned ant-eaters. Another letter to Gunther, headed Devonshire Club, St James's SW and dated 6 May 1887, reads:

You were so kind as to tell me some time ago that if I desired tickets for the Zoological Gardens I need not hesitate to ask you. Can you provide me with three tickets for Sunday week, May 15th-I have a niece in town who would be much pleased by the attention.

Brady left school in 1850 at the age of fifteen to serve as an apprentice to a Mr T. Harvey or R. Richards (sources differ), a chemist in Leeds, for four years. He then went on to study pharmacy in the laboratories of Dr T. Richardson, the forerunner of the Newcastle College of Medicine. On graduating in 1855, he set himself up as a wholesale and retail pharmacist on the corner of St Nicholas' Square in Newcastle. His commercial career prospered from the start,

possibly because pure drugs had previously been in short supply. He was soon able to move to larger premises in Mosley Street, and to diversify into the export trade and into the sale of scientific instruments. In the latter function, he established important contacts with many eminent scientists. His ready acceptance by the inner sanctum of late nineteenth-century foraminiferologists may have been attributable in some measure to his business reputation.

Brady's dynamic energy and organizational ability were both evident in his role as pharmacist. He was largely responsible for the foundation of the British Pharmaceutical Conference and was an active member thereof, serving both as Treasurer (1864–70) and later President (1872–73). He also served on the Council of the Pharmaceutical Society and was a member of the Board of Examiners until 1870 (when ill-health forced him to retire). Further, he did much to promote the scientific education of pharmaceutical chemists and was instrumental in transforming the *Pharmaceutical Journal* (to which he was a regular contributor) from a monthly to a weekly publication. Accolades from professional colleagues were many. Brady was elected an Honorary Member of the American Pharmaceutical Association, the Philadelphia College of Pharmacy, and the Pharmaceutical Societies of St Petersburg and Vienna.

His hunger for intellectual activity also manifested itself in fields other than those directly associated with his work, even early on. He became an enthusiastic member of the Tyneside Naturalists Field Club (incidentally the second oldest in the country) and the Northumberland, Durham, and Newcastle upon Tyne Natural History Society. His first papers on the foraminifera appeared in the 1860s in the transactions of these societies and as a report of the British Association for the Advancement of Science (Brady, 1863, 1864*a*,*b*, 1865*a*,*b*), and essentially concerned those living off the north-east coasts (see also Woodward 1972). Another significant milestone in Brady's early career was the co-publication with Carpenter (Carpenter and Brady 1869) of a monograph of the genera *Loftusia* and *Parkeria* (the latter now known to be non-foraminiferal). This work received extremely favourable reviews from Duncan and Parker, the originals of which are now to be found among the Referees' Reports in the Archives of the Royal Society. It is testimony to Brady's dedication that over thirty subsequent publications on the group (ranging in age from Silurian to Recent) were forthcoming during the course of his working life, notwithstanding the many and varied demands on his time. In recognition of his signal services to natural science he was elected a Fellow of the Geological Society in 1864 and a Fellow of the Royal Society in 1874. In a letter headed Mosley St., Newcastle on Tyne and dated 6 June 1874, Brady wrote to Stokes at the Royal Society acknowledging the latter honour and adding 'I hope to be attending for admission on the 18th Sept.'. This letter now resides among the Miscellaneous Correspondence in the Archives of the Royal Society.

Brady was so successful in business that he was able to sign over his business to his one-time assistant and later partner Nicholas Martin in 1876 (at the age of 41), and to devote the remainder of his fruitful life to the full-time study of the foraminifera. In his work on the group, he was variously associated with most of the leading contemporary authorities, including his sometime co-authors Carpenter (see Plate A), William Kitchen Parker, and Thomas Rupert Jones (see Plate D; see also Murray 1981, 1989 and Hodgkinson 1992). Together with Fortescue William Millett and William Crawford Williamson, these luminaries constituted what was to become known as the 'English

Plate D **Top left:** Brady as a young man, probably *c.* 1868–1870. **Top right:** Brady as an older man. This is the best-known photograph of Brady, and is the one used by Adams (1978) in his recent biography. **Bottom left:** Jones. **Bottom right:** Parker. (From the Brady family photograph album, reproduced by kind permission of the Newcastle upon Tyne City Libraries and Arts.)

School'. The broad species concept of the taxonomically conservative 'English School' contrasted radically with the philosophy of the 'Continental School' (personified by Auguste Emanuel Reuss, Christian Gottfried Ehrenberg (see Plate E), Alcide Dessalines d'Orbigny, and others).

The pinnacle of Brady's achievements as a foraminiferal worker was undoubtedly achieved with the publication of the 'Report on the Foraminifera dredged by HMS *Challenger* during the years 1873–1876' (hereafter '*Challenger* Report'). Brady's work on this colossal project began in 1878 and ended with the submission to the publishers of the final instalment in 1884. The 814 pages of text (written in a delightfully idiosyncratic style far removed from the modern terseness) set new standards of comprehensive presentation of information and attention to detail and make the '*Challenger* Report' an indispensable reference even to this day. The bibliographic section alone occupies 46 pages! Brady's accuracy of observation was particularly exemplary and is perhaps his most enduring legacy. He personally supervised the production of the 116 magnificent colour plates by the skilled draughtsman and lithographer A.T. Hollick, which are of a standard rarely matched before or since (and which were praised fulsomely by Brady in his acknowledgements). Records in a plate proof receipt and despatch book owned by James Chumley of the *Challenger* Office, 45 Frederick Street, Edinburgh (now in the Special Collections Department of Edinburgh University Library) indicate that this in itself was a very time-consuming task, beginning in 1882.

In later life, Brady received many prestigious awards and honours for his contributions to foraminiferal studies, among them a gold medal from the Austrian Emperor Franz Joseph I (see below) and an honorary doctorate from the University of Aberdeen. The letter advising him of the latter honour is among some of the uncatalogued Brady papers in the Library of the Royal Society. It is dated 3 March 1888 and signed H. Alleyne Nicholson and reads:

My Dear Brady, I write a most hurried note in order to catch the early Sunday mail tomorrow morning for the south, that I may have the great pleasure of telling you that the University of Aberdeen has to-day, on my proposal, conferred upon you the degree of LL.D.

He was also appointed as Corresponding Member of the Imperial Geological Institute of Vienna, made an Honorary Member of the Royal Bohemian Museum, Prague, and sat on the 'Committee of Papers' of the Royal Society. His review of 3 June 1889 of a paper by Bateson 'on some varieties of *Cardium edule*' resides among the Referees Reports in the Archives of the Royal Society.

Like many other products of his generation, Brady had a great zest for life. Despite (or perhaps because of) his delicate health (he was troubled by pulmonary disease for many years), he was an avid gentleman-traveller. He journeyed twice around the world, visiting such places as Morocco, India, Ceylon, China, Japan, Java, the Pacific Islands, Australia, New Zealand, and the United States. He seems to have been particularly fascinated by the Orient, and filled the house he had bought for his father's retirement ('Hillfield' at the top of Windmill Hill in Gateshead) with Japanese paintings, vases, and curios.

His interest in the local flora and fauna he encountered, and the native customs he observed on his travels frequently prompted him to write short pieces. One of his letters to Gunther, headed Hillfield and dated 28 October 1878, reads:

During a recent visit to the interior of Morocco, I made a good many observations in respect to the snake performances as practiced by the Clissowa [?].

A subsequent letter, dated 13 November 1878, thanks Gunther for his 'obliging letter just to hand', and laments the lack of 'accurate knowledge on the characters of snakes' displayed earlier, adding: 'had they only been Protozoa, I could have told you more about them—but this comes from having lapsed into a specialist'.

On his last overseas trip, to the Upper Nile in 1889–90, Brady fell seriously ill with oedema of the feet and legs. He spent some time laid up in Cairo before being forced home. On his return, he took up residence in Bournemouth, where he lived as a semi-invalid and was unable to fulfil his final ambition (alluded to in a letter to Gunther dated 22 May 1888) of producing a monograph of the British Foraminifera.

Brady died of pneumonia on Saturday 10 January 1891. The *Newcastle Daily Chronicle* posted a short obituary notice the following Monday, and an account of the funeral the following Thursday. Obituaries were also published in the *Geological Magazine*, in the *Pharmaceutical Journal* and in the 'Notices of Fellows Deceased' in the *Proceedings of the Royal Society* for 1891–92 (the last-named by the metallurgist W.C. Roberts-Austen).

Brady's most fitting epitaph is provided by Dr Michael Foster, who wrote in the issue of *Nature* dated 29 January 1891:

Science has lost a steady and fruitful worker, and many men of science have lost a friend and helpmate whose place they feel no-one else can fill. His wide knowledge of many branches of scientific inquiry and his large acquaintance with scientific men made the hours spent with him always profitable; his sympathy with art and literature, and that special knowledge of men and things which belongs only to the travelled man made him welcome also where science was unknown; while the brave patience with which he bore the many troubles of enfeebled health, his unselfish thoughtfulness for interests other than his own, and a sense of humour which, when needed, led him to desert his usual staid demeanour for the merriment of the moment, endeared him to all his friends.

Readers interested in further details of Brady's life are referred to the two fine and factual biographies known to the author, the one contemporary (by his brother-in-law Thomas Carrick Watson, in Steel 1899), the other modern (Adams 1978).

Brady library

Watson, the executor of Brady's will, wrote to the Treasurer of the Royal Society in a letter (now among the Miscellaneous Correspondence in the Archives of the Royal Society) headed 83 Osborne Road, Newcastle on Tyne and dated 2 July 1891: 'I beg to hand you herewith . . . eight hundred pounds, being the sum bequeathed by the late Henry Bowman Brady to the Royal Society'.

Five hundred pounds was to be placed in the Scientific Relief Fund of the Royal Society in accordance with Clause 10 of the will. The remaining three hundred pounds was to be used for the maintenance and increase of the Brady Library in accordance with Clause 3 of the will, which read:

I bequeath all my books and papers relating to the Protozoa to the Royal Society and I recommend for the greater convenience of reference thereto the said books and papers should be kept together in one place and a distinct collection . . .

Plate E **Left:** C.G. Ehrenberg. **Centre:** A.E. Reuss. **Right:** F. Karrer. (From the Brady family photograph album, reproduced by kind permission of Newcastle upon Tyne City Libraries and Arts.)

The Brady Library, incidentally the only named collection in the entire Library of the Royal Society, now comprises some 180 volumes. Among these are a number of rare and valuable books by such authors as Agassiz, d'Archaiac and Haime, Batsch, de Blainville, Costa, Dujardin, Ehrenberg, Fichtel and Moll, Geinitz, Haeckel, Hantken, Karrer, d'Orbigny, Reuss, Schaudinn, Seguenza, Silvestri, Soldani, Spengler, Stache and Schwager, and Terquem. Also in the collection are a complete early run of the journal *Annals and Magazine of Natural History* and volumes of reprints and unpublished manuscripts and drawings by various workers annotated by Brady himself, some of which are bound under the title 'Memoirs and Papers Relating to the Foraminifera'. Some material has been bound and some added by the Royal Society.

Some letters, manuscripts, and sundry papers are archived separately in the Library of the Royal Society. The letters include one from Edward Heron-Allen to the Library Committee of the Royal Society, headed 33 Hamilton Terrace NW and dated 21 October 1914, which reads:

Gentlemen, By the introduction of Prof. Herdman and with the kind and courteous assistance of the Asst. Librarian Mr. Hastings White, it has been my privilege for some time past to make use of the unique and remarkable collection of works on the Foraminifera bequeathed to your library by the late Dr. H.B. Brady . . .

Other letters in the collection include ones to Brady from G.S. Brady, Carter, Guppy, Halkyard, Hantken, Howchin, Millett, Murray, Robertson, Schwager, and Sherborn.

The manuscripts include one 'On the shallow water and littoral foraminifera of some of the islands of the southern Pacific', probably written *c.* 1890 and never published.

The sundry papers include numerous drawings and tracings of foraminifera by Brady together with a key, plates of Crag foraminifera

by Brady and West, taxonomic notes on Pacific foraminifera (New Caledonia 1884, Fiji 1884–85, and Samoa), and distribution data on the *Challenger* foraminifera.

The *Challenger* data are contained in three black foolscap volumes. Two of these tabulate the distributions of all of the species of foraminifera figured in the '*Challenger* Report' (Brady 1884), species-by-species. The first of these deals with those figured on Plates 1–55 of the '*Challenger* Report' (Miliolidae–Astrorhizidae–Lituolidae–Textularidae–Chilostomellidae) and the second with those figured on Plates 56–115 (Lagenidae–Globigerinidae–Rotalidae–Nummulinidae). The tables are interspersed with notes (e.g. from Revd A.M. Norman on *Psammatodendron*). The third volume lists the distributions of all of the foraminifera figured in the '*Challenger* Report', station-by-station, and adds the date on which the station was sampled, its latitude and longitude, depth, bottom temperature, and substrate data. This volume contains an unsigned note to the effect that 'Brady gives lists from . . . stations not in his Report'.

Brady family photograph album

Additional insights into the Brady's shared characters may be gained by browsing through their family photograph album in the Local Studies Department, Central Library, Newcastle upon Tyne. The frontispiece page features one of the best-known photographs of Brady (that used by Adams 1978; see also Plate D herein) and dedicates the album to his 'affectionate remembrance'. Many of the photographs in this remarkable and fascinating compilation were evidently taken by Brady himself: a number of the earliest ones of the Great Fire of Gateshead are dated October 1854 and signed H.B.B., and it seems reasonable to attribute some later photomicrographs (including ones of pennate and centric diatoms and holothurians) to him also. However, some were also taken by R.B. Bowman (probably a cousin) and some evidently bought as postcards, etc.

Among the photographs present in the album (this is by no means an exhaustive list) are ones of local scenes, beauty spots (home and abroad), still lives, staged set-pieces somewhat in the style of Henry Peach Robinson, family and friends (including the Quaker families Pumphrey and Robinson), literary figures (contemporary English and French authors, and the Romantic Poets), works of art (principally in the classical, neo-classical, renaissance, and Pre-Raphaelite schools, often with a religious theme), architecture (principally eccesiastical, but also industrial and municipal), contemporary figures, politicians, captains of industry (including such diverse figures as the performing artists Ellen Terry and Sarah Siddons, Florence Nightingale, the 'iron master' Crawshay, David Livingstone, the imperialist Rajah Brooke, abolitionist Abraham Lincoln, the Tsar of Russia, the Pope, and Garibaldi), and leading contemporary scientists and naturalists. A touching chord is struck by the inclusion of the obviously much-loved prize-winning dog 'Cato', and a somewhat bizarre one by the inclusion of one 'Crockett the Lion Tamer'.

The *Challenger* foraminifera

The 600 cases of *Challenger* material, including 100 000 'mountable specimens' were originally assumed all to have been deposited in the British Museum (Natural History), in accordance with Admiralty instructions to that effect. However, some material was disposed of during the voyage, some was never returned by the specialists to whom it was sent for description, and some was sent out on subsequent request to various institutions: in particular, sediment samples were very widely dispersed.

The history of dispersal of the *Challenger* material is admirably summarized by Lingwood (1981) and Kempe and Buckley (1987). The following discussion of the location of *Challenger* foraminiferal material is based partly on these accounts, and partly on those of Adams (1960), Adams *et al*. (1980), and Murray and Taplin (1984*a*). McConnell (1990) provides further information on the whereabouts of correspondence and papers pertaining to the *Challenger* expedition and its organizers and participants.

Foraminiferal slides

According to records in the catalogue of '*Challenger* Office specimens sent out 1873–1915' housed in the Palaeontology Library of the British Museum (Natural History), *Challenger* material was sent to Brady in ten batches between 25 November 1882 and 9 September 1887, that despatched on 1 June 1887 being directed to 5 Robert Street, Adelphi, London WC1. This material is only recorded as having been returned in three instances! However, it would seem that little of importance remains unaccounted for.

Most of the specimens of foraminifera figured in the '*Challenger* Report' (Brady 1884) are to be found in picked slides in handsome polished wood cabinets in the Heron–Allen Library in the Palaeontology Department of the British Museum (Natural History) (see Plate F); a few are to be found in the Carpenter collection in the Royal Albert Memorial Museum, Exeter (Murray and Taplin 1984*a*). The slides in the BMNH pertain to the cruises of the *Challenger* herself (1872–76), the *Lightning* (1868), the *Porcupine* (1869), the *Knight Errant* (1879),

and to the Austro-Hungarian and British North Polar expeditions (1872–74 and 1875–76 respectively) (see maps following p. 150). Comparison with authenticated specimens of Brady's hand-writing (for example signed letters to Gunther and to John Hancock, now in the museum that bears his name in Newcastle upon Tyne) suggests that the Victorian copper-plate inscriptions on the undersides of many of these slides are attributable to the man himself.

Unfigured but sorted specimens, and unpicked and unsorted residues are also to be found in slides in the Palaeontology Department of the BMNH (see Plate F). Henry Sidebottom made some attempt to catalogue the former in his unpublished 'MS. index to the collection of type-slides used by H.B. Brady', probably compiled at around the turn of the century. Adams (1960) and Adams *et al*. (1980) list the BMNH register numbers of many of the specimens figured in the '*Challenger* Report' (Brady 1884). They also discuss the status of the specimens of the 97 species described as new in the '*Challenger* Report' (Brady 1884) and the 140 species described as new in earlier works (principally Brady 1878, 1879*a*,*b*, 1881*a*–*c*, 1882) and subsequently figured (in many cases for the first time) in the '*Challenger* Report'. All of these are interpreted as syntypes. Some have been designated as lectotypes and paralectotypes.

Adams (1960) and Adams *et al*. (1980) also give something of the history of the collections. It appears from their accounts that Brady presented the *Porcupine* material to the BMNH in 1885, and the *Knight Errant* material and 612 slides of *Challenger* material in 1888. At least the bulk of the remainder of the *Challenger* material was originally deposited in the University Museum of Zoology, Cambridge, whence, at the instigation of Sir Clive Forster-Cooper and Edward Heron-Allen, it was removed to the BMNH between 1939 and 1959 (see also Joysey 1960). There are no records of any material remaining at Cambridge (R. Symonds, personal communication).

In fact, Brady appears to have presented some *Challenger* material to the BMNH as early as 1885. A letter from Brady to Gunther, headed Savile Club, Piccadilly and dated 8 October 1885, intimated that Brady expected to finish the following day 'sorting and arranging the collection of *Challenger* and other Foraminifera and mounting a suitable series for exhibition [in the BMNH]'. A second letter, dated 9 October confirmed that he had indeed 'finished, as far as I can at the moment, the work I have been engaged upon . . . in connection with the collections'. A third letter, headed White Hart Hotel, Reigate and dated 13 October noted that Brady was 'well pleased that the series of Foraminifera meets with your approval. I hope still, as I may have time, to do a good deal to render it more complete'.

Brady is also known to have presented 435 slides of *Challenger* and *Porcupine* material, and practically all of the Austro-Hungarian North Polar expedition material to his friend Felix Karrer (see Plate E) at the Naturhistorisches Museum, Vienna, in 1887. In recognition of this, Franz von Hauer, a geologist who at the time was director of the Museum, approached the Austrian Emperor Franz Joseph I and arranged for Brady to be presented with a gold medal inscribed 'k.k. österr.-ung. Ehrenzeichen fur Kunst und Wissenschaft' (Insignia of the Royal and Imperial Austro-Hungarian Empire for Art and Science), and bearing His Imperial Majesty's portrait and device (F. Rögl, personal communication). This material is still in the NMH, Vienna.

Additional *Challenger* material was presented to David Robertson, 117 slides of which are now in the Robertson Museum and Aquarium collection, Millport, and Art Gallery and Museum collection, Glasgow

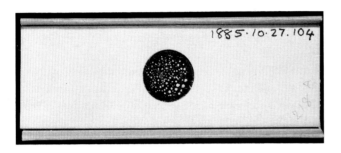

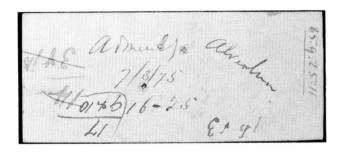

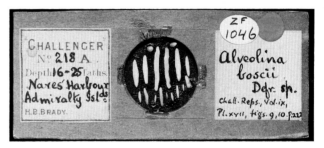

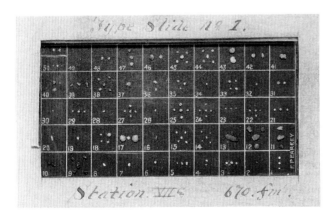

Plate F **Top left:** slide of residue material from *Challenger* Station 150, between Kerguelen and Heard Island. **Upper centre left:** slide of picked residue from *Challenger* Station 218A, Admiralty Islands, with old British Museum (Natural History) register number (1885.10.27.104). **Lower centre left:** reverse of slide shown lower centre right annotated by Brady, with old BMNH register number (1885.9.25.11). **Bottom left:** slide of sorted residue from Station VIIC. The fauna in many of these so-called 'type slides' has been identified by Sidebottom ('MS. index to the collection of type-slides used by H.B. Brady'). **Top right:** slide of the benthonic foraminifer *Keramosphaera murrayi* Brady from *Challenger* Station 157, with new BMNH register number (ZF3617). **Upper centre right:** slide of the planktonic foraminifer *Globigerina aequilateralis* Brady [now identified as *Globigerinella aequilateralis* (Brady)] from a surface tow in the Pacific, with new BMNH register number (ZF1466), figured in the '*Challenger* Report' (Brady 1884; Pl. 80, Fig. 20). **Lower centre right:** slide of the benthonic foraminifer *Alveolina boscii* Defrance, sp. [now *Alveolinella quoyi* (d'Orbigny)] from *Challenger* Station 218A, Admiralty Islands, with new BMNH register number (ZF1046), figured in the '*Challenger* Report' (Brady 1884; Pl. 17, Figs 9–10). Reverse of this slide shown lower centre left. **Bottom right:** Thin-section of the benthonic foraminifer *Orbitolites complanata* var. *laciniata* Brady [now *Marginopora vertebralis* var. *plicata* Dana] from *Challenger* Station 172A, Tongatabu Reef, with new BMNH register number (ZF2037), figured in the '*Challenger* Report' (Brady 1884; Pl. 16, Fig. 11). (All reproduced by courtesy of the Palaeontology Department of the BMNH.)

(F. Woodward, personal communication). The Hancock Museum, Newcastle upon Tyne also has 'a few slides' (P.S. Davis, personal communication), and the Laboratorium voor Paleontologie, Katholieke Universiteit, Leuven, Belgium, has twelve (Hooyberghs and van de Sande 1988).

Raw material

A vast number of sediment sounding samples collected by *Challenger* are now housed in the BMNH. These were originally stored in one collection in the *Challenger* expedition commission's offices in Queen Street in Edinburgh, which was moved in 1890 to nearby Frederick Street. A set of samples from this collection, one from each *Challenger* station, was sent to the Geology Department of the BMNH in 1895 and subsequently transferred to the Zoology Department of that institution in 1922, and to the Mineralogy Department in 1938. The remainder of the Edinburgh collection was transferred in 1904 to Sir John Murray's residence 'Villa Medusa' on the northern outskirts of the city, where it remained until long after Murray's death in an automobile accident in 1914. Here it was examined by Edward Heron-Allen and Arthur Earland in 1919 and found to comprise 9746 samples of 'Marine Deposits' (soundings, dredgings, etc., in bottles and boxes), together with the *Challenger* expedition glass photographic plates, microscopical preparations, and an extensive oceanographic library. This part of the collection was eventually acquired by the Zoology Department of the BMNH in 1921 and by the Mineralogy Department either later in 1921 (a few specimens of phosphates, etc.) or in 1935. Some of it has subsequently been transferred to the Palaeontology Department.

Murray had stipulated in 1914 to Heron-Allen that in the event of his untimely death his collection should be bequeathed either to his two sons, if they wished to carry on his work, or to a reputable institution such as the BMNH, or to the Imperial College of Science (University of London). In the latter event a responsible curator, salaried from a trust fund, was meticulously to catalogue every sample. After the inevitable delay, this condition (subsequently stressed by the Murray family solicitor) has now been fully complied with: the Sir John Murray Collection of zoological, botanical, and geological specimens from the cruise of HMS *Challenger* is catalogued by Buckley *et al.* (1979, 1984) and most of it has been entered onto a computer database.

Revised taxonomy of the *Challenger* foraminifera

Introduction

Format

The format of the succeeding section essentially follows that of Barker (1960), though changes have been made to over half (nearly 500) of his identifications. The bulk of the text comprises reproductions and explanations of the text-figures and plates from the 'Report on the foraminifera dredged by HMS *Challenger*' (Brady 1884). Two supplementary plates refiguring those few forms inadequately figured by Brady are also incorporated. The plate explanations are annotated by locality details, corrected where appropriate, after Nuttall (1927, 1931), and taxonomic notes. Also included here are British Museum (Natural History) register numbers and relevant information on the status of specimens (lectotypes etc.). BMNH register numbers prefixed with the year of accession (1959) refer to the actual specimen(s): those prefixed with the letters ZF (Zoology: Foraminifera) refer to the slide, which often contains more than one specimen and less often more than one species. Four appendices (see below), a list of references (with over 300 entries), and a taxonomic index (with over 3000 entries) are to be found at the back of the book.

Preferred names in the taxonomic section are based upon a comparison of Brady's specimens with material figured in the cited references. Original designations are included here in square brackets so as to allow the interested reader to consult the Ellis and Messina *Catalogue of foraminifera* (1940 *et seq.*) for comparison. Specific determinations are at times subjective; generic assignments almost always to (particularly where strictly dependent on study of internal characteristics, as in the case of many representatives of the orders Nodosariida and Buliminida). The final responsibility for any misidentifications or other errors of commission or of omission rests with me, although I have taken advice from the following acknowledged experts as appropriate: Paul Brönnimann (now sadly deceased), Fred Banner, Mike Charnock, Fred Clark, Stanislaw Geroch, Andy Gooday, Jo Haynes, Mike Kaminski, Robert Knight, Ewa Łuczkowska, Tim Patterson, Stefan Revets, John Whittaker, and Toine Wonders. This assistance is gratefully acknowledged. Philip Tubbs of the BMNH and International Commission on Zoological Nomenclature advised me as to the status of the species listed and in some cases described by d'Orbigny in 1826, and the status and attribution of the genera first formally described as new by Patterson and Richardson in 1988, but validated by the prior descriptions of Loeblich and Tappan in 1987.

A suggested suprageneric classification of the *Challenger* foraminifera, based on the scheme outlined by Haynes (1981), is given in Appendix 1. Summary ecological and stratigraphic ranges of selected *Challenger* foraminifera are given in Appendices 2–3 and 4 respectively.

New names and new species

The new name *Textularia hystrix* (from the Latin for 'porcupine') is proposed for *Textularia aspera* Brady 1882, which is a primary junior homonym of *T. aspera* Ehrenberg 1840. An additional fifty-nine species are recognized as new, but provisionally left in open nomenclature (for example, *Biloculinella* sp. nov. (Plate 3, Figs 1–2)).

Abbreviations

The following abbreviations are adopted throughout:

Sta. = Station;
fm. = fathoms;
BMNH = British Museum (Natural History), now known as The Natural History Museum, London;
ICZN = International Commission on Zoological Nomenclature.

Imperial to metric conversion

To convert fathoms to metres, multiply by 1.83.

Ecological, palaeoecological, and stratigraphic applications

One of the main aims of revising the taxonomy of the *Challenger* foraminifera is to refine the ecological ranges of the species described therein (Appendices 2–3). Refined ecological ranges can be used to give refined palaeoecological interpretations of fossil assemblages containing either extant species or eco-homoeomorphs of extant species, as in Pleistogene (Quaternary) to Neogene of the Gulf of Mexico and offshore Niger Delta, the Neogene of the Gulf of Suez and South-East Asia, the Neogene to Palaeogene of California and offshore Angola, and the Palaeogene to Senonian of the North Sea. Palaeoecologically controlled 'apparent extinctions' of some species can in turn be of regional stratigraphic use, as in the North Sea (Jones and Charnock 1990; see also Appendix 4).

Explanations of plates in the '*Challenger* Report'

Plate 1 (I)

Figures 1–4 *Nodophthalmidium simplex* Cushman and Todd 1947

Fig. 1 × 50. From *Challenger* Sta. 24, off Culebra Island, West Indies (390fm.). British Museum (Natural History) register number ZF2010.
Figs 2–4 × 50. From *Challenger* Sta. 217A, Humboldt Bay, off Papua, Pacific (37fm.). BMNH reg. no. ZF2011.

Referred by Brady (1884) to *Nubecularia tibia* Jones and Parker, by Thalmann (1932) to *Nodobacularia tibia* (Jones and Parker), by Barker (1960) to *Nodophthalmidium* sp. nov., and by Zheng (1988) to *Nodopithalmidium (sic) simplex* Cushman and Todd.

Figures 5–8 *Parrina bradyi* (Millett 1898)

[*Nubecularia bradyi* Millett 1898]
× 50. From *Challenger* Sta. 260A, off Honolulu Reefs, Pacific (40fm.). ZF2007.

Referred by Brady (1884) to *Nubecularia inflata*, by Thalmann first

(1932) to *Parrina inflata* (Brady) and later (1937) to *P. bradyi* var. *Fistulata* Rhumbler (Figs 5–6) and *P. bradyi* var. *sufflata* Rhumbler (Figs 7–8), and by Barker (1960) to *P. bradyi* (Millett) (Figs 5–6) and *P. bradyi* var. *sufflata* Rhumbler (Figs 7–8). Rhumbler's varietal names appear unnecessary. *Nubecularia bradyi* Millett 1898 (a new name for *N. inflata* Brady 1884, not Terquem 1876) is the type-species by original designation of the genus *Parrina* Cusham 1931 (a new name for *Silvestria* Schubert 1921, not Verhoeff 1895 or Brian 1902).

Figures 9–16 *Nubecularia lucifuga* Defrance 1825

Figs 9–11 ×20.
Figs 12–15 ×20. From off Dernah and Burdah, Gulf of Bombah, Tripoli, Mediterranean. ZF2008.
Fig. 16 ×25. From shore sand, Melbourne. ZF2009.

Referred by Brady (1884) and Thalmann (1932) to *Nubecularia lucifuga* Defrance, by Thalmann (1942) and Barker (1960) to *N. lucifuga* Defrance (Figs 9–11 and 13–16) and *N. lucifuga* var. *lapidea* Wiesner (Fig. 12). Wiesner's varietal name appears unnecessary. This is the type-species by original designation of the genus *Nubecularia* Defrance 1825.

Figures 17–18 *Pyrgoella irregularis* (d'Orbigny 1839)

[*Biloculina irregularis* d'Orbigny 1839]
Fig. 17 ×20. From *Challenger* Sta. 23, off Sombrero Island, West Indies (450fm.).
Fig. 18 ×30. From *Challenger* Sta. 85, off the Canaries, Atlantic (1125fm.). ZF1156.

Referred by Brady (1884) to *Biloculina*, and by Thalmann (1932), Barker (1960), Hofker (1976), and van Marle (1991) to *Nummoloculina*. Łuczkowska (pers. comm.) prefers placement in *Pyrgoella*.

Figures 19–20 *Pseudotriloculina cyclostoma* (Reuss 1850)

[*Biloculina cyclostoma* Reuss 1850]
×45. From off Dernah and Burdah, Gulf of Bombah, Tripoli, Mediterranean. ZF2490.

Referred by Brady (1884), Thalmann (1932), and Barker (1960) to *Triloculina cuneata* Karrer, 'biloculine variety', and by Łuczkowska (pers. comm.) to *Sinuloculina cyclostoma* (Reuss). This is the type-species by original designation of the genus *Sinuloculina* Łuczkowska 1972, regarded by Loeblich and Tappan (1987) as a junior synonym of *Pseudotriloculina* Cherif 1970.

Plate 2 (II)

Figures 1–3 *Sigmoilina sigmoidea* (Brady 1884)

[*Planispirina sigmoidea* Brady 1884]
Figs 1–2 ×40. ZF2112. Fig. 3 ×30. ZF2113. From *Challenger* Sta. 120, off Pernambuco, Atlantic (675fm.).

Referred by Brady (1884) to *Planispirina sigmoidea*, by Thalmann (1932), Hofker (1976), and Zheng (1988) to *Sigmoilina sigmoidea* (Brady), and by Barker (1960) to *Nummoloculina contraria* (d'Orbigny) (see also Text-figure 5). This is the type-species by original designation of the genus *Sigmoilina* Schlumberger 1887.

Figure 4 *Pyrgoella sphaera* (d'Orbigny 1839)

[*Bioculina sphaera* d'Orbigny 1839]
×25. From *Challenger* Sta. 246, North Pacific (2050fm.). ZF1167.

Referred by Brady (1884) to *Biloculina*, by Thalmann (1932) to *Planispirina*, and by Barker (1960), Hofker (1976), and Zheng (1988) to *Pyrgoella* (see also Text-figure 1). This is the type-species by original designation of the genus *Pyrgoella* Cushman and White 1936.

Figures 5–6 *Pyrgo lucernula* (Schwager 1866)

[*Biloculina lucernula* Schwager 1866]
Fig. 5 ×25. From *Challenger* Sta. 76, off the Azores, Atlantic (900fm.). ZF1139.
Fig. 6 ×40. From *Porcupine* Sta. 36, off Ireland, Atlantic (725fm.). ZF1141.

Referred by Brady (1884) to *Biloculina bulloides* d'Orbigny, and by Thalmann (1932), Barker (1960), Srinivasan and Sharma (1980), van Marle (1991), and Łuczkowska (pers. comm.) to *Pyrgo lucernula* (Schwager). Srinivasan and Sharma (1980) have recently designated a neotype for this species from Schwager's Kar Nicobar material in the Hindu University, Banaras, India.

Figure 7 *Pyrgo sarsi* (Schlumberger 1891)

[*Biloculina sarsi* Schlumberger 1891]
×25. From *Challenger* Sta. 24, West Indies (390fm.). ZF1159.

Referred by Brady (1884) to *Biloculina ringens* Lamarck, sp., by Thalmann first (1932) to *Pyrgo bradyi* (Schlumberger) and later (1937) to *P. fornasinii* Chapman and Parr 1935 (a new name for *Biloculina bradyi* Schlumberger 1891, not Fornasini 1886), also by Barker (1960) to *P. fornasinii*, and by Zheng (1988) to *P. fornasini* (*sic*). *Pyrgo fornasinii* Chapman and Parr 1935 is regarded by the present author as a junior synonym of *Biloculina sarsi* Schlumberger 1891.

Figure 8 *Sigmopyrgo vespertilio* (Schlumberger 1891)

[*Biloculina vespertilio* Schlumberger 1891]
×25. From *Challenger* Sta. 78, North Atlantic (1000fm.). ZF1160.

Referred by Brady (1884) to *Biloculina ringens* Larmarck, sp., and by Thalmann (1932), Barker (1960), and van Marle (1991) to *Pyrgo vespertilio* (Schlumberger). This is the type-species by original designation of the genus *Sigmopyrgo* Hofker 1983, which differs from *Pyrgo* Defrance 1824 in the initially sigmoiline rather than quinqueloculine coiling mode in the microspheric generation.

Figure 9 *Pyrgo elongata* (d'Orbigny 1826)

[*Biloculina elongata* d'Orbigny, 1826]
×45. From *Challenger* Sta. 24, West Indies (690fm.). ZF1154.

Referred by Brady (1884) to *Biloculina*, and by Thalmann (1932) and Barker (1960) to *Pyrgo*.

Figures 10–11, 15 *Pyrgo murrhina* (Schwager 1866)

[*Biloculina murrhina* Schwager 1866]
Fig. 10 ×50. From *Challenger* Sta. 224, North Pacific (1850fm.). ZF1152.

Fig. 11 × 50. ZF1153. Fig. 15 × 40. ZF1150. From *Challenger* Sta. 323, South Atlantic (1900fm.).

Referred by Brady (1884) to *Biloculina depressa* var. *murrhyna (sic)* Schwager (Figs 10–11) and *B. depressa* d'Orbigny (Fig. 15), by Thalmann (1932) to *Pyrgo murrhyna (sic)* (Schwager) (Figs 10–11) and *P. depressa* (d'Orbigny) (Fig. 15), by Barker (1960) to *P. murrhyna (sic)* (Schwager), and by Srinivasan and Sharma (1980), Zheng (1988) (Fig. 15 only), Hermelin (1989), and van Marle (1991) to *P. murrhina* (Schwager). Srinivasan and Sharma (1980) have recently designated a neotype for *Biloculina murrhina* Schwager from Schwager's Kar Nicobar material in the Hindu University, Banaras, India. The range of variability exhibited by this species clearly embraces such forms as Brady's Figure 15 (*Biloculina bradyi* Fornasini 1886).

Figures 12, 16–17 *Pyrgo depressa* (d'Orbigny 1826)

[*Biloculina depressa* d'Orbigny 1826]
Fig. 12 × 30. From *Challenger* Sta. 352A, off Cape Verde Islands, Antlantic (11fm.). ZF1151.
Fig. 16 × 35. From *Porcupine* Sta. 23, off NW Ireland, Atlantic (630fm.). ZF1146.
Fig. 17 × 40. ZF1149.

Referred by Brady (1884) to *Biloculina*, and by Thalmann (1932), Barker (1960), Hofker (1976), and van Marle (1991) to *Pyrgo*.

Figures 13–14 *Pyrgo laevis* Defrance 1824

Fig. 13 × 30. From *Porcupine* Sta. 28, off NW Ireland, Atlantic (1215fm.). ZF1157.
Fig. 14 × 20. From *Challenger* Sta. 24, West Indies (390fm.). ZF1158.

Referred by Brady (1884) to *Biloculina*, and by Thalmann (1932) and Barker (1960) to *Pyrgo*. This is the type-species by original monotypy of the *genus* Pyrgo Defrance 1824.

Plate 3 (III)

Figures 1–2 *Biloculinella* sp. nov.

× 40. From *Porcupine* Sta. AA, Loch Scavaig, North Atlantic (45–60fm.).
Fig. 1 ZF1144; Fig. 1 ZF1145.

Referred by Brady (1884) to *Biloculina depressa* d'Orbigny, and by Thalmann (1932) and Barker (1960) to *Pyrgo depressa* (d'Orbigny) (see Plate 2, Figs 12, 16–17). The 'biloculine' coiling mode and 'miliolinelline' apertural characteristics indicate placement in the genus *Biloculinella*.

Figure 3 *Pyrgo serrata* (Bailey 1861)

[*Biloculina serrata* Bailey 1861]
× 25. From *Challenger* Sta. 168, off New Zealand, South Pacific (1100fm.). ZF1148.

Referred by Brady (1884) to *Biloculina depressa* var. *serrata*, and by Thalmann (1932), Barker (1960), Zheng (1988), and Hermelin (1989) to *Pyrgo serrata* (Bailey). *Biloculina depressa* var. *serrata* Brady 1884 is both a junior homonym and a junior synonym of *B. serrata* Bailey 1861.

Figures 4–5 *Pyrgo denticulata* (Brady 1884)

[*Biloculina ringens* var. *denticulata* Brady 1884]
Fig. 4a × 40. From *Challenger* Sta. 172, off Tongatabu, Friendly Islands, Pacific (18fm.). ZF1161.
Figs 4b, 5 × 40. From *Challenger* Sta. 260A, off Honolulu Reefs, Pacific (40fm.). ZF1162.

Referred by Brady (1884) to *Biloculina ringens* var. *denticulata*, by Thalmann (1932), Barker (1960), Whittaker and Hodgkinson (1979), and Haig (1988) to *Pyrgo denticulata*, and by Hofker (1976) to *Pyrgoides denticulatus* (Brady). This is the type-species by original designation of the genus *Pyrgoides* Hofker 1976, regarded by Loeblich and Tappan (1987) as a junior synonym of *Pyrgo* Defrance 1824.

Figures 6, 14 *Pyrgo lucernula* (Schwager 1866)

[*Biloculina lucernula* Schwager, 1866].
Fig. 6 × 35. From *Challenger* Sta. 174C, off Fiji, Pacific (210fm.). ZF1168.
Fig. 14 × 20. From *Challenger* Sta. 76, off the Azores, Atlantic (900fm.). ZF1907.

Referred by Brady (1884) to *Biloculina tubulosa* Costa (Fig. 6) and *Miliolina trigonula* Larmarck sp. (Fig. 14), and by Thalmann (1932), Barker (1960), and van Marle (1991) (Fig. 14 only) to *Pyrgo lucernula* (Schwager) (see also Plate 2, Figs 5–6).

Figures 7–8 *Pyrgo denticulata* var. *striolata* (Brady 1884)

[*Biloculina ringens* var. *striolata* Brady 1884]
Fig. 7 × 45. From *Challenger* Sta. 187A, off Booby Island, Torres Strait, Pacific (8fm.). ZF1164.
Fig. 8 × 45. From *Challenger* Sta. 187, off Papua, Pacific (6fm.). ZF1163.

Referred by Brady (1884) to *Biloculina ringens* var. *striolata*, by Thalmann (1932) and Barker (1960) to *Pyrgo denticulata* var. *striolata* (Brady), and by Haig (1988) to *P. striolata* (Brady).

Figure 9 *Pyrgo comata* (Brady 1881)

[*Biloculina comata* Brady 1881]
× 40. From *Challenger* Sta. 24, West Indies (390fm.). ZF1143.

Referred by Brady (1884) to *Biloculina*, and by Thalmann (1932), Barker (1960), Hofker (1976), Zheng (1988), and van Marle (1991) to *Pyrgo*. Van Marle (1991) regards *Biloculina comata* Brady 1881 as a *nomen nudum*, but gives no justification.

Figures 10–13 Miliolid juvenaria

Fig. 10 × 50. From *Challenger* Sta. 205A, Hong Kong Harbour, Pacific (76fm.). ZF1923.
Figs 11–12 × 50. From *Challenger* Sta. 217A, off Papua, Pacific (37fm.). ZF1924.
Fig. 13 × 50.

Referred by Brady (1884) (albeit tentatively) to *Adelosina soldanii* d'Orbigny (Fig. 10), *A. laevigata* d'Orbigny (Fig. 11), and *A. semistriata* d'Orbigny (Fig. 12), by Thalmann (1932) and Barker (1960) to 'Miliolidae juv.' and 'Young Miliolidae' respectively, and by Łuczkowska (pers. comm.) to *Adelosina* spp. No specific determination is possible.

Figures 15–16 *Triloculina trigonula* (Lamarck 1804)

[*Miliolites trigonula* Lamarck 1804]
Fig. 15a × 40. From *Porcupine* Sta. 23, off NW Ireland, Atlantic (630fm.). ZF1909.
Fig. 15b × 40. From *Porcupine* Sta. AA, off Skye, North Atlantic (45–60fm.). ZF1910.
Fig. 16 × 25. ZF1908.

Referred by Brady (1884) to *Miliolina*, and by Thalmann (1932), Barker (1960), Haynes (1973a), Whittaker and Hodgkinson (1979), and Hermelin (1989) to *Triloculina*. This is the type-species by subsequent designation of the genus *Triloculina* d'Orbigny 1826.

Figure 17 *Triloculina tricarinata sensu* Parker, Jones and Brady 1865

× 50. From Shell Cove, Port Jackson, Australia. ZF1906.

Referred by Brady (1884) to *Miliolina tricarinata* d'Orbigny, sp., and by Thalmann (1932), Barker (1960), Hofker (1976), Whittaker and Hodgkinson (1979), Zheng (1988), Hermelin (1989), and van Marle (1991) to *Triloculina tricarinata* d'Orbigny 1826. *Triloculina tricarinata* d'Orbigny 1826 was a *nomen nudum* (Tubbs (ICZN), pers. comm.). The name was validated either by the publication of a figure identified as *Triloculina tricarinata* d'Orbigny (and stated to be 'd'aprés d'Orbigny') by Guérin-Méneville in the seldom-read *Iconographie du Règne Animal de G. Cuvier*, published in parts (livraisons) between 1829 and 1843 (Cowan 1971), or by the description provided by Deshayes in the equally seldom-read *Encyclopédie Méthodique* of 1830. Unfortunately, Guérin-Méneville's figure shows a *Spiroloculina* rather than a *Triloculina*. It seems likely that Guérin-Méneville, who was not a foraminiferal specialist, inadvertently reproduced the previously unpublished figure of '*Spiroloculina tricarinata* d'Orbigny 1826' believing it to be '*Triloculina tricarinata* d'Orbigny 1826'. In view of this confusion, *Triloculina tricarinata* is herein interpreted in the sense of Parker *et al.* (1865). These authors based their concept of the species (their drawing) on the three-dimensional model fashioned by d'Orbigny and supplied along with his 1826 '*Tableau Méthodique . . . .*'. The model was unfortunately in itself not valid for the purpose of taxonomic description (Tubbs (ICZN), pers. comm.).

Plate 4 (IV)

Figures 1–2 *Planispirinoides bucculentus* var. *placentiformis* (Brady 1884)

[*Miliolina bucculenta* var. *placentiformis* Brady 1884]
Fig. 1 × 20. From *Challenger* Sta. 24, West Indies (390fm.). ZF1855.
Fig. 2 × 20. From *Challenger* Sta. 149D, Kerguelen Island, South Pacific (20–60fm.). ZF1857.

Referred by Brady (1884) to *Miliolina bucculenta* var. *placentiformis*, by Thalmann first (1932), to *Triloculina bucculenta* var. *placentiformis* (Brady), later (1933a) to *Planispirina bucculenta* var. *placentiformis* (Brady), and finally (1937) to *Miliolinella bucculenta* var. *placentiformis* (Brady), and by Barker (1960) to *Planispirinoides bucculentus* var.

placentiformis (Brady). Hofker (1976) regarded *P. bucculentus* var. *placentiformis* (Brady) as indistinguishable from *P. bucculentus* (Brady), *s.s.* (see Plate 114, Fig. 3). Most authors, however, maintain the distinction.

Figure 3 *Miliolinella subrotunda* (Montagu 1803)

[*Vermiculum subrotundum* Montagu 1803]
× 50. From *Challenger* Sta. 145, Prince Edward Island, South Pacific (50–150fm.). ZF1858.

Referred by Brady (1884) to *Miliolina circularis* Bornemann, sp., by Thalmann (1932) to *Triloculina subrotunda* (Montagu), by Barker (1960) to *Miliolinella subrotunda* (Montagu), and by Hofker (1976) to *M. circularis* (Bornemann). This is the type-species by original designation of the genus *Miliolinella* Weisner, 1931.

Figures 4–5 *Cribromiliolinella subvalvularis* (Parr 1950)

[*Triloculina subvalvularis* Parr 1950]
Fig. 4 × 16. From *Challenger* Sta. 168, NE coast of New Zealand, Pacific (110fm.). ZF1916.
Fig. 5 × 40. ZF1917.

Referred by Brady (1884) to *Miliolina valvularis* Reuss, sp., by Thalmann (1932) to *Triloculina valvularis* Reuss, and by Barker (1960) to *T. subvalvularis* Parr. This is the type-species by original designation of the genus *Cribromiliolinella* Saidova, 1981.

Figure 6 *Triloculina transversestriata* (Brady 1881)

[*Miliolina transversestriata* Brady 1881]
× 50. From *Challenger* Sta. 185, off Raine Island, Torres Strait, Pacific (155fm.). ZF1905.

Referred by Brady (1884) to *Miliolina*, by Thalmann (1932) and Barker (1960) to *Triloculina*, and by Haig (1988) to *Quinqueloculina*.

Figure 7 *Triloculinella sublineata* (Brady 1884)

[*Miliolina circularis* var. *sublineata* Brady 1884]
× 50. From *Challenger* Sta. 218A, Nares Harbour, Admiralty Islands, Pacific (16–25fm.). ZF1860.

Referred by Brady (1884) to *Miliolina circularis* var. *sublineata*, by Thalmann (1932) to *Triloculina circularis* var. *sublineata* (Brady), by Barker (1960) to *Miliolinella sublineata* (Brady), by Haig (1988) to *Miliola sublineata* (Brady), and by the present author to *Triloculinella sublineata* (Brady). Loeblich and Tappan first (1964) synonymized *Triloculinella* Riccio 1950 with *Miliolinella* Wiesner 1931, but later (1987) reinstated it. Łuczkowska (pers. comm.) nonetheless prefers to retain the present species in *Miliolinella*.

Figure 8 *Pyrgo comata* (Brady 1881)

[*Biloculina comata* Brady 1881]
× 27. From *Challenger* Sta. 24, West Indies (390fm.). ZF1867.

Referred by Brady (1884) to *Miliolina insignis* Brady, by Thalmann (1932) to *Flintina grata* (Terquem), and by Barker (1960) to *Triloculina insignis* (Brady). Appears to the present author to fall within the range of variation exhibited by *Pyrgo comata* (Brady) (see Plate 3, Fig. 9).

Figure 9 *Flintina bradyana* **Cushman 1921**

× 35. From *Challenger* Sta. 233B, Japan (15fm.). ZF1865.

Referred by Brady (1884) to *Miliolina fichteliana* d'Orbigny, sp., by Thalmann (1932), Barker (1960), Zheng (1988), and van Marle (1991) to *Flintina bradyana*, and by Whittaker and Hodgkinson (1979) to *Triloculina bradyana* (Cushman). This is the type-species by original designation of the genus *Flintina* Cushman 1921.

Figure 10 *Triloculina trigonula* var. *striatotrigonula* **(Parr 1941)**

[*Triloculina striatotrigonula* Parr 1941]
× 40. From *Challenger* Sta. 162, Bass Strait, Pacific (38–40fm.). ZF1868.

Referred by Brady (1884) to *Miliolina insignis* Brady, by Thalmann (1932) to *Triloculina insignis* (Brady, and by Barker (1960) to *T. striatotrigonula* Parker and Jones. *Triloculina striatotrigonula* Parker and Jones 1865 was a *nomen nudum* (Hodgkinson 1992). The name was validated by the publication of a description by Parr in 1941 (Hodgkinson 1992). Brady's illustrations constitute the type figures. The striate morphotype illustrated by Brady is herein regarded as no more than subspecifically distinct from *Triloculina trigonula* (Lamarck 1804) *sensu stricto* (see Plate 3, Figs 15–16).

Plate 5 (V)

Figures 1–2 *Edentostomina cultrata* (Brady 1881)

[*Miliolina cultrata* Brady 1881]
× 50. From *Challenger* Sta. 217A, off Papua, Pacific (37fm.). ZF1862.

Referred by Brady (1884) to *Miliolina*, by Thalmann (1932) and Barker (1960) to *Quinqueloculina*, and by Collins (1958), Haig (1988), and Zheng (1988) to *Edentostomina*. This is the type-species by original designation of the genus *Edentostomina* Collins 1958.

Figure 3 *Quinqueloculina tropicalis* Cushman 1924

× 50. From *Challenger* Sta. 217A, off Papua, Pacific (37fm.). ZF1866.

Referred by Brady (1884) to *Miliolina gracilis* d'Orbigny, sp., and by Thalmann (1932), Barker (1960), Whittaker and Hodgkinson (1979), and Zheng (1988) to *Quinqueloculina tropicalis* Cushman. The generic identity of this species is somewhat questionable.

Figure 4 *Triloculinella obliquinodus* Riccio 1950

× 50. From *Challenger* Sta. 174B, off Kandavu, Fiji, Pacific (610fm.). ZF1876.

Referred by Brady (1884) to *Miliolina oblonga* Montagu, sp., by Thalmann (1932) to *Triloculina oblonga* (Montagu), and by Barker (1960) (albeit tentatively) to *Miliolinella oblonga* (Montagu). Brady's species appears distinct from Montagu's and in fact appears almost identical to *Triloculinella obliquinodus* Riccio 1950, the type-species by original designation of *Triloculinella* Riccio 1950. Loeblich and Tappan first (1964) synonymized *Triloculinella* Riccio 1950 with *Miliolinella* Wiesner 1931, but later (1987) reinstated it on the basis of its initially 'cryptoquinqueloculine' chamber arrangement (Bogdanovich 1969).

Figure 5 *Edentostomina* **sp. nov.**

× 50. From *Challenger* Sta. 142A, Simon's Bay, Cape of Good Hope, off South Africa (15–20fm.). ZF1919.

Referred by Brady (1884) to *Miliolina venusta* Karrer, sp., and by Thalmann (1932), Barker (1960), Hermelin (1989), and van Marle (1991) to *Quinqueloculina venusta* Karrer (see below). Regarded by the present author as a new species of *Edentostomina*. Note, however, that Łuczkowska (pers. comm.) prefers placement in *Lachlanella*.

Figure 6 *Quinqueloculina seminulum* **(Linné 1758)**

[*Serpula seminulum* Linné 1758]
Fig. 6a × 50. From *Porcupine* Sta. AA, off Skye, North Atlantic (45–60fm.). ZF1897.
Figs 6b, c × 50. From *Porcupine* Sta. 23, off NW Ireland, Atlantic (630fm.). ZF1898.

Referred by Brady (1884) to *Miliolina seminulum* Linne, sp., by Thalmann (1932) to *Quinqueloculina akneriana* d'Orbigny, and by Barker (1960), Haynes (1973a), and van Marle (1991) to *Q. seminulum* (Linne) (see also Text-figures 2–3). This is the type-species by subsequent designation of the genus *Quinqueloculina* d'Orbigny 1826.

Figure 7 *Quinqueloculina venusta* **Karrer 1868**

× 50. From *Challenger* Sta. 346, South Atlantic (2350fm.). ZF1918.

Referred by Brady (1884) to *Miliolina venusta* Karrer, sp., by Thalmann (1932) to *Quinqueloculina venusta* Karrer, and by Barker (1960) and Hermelin (1989) to *Q. lamarckiana* d'Orbigny.

Figures 8–9 *Quinqueloculina auberiana* **d'Orbigny 1839**

Fig. 8 × 30. From *Challenger* Sta. 24, West Indies (390fm.). ZF1847.
Fig. 9 × 40. From *Challenger* Sta. 24, West Indies (390fm.). ZF1848.

Referred by Brady (1884) to *Miliolina auberiana* d'Orbigny, sp., by Thalmann (1932), Barker (1960) and Hofker (1976) to *Quinqueloculina auberiana* d'Orbigny, and by Whittaker and Hodgkinson (1979) tentatively to *Q. cuvieriana* d'Orbigny.

Figures 10–11 *Sigmamiliolinella australis* **(Parr 1932)**

[*Quinqueloculina australis* Parr 1932]
× 50. From *Challenger* Sta. 162, Bass Strait, Pacific (38–40fm.). ZF1903.

Referred by Brady (1884) to *Miliolina subrotunda* Montagu, sp., by Thalmann (1932) to *Quinqueloculina australis* Parr, by Barker (1960) and van Marle (1991) to *Miliolinella australis* (Parr), by Hofker (1976) to *M. subrotunda* (Montagu) (Fig. 10 only), and by Zheng (1988) to *Sigmamiliolinella australis* (Parr). This is the type-species by original designation of the genus *Sigmamiliolinella* Zheng 1988.

Figure 12 *Quinqueloculina lamarckiana* **d'Orbigny 1839**

× 40. From *Challenger* Sta. 233B, Inland Sea, Japan (15fm.). ZF1863.

Referred by Brady (1884) to *Miliolina cuvieriana* d'Orbigny, sp., by Thalmann (1932), Barker (1960), and Hermelin (1989) to *Quinqueloculina lamarckiana* d'Orbigny, and by Whittaker and Hodgkinson

(1979) to *Q. cuvieriana* d'Orbigny. *Quinqueloculina cuvieriana* d'Orbigny 1839 and *Q. lamarckiana* d'Orbigny 1839 appear synonymous. *Q. lamarckiana* is the senior name only in that it has page precedence. Recommendation 69B(11) of the *International Code of Zoological Nomenclature* (ICZN) (Ride *et al.* 1985) reads that 'all other things being equal, preference should be given to the nominal species cited first in the work, page or line ("position preference")'.

Figures 13–14 *Miliolinella subrotunda* (Montagu 1803)

[*Vermiculum subrotundum* Montagu 1803]
× 50. From *Challenger* Sta. 135, South Atlantic (100–150fm.). ZF1859.

Referred by Brady (1884) to *Miliolina circularis* Bornemann, sp., by Thalmann (1932) to *Triloculina subrotunda* (Montagu), by Barker (1960) to *Miliolinella subrotunda* (Montagu), and by Hofker (1976) to *M. circularis* (Bornemann). This is the type-species by original designation of the genus *Miliolinella* Wiesner 1931.

Figure 15 *Quinqueloculina* sp. (aberrant)

× 12. From *Challenger* Sta. 246, North Pacific (2050fm.). ZF1921.

Referred by Brady (1884) to *Miliolina* sp. ('anomalous specimin . . . too obviously a monstrosity to require a distinctive name'), and by Thalmann (1932) and Barker (1960) to *Quinqueloculina* sp.

Plate 6 (VI)

Figures 1–2 *Pseudomassilina australis* (Cushman 1932)

[*Massilina australis* Cushman 1932]
× 50. From *Challenger* Sta. 187A, Booby Island, Torres Strait, Pacific (8fm.). ZF1896.

Referred by Brady (1884) to *Miliolina secans* d'Orbigny, sp., by Thalmann (1932) and Haynes (1973a) to *Massilina secans* (d'Orbigny), and by Barker (1960) and Whittaker and Hodgkinson (1979) to *Pseudomassilina australis* (Cushman). This is the type-species by original designation of the genus *Pseudomassilina* Lacroix 1938.

Figures 3–5 *Flintinoides labiosa* (d'Orbigny 1839)

[*Triloculina labiosa* d'Orbigny 1839]
Figs 3, 5 × 45. From *Challenger* Sta. 135, off Nightingale Island, Tristan d'Acunha, Atlantic (100–150fm.). ZF1871.
Fig. 4 × 45. From *Challenger* Sta. 306, West of Patagonia, Pacific (345fm.). ZF1870.

Referred by Brady (1884) to *Miliolina*, by Thalmann first (1932) to *Triloculina* and later (1937) to *Miliolinella*, and also by Barker (1960) to *Miliolinella*. This is the type-species by original designation of the genus *Flintinoides* Cherif 1970, which differs from *Miliolinella* Wiesner 1931 in the ultimately planispiral rather than 'massiline' chamber arrangement. It has been neotypified by Le Calvez (1977).

Figures 6–7 *Quinqueloculina bradyana* Cushman 1917

Figs 6–7 × 50. From *Challenger* Sta. 162, Bass Strait, Pacific (38–40fm.). ZF1915.

Referred by Brady (1884) to *Miliolina undosa* Karrer, sp., and by Thalmann (1932) and Barker (1960) to *Quinqueloculina bradyana* Cushman.

Figure 8 *Adelosina pascuaensis* Koutsoukos and Falcetta 1987, var.

Fig. 8 × 50. From *Challenger* Sta. 260A, off Honolulu reefs, Pacific (40fm.). ZF1914.

Referred by Brady (1884) to *Miliolina undosa* Karrer, sp., and by Thalmann (1932) and Barker (1960) to *Quinqueloculina bradyana* Cushman. Regarded by the present author as no more than varietally (i.e. infra-subspecifically) distinct from *Adelosina pascuaensis* Koutsoukos and Falcetta 1987. *Adelosina* d'Orbigny 1826 differs from *Quinqueloculina* d'Orbigny 1826 externally principally in the terminal placement of the aperture and in the generally greater test ornamentation.

Figure 9 *Adelosina bicornis* (Walker and Jacob 1798), emend. Haynes 1973

[*Serpula bicornis* Walker and Jacob 1798]
× 40. From *Porcupine* Sta. AA, off Skye, North Atlantic (45–60fm.). ZF1852.

Referred by Brady (1884) to *Miliolina*, by Thalmann (1932), and Barker (1960), and Haynes (1973a) to *Quinqueloculina*. Loeblich and Tappan (1987) refer this species to *Adelosina*, which differs from *Quinqueloculina* principally in the terminal placement of the aperture (see above). Łuczkowska (pers. comm.) prefers to refer this species to *Lachlanella* Vella 1957.

Figure 10 *Massilina amygdaloides* (Brady 1884)

[*Miliolina amygdaloides* Brady 1884]
× 50. From *Challenger* Sta. 232, South of Japan (345fm.). ZF1846.

Referred by Brady (1884) to *Miliolina*, and by Thalmann (1932), Barker (1960), and van Marle (1991) to *Quinqueloculina*. The generic identity of this species is somewhat questionable.

Figures 11–12 *Adelosina intricata* (Terquem 1878)

[*Quinqueloculina intricata* Terquem 1878]
× 35. From *Porcupine* Sta. AA, off Skye, North Atlantic (45–60fm.). ZF1853.

Referred by Brady (1884) to *Miliolina bicornis* Walker and Boys, sp., and by Thalmann (1932), Barker (1960), and Haynes (1973a) to *Quinqueloculina intricata* Terquem. Łuczkowska (pers. comm.) prefers to refer this species to *Adelosina*, which differs from *Quinqueloculina* principally in the terminal placement of the aperture (see above).

Figures 13–14 *Adelosina pulchella* d'Orbigny 1846

Fig. 13 × 25. From West of Scotland, Atlantic. ZF1879.
Fig. 14 × 35. From *Porcupine* Sta. AA, off Skye, North Atlantic (45–60fm.). ZF1880.

Referred by Brady (1884) to *Miliolina*, and by Thalmann (1932) and Barker (1960) to *Quinqueloculina*. Łuczkowska (pers. comm.) again prefers to refer this species to *Adelosina*, which differs from *Quinqueloculina* principally in the terminal placement of the aperture (see above). Papp and Schmid (1985) synonymize *Adelosina pulchella* d'Orbigny 1846 with *A. schreibersii* d'Orbigny 1846.

Figures 15–20 *Adelosina granulocostata* (Germeraad 1946)

[*Quinqueloculina granulocostata* Germeraad 1946]
Figs 15, 17, 19, 20 × 35. From *Challenger* Sta. 260A, off Honolulu reefs, Pacific (40fm.). ZF1873.
Fig. 16 × 35. From *Challenger* Sta. 218A, Nares Harbour, Admiralty Islands, Pacific (16–25fm.). ZF1872.
Fig. 18 × 35. From *Challenger* Sta. 187A, off Booby Island, Pacific (8fm.). ZF1874.

Referred by Brady (1884) to *Miliolina linneiana* d'Orbigny, sp., by Thalmann (1932) to *Quinqueloculina linneiana* (d'Orbigny), and by Barker (1960), Zheng (1988), and van Marle (1991) to *Quinqueloculina granulocostata* Germeraad. Łuczkowska (pers. comm.) yet again prefers to refer this species to *Adelosina*, which differs from *Quinqueloculina* principally in the placement of the aperture (see above).

Plate 7 (VII)

Figures 1–4 *Ptychomiliola separans* (Brady 1881)

[*Miliolina separans* Brady 1881]
Fig. 1 × 20. From *Challenger* Sta. 182, Torres Strait, Pacific (155fm.). ZF1899.
Figs 2, 3 × 20. From Storm Bay, Tasmania. Fig. 2 ZF1900; Fig. 3 ZF1902.
Fig. 4 × 20. From *Challenger* Sta. 187A, Torres Strait, Pacific (8fm.). ZF1901.

Referred by Brady (1884) to *Miliolina*, and by Thalmann (1932) and Barker (1960) to *Ptychomiliola*. This is the type-species by subsequent designation of the genus *Ptychomiliola* Eimer and Fickert 1899.

Figures 5–6 *Pseudomassilina macilenta* (Brady 1884)

[*Miliolina macilenta* Brady 1884]
× 50. From *Challenger* Sta. 219A, Admiralty Islands, Pacific (17fm.). ZF1875.

Referred by Brady (1884) to *Miliolina*, by Thalmann first (1932) to *Massilina* and later (1942) to *Pseudomassilina*, and also by Barker (1960) and Haig (1988) to *Pseudomassilina*.

Figures 7–12 *Edentostomina rupertiana* (Brady 1881)

[*Miliolina rupertiana* Brady 1881]
Fig. 7 × 30; Fig. 11 × 100. From *Challenger* Sta. 187, South of New Guinea, Pacific (6fm.). Fig. 7 ZF1889; Fig. 11 ZF1890.
Fig. 8 × 30. From *Challenger* Sta. 186, Flinders Passage, Pacific (7–8fm.). ZF1892.
Fig. 9 × 30. From *Challenger* Sta. 188, South of Papua, Pacific (28fm.). ZF1893.
Fig. 10 × 30. From *Challenger* Sta. 187A, Torres Strait, Pacific (8fm.). ZF1894.
Fig. 12 × 100. From *Challenger* Sta. 189, South of Papua, Pacific (28fm.).

Referred by Brady (1884) to *Miliolina*, by Thalmann (1932) and Barker (1960) to *Triloculina*, and by Collins (1958) and Haig (1988)

to *Edentostomina* (see also Text-figure 4). This is the type-species by original designation of the genus *Rupertianella* Loeblich and Tappan 1985 (a new name for *Pseudotriloculina* Rasheed 1971 not Cherif 1970), regarded by the present author as a junior synonym of *Edentostomina* Collins 1958.

Figure 13 *Quinqueloculina boueana* d'Orbigny 1846

× 50. From *Challenger* Sta. 204A, Manila Harbour, Philippines (4fm.). ZF1854.

Referred by Brady (1884) to *Miliolina*, and by Thalmann (1932), Barker (1960), and van Marle (1991) to *Quinqueloculina*. This species has recently been lectotypified by Papp and Schmid (1985).

Figure 14 *Quinqueloculina parkeri* (Brady 1881)

[*Miliolina parkeri* Brady 1881]
× 40. From the Seychelles. ZF1877.

Referred by Brady (1884) to *Miliolina*, and by Thalmann (1932), Barker (1960), Whittaker and Hodgkinson (1979), and Haig (1988) to *Quinqueloculina*. Łuczkowska (pers. comm.) prefers to refer this species to *Lachlanella* Vella 1957.

Figures 15–17 *Pseudohauerina occidentalis involuta* (Cushman 1946)

[*Hauerina involuta* Cushman 1946]
Fig. 15 ZF1568; Fig. 16 ZF1569; Fig. 17 ZF1571. × 50. From *Challenger* Sta. 218A, Nares Harbour, Admiralty Islands, Pacific (16–25fm.).

Referred by Brady (1884) and Thalmann (1932) to *Hauerina ornatissima* (Karrer), by Barker (1960) to *H. involuta* Cushman, and by Ponder (1972) to *Pseudohauerina occidentalis involuta* (Cushman). *Hauerina occidentalis* Cushman 1946 is the type-species by original designation of the genus *Pseudohauerina* Ponder 1972.

Figures 18–22 *Pseudohauerina orientalis* (Cushman 1946)

[*Hauerina orientalis* Cushman 1946]
Figs 18–19 × 50. From Levuka, Ovalau, Fiji, Pacific (12fm.). ZF1570.
Fig. 20 × 50. From *Challenger* Sta. 174C, off Fiji, Pacific (210fm.). ZF1572.
Fig. 21 × 50. From *Challenger* Sta. 218A, Nares Harbour, Admiralty Islands, Pacific (16–25fm.). ZF1573.
Fig. 22 × 50. From *Challenger* Sta. 172, Friendly Islands, Pacific (18fm.). ZF1574.

Referred by Brady (1884) and Thalmann (1932) to *Hauerina ornatissima* (Karrer), by Barker (1960) to *H. orientalis* Cushman (Figs 18–20) and *H.* sp. (Figs 21–22), and by Ponder (1972) to *Pseudohauerina orientalis* (Cushman) (Fig. 20).

Plate 8 (VIII)

Figures 1–4 *Sigmoilopsis schlumbergeri* (Silvestri 1904)

[*Sigmoilina schlumbergeri* Silvestri 1904]
Figs 1–2 × 50. From *Porcupine* Sta. 23, off NW Ireland, Atlantic (630fm.). ZF2099.

Fig. 3 × 50. From *Challenger* Sta. 120, off Pernambuco, Atlantic (675fm.). ZF2101.

Fig. 4 × 50. From *Porcupine* Sta. 37, off Ireland, Atlantic (2435fm.). ZF2100.

Referred by Brady (1884) to *Planispirina celata* Costa, sp., by Thalmann (1932) and Hofker (1976) to *Sigmoilina schlumbergeri* (Silvestri), and by Barker (1960), Schroder (1986), Zheng (1988), Hermelin (1989), and van Marle (1991) to *Sigmoilopsis schlumbergeri* (Silvestri). This is the type-species by original designation of the genus *Sigmoilopsis* Finlay 1947.

Figure 5 *Siphonaperta crassatina* (Brady 1884)

[*Miliolina crassatina* Brady 1884]
× 50. From *Challenger* Sta. 162, Bass Strait, Pacific (38fm.). ZF1861.

Referred by Brady (1884) to *Miliolina crassatina* (a new name for *M. incrassata* Brady 1881, not *Quinqueloculina incrassata* Karrer 1868), by Thalmann (1932) and Barker (1960) to *Flintina crassatina* (Brady), by Haig (1988) to *Quinqueloculina crassatina* (Brady), and by Zheng (1988) to *Siphonaperta crassatina* (Brady). The chamber arrangement appears to be triloculine rather than quinqueloculine as in typical *Siphonaperta*.

Figures 6–7 *Siphonaperta* sp. nov.

Fig. 6 × 50. From *Challenger* Sta. 218A, Nares Harbour, Admiralty Islands, Pacific (16–25fm.). ZF1817.

Fig. 7 × 50. From *Porcupine* Sta. 37, off Ireland, Atlantic (2435fm.). ZF1925.

Referred by Brady (1884) to *Miliolina agglutinans* d'Orbigny, sp., by Thalmann (1932) to *Quinqueloculina agglutinans* d'Orbigny, and by Barker (1960) to *Dentostomina agglutinans* (d'Orbigny). Barker (1960) noted that Brady's figures 'may represent a new species'.

Figures 8–10 *Pseudoflintina triquetra* (Brady 1879)

[*Miliolina triquetra* Brady 1879]
Figs 8–9 ZF1912. Fig. 10 ZF1913. × 30. From *Challenger* Sta. 162, Bass Strait, Pacific (38–40fm.).

Referred by Brady (1884) to *Miliolina*, by Thalmann (1932) and Barker (1960) to *Flintina*, and Zheng (1988) to *Pseudoflintina*. This is the type-species by original designation of the genus *Pseudoflintina* Saidova 1981. Łuczkowska (pers. comm.) prefers placement in *Sigmoilopsis*.

Figure 11 *Proemassilina* sp. nov.

× 50. From *Porcupine* Sta. 47, Faroe Channel, North Atlantic (542fm.). ZF2393.

Referred by Brady (1884) to *Spiroloculina asperula* Karrer, by Thalmann (1932) to *Massilina alveoliniformis* (Millett), and by Barker (1960) to *Ammomassilina alveoliniformis* (Millett) (or *Praemassilina* (*sic*) sp.). Regarded by the present author as a new species of *Proemassilina* Lacroix 1938, a genus which Loeblich and Tappan first (1964) synonymized with *Massilina* Schlumberger 1893 and later (1987) reinstated. Łuczkowska (pers. comm.) again prefers *Sigmoilopsis*.

Figure 12 *Proemassilina arenaria* (Brady 1884)

[*Spiroloculina arenaria* Brady 1884]
× 30. From *Challenger* Sta. 174C, off Fiji, Pacific (210fm.). ZF2392.

Referred by Brady (1884) to *Spiroloculina*, by Thalmann (1932) and van Marle (1991) to *Massilina*, and by Barker (1960) to *Praemassilina* (*sic*). Łuczkowska (pers. comm.) yet again prefers *Sigmoilopsis*

Figure 13 *Ammomassilina alveoliniformis* (Millett 1898)

[*Massilina alveoliniformis* Millett 1898]
× 18. From *Challenger* Sta. 218A, Nares Harbour, Admiralty Islands, Pacific (16–25fm.). ZF2394.

Referred by Brady (1884) to *Spiroloculina asperula* Karrer, by Thalmann (1932) to *Massilina alveoliniformis* Millett, and by Barker (1960) (albeit tentatively) to *Ammomassilina alveoliniformis* (Millett). This is the type-species by original designation of the genus *Ammomassilina* Cushman, 1933. Łuczkowska (pers. comm.) prefers placement in *Sigmoilopsis*.

Figure 14 *Spiroglutina asperula* (Kerrer 1868)

[*Spiroloculina asperula* Karrer 1868]
× 50. From *Challenger* Sta. 217A, off Papua, Pacific (37fm.). ZF2395.

Referred by Brady (1884) to *Spiroloculina asperula* Karrer, by Thalmann (1932) to *Massilina alveoliniformis* (Millett), and by Barker (1960) to *Ammomassilina alveoliniformis* (Millett) (or *Praemassilina* (*sic*) *asperula* Karrer). *Spiroloculina asperula* Karrer 1868 is the type-species by original designation of the genus *Spiroglutina* Mikhalevich 1983, which Loeblich and Tappan (1987) regarded as a junior synonym of *Sigmoilinita* Seiglie 1965. It was referred by Zheng (1988) to *Sigmoilopsis* Finlay 1947. Łuczkowska (pers. comm.) also prefers placement in *Sigmoilopsis*. However, *Sigmoilopsis* lacks a late planispiral portion.

Figures 15–20 *Schlumbergerina alveoliniformis* (Brady 1879)

[*Miliolina alveoliniformis* Brady 1879]
Fig. 15 × 25. From *Challenger* Sta. 218A, Nares Harbour, Admiralty Islands, Pacific (16–25fm.). ZF1818.
Fig. 16 × 25.
Fig. 17 × 25. From *Challenger* Sta. 260A, off Honolulu reefs, Pacific (40fm.). ZF1819.
Fig. 18 × 40; Figs 19–20 × 25. From Tongatabu, Pacific (18fm.).
Fig. 18 ZF1820; Fig. 19 ZF1842; Fig. 20 ZF1821.

Referred by Brady (1884) to *Miliolina*, and by Thalmann (1932) and Barker (1960) to *Schlumbergerina*.

Plate 9 (IX)

Figure 1 *Vertebralina* sp.

× 50. From *Challenger* Sta. 174A, off Fiji, Pacific (255fm.). ZF1922.

Referred by Brady (1884) to '*Miliolina . . . or possibly . . . Articulina*', by Thalmann first (1932) to *Quinqueloculina* and later (1933a) to *Vertebralina*, and also by Barker (1960) to *Vertebralina*. This is a juvenile specimen whose specific identity remains questionable.

Figures 2–3 *Quinqueloculina pseudoreticulata* **Parr 1941**

× 25. From *Challenger* Sta. 188, South of New Guinea, Pacific (28fm.). ZF1883.

Referred by Brady (1884) to *Miliolina reticulata* d'Orbigny, sp., by Thalmann (1932) to *Quinqueloculina reticulata* (d'Orbigny), and by Barker (1960), Zheng (1988), Whittaker and Hodgkinson (1979), and van Marle (1991) to *Q. pseudoreticulata* Parr.

Figure 4 *Adelosina reticulata* **(d'Orbigny 1826)**

[*Triloculina reticulata* d'Orbigny 1826]
× 25. From *Challenger* Sta. 24, West Indies (390fm.). ZF1884.

Referred by Brady (1884) to *Miliolina*, by Thalmann (1932) and later Barker (1960) to *Quinqueloculina*, and by the present author to *Adelosina*. *Adelosina* differs from *Quinqueloculina* principally in the terminal placement of the aperture.

Figures 5–6 *Spiroloculina communis* **Cushman and Todd 1944**

× 30. From *Challenger* Sta. 218A, Nares Harbour, Admiralty Islands, Pacific (16–25fm.). ZF2399. See also 1962:5:20:49.

Referred by Brady (1884) to *Spiroloculina excavata* d'Orbigny, by Thalmann (1932) to *S. grateloupi* d'Orbigny, and by Barker (1960), Haig (1988), Zheng (1988), and van Marle (1991) to *S. communis* Cushman and Todd. This is the type-species by original designation of the genus *Bidentina* Mikhalevich 1988, regarded by the present author as a junior synonym of *Spiroloculina* d'Orbigny 1826.

Figures 7–8 *Spiroloculina robusta* **Brady 1884**

× 20. From *Challenger* Sta. 24, West Indies (390fm.). ZF2414.

Referred by Brady (1884) and Zheng (1988) to *Spiroloculina*, and by Thalmann (1932) and Barker (1960) to *Flintia*. This is the type-species by original designation of the genus *Flintia* Schubert 1911, regarded by the present author as a junior synonym of *Spiroloculina* d'Orbigny 1826.

Figures 9–10 *Massilina carinata* **(Fornasini 1903)**

[*Spiroloculina nitida* var. *carinata* Fornasini 1903]
Fig. 9 × 50. From *Challenger* Sta. 233B, off Japan, Pacific (15fm.). ZF2411.
Fig. 10 × 40. From *Challenger* Sta. 187A, Torres Strait, Pacific (8fm.). ZF2412.

Referred by Brady (1884) to *Spiroloculina nitida* d'Orbigny, by Thalmann (1932) to *S. milletti* Wiesner, by Barker (1960) to *Massilina milletti* (Wiesner), and by Atkinson (1968) and Haynes (1973a) to *M. carinata* (Fornasini). Łuczkowska (pers. comm.) prefers placement in *Spiroloculina*.

Figure 11 *Spiroloculina henbesti* **Thalmann 1955**

× 35. From *Porcupine* Sta. AA, off Skye, North Atlantic (45–60fm.). ZF2413.

Referred by Brady (1884) and Thalmann (1932) to *Spiroloculina planulata* (Lamarck), and by Barker (1960) to *S. henbesti* Petri. This species should in fact be attributed to Thalmann (1955).

Figures 12–14 *Hauerina fragilissima* **(Brady 1884)**

[*Spiroloculina fragilissima* Brady 1884]
Figs 12, 13 × 50. From *Challenger* Sta. 279C, off Tahiti, Pacific (620fm.). Fig. 12 ZF2400; Fig. 13 ZF2401.
Fig. 14 × 50. From *Challenger* Sta. 218A, Nares Harbour, Admiralty Islands, Pacific (16–25fm.). ZF2402.

Referred by Brady (1884) to *Spiroloculina*, and by Thalmann (1932), Barker (1960), Whittaker and Hodgkinson (1979), and Haig (1988) to *Hauerina*.

Figures 15–16 *Spiroloculina rotunda* **d'Orbigny 1826**

Fig. 15 × 50. From *Challenger* Sta. 344, Ascension Island, South Atlantic (420fm.). ZF2408.
Fig. 16 × 30. ZF2409.

Referred by Brady (1884) to *Spiroloculina limbata* d'Orbigny, by Thalmann (1932) to *S. depressa* d'Orbigny, and by Barker (1960) (albeit tentatively) and van Marle (1991) to *S. rotunda* d'Orbigny. Van Marle (1991) regards *Spiroloculina rotunda* d'Orbigny 1826 as a *nomen nudum*. It is true that d'Orbigny's citation (d'Orbigny 1826) is unaccompanied either by a description or a figure, but it is nonetheless valid as it makes reference to the earlier figures of Soldani (1789, 1798).

Figure 17 *Spiroloculina depressa* **d'Orbigny 1826**

× 30. From *Porcupine* Sta. AA, off Skye, North Atlantic (45–60fm.). ZF2410.

Referred by Brady (1884) to *Spiroloculina limbata* d'Orbigny, and by Thalmann (1932), Barker (1960) (albeit tentatively), Haynes (1973a), and van Marle (1991) to *S. depressa* d'Orbigny. Van Marle (1991) regards *Spiroloculina depressa* d'Orbigny 1826 as a *nomen nudum*, but it is valid for the same reasons as those quoted in the case of *S. rotunda* d'Orbigny 1826 (see above). *Spiroloculina depressa* d'Orbigny, 1826 is the type-species by subsequent designation of the genus *Spiroloculina* d'Orbigny 1826.

Plate 10 (X)

Figures 1–2 *Spiroloculina bradyi* **Barker 1960**

× 30. From *Challenger* Sta. 189, South of Papua, Pacific (25–29fm.).

Referred by Brady (1884) and Thalmann (1932) to *Spiroloculina limbata* d'Orbigny, var., by Barker (1960) to *S.* sp. nov. (for which 'the new name *S. bradyi* is proposed'), and by Haig (1988) to *S. bradyi* Barker.

Figures 3–4 *Spiroloculina communis* **Cushman and Todd 1944**

× 40. From *Challenger* Sta. 217A, South of Papua, Pacific (37fm.). ZF2407

Referred by Brady (1884) to *Spiroloculina impressa* Terquem, by Thalmann (1932) to *S. canaliculata* d'Orbigny, and by Barker (1960) to *S. communis* Cushman and Todd. This is the type-species by original designation of the genus *Bidentina* Mikhalevich 1988, regarded by the present author as a junior synonym of *Spiroloculina* d'Orbigny 1826 (see also Plate 9, Figs 5, 6).

Figures 5–6 *Spiroloculina tenuiseptata* Brady 1884

Fig. 5 × 40. From *Challenger* Sta. 191A, Ki Islands, Japan (580fm.). ZF2418.
Fig. 6 × 75. From *Challenger* Sta. 174B, off Fiji, Pacific (610fm.). ZF2419.

Referred by Brady (1884), Thalmann (1932), and Barker (1960) to *Spiroloculina*, and by Zheng (1988) to *Ophthalmidium*.

Figures 7, 8, 11 *Spirosigmoilina tenuis* (Čžjžek 1848)

[*Quinqueloculina tenuis* Čžjžek 1848]
× 50. From *Challenger* Sta. 302, South Pacific (1450fm.). ZF2415.

Referred by Brady (1884) and Thalmann (1932) to *Spiroloculina*, and by Barker (1960) to *Sigmoilina*. This is the type-species by original designation of the genus *Sigmoilinita* Seiglie 1965, regarded by the present author as a junior synonym of *Spirosigmoilina* Parr 1942. Zheng (1988) also referred it to *Spirosigmoilina*, while erroneously retaining *Sigmoilinita* for another species.

Figures 9–10 *Spirosigmoilina pusilla* (Earland 1934)

[*Spiroloculina pusilla* Earland 1934]
× 50. From *Challenger* Sta. 332, South Atlantic (2200fm.). ZF2416.

Referred by Brady (1884) to *Spiroloculina tenuis* Čžjžek, sp., by Thalmann first (1932) to *S. tenuissima* Reuss and later (1937) to *S. pusilla* Earland, by Barker (1960) to *Spirophthalmidium pusillum* (Earland), by Zheng (1988) to *Spirosigmoilina pusilla* (Earland), and by Hermelin (1989) to *Ophthalmidium pusillum* (Earland). Łuczkowska (pers. comm.) prefers placement of the present species in the genus *Sigmoilinita* Seiglie 1965, regarded by the present author as a junior synonym of *Spirosigmoilina* Parr 1942.

Figure 12 *Spiroloculina elevata* Wiesner 1923

× 50. From *Challenger* Sta. 185, Torres Strait, Pacific (155fm.). ZF2389.

Referred by Brady (1884) to *Spiroloculina acutimargo* (see below), by Thalmann (1932) to *S. affixa* Terquem, and by Barker (1960) to *S. elevata* Wiesner. Łuczkowska (pers. comm.) prefers placement in *Inaequalina* Łuczkowska 1971, regarded by the present author as a junior synonym of *Spiroloculina* d'Orbigny 1826.

Figure 13 *Spirophthalmidium acutimargo* (Brady 1884)

[*Spiroloculina acutimargo* Brady 1884]
× 50. From *Challenger* Sta. 120, off Pernambuco, West Atlantic (675fm.). ZF2390.

Referred by Brady (1884) to *Spiroloculina*, by Thalmann (1932) to *Siprophthalmidium (sic)*, by Barker (1960) to *Spirophthalmidium*, and by Łuczkowska (1971) and van Marle (1991) to *Ophthalmidium*. This is the type-species by original designation of the genus *Spiropthalmidium* Cushman 1927 (*nom. imperf.*) [= *Spirophthalmidium* Paalzow, 1932 (nom. corr.)].

Figure 14 *Spirophthalmidium emaciatum* (Haynes 1973)

[*Spiropthalmidium acutimargo* var. *emaciatum* Haynes 1973]
× 50. From *Challenger* Sta. 174A, off Fiji, Pacific (255fm.). ZF2387.

Referred by Brady (1884) to *Spiroloculina acutimargo* (see above), by Thalmann (1932) and Barker (1960) to *S.* sp. abnorm., and by Haynes (1973a) to *Spiropthalmidium (sic) acutimargo* var. *emaciatum*. The generic identity of this species is somewhat questionable.

Figure 15 *Spiroloculina venusta* Cushman and Todd 1944

× 50. From *Challenger* Sta. 185A, Torres Strait, Pacific (155fm.). ZF2388.

Referred by Brady (1884) to *Spiroloculina acutimargo* (see above), by Thalmann (1932) to *S. tenuirostra* Karrer nov. var., and by Barker (1960) to *S. venusta* Cushman and Todd.

Figures 16–17, 22–23 *Spiroloculina angulata* (Cushman 1917)

[*Spiroloculina grata* var. *angulata* Cushman 1917]
Fig. 16 × 50. From *Challenger* Sta. 172, Friendly Islands, Pacific (18fm.). ZF2403.
Fig. 17 × 50. From *Challenger* Sta. 218A, Nares Harbour, Admiralty Islands, Pacific (16–25fm.). ZF2404.
Fig. 22 × 30. From Shell Cove, Port Jackson, Australia. ZF2405.
Fig. 23 × 30. From *Challenger* Sta. 187A, Booby Island, Torres Strait, Pacific (8fm.). ZF2406.

Referred by Brady (1884) to *Spiroloculina grata* Terquem, by Thalmann (1932) to *S. grata* Terquem (Figs 16–17) and *S. antillarum* var. *angulata* Cushman (Figs 22–23), and by Barker (1960) to *S. angulata* Cushman.

Figures 18–20 *Nodobaculariella convexiuscula* (Brady 1884)

[*Spiroloculina (?) convexiuscula* Brady 1884].
× 50. From *Challenger* Sta. 185, Torres Strait, Pacific (155fm.). ZF2396.

Referred by Brady (1884) to *Spiroloculina (?)*, by Thalmann (1932) to *Quinqueloculina*, and by Barker (1960) and Haig (1988) to *Nodobaculariella*.

Figure 21 *Spiroloculina antillarum* d'Orbigny 1839

× 30. From *Challenger* Sta. 122, SE of Pernambuco, Atlantic (350fm.). ZF2391.

Figures 24–26 *Spirosigmoilina bradyi* Collins 1958

Fig. 24 × 75. From *Challenger* Sta. 218A, Nares Harbour, Admiralty Islands, Pacific (16–25fm.). ZF2397.
Figs 25–26 × 75. From *Challenger* Sta. 260A, Hawaii (40fm.). ZF2398.

Referred by Brady (1884) to *Spiroloculina crenata* Karrer, sp., by Thalmann (1932) to *Massilina crenata* Karrer, by Collins (1958) and Haig (1988) to *Spirosigmoilina bradyi*, and by Barker (1960) to *Hauerina speciosa* Karrer.

Plate 11 (XI)

Figures 1–3 *Cornuspira involvens* (Reuss 1849)

[*Operculina involvens* Reuss 1849]

Figs 1a–b × 30. From *Challenger* Sta. 23 or 24, West Indies (450 or 390fm. respectively). Fig. 1a ZF1294; Fig. 1b ZF1297.

Fig. 2 × 30. From *Challenger* Sta. 279C, Tahiti, Pacific (620fm.). ZF1295.

Fig. 3 × 30. From *Challenger* Sta. 149D, Balfour Bay, Kerguelen Islands, South Pacific (20–60fm.). ZF1296.

Referred by Brady (1884), Thalmann (1932), and Barker (1960) to *Cornuspira*, and by Zheng (1988) and van Marle (1991) to *Cyclogyra*. The ICZN have recently suppressed *Cyclogyra* Wood 1842 in favour of *Cornuspira* Schultze 1854 (Melville 1978).

Figure 4 *Cornuspira carinata* (Costa 1856)

[*Operculina carinata* Costa 1856]
× 30. From *Porcupine* Sta. 11, West of Ireland, North Atlantic (1630fm.). ZF1288.

Referred by Brady (1884), Thalmann (1932) and Barker (1960) to *Cornuspira*, and by Hofker (1976) and Zheng (1988) to *Cyclogyra* (see above).

Figures 5–6 *Cornuspira foliacea* (Philippi 1844)

[*Orbis foliaceus* Philippi 1844]
× 15. From *Challenger* Sta. 24, West Indies (390fm.). ZF1290.

Referred by Brady (1884) to *Cornuspira*, and by Thalmann (1932), Barker (1960), and Zheng (1988) to *Cornuspiroides*. This is the type-species by subsequent designation of the genus *Cornuspira* Schultze, 1854.

Figure 7 *Cornuspirella diffusa* (Heron-Allen and Earland, 1913)

[*Cornuspira diffusa* Heron-Allen and Earland, 1913].
× 16. From *Challenger* Sta. 174C, off Fiji, Pacific (210fm.). ZF1291.

Referred by Brady (1884) to *Cornuspira foliacea* Philippi, sp. (see above), by Thalmann first (1932) to *Cornuspiroides foliaceum* (Philippi) and later (1933a) to *Cornuspirella diffusa* (Heron-Allen and Earland), and also by Barker (1960) to *Cornuspirella diffusa* (Heron-Allen and Earland). This is the type-species by original designation of the genus *Cornuspirella* Cushman 1928.

Figures 8–9 *Cornuspiroides primitivus* (Rhumbler 1904)

[*Cornuspira primitiva* Rhumbler 1904]
× 16. From *Challenger* Sta. 24, West Indies (390fm.). ZF1292; Fig. 9 ZF1293.

Referred by Brady (1884) to *Cornuspira foliacea* Philippi, sp. (see above), by Thalmann first (1932) to *Cornuspiroides foliaceum* (Philippi) and later (1933a) to *C. primitivus* (Rhumbler), and also by Barker (1960) to *C. primitivus* Rhumbler. Note that the present species lacks the late flabelliform development of typical *Cornuspiroides*.

Figures 10–11 *Nummoloculina contraria* (d'Orbigny 1846)

[*Biloculina contraria* d'Orbigny 1846]
Fig. 10a × 20. From *Porcupine* Sta. AA, off Skye, North Atlantic (45–60fm.). ZF2105.
Figs 10b–11 × 20. From off the Hebrides, North Atlantic (?70–100fm.). ZF2106.

Referred by Brady (1884) to *Planispirina*, and by Thalmann (1932), Barker (1960), and Zheng (1988) to *Nummoloculina* (see also Text-figure 5). This is the type-species by original designation of the genus *Nummoloculina* Steinmann 1881. It has been lectotypified by Papp and Schmid (1985).

Figures 12–13 *Sigmoihauerina bradyi* (Cushman 1917)

[*Hauerina bradyi* Cushman 1917]
Fig. 12 × 50. From *Challenger* Sta. 187A, Torres Strait, Pacific (8fm.). ZF1566.
Fig. 13 × 50. From *Challenger* Sta. 187, Booby Island, Pacific (6fm.). ZF1567.

Referred by Brady (1884) to *Hauerina compressa* d'Orbigny, and by Thalmann (1932) and Barker (1960) to *H. bradyi* Cushman. This is the type-species by original designation of the genus *Sigmoihauerina* Zheng 1979.

Figures 14–16 *Polysegmentina circanata* (Brady 1881)

[*Hauerina circanata* Brady 1881]
Figs 14–15 × 50. From *Challenger* Sta. 187A, Torres Strait, Pacific (8fm.). ZF1565. Lectotype (Loeblich and Tappan 1964) removed from this slide to ZF3629.
Fig. 16 × 50.

Referred by Brady (1884), Thalmann (1932), Whittaker and Hodgkinson (1979), and Haig (1988) to *Hauerina*, and by Barker (1960) to *Polysegmentina*. This is the type-species by original designation of the genus *Polysegmentina* Cushman 1946, which Loeblich and Tappan first (1964) synonymized with *Hauerina* d'Orbigny 1839 and later (1987) reinstated.

Plate 12 (XII)

Figures 1–4 *Planispirinella exigua* (Brady 1879)

[*Planispirina exigua* Brady 1879]
Figs 1–2, 4 × 50. From *Challenger* Sta. 187A, off Booby Island, Pacific (8fm.). Figs 1–2 ZF2107; Fig. 4 ZF2110. Lectotype (Loeblich and Tappan 1964) ZF3616 ex ZF2110.
Fig. 3 × 50. From *Challenger* Sta. 185, Torres Strait, Pacific (155fm.). ZF2109.

Referred by Brady (1884) and Thalmann (1932, 1933a) to *Planispirina*, and by Barker (1960) and Haig (1988) (Figs 1–2, 4 only) to *Planispirinella* (see also Text-figure 5). This is the type-species by original designation of the genus *Planispirinella* Wiesner 1931.

Figures 5, 7–8 *Cornuloculina inconstans* (Brady 1879)

[*Hauerina inconstans* Brady 1879]
Fig. 5 × 30. From *Challenger* Sta. 120, off Pernambuco, West Atlantic (675fm.). ZF2021.
Figs 7–8 × 50. From *Challenger* Sta. 24, West Indies (390fm.). ZF2021.

Referred by Brady (1884) and Thalmann (1932) to *Ophthalmidium*, by Thalmann (1933a) and Barker (1960) to *Hauerinella*, and by Hofker (1976), Zheng (1988), and van Marle (1991) to *Cornuloculina*. This is the type-species by original designation of the genus

Hauerinella Schubert 1921, and also the type-species by subsequent designation of the genus *Cornuloculina* Burbach 1886. Thus, *Hauerinella* is an isotypic junior synonym of *Cornuloculina*. Van Marle (1991) regards *Hauerina inconstans* Brady 1881 as a *nomen nudum*, but gives no justification.

Figure 6 *Cornuloculina tumidula* (Brady 1884)

[*Ophthalmidium tumidulum* Brady 1884]
× 50. From *Challenger* Sta. 24, West Indies (390fm.). ZF2021.

Referred by Brady (1884) to *Ophthalmidium tumidulum*, by Thalmann first (1932) to *O. inconstans* (Brady) (see above) and later (1933a) to *Hauerinella inconstans* (Brady), and by Barker (1960) to *H. tumidulum* (Brady). As stated above, *Hauerinella* is a junior synonym of *Cornuloculina*.

Figures 9–11 *Vertebralina insignis* Brady 1884

Figs 9–10 × 50. Fig. 11 × 50. ZF2614. From *Challenger* Sta. 172, Friendly Islands, Pacific (18fm.).

Figures 12–13, 22, ?23, 24 *Articulina pacifica* Cushman 1944

Figs 12–13, 23–24 × 50. From *Challenger* Sta. 260A, off Honolulu reefs, Pacific (40fm.). Figs 12–13 ZF1101; Figs 23–24 ZF1100.
Fig. 22 × 50. From *Challenger* Sta. 172, Friendly Islands, Pacific (18fm.). ZF1099.

Referred by Brady (1884) and Thalmann (1932) to *Articulina sulcata* Reuss (Figs 12–13) and *A. sagra* d'Orbigny (Figs 22–24), and by Barker (1960) (Figs 12–13 and questionably Figs 22–24) to *A. pacifica* Cushman. The specimen represented by Fig. 23 appears to possess a quinqueloculine rather than sigmoiline initial chamber arrangement may belong in *Articularia*. Łuczkowska (pers. comm.) prefers *Nodobaculariella*.

Figures 14–16 *Vertebralina striata* d'Orbigny 1826

Fig. 14 × 50. From *Challenger* Sta. 187, South of New Guinea, Pacific (6fm.). ZF2615.
Figs 15–16 × 50. From *Challenger* Sta. 187A, Torres Strait, Pacific (8fm.). ZF2616.

This is the type-species by original monotypy of the genus *Vertebralina* d'Orbigny, 1826.

Figures 17–18 *Articularia sagra* (d'Orbigny 1839)

[*Articulina sagra* d'Orbigny 1839]
× 50. From *Challenger* Sta. 24, West Indies (390fm.). ZF1092.

Referred by Brady (1884) and Thalmann (1932, 1933a) to *Articulina conico-articulata* (Batsch), and by Barker (1960) to *A. sagra* d'Orbigny. Cushman (1944) regarded *Nautilus* (*Orthoceras*) *conico-articulatus* Batsch 1791 as unrecognizable and suggested that the name be allowed to lapse. The quinqueloculine rather than milioline initial chamber arrangement indicates placement in *Articularia* rather than *Articulina*.

Figures 19–21 *Articularia lineata* (Brady 1884)

[*Articulina lineata* Brady 1884]
× 50. From *Challenger* Sta. 33, off Bermuda, Atlantic (435fm.). ZF1098.

Referred by Brady (1884), Thalmann (1932), and Barker (1960) to *Articulina*, and transferred by the present author to *Articularia* (see above).

Plate 13 (XIII)

Figures 1–?2 *Articulina mayori* Cushman 1944

Fig. 1 × 50. From *Challenger* Sta. 33, off Bermuda (435fm.). ZF1091.
Fig. 2 × 50. From *Challenger* Sta. 297A, Tahiti (420fm.). ZF1090.

Referred by Brady (1884) and Thalmann (1932) to *Articulina conico-articulata* (Batsch), and by Barker (1960) to *A. mayori* Cushman. Brady (1884) regarded Figure 2 as representing a 'misshapen' specimen.

Figures 3–5 *Tubinella inornata* (Brady 1884)

[*Articulina funalis* var. *inornata* Brady 1884]
× 30. From *Challenger* Sta. 149D, Balfour Bay, Kerguelen Island, South Pacific (20–60fm.). Fig. 3 ZF1096; Fig. 4 ZF3656 lectotype designated by Loeblich and Tappan (1955); Fig. 5 ZF1096.

Referred by Brady (1884) to *Articulina funalis* var. *inornata*, and by Thalmann (1932) and Barker (1960) to *Tubinella inornata* (Brady). This is the type-species by subsequent designation of the genus *Tubinella* Rhumbler 1906.

Figures 6–11 *Tubinella funalis* (Brady 1884)

[*Articulina funalis* Brady 1884]
Figs 6–7, 9, 11 × 30. From *Challenger* Sta. 145, Prince Edward Island, South Pacific (50–150fm.). Figs 6–7 ZF1095; Fig. 9 ZF1094; Fig. 11 ZF1097. Lectotype designated by Loeblich and Tappan (1964) ex ZF1095 (specimen represented by Fig. 7).
Figs 8, 10 × 30. From *Challenger* Sta. 149D, Kerguelen Island, South Pacific (20–60fm.). ZF1093.

Referred by Brady (1884) to *Articulina*, and by Thalmann (1932), Barker (1960), and Haig (1988) to *Tubinella*. This is the type-species by original designation of the genus *Tubinellina* Wiesner 1931, regarded by the present author as a junior synonym of *Tubinella* Rhumbler 1906. Various aspects of it have been reviewed by Cooke (1978).

Figures 12–13 *Peneroplis bradyi* Cushman 1930

× 50. From *Challenger* Sta. 33, off Bermuda, Atlantic (435fm.). ZF2086.

Referred by Brady (1884) to *Peneroplis pertusus* var. g (*P. laevigatus* Karrer), by Thalmann (1932) to *P. pertusus* (Forskål), by Barker (1960) to *Puteolina bradyi* (Cushman), and by Haig (1988) to *Spirolina pertusus* (Forskål). *Puteolina* Hofker 1952 was regarded by Leoblich and Tappan (1987) as a junior synonym of *Laevipeneroplis* Sulc 1936, which is regarded by the present author as a junior synonym of *Peneroplis* de Montfort 1808.

Figure 14 *Peneroplis carinatus* d'Orbigny 1839

× 50. From *Challenger* Sta. 352A, Cape Verde Islands, Atlantic (11fm.). ZF2084.

Referred by Brady (1884) to *Peneroplis pertusus* var. f (*P. carinatus* d'Orbigny), by Thalmann (1932) and Barker (1960) to *P. carinatus* d'Orbigny, and by Haig (1988) to *Spirolina pertusus* (Forskål).

FIgure 15 *Peneroplis planatus* (Fichtel and Moll 1798)

[*Nautilus planatus* Fichtel and Moll 1798]
× 50. From *Challenger* Sta. 218A, Nares Harbour, Admiralty Islands, Pacific (16–25fm.). ZF2088.

Referred by Brady (1884) to *Peneroplis pertusus* var. a (*P. planatus* Fichtel and Moll, sp.), by Thalmann (1932), Barker (1960), and van Marle (1991) to *P. planatus* (Fichtel and Moll), and by Haig (1988) to *Spirolina pertusus* (Forskål). This is the type-species by original designation of the genus *Peneroplis* de Montfort 1808. The long-lost holotype of this species has been discovered and illustrated by Rögl and Hansen (1984).

Figures 16–17, 23 *Peneroplis pertusus* (Forskål 1775)

[*Nautilus pertusus* Forskål 1775]
Fig. 16 × 40. From *Challenger* Sta. 185A, off Cape York, Torres Strait, Pacific (3–11fm.). ZF2078.
Fig. 17 × 40. From *Challenger* Sta. 218A, Nares Harbour, Admiralty Islands, Pacific (16–25fm.). ZF2082.
Fig. 23 × 200. ZF2079.

Referred by Brady (1884) to *Peneroplis pertusus* var. b, by Thalmann (1932), Barker (1960), and van Marle (1991) to *P. pertusus* (Forskål), and by Haig (1988) to *Spirolina pertusus* (Forskål). Van Marle (1991) regards *Nautilus pertusus* Forskål 1775 as a *nomen nudum*.

Figures 18–19, 22 *Coscinospira arietina* (Batsch 1791)

[*Nautilus* (*Lituus*) *arietinus* Batsch 1791]
Figs 18–19 × 40. From *Challenger* Sta. 218A, Nares Harbour, Admiralty Islands, Pacific (16–25fm.). ZF2081–2082.
Fig. 22 × 40. From the Aegean Sea, Mediterranean.

Referred by Brady (1884) to *Peneroplis pertusus* var. c (*P. arietinus* Batsch, sp), by Thalmann (1932) and Barker (1960) to *Spirolina arietina* (Batsch), and by Haig (1988) to *S. pertusus* (Forskål). *Nautilus* (*Lituus*) *arietinus* Batsch 1791 is the type-species, by synonymy with the originally designated *Cribrospirolina distinctiva* Haman 1972 (the former type-species fixed by original designation), of the genus *Cribrospirolina* Haman 1972, and also the type-species by synonymy with *Coscinospira hemprichii* Ehrenberg 1839 (the former type-species fixed by subsequent designation) of the genus *Coscinospira* Ehrenberg 1839 (Al-Abdul Razzaq and Bhalla 1987; Loeblich and Tappan 1987). *Cribrospirolina* is a junior synonym of *Coscinospira*.

Figures 20–21 *Spirolina cylindracea* (Lamarck 1804)

[*Nautilus* (*Lituus*) *cylindraceus* Lamarck 1804]
× 60. From *Challenger* Sta. 173A, off Fiji, Pacific (12fm.). ZF2085.

Referred by Brady (1884) to *Peneroplis pertusus* var. d (*P. cylindraceus* Lamarck, sp.), by Thalmann (1932) to *P. cylindraceus* (Larmarck), by

Barker (1960) to *Spirolina acicularis* (Batsch), and by Haig (1988) to *S. pertusus* (Forskål). This is the type-species by subsequent designation of the genus *Spirolina* Lamarck 1804.

Figures 24–25 *Monalysidium politum* (Chapman 1900)

[*Spirolina* (*Monalysidium*) *polita* Chapman 1900]
× 75. From *Challenger* Sta. 352A, Cape Verde Islands, Atlantic (11fm.). ZF2087.

Referred by Brady (1884) to *Peneroplis pertusus* Forskål, sp. var. e (*P. lituus* Gmelin, sp?), by Thalmann (1932) and Barker (1960) to *Monalysidium politum* (Chapman) (with the specific identity of Fig. 25 questioned in both cases), and by Haig (1988) to *Spirolina pertusus* (Forskål). The generic identity of the present species is somewhat questionable as it lacks the early planispiral portion of typical *Monalysidium* or *Spirolina*. *Monalysidium* Chapman 1900 differs from *Spirolina* Larmarck 1804 in possessing an everted or phialine apertural lip.

Plate 14 (XIV)

Figures 1–2, 5–6, 10–13 *Archaias angulatus* (Fichtel and Moll 1798)

[*Nautilus angulatus* Fichtel and Moll 1798]
Figs 1, 5 × 30; Figs 10, 11, 12 × 20–30; Fig. 13 × 100. From coral sand, Bermuda, Atlantic. Figs 1, 5 ZF2023; Fig. 12 ZF2030; Fig. 13 ZF2032.
Fig. 2 × 30. From *Challenger* Sta. 24, West Indies (390fm.). ZF2024.
Fig. 6 × 30. From *Challenger* Sta. 122, off Pernambuco, West Atlantic (350fm.). ZF2027.

Referred by Brady (1884) to *Orbiculina adunca* Fichtel and Moll, sp. (Figs 1–2 being regarded as the 'very young, or nautiloid condition; *Nautilus angulatus*, Fichtel and Moll', Figs 5–6 as the 'young or intermediate stage; *Nautilus orbiculus*, Fichtel and Moll' and Figs 10–13 as '*Nautilus aduncus*, Fichtel and Moll'), by Thalmann (1932) to *Archaias aduncus* (Fichtel and Moll), and by Barker (1960) and van Marle (1991) to *A. angulatus* (Fichtel and Moll). This is the type-species by original designation of the genus *Archaias* de Montfort 1808. The long-lost holotype of this species has been discovered and illustrated by Rögl and Hansen (1984). *Nautilus aduncus* Fichtel and Moll 1798 and *N. orbiculus* Fichtel and Moll 1798 are junior synonyms of *N. angulatus* Fichtel and Moll 1798 (Rögl and Hansen 1984).

Figures 3–4 *Peneroplis proteus* d'Orbigny 1839

Fig. 3 × 30. From *Challenger* Sta. 24, West Indies (390fm.). ZF2025.
Fig. 4 × 30. From coral sand, Bermuda, Atlantic. ZF2026.

Referred by Brady (1884) to *Orbiculina adunca* Fichtel and Moll, sp., var. 'analogous to the "Spiroline" modifications of *Peneroplis*', by Thalmann (1932) to *Peneroplis proteus* d'Orbigny, and by Barker (1960) to *Puteolina proteus* (d'Orbigny). *Peneroplis proteus* d'Orbigny 1839 is the type-species of *Puteolina* Hofker 1952, regarded by the present author as a junior synonym of *Peneroplis* de Montfort 1808.

Figures 7–9 *Cyclorbiculina compressa* (d'Orbigny 1839)

[*Orbiculina compressa* d'Orbigny 1839]
Fig. 7 × 20. From *Challenger* Sta. 24, West Indies (390fm.). ZF2025.
Figs 8, 9 × 20. From *Challenger* Sta. 352A, St Vincent, Cape Verde Islands, Atlantic (11fm.). Fig. 8 ZF2029; Fig. 9 ZF2028.

Referred by Brady (1884) to *Orbiculina adunca* Fichtel and Moll, sp. (Figs 7–8 being regarded as 'flabelliform adult shells; *Nautilus aduncus* FIchtel and Moll' and Fig. 9 as an 'adult . . ., discoidal form; *Orbiculina compressa* d'Orbigny'), by Thalmann first (1932) to *Archaias aduncus* (Fichtel and Moll) and later (1942) to *Cyclorbiculina compressa* (d'Orbigny), and by Barker (1960) to *A. compressus* (d'Orbigny). This is the type-species by original designation of the genus *Cyclorbiculina* Silvestri 1937.

Plate 15 (XV)

Figures 1–3, 5 *Parasorites marginalis* (Lamarck 1816)

[*Orbulites marginalis* Lamarck 1816]
Figs 1–3 × 30. From *Challenger* Sta. 260, Honolulu reefs, Pacific (40fm.). ZF2040.
Fig. 5 × 30. From *Challenger* Sta. 173A, off Fiji, Pacific (12fm.). ZF2042.

Referred by Brady (1884) to *Orbitolites marginalis* Larmarck, sp., by Thalmann (1932) to *Praesorites orbitolitoides* Hofker (see below), by Barker (1960) to *Sorites marginalis* (Lamarck) (*S. hofkeri* Lacroix?), and also by van Marle (1991) to *S. marginalis* (Larmarck). Van Marle (1991) regards *Orbulites marginalis* Larmarck 1816 as a *nomen nudum*.

Figures 4 *Parasorites orbitolitoides* (Hofker 1930)

[*Praesorites orbitolitoides* Hofker 1930]
× 30. From *Challenger* Sta. 33, off Bermuda, Atlantic, 4 April 1873 (435fm.). ZF2041.

Referred by Brady (1884) to *Orbitolites marginalis* Larmarck, sp. (see above), and by Thalmann (1932) and Barker (1960) to *Praesorites orbitolitoides* Hofker. *Praesorites orbitolitoides* Hofker 1930 is the type-species by original designation of the genus *Parasorites* Seiglie, Grove and Rivera 1976. *Praesorites* Douvillé 1902 is a junior synonym of *Broeckina* Munier-Chalmas 1882.

Figures 6–7 *Discospirina italica* (Costa 1856)

[*Pavonina italica* Costa 1856]
Figs 6a, c–d, 7 × 20. From *Challenger* Sta. 44, off Cape Hatteras, North Atlantic (17fm.). ZF2043.
Fig. 6b × 20. From *Porcupine* Sta. 24, West of Ireland, Atlantic. ZF2044.

Referred by Brady (1884) to *Orbitolites tenuissimus* Carpenter, and by Thalmann (1932) and Barker (1960) to *Discospirina tenuissima* (Carpenter). *Orbitolites tenuissimus* Carpenter 1870 is a junior synonym of *Pavonina italica* Costa 1856, the type-species by original designation of the genus *Discospirina* Munier-Chalmas 1902. It is also the type-species by original designation of the genus *Krumbachia* Wiesner 1920, which is therefore an isotypic junior synonym of *Discospirina*.

Plate 16 (XVI)

Figures 1–6 *Marginopora vertebralis* Quoy and Gaimard 1830

Figs 1, 4 × 30. From *Challenger* Sta. 260A, Honolulu reefs, Pacific (40fm.).
Figs 2–3 × 30. From the Loo Choo Islands, Formosa, Pacific.
Figs 5, 6 × 20. From *Challenger* Sta. 172, off Tongatabu, Friendly Islands (18fm.). Fig. 5 ZF2033; Fig. 6 ZF2036.

Referred by Brady (1884) and Thalmann (1932) to *Orbitolites complanata/us* (Larmarck) (Figs 1–4) and *O. duplex* Carpenter (Figs 5–6), and by Barker (1960) to *Marginopora vertebralis* Quoy and Gaimard. This is the type-species by original monotypy of the genus *Marginopora* Quoy and Gaimard 1830.

Figure 7 *Amphisorus hemprichii* Ehrenberg 1839

× 11. From *Challenger* Sta. 173A, off Fiji, Pacific (12fm.). ZF2039.

Referred by Brady (1884) to *Orbitolites duplex* Carpenter, by Thalmann (1932) to *Amphisorus hemprichii* Ehrenberg, by Collins (1958) to *A. duplex* (Carpenter), and by Barker (1960) to '*Orbitolites duplex* Carpenter' (see also Text-figure 7). *Orbitolites duplex* Carpenter 1883 is a junior synonym of *Amphisorus hemprichii* Ehrenberg 1839, the type-species by original designation of the genus *Amphisorus* Ehrenberg 1839. It is also the type-species by original designation of the genus *Bradyella* Munier-Chalmas 1902, which is therefore an isotypic junior synonym of *Amphisorus*.

Figures 8–11 *Marginopora vertebralis* var. *plicata* Dana 1846

Figs 8–9 × 1; Fig. 10 × 3. From *Challenger* Sta. 172, Friendly Islands, Pacific (18fm.). 1959.5.5.772–774.
Fig. 11 × 12. From *Challenger* Sta. 172A?, Tongatabu Reef, Pacific (?18fm.). ZF2037.

Referred by Brady (1884) to *Orbitolites complanata* var. *laciniata* Brady, by Thalmann (1932) to *O. complanatus* var. *plicatus* (Dana), and by Barker (1960) to *Marginopora vertebralis* var. *plicata* Dana. *Orbitolites laciniatus* Brady 1881 was lectotypified by Smout (1936). It is regarded by the present author as a junior synonym of *Marginopora vertebralis* var. *plicata* Dana 1846.

Plate 17 (XVII)

Figures 1–6 *Marginopora vertebralis* Quoy and Gaimard 1830

Figs 1, 3, 5 ZF2034; Figs 2, 4 ZF2035; Fig. 6 ZF2036. × 12. From the Gulf of Suez.

Referred by Brady (1884) to *Orbitolites complanata* Laramrck, sp. ('monstrous specimens, showing the results of redundant and inequilateral growth'), by Thalmann first (1932) to *O. complanatus* (Lamarck) and later (1933a) to *O. duplex* Carpenter, and by Barker (1960) to *Marginopora vertebralis* Quoy and Gaimard (see also Plate 16, Figs 8–11).

Figures 7–12 *Alveolinella quoii* (d'Orbigny 1826)

[*Alveolina quoii* d'Orbigny 1826]
Figs 7–8 × 15.

Figs 9–10 ×15. From *Challenger* Sta. 218A, Nares Harbour, Admiralty Islands, Pacific (16–25fm.). ZF1046.
Figs 11, 12 × 25. From the Admiralty Islands, Pacific (16fm.). Fig. 11 ZF1050; Fig. 12 ZF1047.

Referred by Brady (1884) to *Alveolina boscii* Defrance, sp., by Thalmann first (1932) to *Alveolinella boscii* (Defrance) and later (1933a) to *A. quoyi* (d'Orbigny), and also by Barker (1960) and Whittaker and Hodgkinson (1979) to *A. quoyi* (d'Orbigny). This is the type-species by original designation of the genus *Alveolinella* Douvillé 1907.

Figures 13–15 *Borelis melo* (Fichtel and Moll 1798)

[*Nautilus melo* Fichtel and Moll 1798]
Fig. 13 ZF1056; Fig. 14 ZF1048. × 50. Coral sand, Bermuda.
Fig. 15 × 50. From off Ascension Island, Atlantic.

Referred by Brady (1884) to *Alveolina melo* Fichtel and Moll, sp., by Thalmann first (1932) to *Neoalveolina melo* (Fichtel and Moll) and later (1933a) to *N. bradyi* (Silvestri), and by Barker (1960) to *N. melo* (Fichtel and Moll). *Nautilus melo* Fichtel and Moll 1798 is the type-species by equivalence with the originally designated *B. melonoides* de Montfort 1808 of the genus *Borelis* de Montfort 1808 (Loeblich and Tappan 1964, 1987). *Alveolina bradyi* Silvestri 1927 is a junior synonym of *N. melo* Fichtel and Moll 1798. As *A. bradyi* Silvestri is the type-species by original designation of the genus *Neoalveolina* Silvestri 1928, *Neoalveolina* Silvestri 1928 is a junior synonym of *Borelis* de Montfort 1808. Smout (1963) has designated a neotype for *Nautilus melo* Fichtel and Moll 1798 from the Middle Miocene (Tortonian) of Transylvania. The neotype subsequently designated by Rögl and Hansen (1984) from the Middle Miocene (Badenian) of the Vienna Basin is therefore invalid (Loeblich and Tappan 1987; Tubbs 1990).

Plate 18 (XVIII)

Figures 1–8 *Psammosphaera fusca* Schulze 1875

Fig. 1 × 20. From *Porcupine* Sta. 16, West of Ireland, Atlantic (?). ZF2190.
Figs 2–3 ×40. From *Challenger* Sta. 286, South Pacific (2335fm.). ZF2191.
Fig. 4 ZF2192; Fig. 7 ZF2189. × 40. From *Challenger* Sta. 122, SE of Pernambuco, West Atlantic (350fm.).
Figs 5, 8 ×40. From *Challenger* Sta. 323, South Atlantic (1900fm.). 1959.5.11.59–60.
Fig. 6 ×40. From *Challenger* Sta. 23, West Indies (450fm.). ZF2193.

Referred by Brady (1884) to *Psammosphaera fusca* Schulze, by Thalmann (1932) and Barker (1960) to *P. fusca* Schulze (Figs 1, 5–8) and *P. parva* Flint (Figs 2–4), by Hofker (1972, 1976) to *P. fusca* Schulze (Figs 1, 5–6) and *P. parva* Flint (Fig. 4), by Haynes (1973a) to *P. parva* Flint (Figs 2–4), and by Zheng (1988) to *P. fusca* Schulze (Figs 1–8). *Psammosphaera parva* Flint 1899 was originally distinguished from *P. fusca* Schulze 1875 on the basis of its spicular construction. It is regarded by the present author as a junior synonym of *P. fusca*, which is known to be non-selective and to employ a wide variety of materials in its test construction (see, for instance, Schröder 1986). *P. fusca* is the type-species by original designation of the genus *Psammosphaera* Schulze 1875.

Figures 9–10 *Sorosphaera confusa* Brady 1879

Fig. 9 ×15. From *Porcupine* Sta. 23, off NW Ireland, Atlantic (630fm.). 1959.5.11.402.
Fig. 10 ×15. From *Challenger* Sta. 76, off the Azores, Atlantic (900fm.). ZF2364.

Referred by Brady (1884), Thalmann (1932), and Barker (1960) to *Sorosphaera confusa* Brady. This is the type-species by original monotypy of the genus *Sorosphaera* Brady 1879.

Figures 11–15, ?17 *Saccammina sphaerica* Brady 1871

Figs 11, 13 ×15. From the North Atlantic ('deep water'), ZF2340.
Figs 12, 14 ×15.
Fig. 15 ×15. From the Hardanger Fjord, Norway (400–500fm.). 1959.5.11.351.
Fig. 17 ×15. From *Challenger* Sta. 149D, Kerguelen Island, South Pacific (20–60fm.). ZF2342.

Referred by Brady (1884), Thalmann (1932), Barker (1960), and van Marle (1991) to *Saccammina sphaerica* M. Sars, and by Hofker (1972), Schröder (1986), Zheng (1988), and Charnock and Jones (1990) to *S. sphaerica* Brady. The latter attribution is the correct one since *Saccammina sphaerica* M. Sars 1868 (1869) was a *nomen nudum*. *Saccammina sphaerica* Brady 1871 is the type-species by subsequent designation of the genus *Saccammina* Carpenter 1869 (Loeblich and Tappan 1964, 1987).

Figure 16 *Reophax bradyi* Brönnimann and Whittaker 1980

× 15. From *Porcupine* Sta. 23, off NW Ireland, Atlantic (630fm.). ZF2341.

Referred by Brady (1884) to *Saccammina sphaerica* M. Sars, by Thalmann (1932) and Barker (1960) to *S. sphaerica* var. *catenulata* Cushman, and by the present author to *Reophax bradyi* Brönnimann and Whittaker (see also Plate 30, Fig. 12).

Figures 18–19 *Saccammina socialis* Brady 1884

Fig. 18 × 20. From *Challenger* Sta. 246, North Pacific (2050fm.).
Fig. 19 × 20. From *Challenger* Sta. 246, North Pacific (2050fm.). ZF2339.

Plate 19 (XIX)

Figures 1–4 *Astrorhiza limnicola* Sandahl 1858

Figs 1, 3 ×8; Fig. 4 × 100. From off the coast of Norway. ZF1122.
Fig. 2 × 8. From off Torquay, England. ZF1121.

Referred by Brady (1884), Thalmann (1932), Barker (1960), and Hofker (1972) to *Astrorhiza limicola* Sandahl (*sic*). *Astrorhiza limnicola* Sandahl 1858 is the type-species by original monotypy of the genus *Astrorhiza* Sandahl 1858.

Figures 5, 6?, 7–10 *Astrorhiza arenaria* Norman 1876

Fig. 5 × 8.
Fig. 6 × 8. From *Porcupine* Sta. 64, Faroe Channel, North Atlantic (640fm.). ZF1116.
Fig. 7 × 8.

Figs 8–9 × 8; Fig. 10 × 100. From *Porcupine* Sta. 47, Faroe Channel, North Atlantic (542fm.). 1959.5.5.123–126.

Referred by Brady (1884), Thalmann (1932), Barker (1960), Hofker (1972), and Zheng (1988) to *Astrorhiza arenaria* Norman. Schröder (1986) attributed the authorship of this species to Carpenter.

Plate 20 (XX)

Figures 1–9 *Astrorhiza crassatina* Brady 1881

Figs 1–2, 4, 6–8 × 8.
Figs 3, 5 × 8; Fig. 9 × 100. From *Porcupine* Sta. 64, Faroe Channel, North Atlantic (?). ZF1117.

Referred by Brady (1884), Thalmann (1932), and Schröder (1986) to *Astrorhiza*, and by Barker (1960) and Hofker (1972) to *Psammosiphonella*. *Psammosiphonella* Avnimelech 1952 was synonymized by Loeblich and Tappan (1964, 1987) with *Bathysiphon* M. Sars 1872, and by Charnock and Jones (1990) with *Rhabdammina* M. Sars 1869.

Figures 10–23 *Astrorhiza granulosa* (Brady 1879)

[*Marsipella granulosa* Brady 1879]
Figs 10–13 × 8. ZF1115. Figs 14, 17, 19, 20 × 8. ZF1119. Figs 15–16, 21–22 × 8. Fig. 18 × 8. ZF1120 Fig. 23 × 100. From *Challenger* Sta. 78, SE of the Azores, Atlantic (100fm.).

Figures 10–13 were referred by Brady (1884), Thalmann (1932), Barker (1960), and Schröder (1986) to *Astrorhiza angulosa* Brady, which species was referred by Saidova (1975) to *Astrorhizinulla*. Figures 14–23 were referred by Brady (1884), Thalmann (1932), and Zheng (1988) to *Astrorhiza granulosa* (Brady), and by Barker (1960) to *Amphitremoidea* (sic) *granulosa* (Brady), which species was again referred by Saidova (1975) to *Astrorhizinulla*. All the figures are regarded by the present author as representing *Astrorhiza granulosa* (Brady 1879), with *A. angulosa* Brady 1881 regarded as synonymous. *Astrorhizinulla* Saidova 1975 is regarded as synonymous with *Astrorhiza* Sandahl 1858. *Amphitremoida* Eisenack 1938 [*Amphitremoidea* Thalmann 1941, *nom. van.*] is an Ordovician genus.

Plate 21 (XXI)

Figures 1–8, 10–13 *Rhabdammina abyssorum* M. Sars 1869

Fig. 1 × 10. From *Challenger* Sta. 78, SE of the Azores, Atlantic (1000fm.). ZF2292.
Figs 2–3 × 10.
Fig. 4 × 10. ZF2293. Fig. 7 × 10. ZF2294. From *Challenger* Sta. 195, Banda Sea, Pacific (1425fm.).
Figs 5–6 × 10.
Figs 8, 10 × 10. Figs 9, 11 × 10. ZF2295. Fig. 12 × 60. From *Challenger* Sta. 218, North of Papua, Pacific (1070fm.).
Fig. 13 × 60. From *Porcupine* Sta. 23, off NW Ireland, Atlantic (630fm.). 1959.5.11.185.

Referred by Brady (1884), Thalmann (1932), Barker (1960), Hofker (1972, 1976) (Figs 1–3 only), Schröder (1986), and Charnock and Jones (1990) to *Rhabdammina abyssorum* M. Sars. This is the type-species by original designation of the genus *Rhabdammina* M. Sars 1869.

Figure 9 *Rhabdammina major* de Folin 1887

× 10.

Referred by Brady (1884), Thalmann (1932), and Barker (1960) to *Rhabdammina abyssorum* M. Sars, 'variety with branched arms (*Rhabdammina irregularis* Carpenter?)', by Schröder (1986) to *R. irregularis* Carpenter, and by Gooday (1986) to *R. major* de Folin.

Plate 22 (XXII)

Figures 1–3, 5–6 *Rhabdammina linearis* Brady 1879

Fig. 1 × 10. Figs 2–3 × 10, Fig. 6 × 60. ZF2305. Fig. 5 × 10. ZF2307. From *Challenger* Sta. 323, South Atlantic (1900fm.).

Referred by Brady (1884), Thalmann (1932), Hofker (1972, 1976), and Schröder (1986) to *Rhabdammina*, and by Barker (1960) to *Oculosiphon*. This is the type-species by original designation of the genus *Oculosiphon* Avnimelech 1952, regarded by the present author as a junior synonym of *Rhabdammina* M. Sars 1869.

Figure 4 *Rhabdammina neglecta* Gooday 1986

× 10. From *Porcupine* Sta. 16, West of Ireland (?). ZF2306.

Referred by Brady (1884), Thalmann (1932), Hofker (1976), and Shröder (1986) to *Rhabdammina linearis* Brady, by Barker (1960) to *Oculosiphon linearis* (Brady) (see above), and by Gooday (1986) to *Rhabdammina neglecta*.

Figures 7–10 *Rhabdammina discreta* Brady 1881

Fig. 7 × 10. From the Faroe Channel. ZF2299.
Fig. 8 × 10. From *Porcupine* Sta. 4, North Atlantic. ZF2300.
Fig. 9 × 10. From *Porcupine* Sta. 23, off NW Ireland (630fm.). ZF2301.
Fig. 10 × 8. From *Valorous* Sta. 5, off Sydney, Australia (410fm.). ZF2302.

Referred by Brady (1884), Thalmann (1932), Hofker (1972) (erroneously citing Figs 11–13), Schröder (1986), and Charnock and Jones (1990) to *Rhabdammina*, and by Barker (1960) to *Psammosiphonella*. *Psammosiphonella* Avnimelech 1952 was synonymized by Loeblich and Tappan (1964, 1987) with *Bathysiphon* M. Sars 1872, and by Charnock and Jones (1990) with *Rhabdammina* M. Sars 1869.

Figures 11–13 *Rhabdammina cornuta* (Brady 1879)

[*Astrorhiza cornuta* Brady 1879]
Fig. 11 × 15. ZF2296. Fig. 12 × 13. ZF2297. From *Challenger* Sta. 122, SE of Pernambuco, West Atlantic (350fm.).
Fig. 13 × 15. From *Porcupine* Sta. 28, off NW Ireland, Atlantic (1215fm.). ZF2298.

Referred by Brady (1884), Thalmann (1932), Barker (1960), and Schröder (1986) to *Rhabdammina cornuta* (Brady). This is the type-species by original designation of *Astrorhiza* (*Astrorhizoides*) Shchedrina 1969, regarded by the present author as a junior synonym of *Rhabdammina* M. Sars 1869.

Figures 14–18 *Jaculella acuta* Brady 1879

Fig. 14 ×12 1966.2.11.5. Lectotype designated by Loeblich and Tappan (1964) (ex Slide No. ZF1602). Fig. 15 ×12. ZF1602. Figs 16, 18 ×12. 1959.5.5.602–603. Fig. 17 ×12. ZF1603. From *Challenger* Sta. 122, SE of Pernambuco, West Atlantic (350fm.).

Referred by Brady (1884), Thalmann (1932), Barker (1960), Schröder (1986), and Zheng (1988) to *Jaculella acuta* Brady. This is the type-species by original monotypy of the genus *Jaculella* Brady, 1879.

Figures 19–22 *Jaculella obtusa* Brady 1882

×12. From *Knight Errant* Sta. 7, Faroe Channel, North Atlantic (530fm.). ZF1604.

Referred by Brady (1884), Barker (1960), Hofker (1972), and Zheng (1988) to *Jaculella obtusa* Brady, and by Thalmann (1932) to *J. acuta* Brady.

Plate 23 (XXIII)

Figures 1–3, 5–6 *Hyperammina friabilis* Brady 1884

Fig. 1 ×10. From *Challenger* Sta. 122, SE of Pernambuco, West Atlantic (350fm.). ZF1594.
Figs 2, 6 ×10. From *Porcupine* Sta. 47, Faroe Channel, North Atlantic (542fm.). 1959.5.5.579–580.
Fig. 3 ×10. From *Challenger* Sta. 195, Banda Sea, Pacific (1425fm.). ZF1593.
Fig. 5 ×20.

Referred by Brady (1884), Thalmann (1932, 1933a), Barker (1960), Hofker (1972, 1976), Zheng (1988), and Charnock and Jones (1990) to *Hyperammina friabilis*. This is the type-species by subsequent designation of the genus *Hyperammina* Eimer and Fickert 1899, regarded by Loeblich and Tappan (1964, 1987) as a junior synonym of *Hyperammina* Brady 1878.

Figures 4, 7 *Hyperammina cylindrica* Parr 1950

Fig. 4 ×10. From off Cumbrae, North Atlantic (60–65fm.). ZF1590.
Fig. 7 ×10. From *Challenger* Sta. 323, South Atlantic (1900fm.). ZF1591.

Referred by Brady (1884), Thalmann (1932, 1933a), and Hofker (1976) to *Hyperammina elongata* Brady (see below), and by Barker (1960), Zheng (1988), and Charnock and Jones (1990) to *H. cylindrica* Parr.

Figure 8 *Hyperammina elongata* Brady 1878

×10. From *Challenger* Sta. 24, West Indies (390fm.). ZF1592.

Referred by Brady (1884), Thalmann (1932), Barker (1960), Hofker (1972, 1976), and Schröder (1986) to *Hyperammina elongata* Brady, and by Zheng (1988) to *H. laevigata* (J.B. Wright) (see below). *Hyperammina elongata* Brady 1878 is the type-species by original monotypy of the genus *Hyperammina* Brady 1878. It is also the type-species by original monotypy of the genus *Bactrammina* Eimer and Fickert 1899. *Bactrammina* is an isotypic junior synonym of *Hyperammina*.

Figures 9–10 *Hyperammina laevigata* (J.B. Wright 1891)

[*Hyperammina elongata* var. *laevigata* J.B. Wright 1891]
Magnification not given. From *Challenger* Sta. 323, South Atlantic (1900fm.). ZF1591.

Referred by Brady (1884) and Thalmann (1933a) to *Hyperammina elongata* (see above), and by Thalmann (1932, 1942), and Barker (1960), Hofker (1972), Zheng (1988), and Charnock and Jones (1990) to *H. laevigata* (Wright). Inadvertently referred by Hofker (1976) to both *Hyperammina elongata* Brady and *H. laevigata* (Wright).

Figures 11–14 *Archimerismus subnodosus* (Brady 1884)

[*Hyperammina subnodosa* Brady 1884]
×10.

Referred by Brady (1884), Thalmann (1932), Barker (1960), and Hofker (1972) to *Hyperammina*, and by Loeblich and Tappan (1984) to *Archimerismus*. This is the type-species by original designation of *Archimerismus* Loeblich and Tappan 1984, which differs from *Hyperammina* Brady 1878 in the partial internal subdivision of the test.

Figures 15–19 *Saccorhiza ramosa* (Brady 1879)

[*Hyperammina ramosa* Brady 1879]
Figs 15, 19 ×16. ZF1595. Figs 17–18 ×16. From *Porcupine* Sta. 23, off NW Ireland (630fm.). 1959.5.5.582–585.
Fig. 16 ×16. From *Challenger* Sta. 246, North Pacific (2050fm.). ZF3612 (ex ZF1596). Lectotype designated by Loeblich and Tappan (1964).

Referred by Brady (1884) to *Hyperammina*, by Thalmann (1932), Barker (1960), Schröder (1884), and Zheng (1988) to *Saccorhiza*, and by Hofker (1972) to *Hyperammina* (*Saccorhiza*). This is the type-species by original monotypy of the genus *Saccorhiza* Eimer and Fickert 1899. *Hyperammina* (*Saccorhiza*) *caribbeana* Hofker 1972 appears synonymous.

Plate 24 (XXIV)

Figures 1–5 *Tolypammina vagans* (Brady 1879)

[*Hyperammina vagans* Brady 1879]
Figs 1–2 ×15. From *Challenger* Sta. 120, off Pernambuco, West Atlantic (675fm.). ZF1597.
Fig. 3 ×15. ZF1599. Figs 4–5 ×15. ZF1600. From *Porcupine* Sta. 23, off NW Ireland, Atlantic (630fm.).

Referred by Brady (1884) to *Hyperammina*, by Rhumbler (1895), Thalmann (1933a), Barker (1960), Schröder (1986), Zheng (1988), and Charnock and Jones (1990) to *Tolypammina*, by Thalmann (1932) to *Girvanella*, and by Hofker (1972, 1976) to *Hyperammina* (*Tolypammina*). This is the type-species by original monotypy of the genus *Tolypammina* Rhumbler 1895. *Girvanella* Nicholson and Etheridge, 1878 has been shown to be an algal genus.

Figures 6–9 *Tolypammina schaudinni* Rhumbler 1904

Fig. 6 ×15. From *Porcupine* Sta. 23, off NW Ireland, Atlantic (630fm.). ZF1601.

Figs 7–9 ×15. From *Challenger* Sta. 244, North Pacific (2900fm.). ZF1598.

Referred by Brady (1884) to *Hyperammina vagans* Brady (see above), by Thalmann (1932) to *Girvanella vagans* (Brady) (Fig. 6) and *G. schaudinni* (Rhumbler) (Figs 7–9), by Thalmann (1933a) and Barker (1960) to *Tolypammina schaudinni* Rhumbler, and by Hofker (1976) to *Hyperammina* (*Tolypammina*) *vagans* (Brady). *Girvanella* is an alga.

Figures 10–19 *Marsipella elongata* Norman 1878

Fig. 10 ×15. ZF1813. Fig. 11 ×15. ZF1815.From *Porcupine* Sta. 47, Faroe Channel, North Atlantic (542fm.).
Figs 12–14 ×15. From *Porcupine* Sta. 23, off NW Ireland, Atlantic (630fm.). 1959.5.5.622–624.
Fig. 15 ×15. Figs 16–17 ×15. ZF1816. From *Challenger* Sta. 122, SE of Pernambuco, Atlantic (350fm.).
Figs 18–19 ×60. From *Porcupine* Sta. 24, off Ireland, Atlantic (?). ZF1814.

Referred by Brady (1884), Thalmann (1932), Barker (1960), Hofker (1972, 1976), Haynes (1973a) and Schröder (1986) to *Marsipella elongata* Norman. This is the type-species by original monotypy of the genus *Marsipella* Norman 1878.

Figures 20–22 *Marsipella cylindrica* Brady 1882

×15. From *Knight Errant* Sta. 7, Faroe Channel, North Atlantic (530fm.). ZF1811.

Referred by Brady (1884), Barker (1960), Hofker (1972, 1976), and Zheng (1988) to *Marsipella*. This is the type-species by original designation of the genus *Pseudomarsipella* Saidova 1975, regarded by Loeblich and Tappan (1987) as a subjective isotypic junior synonym of *Rhabdamminella* de Folin, 1887. *Rhabdamminella* is regarded by the present author as a junior synonym of *Marsipella* Norman 1878, its type-species by original monotypy, *R. prismaeginosa* de Folin 1887, being a probable junior synonym of the present species.

Plate 25 (XXV)

Figures 1–6 *Pilulina jeffreysii* Carpenter 1875

Figs 1a, 2, 3 ×12. From *Porcupine* Sta. 28, off NW Ireland, Atlantic (1215fm.). ZF2089.
Fig. 1b 1959.5.5.960.
Fig. 4 ×100.
Fig. 5 ZF2091; Fig. 6 ZF2090 ×100. From *Porcupine* Sta. 21, West of Ireland, Atlantic (1476fm.).

Referred by Brady (1884), Thalmann (1932), Barker (1960), and Hofker (1972) to *Pilulina jeffreysii* Carpenter. This is the type-species by subsequent designation of the genus *Pilulina* Carpenter 1870.

Figure 7 *Technitella melo* Norman 1878

×50. From *Challenger* Sta. 344, Ascension Island, Atlantic. ZF2429.

Referred by Brady (1884), Thalmann (1932), and Hofker (1972, 1976) to *Technitella melo* Norman, and by Thalmann (1937) and Barker (1960) to *T. bradyi* Earland.

Figures 8–10 *Technitella legumen* Norman 1878

Fig. 8 ×5; Fig. 12 ×100. From *Challenger* Sta. 276, South Pacific (2350fm.). ZF2428.
Fig. 9 ×40.
Fig. 10 ×30. From *Challenger* Sta. 149I, Kerguelen Island, South Pacific (120fm.). ZF2427.

Referred by Brady (1884), Thalmann (1932), Barker (1960), Hofker (1972) (Figs 8–10 only), Schröder (1986), and Charnock and Jones (1990) to *Technitella legumen* Norman. This is the type-species by subsequent designation of the genus *Technitella* Norman 1878. This genus has been reinterpreted by Haman (1967, 1971), so as to include both free and attached forms. The latter differ from *Halyphysema* Bowerbank 1862 in lacking an internally subdivided basal expansion.

Figure 11 *Technitella* sp. nov.

×25. From off Cumbrae, North Atlantic (60fm.). 1959.5.11.494.

Referred by Brady (1884), Thalmann (1932), Barker (1960), and Schröder (1986) to *Technitella legumen* Norman (see above). Regarded by the present author as a new species, the generic identity of which is somewhat questionable.

Figures 13–14 *Technitella raphanus* Brady 1884

×25. From *Challenger* Sta. 174C, off Kandavu, Fiji (210fm.). ZF2430.

Referred by Brady (1884) and Thalmann (1932) to *Technitella raphanus*. Barker (1960) inadvertantly omitted the caption for this species.

Figures 15–17 *Storthosphaera albida* Schulze 1875

Fig. 15 ×20. Fig. 17 ×20. ZF2426. From Kors Fjord, Norway, North Atlantic.
Fig. 16 ×20. From Bukenfjord, Norway, North Atlantic (365fm.). ZF2426.

Referred by Brady (1884), Thalmann (1932), and Barker (1960) to *Storthosphaera albida* Schulze. This is the type-species by original designation of the genus *Storthosphaera* Schulze 1875.

Figures 18–20 *Pelosina rotundata* Brady 1879

Figs 18, 20 ×20. From *Porcupine* Sta. 23, off NW Ireland, Atlantic (630fm.). ZF2076.
Fig. 19 ×20.

Referred by Brady (1884), Thalmann (1932), Barker (1960), and Hofker (1972) to *Pelosina rotundata* Brady.

Plate 26 (XXVI)

Figures 1–6 *Pelosina cylindrica* Brady 1884

Figs 1, 4–6 ×9. From *Challenger* Sta. 246, North Pacific (2050fm.). 1959.5.5.907–910.
Figs 2–3 ×9. From off Gomera, the Canaries, Atlantic. ZF2073.

Referred by Brady (1884), Thalmann (1932), Barker (1960), and Schröder (1986) (Figs 1–3 only) to *Pelosina cylindrica* Brady.

Figures 7–9 *Pelosina variabilis* Brady 1879

× 9. From off the NE coast of New Zealand (1100fm.). ZF2077.

Referred by Brady (1884), Thalmann (1932), Barker (1960), Hofker (1972, 1976), Schröder (1986), and Zheng (1988) to *Pelosina variabilis* Brady. This is the type-species by subsequent designation of the genus *Pelosina* Brady 1879.

Figures 10–14 *Hippocrepina indivisa* Parker in Dawson 1870

× 45. From Gaspé Bay, Mouth of St Lawrence River, North Atlantic. 1959.5.5.548–552.

Referred by Brady (1884), Thalmann (1932), and Barker (1960) to *Hippocrepina indivisa* Parker. This is the type-species by original designation of the genus *Hippocrepina* Parker 1870. It has recently been lectotypified by Hodgkinson (1992).

Figures 15, 17–20 *Bathysiphon filiformis* M. Sars 1872

Fig. 15 × 3. 87.8.31.1. From the Hardanger Fjord, Norway (collected by the Revd A.M. Norman in 1879). Neotype designated by Gooday (1988).
Fig. 17 × 60. ZF1123. Figs 18–19 × 60; Fig. 20 × 200. From Norway.

Referred by Brady (1884), Thalmann (1932), Barker (1960), Hofker (1972), Gooday (1988), Zheng (1988), and van Marle (1991) to *Bathysiphon filiformis* M. Sars. This is the type-species by original designation of the genus *Bathysiphon* M. Sars 1872.

Figure 16 *Bathysiphon capillaris* De Folin 1886

× 3. No locality details given.

Referred by Brady (1884), Thalmann (1932), Barker (1960), Hofker (1972), Zheng (1988), and van Marle (1991) to *Bathysiphon filiformis* M. Sars, by Gooday (1988) tentatively to *B. capillare* (sic) de Folin, and by Charnock and Jones (1990) to *B. capillaris* de Folin.

Plate 27 (XXVII)

Figures 1–2, 4–11 *Aschemonella scabra* Brady 1879

Figs 1, 5, 7 × 15. ZF1102. Lectotype designated by Loeblich and Tappan (1964) ex this slide. Fig. 2 × 15. ZF1103. Fig. 4 × 15. From *Challenger* Sta. 244, North Pacific (2900fm.). 1959.5.5.104.
Figs 6, 10 × 15. ZF1104. Fig. 8 × 15. ZF1105. Fig. 9 × 15. ZF1106. From *Challenger* Sta. 323, South Atlantic (1900fm.).
Fig. 11 × 10. From *Challenger* Sta. 28 (?), North Atlantic. ZF1108.

Referred by Brady (1884) to *Aschemonella catenata* Norman, sp. (see below), and by Thalmann (1932), Barker (1960), Schröder (1986), and Zheng (1988) to *A. scabra* Brady. This is the type-species by original monotypy of the genus *Aschemonella* Brady, 1879, now known to be a xenophyophorean.

Figure 3 *Aschemonella catenata* (Norman 1876)

[*Astrorhiza catenata* Norman 1876]
× 15. From *Valorous* Sta. 8 (lat. 62° 6′N., long. 55° 56′W.), Davis Strait (1350fm.). ZF1107.

Referred by Brady (1884) and Barker (1960) to *Aschemonella catenata* (Norman), and by Thalmann (1932) to *A. scabra* Brady (see above). There must be some doubt as to whether this species is a foraminifer or a xenophyophore.

Figures 12–15 *Aschemonella ramuliformis* Brady 1884

Fig. 12 × 15. ZF1112. Figs 13, 15 × 15. ZF1113. Fig. 14 × 15. ZF1111. From *Challenger* Sta. 244, North Pacific (2900fm.).

Referred by Brady (1884), Thalmann (1932), Barker (1960) and Schröder *et al.* (1989) to *Aschemonella ramuliformis*, and by Schröder (1986) to *A. ramulifera* (sic) Brady. There is some doubt as to whether this species is a foraminifer or a xenophyophore (Gooday and Nott 1982).

Plate 27A (XXVIIA)

Figures 1–2 *Catena piriformis* Schröder, Medioli and Scott 1989

Fig. 1 × 20. ZF1109. Fig. 2 × 12. ZF1110. From *Challenger* Sta. 244, North Pacific (2900fm.).

Referred by Brady (1884) and Barker (1960) to *Aschemonella catenata* (Norman) (see Plate 27, Fig. 3; see also below), by Thalmann (1932) to *A. scabra* Brady (see Plate 27, Figs 1–2, 4–11), and by Gooday (1983) to 'tectinous chains' (Fig. 1 being an 'agglutinated' form and Fig. 2 'typical'). Gooday's 'tectinous chains' ('strange foraminifera . . . probably distinct at family level') were tentatively referred by Schröder *et al.* (1989) to *Catena piriformis*. This is the type-species by original designation of the baculellid (komokiacean) genus *Catena* Schröder, Medioli and Scott 1989.

Figure 3 *Aschemonella catenata* (Norman 1876)

[*Astrorhiza catenata* Norman 1876]
× 20. From *Challenger* Sta. 244, North Pacific (2900fm.). ZF1114.

Referred by Brady (1884) and Barker (1960) to *Aschemonella catenata* (Norman), and by Thalmann (1932) to *A. scabra* Brady (see Plate 27, Figs 1–2, 4–11). The identity of the present specimens is questionable.

Figures 4–5 *Halyphysema tumanowiczii* Bowerbank 1862

Fig. 4 × 20.
Fig. 5 × 50. From off the coast of Devon, England (adherent on *Laminaria*). 1959.5.5.411.

Referred by Brady (1884), Thalmann (1932) and Barker (1960) to *Haliphysema* (sic) *tumanowiczii* Bowerbank, and by Haynes (1973a) to *Halyphysema tumanowiczii* Bowerbank. This is the type-species by original designation of the genus *Halyphysema* Bowerbank 1862, originally believed to be a sponge. *Squamulina scopula* Carter 1870 is synonymous.

Figure 6 *Halyphysema ramulosa* Bowerbank 1866

× 20. Recorded in Brady's text from the British Isles, Florida and Mauritius.

Referred by Brady (1884), Thalmann (1932) and Barker (1960) to *Haliphysema* (sic).

Figures 7–9 *Dendrophrya erecta* T.S. Wright 1861

Figs 7–8 × 30; Fig. 9 × 150.

Referred by Brady (1884), Thalmann (1932) and Barker (1960) to *Dendrophrya erecta* T.S. Wright. This is the type-species by subsequent designation of the genus *Dendrophrya* T.S. Wright 1861.

Figures 10–12 *Dendrophrya radiata* T.S. Wright 1861

× 45.

Plate 28 (XXVIII)

Figures 1–11 *Rhizammina algaeformis* Brady 1879 *s.l.*

Fig. 1 × 1. Fig. 2 × 8. ZF2311. Fig. 6 × 40. ZF2312. From *Challenger* Sta. 299, South Pacific (2160fm.).
Figs 3–5 × 8. Fig. 7 × 40. Fig. 8 × ?. ZF2313. Fig. 9 × 40. ZF2315. Fig. 10 × 40. Fig. 11 × 40. ZF2315. Probably (from text) from *Porcupine* Sta. 37, North Atlantic (2435fm.).

Referred by Brady (1884), Thalmann (1932), Barker (1960), Schröder (1986), and Zheng (1988) to *Rhizammina algaeformis* Brady. This is the type-species by original monotypy of the genus *Rhizammina* Brady, 1879. Its morphology, internal organization, and taxonomic position have recently been discussed at length by Cartwright *et al.* (1989).

Figures 12–13 *Dendrophyra arborescens* (Norman 1881)

[*Psammatodendron arborescens* Norman 1881]
Fig. 12 × 20; Fig. 13 × 100. From a *Valorous* Station off Holstenborg, Greenland. 1960.4.25.1.

Referred by Brady (1884) to *Hyperammina*, by Thalmann (1932) and Barker (1960) to *Psammatodendron*, and by Schröder (1986) to *Dendrophyra* (see also Text-figure 10). This is the type-species by original designation of the genus *Psammatodendron* Norman 1878, which is regarded by the present author as a junior synonym of *Dendrophyra* T.S. Wright 1861.

Figures 14–15 *Sagenina frondescens* (Brady 1879)

[*Sagenella frondescens* Brady 1879]
× 10. From *Challenger* Sta. 218A, Nares Harbour, Admiralty Islands, Pacific (16–25fm.). ZF2543.

Referred by Brady (1884) to *Sagenella*, and by Thalmann (1932) and Barker (1960) to *Sagenina*. This is the type-species by original designation of the genus *Sagenina* Chapman 1900, a new name for *Sagenella* Brady 1879, not Hall 1851 (Polyzoa).

Plate 29 (XXIX)

Figures 1–3 Pogonophora indet.

× 20. From *Challenger* Sta. 299, South Pacific (2160fm.). ZF2316.

Referred by Brady (1884), Thalmann (1932), and Barker (1960) to 'chitinous rhizopod tubes, probably related to *Rhizammina*'.

Figure 4 Affinities Uncertain

× 30.

Referred by Brady (1884), Thalmann (1932), and Barker (1960) to 'chitinous rhizopod tubes with sarcode filaments extended'.

Figures 5–7 *Testulosiphon indivisus* (Brady 1884)

[*Rhizammina indivisa* Brady 1884]
× 12. Recorded in Brady's text from the Faroe Channel, Cape of Good Hope and Fiji.

Referred by Brady (1884), Thalmann (1932), Schröder (1986), and Zheng (1988) to *Rhizammina*, and by Barker (1960) to *Testulosiphon*. This is the type-species by original designation of the genus *Testulosiphon* Avnimelech 1952.

Figures 8–18 *Botellina labyrinthica* Brady 1881

Figs 8–10 1959.5.5.178–180; Figs 11–12 ZF1234; Figs 13–18 1959.5.5.184–189 × 10. From *Porcupine* Sta. 51, Faroe Channel, North Atlantic (440fm.).

Referred by Brady (1884), Thalmann (1932), and Barker (1960) to *Botellina*, and by Hofker (1972) to *Hyperammina*. This is the type-species by subsequent designation of the genus *Botellina* Carpenter, Jeffreys and Thompson 1870. *Botellina* Carpenter, Jeffreys and Thompson 1870 differs from *Hyperammina* Brady 1878 in possessing a labyrinthic interior to the second chamber (the proloculus is non-labyrinthic).

Plate 30 (XXX)

Figures 1–3 *Lagenammina difflugiformis* (Brady 1879)

[*Reophax difflugiformis* Brady 1879]
× 50. From *Challenger* Sta. 323, South Atlantic (1900fm.). ZF2267.

Referred by Brady (1884), Barker (1960) and Schröder (1986) to *Reophax difflugiformis* Brady, by Thalmann (1932) to *Proteonina difflugiformis* var. *lagenarium* (Berthelin) (Fig. 1) and *P. difflugiformis* (Brady) (Figs 2–3), also by Collins (1958) to *P. difflugiformis* (Brady) (Fig. 2), by Skinner (1961) to *Reophax difflugiformis* Brady, and by Charnock and Jones (1990) to *Lagenammina difflugiformis* (Brady). This is the type-species by original designation of the genus *Proteonella* Lukina 1969, regarded by the present author as a junior synonym of *Lagenammina* Rhumbler 1911.

Figure 4 *Lagenammina spiculata* (Skinner 1961)

[*Reophax difflugiformis spiculata* Skinner 1961]
× 50. From *Challenger* Sta. 276, Tahiti, Pacific (2350fm.). ZF2268.

Referred by Brady (1884) and Barker (1960) to *Reophax difflugiformis*

Brady (see above), by Thalmann (1932) and Collins (1958) to *Proteonina difflugiformis* (Brady), and by Skinner (1961) to *Reophax difflugiformis spiculata* subsp. nov.

Figure 5 *Lagenammina arenulata* (Skinner 1961)

[*Reophax difflugiformis arenulata* Skinner 1961]
× 50. From *Porcupine* Sta. 47, Faroe Channel, North Atlantic (542fm.). ZF2269.

Referred by Brady (1884) to *Reophax difflugiformis* Brady (see above), by Thalmann (1932) to *Proteonina difflugiformis* var. *lagenarium* (Berthelin), by Barker (1960) to *Reophax atlantica* (Cushman), by Skinner (1961) to *R. difflugiformis arenulata* subsp. nov., and by Haynes (1973a) to *Lagenammina arenulata* (Skinner). This species differs from *L. atlantica* (Cushman 1941) in lacking a distinct apertural neck.

Figure 6 *Lagenammina ampullacea* (Brady 1881)

[*Reophax ampullacea* Brady 1881]
× 50. From *Challenger* Sta. 149I, Kerguelen Island, South Pacific (120fm.). ZF2260.

Referred by Brady (1884) to *Reophax*, by Thalmann (1932) to *Proteonina*, by Barker (1960) to *Reophax*, and by Charnock and Jones (1990) to *Lagenammina*. The generic identity of this species is questionable.

Figures 7–10, ?11 *Reophax fusiformis* (Williamson 1858)

[*Proteonina fusiformis* Williamson 1858]
Figs 7–8 × 60. From Loch Fyne, Inverary, Scotland, North Atlantic (30–50fm.). ZF2273.
Figs 9–10 × 60. From off Ferry House, Cumbrae, Scotland, Atlantic (6fm.). ZF2274.
Fig. 11 × 60. From *Challenger* Sta. 209, Philippines, Pacific. ZF2275.

Referred by Brady (1884), Barker (1960), Haynes (1973a), Schröder (1986), and Charnock and Jones (1990) to *Reophax,* and by Thalmann (1932) to *Proteonina*. Most recent authorities regard *Proteonina* Williamson 1858 as a junior synonym of *Reophax* de Montfort 1808. Haynes (1973a), however, has argued that the incomplete septal development constitutes a sufficient basis for its retention ('possibly as a subgenus of *Reophax*').

Figure 12 *Reophax bradyi* Brönnimann and Whittaker 1980

× 30. From off the Hebrides, West of Scotland, Atlantic. ZF2286.

Referred by Brady (1884), Thalmann (1932), Barker (1960), and Hofker (1972, 1976) to *Reophax scorpiurus* de Montfort, and by Brönnimann and Whittaker (1980a) to *R. bradyi*. This species differs from *R. scorpiurus* de Montfort (as neotypified by Brönnimann and Whittaker, 1980a) in the more robust construction embodying more strongly overlapping, bulbous chambers.

Figure 13 *Reophax agglutinatus* Cushman 1913

× 30.

Referred by Brady (1884) and Thalmann (1932) to *Reophax scorpiurus* de Montfort, and by Barker (1960) to *R. agglutinatus*

Cushman. Any relationship with *Reophax scorpiurus* de Montfort *s.s.* (see above) was ruled out by Brönnimann and Whittaker (1980a).

Figure 14 *Reophax* sp. nov. (1)

× 50. From *Challenger* Sta. 323, South Atlantic (1900fm.). ZF2285.

Referred by Brady (1884) and Barker (1960) to *Reophax scorpiurus* de Montfort, and by Thalmann (1932) to *R. agglutinatus* Cushman. Any relationship with *Reophax scorpiurus* de Montfort *s.s.* (see above) was ruled out by Brönnimann and Whittaker (1980a). *R. agglutinatus* Cushman (see above) also appears distinct.

Figures 15–17 *Reophax* sp. nov. (2)

Fig. 15 × 40. From *Challenger* Sta. 260A, Honolulu reefs, Pacific (40fm.). ZF2288.
Fig. 16 × 20. ZF2287. Fig. 17 × 30. ZF2289. From *Challenger* Sta. 185, Torres Strait, Pacific (155fm.).

Referred by Brady (1884), Thalmann (1932), Barker (1960), Hofker (1972, 1976) and Schröder (1986) to *Reophax scorpiurus* de Montfort, any relationship with which was ruled out by Brönnimann and Whittaker (1980a).

Figures 18–20 *Hormosina pilulifera* (Brady 1884)

[*Reophax pilulifera* Brady 1884]
Fig. 18 × 25. From *Porcupine* Sta. 31, North Atlantic (1360fm.). ZF2284 or ZF4878-ZF4780.
Fig. 19 × 10; Fig. 20 × 25. ZF2284.

Referred by Brady (1884), Thalmann (1932), Barker (1960), Schröder (1986), and Zheng (1988) to *Reophax,* and by Charnock and Jones (1990) to *Hormosina*. Figure 18 was erroneously assigned by van Marle (1991) to *Reophax agglutinatus* Cushman (see above).

Figures 21–22 *Reophax dentaliniformis* Brady 1881

Fig. 21 × 35. From *Challenger* Sta. 300, North of Juan Fernandez Island, East Pacific (1375fm.). ZF2265.
Fig. 22 × 40. From *Porcupine* Sta. 23, off NW Ireland, Atlantic (630fm.). ZF2266.

Referred by Brady (1884), Barker (1960), Schröder (1986), and Zheng (1988) to *Reophax dentaliniformis* Brady, and by Thalmann (1932) erroneously to *R. scorpiurus* de Montfort. This is the type-species by subsequent designation of the genus *Nodulina* Rhumbler 1895, regarded by Brönnimann and Whittaker (1980a) as a junior synonym of *Hormosina* Brady 1879, and by the present author as a junior synonym of *Reophax* de Montfort 1808.

Figures 23–24 *Hormosina bacillaris* (Brady 1881)

[*Reophax bacillaris* Brady 1881]
Fig. 23 × 20. ZF2261. Fig. 24 × 17. ZF2262. From *Valorous* Sta. 8, North Atlantic (1750fm.).

Referred by Brady (1884), Thalmann (1932), Barker (1960), and Schröder (1986) to *Reophax,* and by Charnock and Jones (1990) to *Hormosina*.

Plate 31 (XXXI)

Figures 1–2, ?5 *Reophax gaussicus* (Rhumbler 1913)

[*Nodosinella gaussica* Rhumbler 1913].
Fig. 1–2 × 7. From *Challenger* Sta. 156, Antarctic (?).
Fig. 5 × 10. From *Challenger* Sta. 195, off Amboyna, Pacific (1425fm.). ZF2280.

Referred by Brady (1884), Thalmann (1932), and Barker (1960) to *Reophax nodulos-a* Brady, by Schröder (1986) and van Marle (1991) to *R. nodulos-us* Brady, by Zheng (1988) to *Nodosinum nodulosus* (Brady), and by Loeblich and Tappan (1987) to *Nodosinum gaussicum* (Rhumbler) (Figs 1–2, ?5). This is the type-species by original designation of the genus *Nodosinum* Hofker 1930, regarded herein as a junior synonym of *Reophax* de Montfort 1808.

Figures 3–4 *Reophax mortenseni* Hofker 1972

× 10. From *Challenger* Sta. 323, South Atlantic (1425fm.). ZF2281.

Referred by Brady (1884), Thalmann (1932), and Barker (1960) to *Reophax nodulos-a* Brady, by Schröder (1986) and van Marle (1991) to *R. nodulos-us* Brady, by Zheng (1988) to *Nodosinum nodulosus* (Brady), and by the present author to *Reophax mortenseni* Hofker 1972. Hofker (1972) stated in the original description of this species that 'Brady figured this species or one very close to it . . . as *Reophax nodulosus*'.

Figures 6–9 *Reophax nodulosus* Brady 1879

Fig. 6 × 10. ZF2281. Figs 7–9 × 10. ZF2282. From *Challenger* Sta. 323, South Atlantic (1900fm.). Fig. 7 lectotype (Loeblich and Tappan 1987).

Referred by Brady (1884) and Barker (1960) to *Reophax nodulos-a* Brady, by Schröder (1986), Charnock and Jones (1990), and van Marle (1991) to *R. nodulos-us* Brady, and by Zheng (1988) to *Nodosinum nodulosus* (Brady). This is the type-species by original designation of the genus *Pseudonodosinella* Saidova 1970, regarded by the present author as a junior synonym of *Reophax* de Montfort 1808.

Figures 10–15 *Hormosinella guttifera* (Brady 1881)

[*Reophax guttifera* Brady 1881]
× 50. From *Challenger* Sta. 323, South Atlantic (1425fm.). ZF2276.

Referred by Brady (1884), Thalmann (1932), Barker (1960), Schröder (1986), Zheng (1988), and van Marle (1991) to *Reophax*, and by Charnock and Jones (1990) (following Shchedrina 1969) to *Hormosinella*. Van Marle (1991) regards *Reophax guttifera* Brady 1881 as a *nomen nudum*, but gives no justification.

Figures 16–17 *Reophax spiculifer* Brady 1879

Fig. 16 × 50. From *Challenger* Sta. 279C, Tahiti, Pacific (620fm.). ZF2290.
Fig. 17 × 50. From *Challenger* Sta. 174A, Fiji, Pacific (255fm.). ZF2291.

Referred by Brady (1884) and Thalmann (1932) to *Reophax spiculifer-a,* and by Barker (1960), Hofker (1972, 1976), and Zheng (1988) to *R. spiculifer*. The masculine ending is the correct one.

Figures 18–22 *Hormosinella distans* (Brady 1881)

[*Lituola* (*Reophax*) *distans* Brady 1881]
Figs 18–19, 22 × 15. ZF2270. Figs 20–21 × 15. ZF2271. From *Challenger* Sta. 300, North of Juan Fernandez Island, East Pacific (1375fm.).

Referred by Brady (1884), Thalmann (1932), Barker (1960), Schröder (1986), and Zheng (1988) to *Reophax*, and by Charnock and Jones (1990) to *Hormosinella*. This is the type-species by original designation of the genera *Hormosinella* Shchedrina 1969 and *Cadminus* Saidova 1970. *Cadminus* is an isotypic junior synonym of *Hormosinella*.

Figures 23–26 *Subreophax aduncus* (Brady 1882)

[*Reophax adunca* Brady 1882]
Fig. 23 × 30. From *Porcupine* Sta. 28, off NW Ireland, Atlantic (1215fm.). ZF2256.
Fig. 24 × 30. From *Challenger* Sta. 323, South Atlantic (1900fm.). ZF2257.
Fig. 25 × 30. From *Challenger* Sta. 300, North of Juan Fernandez Island, East Pacific. ZF2258.
Fig. 26 × 30. From *Challenger* Sta. 160, Southern Ocean (2600fm.). ZF2259.

Referred by Brady (1884), Barker (1960), and Zheng (1988) to *Reophax*, by Hofker (1972, 1976) to *Hyperammina*, and by Schröder (1986) (as *S. adunc-a* (Brady)) and Charnock and Jones (1990) to *Subreophax*. This is the type-species by original designation of the genus *Subreophax* Saidova 1975.

Plate 32 (XXXII)

Figures 1–4 *Nodellum membranaceum* (Brady 1879)

[*Reophax membranaceum* Brady 1879]
Fig. 1–2 × 50. ZF2271. Fig. 3 × 50. From tow net. ZF2278. Fig. 4 × 50. ZF2279. From *Challenger* Sta. 323, South Atlantic (1900fm.). 1960.5.3.1.

Referred by Brady (1884) and Thalmann (1932) to *Reophax*, and by Thalmann (1933*a*), Barker (1960) and Schröder (1986) to *Nodellum*. This is the type-species by original designation of the genus *Nodellum* Rhumbler 1913.

Figures 5–6 *Loeblichopsis sabulosa* (Brady 1881)

[*Reophax sabulosa* Brady 1881]
× 9. From the 'cold area', Faroe Channel, North Atlantic.

Referred by Brady (1884), Thalmann (1932, 1933*a*), and Barker (1960) to *Reophax,* and by Hofker (1969) to *Loeblichopsis*.

Figures 7–9 *Loeblichopsis cylindrica* (Brady 1884)

[*Reophax cylindrica* Brady 1884]
Fig. 7 × 20. From a *Valorous* Station in the North Atlantic (1750fm.). ZF2264.

Fig. 8 × 20. Fig. 9 × 20. ZF2263. From *Challenger* Sta. 144, Prince Edward Island, South Pacific (1570fm.).

Referred by Brady (1884), Thalmann (1932, 1933a), Barker (1960), and Schröder (1986) to *Reophax*, and by Hofker (1969) to *Loeblichopsis*. This is the type-species by original designation of the genus *Loeblichopsis* Hofker 1969.

Figures 10–11 *Protoschista findens* (Parker 1870)

[*Lituola findens* Parker 1870]
× 40. From Gaspé Bay, Mouth of St. Lawrence River, North Atlantic (18–20fm.). ZF2272.

Referred by Brady (1884) and Thalmann (1932) to *Reophax*, and by Thalmann (1933a) and Barker (1960) to *Protoschista*. This is the type-species by original monotypy of the genus *Protoschista* Eimer and Fickert 1899. It has been lectotypified by Hodgkinson (1992).

Figures 12–18 *Liebusella soldanii* (Jones and Parker 1860)

[*Lituola nautiloidea* var. *soldanii* Jones and Parker 1860]
Fig. 12 ×12. From *Challenger* Sta. 174C, Fiji, Pacific (210fm.). 1959.5.5.513.
Fig. 13 ×15. From *Challenger* Sta. 33, off Bermuda, Atlantic (390fm.). ZF1558.
Figs 14, 16 × 15. ZF1560. Fig. 18 × 15. From *Challenger* Sta. 33, off Bermuda, Atlantic (435fm.).
Fig. 15 Magnification not given. From *Challenger* Sta. 33, off Bermuda, Atlantic. ZF1558.
Fig. 17 × 15. From *Challenger* Sta. 24, N of St Thomas's, West Indies (390fm.). ZF1561.

Referred by Brady (1884) to *Haplostiche soldanii* Jones and Parker, sp., by Thalmann first (1932) to *H. dubia* (d'Orbigny), later (1937) to *Liebusella dubia* (d'Orbigny), and later still (1942) to *L. soldanii* (Jones and Parker) (Figs 13–18), and by Barker (1960) to *L. soldanii* (Jones and Parker) (Figs 12, 14–18) and *L. soldanii* var. *intermedia* (van den Broeck) (Fig. 13). Figs 12 and 14–18 are interpreted by the present author as microspheric, and Fig. 13 as macrospheric, *Liebusella soldanii* (Jones and Parker). This is the type-species by original designation of the genus *Liebusella* Cushman 1933. It has been lectotypified by Hodgkinson (1992). As intimated by Barker (1960), its relationship with *Cylindroclavulina bradyi* (Cushman) (see Plate 48, Figs 32–38) requires further investigation.

Figures 19–20, 24–26 *Ammobaculites agglutinans* (d'Orbigny 1846)

[*Spirolina agglutinans* d'Orbigny 1846]
Fig. 19 × 25, Figs 25–26 × 40. ZF1518. Fig. 24 × 40. ZF1522. From *Challenger* Sta. 216, North of Papua.
Fig. 20 ×40. From *Porcupine* Sta. 23, off NW Ireland, Atlantic (630fm.). ZF1519.

Referred by Brady (1884) to *Haplophragmium agglutinans* d'Orbigny, sp., and by Thalmann (1932), Barker (1960) (albeit tentatively in the case of the Pacific specimens), Schröder (1986), Kaminski and Kuhnt (1991), and van Marle (1991) to *Ammobaculites agglutinans* (d'Orbigny). This is the type-species by original designation of the genus *Ammobaculites* Cushman 1910. It has been lectotypified by Papp and Schmid (1985).

Figures 21?, 22–23 *Ammobaculites filiformis* (Earland 1934)

[*Ammobaculites agglutinans* var. *filiformis* Earland 1934]
Fig. 21 × 50. From *Challenger* Sta. 216, North of Papua (2750fm.). ZF1520.
Figs 22–23 ×50. From *Challenger* Sta. 5, Canaries, Atlantic (2740fm.). ZF1521.

Referred by Brady (1884) to *Haplophragmium agglutinans* d'Orbigny, sp., by Thalmann first (1932) to *Ammobaculites agglutinans* (d'Orbigny) (Figs 21, 23) and *A. reophaciformis* Cushman (Fig. 22), and later (1937) to *A. agglutinans* var. *filiformis* Earland (Figs 22–23), by Barker (1960) and Schröder (1986) to *A. agglutinans* (d'Orbigny) (Fig. 21) and *A. agglutinans* var. *filiformis* Earland (Figs 22–23), by van Marle (1991) to *A. agglutinans* (d'Orbigny) (Fig. 21 only), and by the present author to *A. filiformis* (Earland). The erstwhile variety *filiformis* was raised to the rank of species by Echols (1971).

Plate 33 (XXXIII)

Figures 1–4 *Ammoscalaria pseudospiralis* (Williamson 1858)

[*Proteonina pseudospiralis* Williamson 1858]
Fig. 1 × 35. From *Porcupine* Sta. 1, West of Ireland, Atlantic (370fm.). ZF1549.
Figs 2–4 × 35. From *Porcupine* Sta. AA, Loch Scavaig, Skye, Atlantic (45–60fm.). ZF1548.

Referred by Brady (1884) to *Haplophragmium pseudospirale* Williamson, sp., by Thalmann (1932) to *Ammobaculites prostomum* Hofker, and by Barker (1960) and Zheng (1988) to *Ammoscalaria pseudospiralis* (Williamson).

Figures 5, 8, 12 *Ammobaculites calcareus* (Brady 1884)

[*Haplophragmium calcareum* Brady 1881]
Figs 5, 8 1959.5.5.418–419; Fig. 12 1959.5.5.439. × 18. From *Challenger* Sta. 23, West Indies (390fm.).

Referred by Brady (1884) to *Haplophragmium calcareum*, by Thalmann first (1932) to *Ammobaculites calcar-eous* (sic) (Brady and later (1933a) to *A. calcar-eus* (Brady), and by Barker (1960) and Zheng (1988) also to *A. calcar-eus* (Brady).

Figure 6 *Ammobaculites cylindricus* Cushman 1910

× 18. From *Challenger* Sta. 24, West Indies (390fm.). 1959.5.5.418.

Referred by Brady (1884) to *Haplophragmium calcareum* (see above), by Thalmann (1932) and Barker (1960) (albeit tentatively) to *Ammobaculites cylindricus* Cushman, and by Zheng (1988) to *A. calcareus* (Brady).

Figures 7, 9–11 Ammobaculites sp. nov.

Figs 7, 9, 11 × 18. ZF1525. Fig. 10 × 18. From *Challenger* Sta. 185, off Raine Island, Torres Strait, Pacific (155fm.).

Referred by Brady (1884) to *Haplophragmium calcareum*, by Thalmann first (1932) to *Ammobaculites calcar-eous* (*sic*) (Brady) and later (1933a) to *A. calcar-eus* (Brady), and also by Barker (1960) and Zheng (1988) to *A. calcar-eus* (Brady) (see above). Barker (1960) noted that 'figs. 7 and 9–11 may represent a different species from figs. 5, 8 and 12'.

Figures 13–16 Ammoscalaria tenuimargo (Brady 1882)

[*Haplophragmium tenuimargo* Brady 1882]
Figs 13–14 × 25. ZF1554. Figs 15–16 × 25. ZF1555. From *Challenger* Sta. 323, South Atlantic (1900fm.).

Referred by Brady (1884) to *Haplophragmium*, by Thalmann (1932) to *Ammobaculites*, and by Barker (1960) and Zheng (1988) to *Ammoscalaria*. This is the type-species by original designation of the genus *Ammoscalaria* Høglund 1947.

Figures 17–19 Ammotium cassis (Parker in Dawson 1870)

[*Lituola cassis* Parker in Dawson 1870]
× 35. From Gaspé Bay, Mouth of St. Lawrence River, North Atlantic (16fm.). ZF1530.

Referred by Brady (1884) to *Haplophragmium*, by Thalmann (1932) to *Ammobaculites*, and by Barker (1960) to *Ammotium*. This is the type-species by original designation of the genus *Ammotium* Loeblich and Tappan, 1953. It has been lectotypified by Hodgkinson (1992).

Figures 20–25 Eratidus foliaceus (Brady 1881)

[*Haplophragmium foliaceum* Brady 1881]
Figs 20–21, 23 × 40. ZF1533. Fig. 24 × 40. ZF1535. From *Challenger* Sta. 323, South Atlantic (1900fm.).

Referred by Brady (1884) to *Haplophragmium*, by Thalmann (1932) and van Marle (1991) to *Ammobaculites*, by Barker (1960) and Schröder (1986) to *Ammomarginulina*, and by Charnock and Jones (1990) to *Eratidus*. Van Marle (1991) regards *Haplophragmium foliaceum* Brady 1881 as a *nomen nudum*, but gives no justification. This is the type-species by original designation of the genus *Eratidus* Saidova 1975.

Figures 26–28 Discammina compressa (Goes 1882)

[*Lituolina irregularis* var. *compressa* Goes 1882]
Fig. 26 × 25. From *Challenger* Sta. 24, off Culebra Island, West Indies (390fm.). ZF1531.
Figs 27–28 × 25. From *Challenger* Sta. 23, off Sombrero Island, West Indies (450fm.). ZF1532.

Referred by Brady (1884) to *Haplophragmium emaciatum*, by Thalmann (1932) to *Haplophragmoides emaciatus* (Brady), and by Barker (1960), Hofker (1976), Schröder (1986), and Zheng (1988) to *Discammina compressa* (Goes). This is the type-species, by synonymy with the originally designated *D. fallax* Lacroix 1932, of the genus *Discammina* Lacroix 1932.

Plate 34 (XXXIV)

Figures 1–4 Glaphyrammina americana (Cushman 1910)

[*Ammobaculites americanus* Cushman 1910]
Figs 1–2, 3b × 20. ZF1536. Figs 3a, 4. ZF1537. From *Challenger* Sta. 323, South Atlantic (1900fm.).

Referred by Brady (1884) to *Haplophragmium fontinense* Terquem, and by Thalmann (1932) and Barker (1960) to *Ammobaculites americanus* Cushman. This is the type-species by original designation of the genus *Glaphyrammina* Loeblich and Tappan 1984, which differs from *Ammobaculites* Cushman 1910 in the absence of septa.

Figures 5–6 Evolutinella rotulata (Brady 1881)

[*Haplophragmium rotulatum* Brady 1881]
× 50. From *Challenger* Sta. 5, North Atlantic (2740fm.). ZF1550.

Referred by Brady (1884) to *Haplophragmium*, by Thalmann (1932), Barker (1960) and Schröder (1986) to *Haplophragmoides*, and by Charnock and Jones (1990) tentatively to *Evolutinella*. *Evolutinella* Myatliuk 1971 differs from *Haplophragmoides* Cushman 1910 in its evolute test form.

Figure 7 Haplophragmoides sp. nov. (1)

× 20. From *Challenger* Sta. 300, North of Juan Fernandez Island, East Pacific (1375fm.). ZF1541.

Referred by Brady (1884) to the fossil species *Haplophragmium latidorsatum* Bornemann, sp., by Thalmann first (1932) to *Haplophragmoides latidorsatus* (Bornemann) and later (1933a) to *H. subglobosus* (G.O. Sars), by Barker (1960) to *Alveolophragmium subglobosum* (G.O. Sars), by Zheng (1988) to *Cribrostomoides subglobos-um* (G.O. Sars), and by van Marle (1991) to *C. subglobos-us* (M. Sars) (see below).

Figures 8–10 Cribrostomoides subglobosus (Cushman 1910)

[*Haplophragmoides subglobosum* Cushman 1910]
Fig. 8 × 20. From *Challenger* Sta. 24, off Culebra Island, West Indies (390fm.). ZF1542.
Fig. 9 × 15. From *Challenger* Sta. 246, North Pacific (2050fm.). ZF1543.
Fig. 10 Magnification not given. From *Challenger* Sta. 24, N of St Thomas's, West Indies (390fm.). ZF1545.

Referred by Brady (1884) to the fossil species *Haplophragmium latidorsatum* Bornemann, sp., by Thalmann first (1932) to *Haplophragmoides latidorsatus* (Bornemann) (Figs 8, 10) and *Cribrostomoides bradyi* Cushman (Fig. 9), and later (1933a) to *H. subglobosus* (G.O. Sars) (Figs 8, 10) and *C. bradyi* Cushman (Fig. 9), by Barker (1960) to *Alveolophragmium subglobosum* (G.O. Sars) (Figs 8, 10) and '*Cribrostomoides bradyi* Cushman' (Fig. 9), by Hofker (1976) to *C. bradyi* Cushman (Fig. 9), by Charnock and Jones (1990) and Jones *et al.* (in press) to *C. subglobosus* (Cushman) (Figs 8–10), and by van Marle (1991) to *C. subglobosus* (M. Sars). *Haplophragmoides subglobosus* Cushman 1910 is the type-species, by synonymy with the originally designated *Cribrostomoides bradyi* Cushman 1910, of the genus *Cribrostomoides* Cushman 1910 (see Jones *et al.* (1993)).

Figures 11–13 *Veleroninoides scitulus* (Brady 1881)

[*Haplophragmium scitulum* Brady 1881]
Figs 11–12 × 40. ZF1551–1552. Fig. 13 × 40. ZF1553. From *Knight Errant* Sta. 7, Faroe Channel, North Atlantic (530fm.).

Referred by Brady (1884) to *Haplophragmium*, by Thalmann (1932) to *Haplophragmoides*, by Barker (1960) to *Alveolophragmium*, by Schröder (1986) to *Recurvoides*, by Zheng (1988) and van Marle (1991) to *Cribrostomoides*, by Charnock and Jones (1990) to *Labrospira*, and by the present author (following Jones *et al.* (1993)) to *Veleroninoides*. Van Marle (1991) regards *Haplophragmium scitulum* Brady 1881 as a *nomen nudum*, but gives no justification.

Figure 14 *Haplophragmoides* sp. nov. (2)

× 40. From *Challenger* Sta. 174C, off Kandavu, Fiji, Pacific (210fm.). ZF1544.

Referred by Brady (1884) to the fossil species *Haplophragmium latidorsatum* Bornemann, sp., by Thalmann first (1932) tentatively to *Haplophragmoides latidorsatus* (Bornemann) and later (1933*a*) tentatively to *H. subglobosus* (G.O. Sars), and by Barker (1960) tentatively to *Alveolophragmium subglobosum* (G.O. Sars) (see above). The relationship with *Haplophragmoides sphaeriloculum* Cushman 1910 requires investigation.

Figures 15–18 *Adercotryma glomeratum* (Brady 1878)

[*Lituola glomerata* Brady 1878]
× 100. From *Challenger* Sta. 149I, Kerguelen Island, Pacific (120fm.). ZF1540.

Referred by Brady (1884) to *Haplophragmium*, by Thalmann (1932) to *Haplophragmoides*, and by Barker (1960), Schröder (1986), Brönnimann and Whittaker (1987), Charnock and Jones (1990), and van Marle (1991) to *Adercotryma*. This is the type-species by original designation of the genus *Adercotryma* Loeblich and Tappan 1952. Brönnimann and Whittaker (1987) have designated a lectotype for it, believed to correspond to that figured as Fig. 1b by Brady in 1878. These authors also point out that the correct ending is the neuter *glomerat-um* rather than the feminine *glomerat-a*.

Plate 35 (XXXV)

Figures 1–3, 5 *Veleroninoides jeffreysii* (Williamson 1858)

[*Nonionina jeffreysii* Williamson 1858]
Fig. 1 × 50. From *Challenger* Sta. 135, off Tristan d'Acunha, South Atlantic (100–150fm.). ZF1527.
Figs 2, 5 × 50. From Loch Alsh, Hebrides, Scotland, North Atlantic. ZF1527.
Fig. 3 × 50. From *Challenger* Sta. 145, Prince Edward Island, South Paicific (30–150fm.). ZF1528.

Referred by Brady (1884) to *Haplophragmium canariensis* d'Orbigny, sp., by Thalmann (1932) and Barker (1960) to *Haplophragmoides canariensis* (d'Orbigny), and by Haynes (1973*a*) to *Cribrostomoides jeffreysii* (Williamson). Figure 5 evidently represents an aberrant specimen. *Veleroninoides* Saidova 1981 differs from *Cribrostomoides* Cushman 1910 principally in its evolute planispiral rather than

involute streptospiral coiling, and from *Haplophragmoides* Cushman 1910 in its interio-areal rather than interio-marginal aperture.

Figure 4 *Veleroninoides crassimargo* (Norman 1892)

[*Haplophragmium crassimargo* Norman 1892]
× 30. From Dogger Bank, off Scarborough, North Sea (40–50fm.). ZF1529.

Referred by Brady (1884) to *Haplophragmium canariensis* d'Orbigny, sp., by Thalmann (1932) to *Haplophragmoides canariensis* (d'Orbigny), by Barker (1960) to *Alveolophragmium crassimargo* (Norman), by Schröder (1986) and Zheng (1988) to *Cribrostomoides crassimargo* (Norman), and by the present author to *Veleroninoides crassimargo* (Norman). This is the type-species by original designation of the genus *Labrospira* Høglund 1947. As shown by Jones *et al.* (1993*)*, *Labrospira* Høglund 1947 is a junior synonym of *Cribrostomoides* Cushman 1910. *Alveolophragmium* Shchedrina 1936 is an alveolar genus.

Figure 6 *Portatrochammina bipolaris* Brönnimann and Whittaker 1980

× 120. From Franz Josef Land, North Atlantic (113fm.). ZF1546.

Referred by Brady (1884) to *Haplophragmium nanum* Brady, by Thalmann (1932), and Barker (1960) to *Trochammina nana* (Brady) (see below), and by Brönnimann and Whittaker (1980*b*) to *Portatrochammina bipolaris*. Brönnimann and Whittaker (1980*b*) could not locate the specimen figured in the '*Challenger* Report', but chose the holotype for their *P. bipolaris* from a slide labelled as containing specimens figured by Brady. This clearly exhibits the overlapping umbilical flaps characteristic of the genus *Portatrochammina* Echols 1971, which were seemingly overlooked by Brady.

Figures 7–8 *Trochammina nana* (Brady 1881)

[*Haplophragmium nanum* Brady 1881]
× 60. From *Challenger* Sta. 346, South Atlantic (2350fm.). ZF1547.

Referred by Brady (1884) to *Haplophragmium nanum* Brady, and by Thalmann (1932), Barker (1960) and Brönnimann and Whittaker (1980*b*) to *Trochammina nana* (Brady). Brönnimann and Whittaker (1980*b*) have redescribed this species and designated a lectotype for it believed to be the specimen corresponding to Figure 8 of Brady (1884). These authors also erected the new species *Trochammina pintoi* for unfigured forms erroneously accommodated in *Haplophragmium nanum* Brady by Brady in 1881 and 1884.

Figure 9 *Recurvoides turbinatus* (Brady 1881)

[*Haplophragmium turbinatum* Brady 1881]
× 40. From *Challenger* Sta. 346, South Atlantic (2350fm.). ZF1557.

Referred by Brady (1884) to *Haplophragmium*, by Thalmann (1932) to *Trochammina*, and by Thalmann (1937), Barker (1960), and Charnock and Jones (1990) to *Recurvoides*.

Figure 10 *Paratrochammina challengeri* Brönnimann and Whittaker 1988

× 25. From *Challenger* Sta. 323, South Atlantic (1900fm.). ZF1538.

Referred by Brady (1884) to *Haplophragmium globigeriniforme* Parker and Jones, sp., by Thalmann (1932), Hofker (1976), and Zheng (1988) to *Trochammina globigeriniformis* (Parker and Jones), by Thalmann (1933*a*) and Barker (1960) to *Ammoglobigerina globigeriniformis* (Parker and Jones), by Schröder (1986) to *Trochammina* cf. *globigeriniformis* (Parker and Jones), by Brönnimann and Whittaker (1988) to *Paratrochammina challengeri*, and by Charnock and Jones (1990) to *Trochamminopsis challengeri* (Brönnimann and Whittaker). For a fuller discussion of the complex taxonomic history of this species, the reader is referred to Brönnimann and Whittaker (1988).

Figure 11 Trochamminid indet.

×25. From *Challenger* Sta. 78, SE of the Azores, Atlantic (1000fm.). 1959.5.5.438.

Referred by Brady (1884) to *Haplophragmium globigeriniforme* Parker and Jones, sp., by Thalmann (1932), Hofker (1976), and Zheng (1988) to *Trochammina globigeriniformis* (Parker and Jones), and by Thalmann (1933*a*) and Barker (1960) to *Ammoglobigerina globigeriniformis* (Parker and Jones).

Figures 12–15 *Globotextularia anceps* (Brady 1884)

[*Haplophragmium anceps* Brady 1884]
Fig. 12 ×25.
Fig. 13 ×20. From *Challenger* Sta. 332, South Pacific (2200fm.). ZF1524.
Fig. 14 ×20. 1966.2.11.4. (ex ZF1523). Lectotype designated by Loeblich and Tappan (1964). Fig. 15 ×20. ZF1523. From *Challenger* Sta. 296, South Pacific (1825fm.).

Referred by Brady (1884) to *Haplophragmium*, and by Thalmann (1932), Barker (1960), and Zheng (1988) to *Globotextularia*. This is the type-species by original designation of the genus *Globotextularia* Eimer and Fickert 1899.

Figures 16–17 *Tholosina bulla* (Brady 1881)

[*Placopsilina bulla* Brady 1881]
Fig. 16 ×20. ZF2092. Fig. 17 ×20. ZF2093. From *Challenger* Sta. 323, South Atlantic (1900fm.).

Referred by Brady (1884) and later Thalmann (1932) to *Placopsilina*, and by Thalmann (1933*a*) and Barker (1960) to *Tholosina*. This is the type-species by subsequent designation of the genus *Tholosina* Rhumbler 1895.

Figures 18–19 *Tholosina vesicularis* (Brady 1879)

[*Placopsilina vesicularis* Brady 1879]
Fig. 18 ×12. ZF2097. Fig. 19 ×12. ZF2098. From *Porcupine* Sta. 23, off NW Ireland, Atlantic (630fm.).

Referred by Brady (1884) and Thalmann (1932) to *Placopsilina vesicularis* Brady, by Thalmann (1933*a*) and Barker (1960) to *Tholosina vesicularis* (Brady), and by Thalmann (1942) to *T. vesicularis* var. *erecta* Heron-Allen and Earland.

Plate 36 (XXXVI)

Figure 1 *Placopsilina bradyi* Cushman and McCulloch 1939

×15. From *Challenger* Sta. 195A, off Amboyna, East Indies (15–20fm.). ZF2094.

Referred by Brady (1884) and Thalmann (1932, 1933*a*) to the fossil species *Placopsilina cenomana* d'Orbigny, and by Thalmann (1942), Barker (1960), and Zheng (1988) to *P. bradyi* Cushman and McCulloch.

Figures 2–3 *Placopsilina confusa* Cushman 1920

Fig. 2 ×30. From *Porcupine* Sta. 12, West of Ireland, Atlantic (670fm.). ZF2095.
Fig. 3 ×50. From *Challenger* Sta. 122, SE of Pernambuco, West Atlantic (350fm.). ZF2096.

Referred by Brady (1884) to *Placopsilina cenomana* d'Orbigny, by Thalmann first (1932) to *P. cenomana* d'Orbigny (Fig. 2) and *P. confusa* Cushman (Fig. 3), later (1933*a*) to *P. cenomana* d'Orbigny (Fig. 2 only), later still (1937) to *P. confusa* Cushman (Figs 2–3), and finally (1942) to *P. bradyi* Cushman and McCulloch (Figs 2–3), and by Barker (1960) to *P. confusa* Cushman.

Figures 4–6 *Bdelloidina aggregata* Carter 1877

Fig. 4 ×15. 1959.5.5.141. Fig. 5 ×15. ZF1124. Fig. 6 ×30. From *Challenger* Sta. 218A, Nares Harbour, Admiralty Islands, Pacific (16–25fm.).

Referred by Brady (1884), Thalmann (1932, 1933*a*), and Barker (1960) to *Bdelloidina aggregata* Carter. This is the type-species by original designation of the genus *Bdelloidina* Carter 1877. *B. aggregata* var. *bradyi* Elias 1950 appears superfluous.

Figures 7–18 *Thurammina papillata* Brady 1879

Fig. 7 ×30. From *Challenger* Sta. 122, SE of Pernambuco, West Atlantic (350fm.). 1959.9.18.923.
Fig. 8 ×50. ZF2483. Fig. 15 ×50. ZF2486. From *Porcupine* Sta. 23, West of Ireland, Atlantic (630fm.).
Figs 9–11 ×50. From *Challenger* Sta. 120, off Pernambuco, West Atlantic (675fm.). ZF2484.
Figs 12, 14 ×50. From *Challenger* Sta. 160, Southern Ocean (2600fm.).
Fig. 13 Magnification not given. From *Challenger* Sta. 323, South Atlantic (1900fm.). ZF2485.
Figs 16–18 ×100. From *Porcupine* Sta. 24, off Ireland. ZF2487.

Referred by Brady (1884), Thalmann (1932), Barker (1960), and Schröder (1986) to *Thurammina papillata* Brady, by Thalmann (1933*a*) to *T. castanea* Heron-Allen and Earland, and by Hofker (1972) to *Psammosphaera* (*Thurammina*) *papillata* (Brady). This is the type-species by subsequent designation of the genus *Thurammina* Brady 1879. It is interpreted in a broad sense by most modern authors (e.g. Barker 1960).

Plate 37 (XXXVII)

Figure 1 *Thurammina compressa* Brady 1879

×50. From *Porcupine* Sta. 23, West of Ireland, Atlantic (630fm.). ZF2481.

Referred by Brady (1884), Thalmann (1932), and Barker (1960) to *Thurammina compressa* Brady.

Figures 2–7 *Thurammina albicans* Brady 1879

Figs 2–4 × 50. From *Challenger* Sta. 323, South Atlantic (1900fm.). ZF2479.
Figs 5–6 × 50. ZF2478. Figs 7 × 100. ZF2480. From *Challenger* Sta. 323, South Atlantic (1900fm.).

Referred by Brady (1884), Thalmann (1932), and Barker (1960) to *Thurammina albicans* Brady.

Figures 8–16 *Cyclammina cancellata* Brady 1879

Fig. 8 × 15. From a *Porcupine* Station in the North Atlantic. 1959.5.5.303.
Fig. 9 × 15. From *Challenger* Sta. 168, off New Zealand (1100fm.). 1964.12.9.32 (ex ZF1361). Lectotype designated by Banner (1966).
Fig. 10 × 15.
Figs 11, 15 × 15. From a *Porcupine* Station in the North Atlantic (20/17/1897).
Fig. 12 × 15. 1959.5.5.296. Fig. 14 × 15. ZF1359 or ZF1360. Fig. 16 × 30. ZF1359. From *Challenger* Sta. 24, off Culebra Island, West Indies (390fm.).
Fig. 13 × 15. ZF1360–1361.

Referred by Brady (1884), Thalmann (1932), Leroy (1944*a*), Barker (1960), Hofker (1976), and Hermelin (1989) to *Cyclammina cancellata* Brady, and by Zheng (1988) to *C. cancellata* Brady (Figs 8, 10–16) and *C. compressa* Cushman (Fig. 9: the lectotype of *C. cancellata*). *C. cancellata* Brady 1879 is the type-species by original monotypy of the genus *Cyclammina* Brady 1879.

Figures 17–19 *Cyclammina rotundidorsata* (Hantken 1876)

[*Haplophragmium rotundidorsatum* Hantken 1876]
Fig. 17 × 15. Fig. 18 × 15. ZF1363. Fig. 19 × 15. ZF1364. From *Challenger* Sta. 323, South Atlantic (1900fm.).

Referred by Brady (1884) and Thalmann (1932) to *Cyclammina orbicularis* Brady, by Thalmann (1933*a*) and Barker (1960) to *C. orbicularis* Brady (Figs 17–18) and *C. orbicularis* var. *sellina* Rhumbler (Fig. 19), and by Charnock and Jones (1990), based on a comparative study of type material, to *Cyclammina* (*Reticulophragmium*) *rotundidorsata* (Hantken).

Figures 20–23 *Cyclammina pusilla* Brady 1881

Figs 20–22 × 15; Fig. 23 × 30. From *Challenger* Sta. 323, South Atlantic (1900fm.). ZF1365.

Referred by Brady (1884), Thalmann (1932), Barker (1960), Schröder (1986), and Zheng (1988) to *Cyclammina pusilla* Brady. The relationship with *Cyclammina trullissata* (Brady) (see Plate 40, Figs 13, 16) requires investigation.

Plate 38 (XXXVIII)

Figures 1, 2?, 3 *Ammodiscus anguillae* Høglund 1947

Figs 1, 3 × 15. From *Challenger* Sta. 24, off Culebra Island, West Indies (390fm.). ZF1078.
Fig. 2 × 25. From *Challenger* Sta. 332, South Pacific (2200fm.). ZF1059.

Referred by Brady (1884), Thalmann (1932), Schröder (1986), and van Marle (1991) (Figs 1, 3 only) to *Ammodiscus incertus* (d'Orbigny), and by Barker (1960) to *Involutina anguillae* (Høglund) (Figs 1, 3) and *Cornuspira incertà* (d'Orbigny) (?) (Fig. 2). Loeblich and Tappan first (1954) synonymized the agglutinated *Ammodiscus* Reuss 1862 with the hyaline genus *Spirillina* Ehrenberg 1843, but later (1961, 1964, 1987) reinstated it. *Involutina* Terquem 1862 is an aragonitic genus and *Cornuspira* Schultze 1854 a porcelaneous one.

Figures 4–6 *Ammodiscus tenuis* (Brady 1881)

[*Trochammina* (*Ammodiscus*) *tenuis* Brady 1881]
Fig. 4 × 25. From *Challenger* Sta. 46, North Atlantic (1350fm.). ZF1064.
Figs 5–6 × 15. From *Challenger* Sta. 168, off NE New Zealand (1100fm.). 1959.5.5.63–73.

Referred by Brady (1884) and Charnock and Jones (1990) to *Ammodiscus tenuis* (Brady), by Thalmann (1932) to *A. incertus* (d'Orbigny), and by Barker (1960) to *Involutina intermedia* (Høglund) (?) (immature specimen) (Fig. 4) and *I. tenuis* (Brady) (Figs 5–6) (see also discussion above).

Figures 7–9 *Glomospira gordialis* (Jones and Parker 1860)

[*Trochammina squamata* var. *gordialis* Jones and Parker 1860]
Figs 7–8 × 70. From *Challenger* Sta. 5. Canaries, Atlantic (2740fm.). ZF1057.
Fig. 9 × 70. From *Challenger* Sta. 174B, off Kandavu, Fiji, Pacific (610fm.). ZF1058.

Referred by Brady (1884) to *Ammodiscus*, and by Thalmann (1932), Barker (1960), Schröder (1986), Zheng (1988), and Charnock and Jones (1990) to *Glomospira*. This is the type-species by original monotypy of the genus *Glomospira* Rzehak 1885. It has recently been lectotypified (more correctly, neotypified (Hodgkinson 1992)) by Berggren and Kaminski (1990).

Figures 10–16 *Usbekistania charoides* (Jones and Parker 1860)

[*Trochammina squamata* var. *charoides* Jones and Parker 1860]
Figs 10–12, 15 × 70. From *Challenger* Sta. 46, North Atlantic (1350fm.). ZF1052.
Figs 13–14 × 70. From *Challenger* Sta. 323, South Atlantic (1900fm.). ZF1053.
Fig. 16 × 70. From *Challenger* Sta. 218, North of Papua (1070fm.). ZF1054.

Referred by Brady (1884) to *Ammodiscus*, by Thalmann (1932), Barker (1960), Schröder (1986), Zheng (1988), and van Marle (1991) to *Glomospira*, and by Charnock and Jones (1990) to *Usbekistania*. This is the type-species by original designation of the genus *Repmanina* Suleymanov 1966, herein regarded as a junior synonym of *Usbekistania* Suleymanov 1960. It has recently been lectotypified by Berggren and Kaminski (1990). It is also, by synonymy with the originally designated *Glomospirella* (*Usbekistania*) *mubarakensis* Suleymanov 1960, the type-species of *Usbekistania* Suleymanov 1960. *Usbekistania* differs from the irregularly coiled *Glomospira* in being strictly streptospiral, that is in being characterized by regular changes in the axis of coiling.

Figures 17–19 *Turritellella shoneana* **(Siddall 1878)**

[*Trochammina shoneana* Siddall 1878]
Figs 17, 18b × 100. ZF1060. Fig. 18a × 100. ZF1061. Fig. 19a × 100. ZF1062. Fig. 19b × 100. ZF1063. From *Challenger* Sta. 149I, Kerguelen Island, Pacific (120fm.).

Referred by Brady (1884) to *Ammodiscus,* and by Thalmann (1932) and Barker (1960) to *Turritellella.* This is the type-species by subsequent designation of the genus *Turritellella* Rhumbler 1904.

Figures 20–22 *Turritellella spectabilis* **(Brady 1881)**

[*Ammodiscus spectabilis* Brady 1881]
× 12. From a *Porcupine* Station in the North Atlantic (358fm.).

Referred by Brady (1884) to *Ammodiscus,* and by Thalmann (1932) and Barker (1960) to *Turritellella.*

Plate 39 (XXXIX)

Figures 1–4, 6 *Hormosina globulifera* **Brady 1879**

Fig. 1 × 15. ZF1581. Fig. 2 × 15. ZF1582. From *Challenger* Sta. 246, North Atlantic (2050fm.).
Figs 3–4 × 15. ZF1584. From *Porcupine* Sta. 23, off NW Ireland, North Atlantic (630fm.).
Fig. 6 × 15. From *Challenger* Sta. 218, North of Papua, Pacific (1070fm.). ZF1584.

Referred by Brady (1884), Thalmann (1932), Barker (1960), Hofker (1972) (Figs 1–4 only), Schröder (1986), Charnock and Jones (1990), and van Marle (1991) to *Hormosina globulifera* Brady. This is the type-species by subsequent designation of the genus *Hormosina* Brady 1879.

Figure 5 *Ginesina* **sp. nov.**

× 15. From *Challenger* Sta. 246, North Pacific (2050fm.). ZF1582.

Referred by Brady (1884), Thalmann (1932), Barker (1960), Hofker (1972), Schröder (1986), and van Marle (1991) to *Hormosina globulifera* Brady (see above). Regarded by the present author as new species, the generic identity of which is somewhat questionable.

Figures 7–9 *Hormosinella ovicula* **(Brady 1879)**

[*Hormosina ovicula* Brady 1879]
Fig. 7 × 15. From *Challenger* Sta. 241, North Pacific (2300fm.). ZF1588.
Figs 8a, 9 × 15. From *Challenger* Sta. 246, North Pacific (2050fm.). ZF1587.
Fig. 8b × 15. From *Challenger* Sta. 323, South Atlantic (1900fm.). ZF1589.

Referred by Brady (1884), Barker (1960), and Hofker (1972, 1976) to *Hormosina,* and by Schröder (1986) to *Reophax.* This is the type-species by original designation of the genus *Reophanus* Saidova 1970. Loeblich and Tappan (1987) distinguished this genus from *Hormosinella* Shchedrina 1969 on the basis of its supposed mono- rather than multi-layered wall. Note, however, that this is not proven, for while Mendelson (1982) stated the wall of *Hormosina ovicula* Brady to be

mono-layered, Hofker (1972) had earlier stated it to be multi-layered. In view of this, and pending further comparative study, *Reophanus* is regarded as a junior synonym of *Hormosinella* (see also Charnock and Jones 1990).

Figures 10–13 *Subreophax monile* **(Brady 1881)**

[*Trochammina (Hormosina) monile* Brady 1881]
× 10. From *Challenger* Sta. 122, SE of Pernambuco, West Atlantic (350fm.). ZF1585.

Referred by Brady (1884), Thalmann (1932), Barker (1960), and Zheng (1988) to *Hormosina.*

Figures 14–18 *Hormosinella carpenteri* **(Brady 1881)**

[*Hormosina carpenteri* Brady 1881]
Figs 14–16 × 10. From *Challenger* Sta. 78, East of the Azores, Atlantic (1000fm.). ZF1579.
Figs 17–18 × 10. From *Challenger* Sta. 76, off the Azores, Atlantic (900fm.). 1959.5.5.567–568.

Referred by Brady (1884), Barker (1960), and Schröder (1986) to *Hormosina,* by Hofker (1972) to *Hyperammina,* and by Charnock and Jones (1990) (following Shchedrina 1969) to *Hormosinella.*

Figures 19–23 *Hormosina normani* **Brady 1881**

× 10. From various stations in the North Atlantic (1750fm.).

Referred by Brady (1884), Barker (1960), Hofker (1972), and Schröder (1986) to *Hormosina normani* Brady.

Plate 40 (XL)

Figures 1–2 *Trochamminoides olszewskii* **(Grzybowski 1898)**

[*Trochammina olszewskii* Grzybowski 1898]
× 25. From *Challenger* Sta. 120, off Pernambuco, West Atlantic (675fm.). ZF2510.

Referred by Brady (1884) to *Trochammina proteus* Karrer, and by Thalmann (1932), Barker (1960), and Zheng (1988) to *Trochamminoides proteus* (Karrer). Referred herein to *Trochamminoides olszewskii* (Grzybowski) on account of its proportionately more tangentially elongate later chambers.

Figure 3 *Trochamminoides grzybowskii* **Kaminski and Geroch 1992**

× 25. From *Challenger* Sta. 24, off Culebra Island, West Indies (390fm.). ZF2511.

Referred by Brady (1884) to *Trochammina proteus* Karrer, and by Thalmann (1932), Barker (1960), and Zheng (1988) to *Trochamminoides proteus* (Karrer). Brady's form appears distinct from Karrer's. It is practically identical to *Trochamminoides grzybowskii* Kaminski and Geroch, the replacement name for *Trochamminoides elegans* Grzybowski 1898, not Egger 1895.

Figures 4–7 *Lituotuba lituiformis* **(Brady 1879)**

[*Trochammina lituiformis* Brady 1879]
Fig. 4 × 20. From *Challenger* Sta. 24, off the Azores, Atlantic (900fm.). ZF2503.

Figs 5–6 ×20. From *Challenger* Sta. 120, off Pernambuco, West Atlantic (675fm.). ZF2504.

Fig. 7 × 20. From *Challenger* Sta. 24, off Culebra Island, West Indies (390fm.). ZF2505.

Referred by Brady (1884) to *Trochammina*, and by Thalmann (1932), Barker (1960), Hofker (1972, 1976), Schröder (1986), Zheng (1988), and Charnock and Jones (1990) to *Lituotuba*. This is the type-species by subsequent designation of the genus *Lituotuba* Rhumbler 1895.

Figures 8–12 *Conglophragmium coronatum* (Brady 1879)

[*Trochammina coronata* Brady 1879]
Fig. 8 × 30. ZF2497. Fig. 9 × 30. ZF2498. Fig. 10 × 20. ZF2499. From *Challenger* Sta. 23, off Sombrero Island, West Indies (450fm.).
Fig. 11 ×20. From *Challenger* Sta. 120, off Pernambuco, West Atlantic (675fm.). ZF2500.
Fig. 12 ×20. From *Challenger* Sta. 24, off Culebra Island, West Indies (390fm.). 1959.5.11.592.

Figures 8–9 were referred by Brady (1884), Thalmann (1932) and Barker (1960) to *Trochammina conglobata*. Figures 10–12 were referred by Brady (1884) to *Trochammina coronata* Brady, by Thalmann (1932) and Barker (1960) to *Haplophragmoides coronatus* (Brady), and by Zheng (1988) to *Trochamminoides coronatum* (Brady). All four figures were referred by Charnock and Jones (1990) to *Conglophragmium coronatum* (Brady). This is the type-species, by synonymy with the originally designated *Trochammina conglobata* Brady 1884, of the genus *Conglophragmium* Bermúdez and Rivero 1963, and incidentally of its isotypic junior synonym *Conglobatoides* Saidova 1981.

Figures 13, 16 *Cyclammina trullissata* (Brady 1879)

[*Trochammina trullissata* Brady 1879]
Fig. 13 ×30, Fig. 16 ×60. From *Challenger* Sta. 24, West Indies (390fm.). ZF2518.

Referred by Brady (1884) to *Trochammina trullissata* Brady, by Thalmann (1932) to *Cyclammina bradyi* Cushman (Fig. 13) and *Haplophragmoides trullissatus* (Brady) (Fig. 16), by Barker (1960) (Fig. 13 only) and Schröder (1986) to *Cyclammina trullissata* (Brady), and by Zheng (1988) to *Haplophragmoides trullisatum* (Brady) (Fig. 13 only). The relationship with *Cyclammina pusilla* Brady (see Plate 37, Figs 20–23) requires investigation.

Figures 14–15 *Veleroninoides wiesneri* (Parr 1950)

[*Labrospira wiesneri* Parr 1950]
Fig. 14 ×30. ZF2519. Fig. 15 × 50. ZF2520. From *Challenger* Sta. 323, South Atlantic (1900fm.).

Referred by Brady (1884) to *Trochammina trullissata* Brady (see above), by Thalmann (1932) to *Haplophragmoides trullissatus* (Brady), by Barker (1960) to *Alveolophragmium wiesneri* (Parr), and by Schröder (1986) and Zheng (1988) (Fig. 14 only) to *Cribrostomoides wiesneri* (Parr). As shown by Jones *et al.* (1993), *Labrospira* Høglund, 1947 is a junior synonym of *Cribrostomoides* Cushman 1910. The present species is closer to *Veleroninoides* Saidova 1981, though the wall texture is atypical, and recalls that of *Buzasina*

Loeblich and Tappan 1985 (see below). *Labrospira arctica* Parker 1952 appears synonymous. *Alveolophragmium* Shchedrina 1936 is an alveolar genus.

Figures 17–18 *Buzasina ringens* (Brady 1879)

[*Trochammina ringens* Brady 1879]
Fig. 17 ×45. From *Challenger* Sta. 323, South Atlantic (1900fm.). ZF2512.
Fig. 18a × 30. From *Challenger* Sta. 98, off the West coast of Africa (1750fm.). ZF2513.
Fig. 18b × 30. From *Challenger* Sta. 70, North Atlantic (1675fm.). ZF2514.

Referred by Brady (1884) to *Trochammina*, by Thalmann first (1932) to *Haplophragmoides* and later (1933a) to *Ammochilostoma*, by Barker (1960) to *Alveolophragmium*, by Loeblich and Tappan (1985a) to *Buzasina*, and by Schröder (1986) and Zheng (1988) to *Cribrostomoides*. This is the type-species by original designation of the genus *Buzasina* Loeblich and Tappan 1985.

Figures 19–23 *Buzasina galeata* (Brady 1881)

[*Trochammina galeata* Brady 1881]
Figs 19–20, 23 ×50. From *Challenger* Sta. 346, South Atlantic (2350fm.). ZF2501.
Figs 21–22 ×50. From *Challenger* Sta. 332, South Atlantic (2200fm.). ZF2502.

Referred by Brady (1884) to *Trochammina*, by Thalmann (1932) to *Ammochilostoma*, by Barker (1960) and Schröder (1986) to *Cystammina*, and by Charnock and Jones (1990) to *Buzasina*. This is the type-species by original designation of the genus *Cystamminella* Lukina 1980 (not *Cystamminella* Myatliuk 1966), regarded by the present author as a junior synonym of *Buzasina* Loeblich and Tappan 1985 (see above).

Plate 41 (XLI)

Figure 1 *Cystammina pauciloculata* (Brady 1879)

[*Trochammina pauciloculata* Brady 1879]
From *Challenger* Sta. 300, North of Juan Fernandez Island (1375fm.). ZF2508. Lectotype designated by Loeblich and Tappan (1964)

Referred by Brady (1884) to *Trochammina*, by Thalmann (1932) to *Ammochilostoma*, and by Thalmann (1933a), Barker (1960), Schröder (1986), Zheng (1988), and Charnock and Jones (1990) to *Cystammina*. This is the type-species by subsequent designation of the genus *Cystammina* Neumayr 1889. *Ammochilostoma* Eimer and Fickert 1899 is an isotypic junior synonym of *Cystammina* Neumayr 1889.

Figure 2 *Cystammina* sp. nov.

From *Challenger* Sta. 279C, Tahiti, Pacific (620fm.). ZF2509.

Referred by Brady (1884) to *Trochammina pauciloculata* Brady, by Thalmann (1932) to *Ammochilostoma pauciloculata* (Brady), by Thalmann (1933a), Barker (1960), Schröder (1986), Zheng (1988), and Charnock and Jones (1990) to *Cystammina pauciloculata* (Brady) (see above), and by the present author to *C.* sp. nov. This species differs from *Cystammina pauciloculata* (Brady) (see above) principally in possessing an interio-marginal rather than interio-areal aperture. It may not be congeneric.

Figure 3 *Tritaxis challengeri* (Hedley, Hurdle and Burdett 1964)

[*Trochammina challengeri* Hedley, Hurdle and Burdett 1964]
× 50. From *Challenger* Sta. 24, off Culebra Island, West Indies (390fm.). ZF2516.

Referred by Brady (1884), Thalmann (1932), Barker (1960), and van Marle (1991) to *Trochammina squamata* Jones and Parker, and by Hedley *et al.* (1964) and Zheng (1988) to *T. challengeri*.

Figure 4 *Trochammina inflata* (Montagu 1808)

[*Nautilus inflatus* Montagu 1808]
× 50. From Almouth, Northumberland, England (brackish water).

Referred by Brady (1884), Thalmann (1932), Barker (1960), and Haynes (1973a) to *Trochammina inflata* (Montagu). This is the type-species by original monotypy of the genus *Trochammina* Parker and Jones. Brönnimann and Whittaker (1984a) have designated a neotype for this species from the brackish waters of Broadsands, Torbay, Devon (ex Milton Collection). Brady's figured specimen is closely comparable.

Figures 5–6 *Polystomammina nitida* (Brady 1881)

[*Trochammina nitida* Brady 1881]
Fig. 5 × 50. From *Challenger* Sta. 135, Nightingale Island, Tristan d'Acunha, Atlantic (100–150fm.). ZF2506.
Fig. 6 × 50. From *Challenger* Sta. 145, Prince Edward Island, South Pacific (50–150fm.). ZF2507.

Referred by Brady (1884), Thalmann (1932), Barker (1960), and van Marle (1991) to *Trochammina*, and by Seiglie (1965) to *Polystomammina*. *Trochammina nitida* Brady 1881 is the type-species by original designation of the genus *Polystomammina* Seiglie 1965, though van Marle (1991) regards it as a *nomen nudum* (for no apparent reason). Its apertural characteristics (a curved, slit-like primary aperture in the apertural face and a series of arched, posteriorly-opening secondary supplementary apertures at the umbilical tips of each chamber, or at the proximal sides of the umbilical chamber extensions) are well illustrated by Seiglie (1965) (having evidently been overlooked by Brady).

Figures 7–10 *Carterina spiculotesta sensu* Brady 1884

Figs 7–8 × 50; Figs 9–10 × 100. From the Gulf of Suez (40fm.). 1959.5.5.241–245.

Referred by Brady (1884), Thalmann (1932) and Barker (1960) to *Carterina spiculotesta* (Carter), and by Brönnimann and Whittaker (1983) to *C. spiculotesta sensu* Brady. *Rotalia spiculotesta* Carter 1877, which may be distinct, is the type-species by original monotypy of the genus *Carterina* Brady 1884. The 'Carterina problem' is extensively reviewed by Brönnimann and Whittaker (1983), and Loeblich and Tappan (1987).

Figure 11 *Hemisphaerammina bradyi* Loeblich and Tappan 1957

× 25. From off Redcliff, Durham, England (30fm.). ZF2626.

Referred by Brady (1884) and Thalmann (1932) to *Webbina hemisphaerica* (Jones, Parker and Brady), and by Barker (1960) and Haynes (1973a) to *Hemisphaerammina bradyi* Loeblich and Tappan.

Figures 12–16 *Ammolagena clavata* (Jones and Parker 1860)

[*Trochammina irregularis* var. *clavata* Jones and Parker 1860]
Figs 12–13 × 25. From *Challenger* Sta. 122, SE of Pernambuco, West Atlantic (350fm.). ZF2624.
Fig. 14 × 25. From *Challenger* Sta. 323, South Atlantic (1900fm.). ZF2623.
Fig. 15 × 25.
Fig. 16 × 25. From *Challenger* Sta. 164A, off Sydney, Australia (410fm.). ZF2625.

Referred by Brady (1884) to *Webbina*, by Thalmann first (1932) to *Webbina* and later (1933a) to *Ammolagena*, and also by Barker (1960), Hofker (1972, 1976), Schröder (1986), Zheng (1988), Charnock and Jones (1990), and van Marle (1991) to *Ammolagena*. This is the type-species by original monotypy of the genus *Ammolagena* Eimer and Fickert 1899. The existing diagnosis of that genus needs to be emended to take into account the development of supplementary apertures (see Brady's Fig. 15). *Webbina* is a polymorphinid genus.

Plate 42 (XLII)

Figures 1–2 *Bolivinella folia* (Parker and Jones 1865)

[*Textularia agglutinans* var. *folium* Parker and Jones 1865]
× 60. From *Challenger* Sta. 162, Bass Strait, Pacific (38–40fm.). ZF2450.

Referred by Brady (1884) to *Textularia folium* Parker and Jones, and by Thalmann (1932), Barker (1960), and Hayward and Brazier (1980) to *Bolivinella folia* (Parker and Jones). This is the type-species by original designation of the genus *Bolivinella* Cushman 1927. It has been lectotypified by Loeblich and Tappan (1964).

Figures 3, 5 *Bolivinella philippinensis* McCulloch 1977

× 60. From *Challenger* Sta. 174A, Fiji, Pacific (255fm.). ZF2451.

Referred by Brady (1884) to *Textularia folium* Parker and Jones, by Thalmann (1932), Barker (1960), and van Marle (1991) to *Bolivinella elegans* Parr, and by Hayward and Brazier (1980) to *B. philippinensis* McCulloch. *B. philippinensis* McCulloch differs from *B. folia* (Parker and Jones) (see above) in its smaller size, its lower number of generally higher chambers, its less regularly beaded sutural costae, and its more pronounced spinosity. Whether these essentially ornamental differences should be used as the basis for specific distinction is somewhat debatable.

Figure 4 *Bolivinella elegans* Parr 1932

× 60. From *Challenger* Sta. 174A, Fiji, Pacific (255fm.). ZF2451.

Referred by Brady (1884) to *Textularia folium* Parker and Jones, and by Thalmann (1932), Barker (1960), Hayward and Brazier (1980), and van Marle (1991) to *Bolivinella elegans* Parr. *B. elegans* Parr differs from *B. folia* (Parker and Jones) (see above) in its smaller size, its generally more flabelliform test, its single rather than double median costa, and its essentially unbeaded sutural costae. Whether these differences should be used for specific distinction is again somewhat debatable.

Figure 6 *Patellinella inconspicua* (Brady 1884)

[*Textularia inconspicua* Brady 1884]
× 70. From *Challenger* Sta. 162, Bass Strait, Pacific (38–40fm.). ZF2456.

Referred by Brady (1884) to *Textularia*, and by Thalmann (1932) and Barker (1960) to *Patellinella*. This is the type-species by original designation of the genus *Patellinella* Cushman 1928.

Figure 7 *Sagrinella jugosa* (Brady 1884)

[*Textularia jugosa* Brady 1884]
× 70. From *Challenger* Sta. 185, Torres Strait, Pacific (155fm.). ZF2454.

Referred by Brady (1884) and Thalmann (1932) to *Textularia*, and by Barker (1960) and van Marle (1991) to *Patellinella*. This is the type-species, by synonymy with the originally designated *S. guinai* Saidova 1975, of the genus *Sagrinella* Saidova 1975 (Revets, pers. comm.).

Figures 8–12 *Bolivinita quadrilatera* (Schwager 1866)

[*Textilaria quadrilatera* Schwager 1866]
Fig. 8 × 35. From *Challenger* Sta. 164A, off Sydney, Australia (410fm.). ZF2457.
Figs 9–12 × 35. From *Challenger* Sta. 279C, off Tahiti, Pacific (620fm.). ZF2458.

Referred by Brady (1884) to *Textularia*, and by Thalmann (1932), Leroy (1944b), Barker (1960), Belford (1966), Srinivasan and Sharma (1980), and van Marle (1991) to *Bolivinita*. This is the type-species by original designation of the genus *Bolivinita* Cushman 1927. It has recently been neotypified by Srinivasan and Sharma (1980).

Figures 13–14 *Siphotextularia concava* (Karrer 1868)

[*Plecanium concavum* Karrer 1868]
Fig. 13 × 35. From *Challenger* Sta. 219A, Admiralty Islands, Pacific (17fm.). ZF2441.
Fig. 14 × 35. From *Challenger* Sta. 308, West of Patagonia, East Pacific (175fm.). ZF2443.

Referred by Brady (1884), Thalmann (1932), and van Marle (1991) to *Textularia*, and by Barker (1960) to *Siphotextularia*. Barker (1960) was in some doubt as to whether Brady's Recent and Karrer's fossil specimens were conspecific.

Figures 15–16 *Spirorutilus carinatus* (d'Orbigny 1846)

[*Textularia carinata* d'Orbigny 1846]
× 30. From *Challenger* Sta. 209, Philippine Islands (95–100fm.). ZF2440.

Referred by Brady (1884) to *Textularia carinata* d'Orbigny, by Thalmann (1932) and Barker (1960) to *T. pseudocarinata* Cushman, by Zheng (1988) to *Spirorutilis* (sic) *pseudocarinata* (Cushman), and by

Charnock and Jones (1990) to *Spiroplectammina* (*Spiroplectinella*) *carinata* (d'Orbigny) (with *Textularia pseudocarinata* Cushman in synonymy). This is the type-species by original designation of the genus *Spirorutilus* Hofker 1976, regarded by Banner and Pereira (1981), Loeblich and Tappan (1987), and Charnock and Jones (1990) as a junior synonym of *Spiroplectinella* Kisel'man 1970, but by Hottinger *et al.* (1990) as a valid genus (differing from *Spiroplectinella* in its canaliculate wall structure). It has been lectotypified by Papp and Schmid (1985), who refer it to *Spiroplectinella*.

Figures 17–18 *Spiroplectinella wrightii* (Silvestri 1903)

[*Spiroplecta wrightii* Silvestri 1903]
× 35. From a *Porcupine* Station in the Atlantic (no locality details given). ZF2460.

Referred by Brady (1884), Thalmann (1932), Barker (1960), and van Marle (1991) to *Textularia sagittula* Defrance, by Haynes (1973a) to *Spiroplectammina wrightii* (Silvestri), and by Zheng (1988) to *Spirorutilis* (sic) *wrightii* (Silvestri). This is the type-species by original designation of the genus *Spiroplectinella* Kisel'man 1972.

Figures 19–22 *Spirotextularia fistulosa* (Brady 1884)

[*Textularia sagittula* var. *fistulosa* Brady 1884]
× 25. From *Challenger* Sta. 174C, off Kandavu, Fiji, Pacific (210fm.). ZF2461.

Referred by Brady (1884), Thalmann (1932), and Barker (1960) to *Textularia sagittula* var. *fistulosa*, and by Zheng (1988) to *Spirorutilis* (sic) *fistulosa* (Brady). This is the type-species by original designation of the genus *Spirotextularia* Saidova 1975.

Figures 23–24 *Septotextularia rugosa* Cheng and Zheng 1978

× 25. From *Challenger* Sta. 218A, Nares Harbour, Admiralty Islands, Pacific (16–25fm.). ZF2459.

Referred by Brady (1884) to *Textularia rugosa* Reuss, sp., by Thalmann (1932) to *Gaudryina rugulosa* Cushman, by Thalmann (1942) and Barker (1960) to *Gaudryina* (*Siphogaudryina*) *rugulosa* Cushman (Fig. 23) and *Textularia indentata* (Fig. 24), and by Loeblich and Tappan (1985b) to *Septotextularia rugosa* Cheng and Zheng. This is the type-species by original designation of the genus *Septotextularia* Cheng and Zheng 1978. Its complex taxonomic history is discussed at some length by Loeblich and Tappan (1985b).

Figures 25–29 *Siphoniferoides siphoniferus* (Brady 1881)

[*Textularia siphonifera* Brady 1881]
Figs 25, 28a, 29 × 30. From *Challenger* Sta. 260A, Honolulu, Pacific (40fm.). ZF2464.
Fig. 26 × 30. From *Challenger* Sta. 218A, Nares Harbour, Admiralty Islands, Pacific (15–25fm.). ZF2463.
Figs 27, 28b × 30. From *Challenger* Sta. 219A, Admiralty Islands, Pacific (17fm.). ZF2466.

Referred by Brady (1884) to *Textularia*, by Thalmann first (1932) to *Gaudryina* and later (1942) to *G.* (*Siphogaudryina*), and also by Barker (1960) to *G.* (*Siphogaudryina*). This is the type-species by original designation of the genus *Siphoniferoides* Saidova 1981.

Plate 43 (XLIII)

Figures 1–3 *Textularia agglutinans* d'Orbigny 1839

Fig. 1 × 25. From *Challenger* Sta. 135, Tristan d'Acunha, Atlantic (100–150fm.). ZF2431.
Fig. 2 × 25. From *Challenger* Sta. 33, off Bermuda, Atlantic (435fm.). ZF2432.
Fig. 3 × 25. From off the Shetlands, North Atlantic (70–100fm.). 1959.5.11.495.

Referred by Brady (1884), Thalmann (1932), Barker (1960), Zheng (1988) (Fig. 1 only), and van Marle (1991) to *Textularia*. This species has also been referred by various authors to *Textilina* Norvang 1966 (regarded by Banner and Pereira (1981), Loeblich and Tappan (1985*b*, 1987), and the present author as a junior synonym of *Textularia* Defrance 1824).

Figure 4 *Textularia porrecta* (Brady 1884)

[*Textularia agglutinans* var. *porrecta* Brady 1884]
× 25. From *Challenger* Sta. 185, off Raine Island, Torres Strait, Pacific (155fm.). ZF2434.

Referred by Brady (1884) to *Textularia agglutnans* var. *porrecta*, and by Thalmann (1932), Leroy (1944*b*), Barker (1960), and Zheng (1988) to *T. porrecta* (Brady).

Figures 5–8 *Planctostoma luculenta* (Brady 1884)

[*Textularia luculenta* Brady 1884]
Figs 5, 7 × 20. From *Challenger* Sta. 23, off Sombrero Island, West Indies (450fm.). ZF2455.
Figs 6, 8 × 20. From *Challenger* Sta. 24, off Culebra Island, West Indies (390fm.). 1959.5.11.500–501.

Referred by Brady (1884) and Thalmann (1932) to *Textularia*, and by Barker (1960) to *Planctostoma*. This is the type-species by original designation of the genus *Planctostoma* Loeblich and Tappan 1955.

Figures 9–10 *Textularia pseudogramen* Chapman and Parr 1937

Fig. 9 × 30. From *Challenger* Sta. 162, Bass Strait, Pacific (38–40fm.). ZF2452.
Fig. 10 × 18. From *Challenger* Sta. 260A, Honolulu reefs, Pacific (40fm.). ZF2453.

Referred by Brady (1884) and Thalmann (1932) to the fossil species *Textularia gramen* d'Orbigny, and by Barker (1960) and Zheng (1988) to *T. pseudogramen* Chapman and Parr.

Figure 11 *Siphotextularia rolshauseni* Phleger and Parker 1951

× 30. From *Challenger* Sta. 33, off Bermuda, Atlantic (435fm.). ZF2442.

Referred by Brady (1884) and Thalmann (1932) to *Textularia concava* (Karrer) (see Plate 42), and by Barker (1960) to *Siphotextularia* sp. nov. Regarded by the present author as falling within the range of variability exhibited by *S. rolshauseni* Phleger and Parker (see, for instance, Phleger *et al.* 1953).

Figure 12 *Cribrogoesella bradyi* Cushman 1935

× 25. From *Challenger* Sta. 24, West Indies (390fm.). ZF2433.

Referred by Brady (1884) and Thalmann (1932) to *Textularia agglutinans* d'Orbigny, and by Thalmann (1937) and Barker (1960) to *Cribogoesella bradyi* Cushman.

Figures 13–14 *Sahulia conica* (d'Orbigny 1839)

[*Textularia conica* d'Orbigny 1839]
Fig. 13 × 40. From *Challenger* Sta. 172, off Tongatabu, Friendly Islands, Pacific (18fm.). ZF2445.
Fig. 14 × 40. From *Challenger* Sta. 185, off Raine Island, Torres Strait, Pacific (155fm.). ZF2446.

Referred by Brady (1884), Thalmann (1932), and Barker (1960) to *Textularia*, and by Loeblich and Tappan (1985*b*) to *Sahulia*.

Figures 15–16, 18–19 *Sahulia barkeri* (Hofker 1978)

[*Textularia barkeri* Hofker 1978]
Figs 15, 18 × 40. From *Challenger* Sta. 188, New Guinea, Pacific (28fm.). ZF2470.
Fig. 16 × 40. From *Challenger* Sta. 219A, Admiralty Islands (17fm.). ZF2471.
Fig. 19 × 25. From *Challenger* Sta. 24, off Culebra Island, West Indies (24fm.). ZF2473.

Referred by Brady (1884) to *Textularia trochus* d'Orbigny, by Thalmann first (1932) to *T. pseudotrochus* Cushman and later (1942) to *T. orbica* Lalicker and McCulloch, also by Collins (1958) to *T. orbica* (Figs 15–16, 18 only), by Barker (1960) to *T.* sp. nov., by Hofker (1978) to *T. barkeri*, by Loeblich and Tappan (1985*b*) to *Sahulia patelliformis*, and by the present author to *S. barkeri* (Hofker). This is the type-species, by synonymy with the originally designated *S. patelliformis* Loeblich and Tappan 1985, of the genus *Sahulia* Loeblich and Tappan 1985.

Figure 17 *Textulariella barrettii* (Jones and Parker 1876)

[*Textularia barrettii* Jones and Parker 1876]
× 25. From *Challenger* Sta. 24, off Culebra Island, West Indies (390fm.). ZF2472.

Referred by Brady (1884) to *Textularia trochus* d'Orbigny, by Thalmann first (1932) to *T. goesi* Cushman and later (1942) to *T. orbica* Lalicker and McCulloch, and by Barker (1960) to *Textulariella barrettii* (Jones and Parker). This is the type-species by original designation of the genus *Textulariella* Cushman 1927. It has been lectotypified by Loeblich and Tappan (1964).

Plate 44 (XLIV)

Figures 1–2 *Dorothia goesi* (Cushman 1911)

[*Textularia goesi* Cushman 1911]
× 35. From *Challenger* Sta. 174C, off Kandavu, Fiji, Pacific (210fm.). ZF2474.

Referred by Brady (1884) and Thalmann (1932) to *Textularia trochus* d'Orbigny, and by Thalmann (1933*a*), Barker (1960), and van Marle

(1991) to *T. goesi* Cushman. The generic identity of this species is somewhat questionable.

Figures 3, 6–8 *Textulariella barrettii* (Jones and Parker 1876)

[*Textularia barrettii* Jones and Parker 1876]
Fig. 3 ×25. From *Challenger* Sta. 122, SE of Pernambuco, South Atlantic (350fm.). ZF2437.
Fig. 6 ×18; Fig. 8 ×30. From *Challenger* Sta. 33, off Bermuda, Atlantic (435fm.). ZF2438.
Fig. 7 ×20. From *Challenger* Sta. 23, N of St Thomas's, West Indies (390fm.). ZF2439.

Referred by Brady (1884) to *Textularia trochus* d'Orbigny (Fig. 3) and *T. barrettii* Jones and Parker (Figs 6–8), by Thalmann first (1932) to *Textularia trochus* d'Orbigny (Fig. 3) and *Textulariella barrettii* (Jones and Parker) (Figs 6–8), and later (1933a) to *Textularia goesi* Cushman (Fig. 3) and *Textulariella barrettii* (Jones and Parker) (Figs 6–8), and by Barker (1960) to *Textulariella barrettii* (Jones and Parker) (Figs 3, ?6–8). This is the type-species by original designation of the genus *Textulariella* Cushman 1927. It has been lectotypified by Loeblich and Tappan (1964). These individuals are probably microspheric.

Figures 4–5 *Dorothia pseudoturris* (Cushman 1922)

[*Textularia pseudoturris* Cushman 1922]
Fig. 4 ×18. ZF2475. Fig. 5 ×15. ZF2476. From *Challenger* Sta. 24, off Culebra Island, West Indies (390fm.).

Referred by Brady (1884) to *Textularia turris* d'Orbigny, by Thalmann first (1932) to *T. pseudoturris* Cushman and later (1942) to *Dorothia pseudoturris* (Cushman), and also by Barker (1960) to *Dorothia pseudoturris* (Cushman). Desai and Banner (1987) have illustrated the typically canaliculate ultrastucture of this species.

Figures 9–11 *Textularia hystrix* nom. nov.

Figs 9, 11 ×20–25. From a *Porcupine* station in the North Atlantic. ZF2436.
Fig. 10 ×20–25. From *Knight Errant* Sta. 7, Faroe Channel, North Atlantic (530fm.). ZF2437.

Referred by Brady (1884) to *Textularia aspera* Brady, by Thalmann (1932) erroneously to *T. tuberosa* d'Orbigny, and by Barker (1960) to *Tectularia* (sic) *aspera* Brady. *Textularia aspera* Brady 1882 is a primary junior homonym of *T. aspera* Ehrenberg 1840. It is herein renamed *T. hystrix* ('hystrix' being the Latin for 'porcupine').

Figures 12–13 *Dorothia scabra* (Brady 1884)

[*Gaudryina scabra* Brady 1884]
×20–25. From *Challenger* Sta. 23, off Sombrero Island, West Indies (450fm.). ZF2435.

Referred by Brady (1884) to *Textularia aspera* Brady, by Thalmann (1932) to *T. tuberosa* d'Orbigny, and by Barker (1960), Zheng (1988), and Charnock and Jones (1990) to *Dorothia scabra* (Brady). Hofker (1976) interpreted triserial to biserial forms, like those figured on this plate, as megalospheric individuals and trochospiral to biserial forms, like those on Plate 46, as microspheric (note, however, that he did not specifically cite Brady's figures). Hofker (1976) also noted that the wall of this species was canaliculate as in the type-species of *Dorothia*,

D. bulletta (Carsey). This has been confirmed by Loeblich and Tappan (1985b).

Figures 14–18 *Bigenerina nodosaria* d'Orbigny 1826.

Figs 14–17 ×35. From *Porcupine* Sta. 10, West of Ireland, North Atlantic. ZF1132.
Fig. 18 ×35. From Sta. 6, off Ireland, North Atlantic. ZF1138.

Referred by Brady (1884), Thalmann (1932), Barker (1960), and Zheng (1988) to *Bigenerina nodosaria* d'Orbigny. This is the type-species by subsequent designation of the genus *Bigenerina* d'Orbigny 1826.

Figures 19–24 *Bigenerina cylindrica* Cushman 1922

Fig. 19 ×35. From Sta. 18, West of Ireland, North Atlantic (183fm.). ZF1129.
Figs 20–21 ×35; Fig. 22. From *Porcupine* Sta. AA, Loch Scavaig, Skye, Scotland, North Atlantic (45–60fm.). ZF1130.
Fig. 23 ×60. ZF1131.
Fig. 24 ×35. ZF1128.

Referred by Brady (1884) to *Bigenerina digitata* d'Orbigny, by Thalmann (1932) to *Clavulina cylindrica* d'Orbigny, and by Barker (1960) to *Bigenerina cylindrica* Cushman.

Plate 45 (XLV)

Figures 1–8 *Vulvulina pennatula* (Batsch 1798)

[*Nautilus (Orthoceras) pennatula* Batsch 1798]
Figs 1, 3 ×20. ZF1126. Fig. 5 k 20. ZF1134. Fig. 7 ×20. ZF 135. From *Challenger* Sta. 24, off Culebra Island, West Indies (390fm.).
Figs 2, 4 ×20. From *Challenger* Sta. 23, off Sombrero Island, West Indies (450fm.). ZF1125.
Fig. 6 ×20. From *Porcupine* Sta. 27, NW of Ireland, North Atlantic. ZF1133.
Fig. 8 ×20. From *Challenger* Sta. 33, off Bermuda, Atlantic (435fm.). ZF1136.

Figures 1–4 were refered by Brady (1884) to *Bigenerina capreolus* d'Orbigny, by Thalmann first (1932) to *Vulvulina capreolus* d'Orbigny and later (1937) to *Vulvulina pennatula* (Batsch) (Figs 1–2) and *Textularia corrugata* Costa (Figs 3–4), and by van Marle (1991) to *Vulvulina pennatula* (Batsch). Figures 5–8 were referred by Brady (1884) to *Bigenerina pennatula* Batsch, sp. and by Thalmann (1932) to *Vulvulina pennatula* (Batsch). All eight figures were subsequently referred by Barker (1960) and Charnock and Jones (1990) to *Vulvulina pennatula* (Batsch) (with *V. capreolus* d'Orbigny synonymized). Figures 1–4 probably represent the microspheric generation of this species and Figures 5–8 the megalospheric. This is the type-species, by synonymy with the originally designated *Vulvulina capreolus* d'Orbigny 1826, of the genus *Vulvulina* d'Orbigny 1826.

Figures 9–16 *Cribrogoesella robusta* (Brady 1881)

[*Bigenerina robusta* Brady 1881]
Figs 9, 14–16 ×14. From *Challenger* Sta. 24, West Indies (390fm.). 1959.5.5.145–147.
Figs 10–13 ×14. From *Challenger* Sta. 122, SE of Pernambuco, West Atlantic (350fm.). ZF1137.

Referred by Brady (1884) and Thalmann (1932) to *Bigenerina*, and by Thalmann (1942) and Barker (1960) to *Cribrogoesella*. This is the type-species by original designation of the genus *Cribrogoesella* Cushman 1935.

Figures 17–21 *Pavonina flabelliformis* d'Orbigny 1826.

Fig. 17 × 50. From *Challenger* Sta. 260A, Honolulu, Pacific (40fm.). ZF2070.
Fig. 18 × 60. From *Challenger* Sta. 24, West Indies (390fm.). ZF2071.
Figs 19–20 × 60. From Tamatave, Madagascar, Indian Ocean (shore sand). 1959.5.5.900–901.
Fig. 21 × 60. From off Calpentyn, Ceylon (2fm.). ZF2072.

Referred by Brady (1884) and van Marle (1991) (Figs 17, 19–21 only) to *Pavonina flabelliformis* d'Orbigny, and by Thalmann (1932) and Barker (1960) to *P. flabelliformis* d'Orbigny (Figs 17, 19–21) and *P. atlantica* Cushman (Fig. 18). Cushman's species seems superfluous. *Pavonina flabelliformis* d'Orbigny 1826 is the type-species by original monotypy of the genus *Pavonina* d'Orbigny, 1826 and also the type-species by original designation of the genus *Valvopavonina* Hofker 1951. *Valvopavonina* Hofker 1951 is an isotypic junior synonym of *Pavonina* d'Orbigny 1826. *Pavonina* has recently been reviewed by Revets (1991).

Figures 22–23 *Spiroplectella earlandi* Barker 1960

× 60. From *Challenger* Sta. 185, off Raine Island, Torres Strait, Pacific (155fm.). ZF2421.

Referred by Brady (1884) to *Spiroplecta annectens* Parker and Jones, sp., by Thalmann first (1932) to *Spiroplectinata annectens* (Parker and Jones) and later (1937) to *Spiroplectella annectens* (Brady) (sic), by Barker (1960) to *S. earlandi* nom. nov. (sic), and by van Marle (1991) to *Bolivinopsis cubensis* (Cushman and Bermudez). Barker's 'new name' was in fact a new species, Brady's *annectens* being clearly different from Parker and Jones's *annectens,* as shown by Earland (1934) (who referred it to *Spiroplectella annectens* (Brady)). The attribution to *Spiroplectella* Earland 1934 appears correct. Loeblich and Tappan first (1964) synonymized *Spiroplectella* Earland 1934 with *Ammobaculoides* Plummer 1932, but later (1987) reinstated it. In terms of specific identification, Seiglie and Barker (1987) suspected and van Marle (1991) stated Brady's specimens to be referable to *Spiroplectoides cubensis* Cushman and Bermudez 1937, the type-species by original designation of the genus *Duquepsammia* Seiglie and Baker 1987. This can be ruled out as detailed examination of Brady's specimens reveals that they lack the diagnostic internal subdivision of *Duquepsammia*.

Figure 24 *Spiroplectammina* sp. nov.

× 60. From *Challenger* Sta. 185, off Raine Island, Torres Strait, Pacific (155fm.). ZF2420.

Referred by Brady (1884) to *Spiroplecta americana* Ehrenberg, and by Thalmann (1932) and Barker (1960) to *Spiroplectammina* sp. nov.

Figures 25–27 *Spiroplectammina biformis* (Parker and Jones 1865)

[*Textularia agglutinans* var. *biformis* Parker and Jones 1865]
Fig. 25 × 100. From *Challenger* Sta. 285, South Pacific (2375fm.). ZF2422.

Figs 26–27 × 100. From *Challenger* Sta. 323, South Atlantic (1900fm.). ZF2423.

Referred by Brady (1884) to *Spiroplecta biformis* Parker and Jones, sp. and by Thalmann (1932), Barker (1960), and Schröder (1986) to *Spiroplectammina biformis* (Parker and Jones). This is the type-species by original designation of the genus *Spiroplectammina* Cushman 1927. It has been lectotypified by Loeblich and Tappan (1964).

Plate 46 (XLVI)

Figures 1–4 *Karreriella bradyi* (Cushman 1911)

[*Gaudryina bradyi* Cushman 1911]
Figs 1, 2b, 3 × 30. From *Porcupine* Sta. 42, SW of Ireland, Atlantic (862fm.). ZF1453.
Fig. 2a × 30. From *Challenger* Sta. 224, North Pacific (1850fm.). ZF1452.
Fig. 4 × 30. From *Challenger* Sta. 300, North of Juan Fernandez Island, Chile (1375fm.).

Referred by Brady (1884) to *Gaudryina pupoides* d'Orbigny, by Thalmann first (1932) to *G. bradyi* Cushman and later (1942) to *Karreriella bradyi* (Cushman), and also by Barker (1960), Schröder (1986), Zheng (1988), Hermelin (1989), Charnock and Jones (1990), and van Marle (1991) to *K. bradyi* (Cushman). The ultrastructure of this species has been illustrated by Jones (1984c) and Weston (1984).

Figure 5 *Siphotextularia philippinensis* (Keijzer 1953)

[*Textularia philippinensis* Keijzer 1953]
× 25.

Referred by Brady (1884) to *Gaudryina pupoides* var. *chilostoma* Reuss, by Thalmann first (1932) to *G. chilostoma* (Reuss) and later (1942) to *Karreriella chilostoma* (Reuss), by Barker (1960) to *Textularia philippinensis* Keijzer, and by Zheng (1988) to *Siphotextularia philippinensis* (Keijzer).

Figure 6 *Karreriella chilostoma* (Reuss 1852)

[*Textularia chilostoma* Reuss 1852]
× 40. From *Challenger* Sta. 279C, off Tahiti, Pacific (620fm.). ZF1454.

Referred by Brady (1884) to *Gaudryina pupoides* var. *chilostoma* Reuss, by Thalmann first (1932) to *G. chilostoma* (Reuss) and later (1942) to *Karreriella chilostoma* (Reuss), and by Barker (1960) tentatively to *Textularia philippinensis* Keijzer (see above). This species was referred by Loeblich and Tappan (1986a) to their new genus *Meidamonella*. *Meidamonella* Loeblich and Tappan 1986 is regarded herein as a junior synonym of *Karreriella* Cushman 1933.

Figure 7 *Dorothia scabra* (Brady 1884)

[*Gaudryina scabra* Brady 1884]
× 20. From *Challenger* Sta. 24, off Culebra Island, West Indies (390fm.). ZF1458.

Referred by Brady (1884) and Thalmann (1932) to *Gaudryina*, and by

Barker (1960), Zheng (1988), Charnock and Jones (1990), and van Marle (1991) to *Dorothia* (see also Plate 44, Figs 12–13).

Figures 8–11 *Karreriella novangliae* (Cushman 1922)

[*Gaudryina baccata* var. *novangliae* Cushman 1922]
Figs 8, 10 × 20. ZF1449. Fig. 9 × 20. ZF1448. From *Porcupine* Sta. 28, W of Ireland, North Atlantic (1215fm.).
Fig. 11 × 60. From *Challenger* Sta. 174C, off Kandavu, Fiji, Pacific (210fm.). ZF1450.

Referred by Brady (1884) to *Gaudryina baccata* Schwager, by Thalmann first (1932) to *G. baccata* var. *novangliae* Cushman and later (1937) to *Karreriella novangliae* (Cushman), and also by Barker (1960) and Schröder (1986) to *K. novangliae* (Cushman). This is the type-species by original designation of the genus *Meidamonella* Loeblich and Tappan 1986, regarded by the present author as a junior synonym of *Karreriella* Cushman 1933. Srinivasan and Sharma (1980) argue that some of Brady's specimens (not those figured) are true *Karreriella baccata* (Schwager).

Figure 12 *Dorothia pseudofiliformis* (Cushman 1911)

[*Gaudryina pseudofiliformis* Cushman 1911]
× 50. From *Challenger* Sta. 24, off Culebra Island, West Indies (390fm.). ZF1451.

Referred by Brady (1884) to *Gaudryina filiformis* Berthelin, by Thalmann first (1932) to *G. pseudofiliformis* Cushman and later (1942) to *Karreriella pseudofiliformis* (Cushman), by Barker (1960) and Zheng (1988) to *Dorothia exilis* Cushman, and by the present author to *D. pseudofiliformis* (Cushman). *Gaudryina pseudofiliformis* Cushman 1911 is the type-species, by synonymy with the originally designated *Dorothia exilis* Cushman 1936, of the genus *Prolixoplecta* Loeblich and Tappan 1985, regarded by the present author as a junior synonym of *Dorothia* Plummer 1931.

Figure 13 *Dorothia bradyana* Cushman 1936

× 20. From *Challenger* Sta. 24, off Culebra Island, West Indies (390fm.). ZF1462.

Referred by Brady (1884) and Thalmann (1932) to *Gaudryina subrotundata* Schwager, and by Thalmann (1937), Barker (1960), Desai and Banner (1987), and van Marle (1991) to *Dorothia bradyana* Cushman. Desai and Banner (1987) illustrate the typically canaliculate ultrastructure of this species.

Figures 14–16 *Pseudogaudryina* sp. nov. (1)

Fig. 14 × 35. From *Challenger* Sta. 185, off Raine Island, Torres Strait, Pacific (155fm.). ZF1455.
Fig. 15 × 20. From *Challenger* Sta. 174C, off Kandavu, Fiji, Pacific (210fm.). ZF1456.
Fig. 16 × 25. From *Challenger* Sta. 162, Bass Strait, Pacific (38–40fm.). ZF1457.

Referred by Brady (1884) and Thalmann (1932) to *Gaudryina rugosa* d'Orbigny, and by Barker (1960) to *G. (Pseudogaudryina)* sp. nov.

Figures 17–19 *Karrerulina conversa* (Grzybowski 1901)

[*Gaudryina conversa* Grzybowski 1901]
Fig. 17 × 40. From *Challenger* Sta. 218, North of Papua, Pacific (1070fm.). ZF1460.
Figs 18–19 × 40. From *Challenger* Sta. 346, South Atlantic (2350fm.). ZF1461.

Referred by Brady (1884) to *Gaudryina siphonella* Reuss, by Thalmann (1932) to *G. apicularis* Cushman, by Thalmann (1942) and Barker (1960) to *Karreriella (Karrerulina) apicularis* (Cushman), by Collins (1958) to *Karriella (sic) (Karrerulina) apicularis* (Cushman), by Schröder (1986) to *Karreriella apicularis* (Cushman), and by Charnock and Jones (1990) to *Karrerulina conversa* (Grzybowski). This is the type-species, by synonymy with the originally designated *Gaudryina apicularis* Cushman 1911, of the genus *Karrerulina* Finlay 1940. *Gaudryinoides* Saidova 1975 (not Geodakchan 1969) is an junior synonym of *Karrerulina* Finlay 1940.

Figures 20–21 *Chrysalidinella dimorpha* (Brady 1881)

[*Chrysalidina dimorpha* Brady 1881]
Fig. 20 × 70. From *Challenger* Sta. 205A, Hong Kong, Pacific (7fm.). ZF1273.
Fig. 21 × 70. From *Challenger* Sta. 260A, Honolulu, Pacific (40fm.).

Referred by Brady (1884) to *Chrysalidina*, and by Thalmann (1932), Barker (1960), Whittaker and Hodgkinson (1979), and Revets (1991) to *Chrysalidinella*. This is the type-species by original designation of the genus *Chrysalidinella* Schubert 1908, which has been reviewed by Revets (1991). Revets (1991) has shown that some *Challenger* specimens (from Type Slide TS C16) are attributable to *Chrysalidinella pacifica* (Uchio). These specimens had earlier been identified by Sidebottom as *Chrysalidina dimorpha*.

Plate 47 (XLVII)

Figures 1–3 *Reussella spinulosa* (Reuss 1850)

[*Verneuilina spinulosa* Reuss 1850]
Fig. 1 × 70. From *Challenger* Sta. 219A, Admiralty Islands, Pacific (17fm.). ZF2608.
Figs 2–3 × 70. From *Challenger* Sta. 217A, Papua, Pacific (37fm.). ZF2609.

Referred by Brady (1884) to *Verneuilina spinulosa* Reuss, by Thalmann (1932) to *Reussia spinulosa* (Reuss), by Thalmann (1933a) and Leroy (1944b) to *Reussella spinulosa* (Reuss), by Barker (1960) to *R. simplex* (Cushman)? (Fig. 1) and *R. aculeata* Cushman (Figs 2?, 3), by Belford (1966) to *R. aculeata* Cushman (Figs 2? 3), and by van Marle (1991) to *R. simplex* (Cushman) (Fig. 1). *Verneuilina spinulosa* Reuss 1850 is the type-species by original designation of the genus *Reussella* Galloway 1933, which has been reviewed by Revets (1991). *Reussella aculeata* Cushman appears superfluous.

Figures 4–7 *Eggerella bradyi* (Cushman 1911)

[*Verneuilina bradyi* Cushman 1911]
Fig. 4 × 35. ZF2604. Fig. 5 × 25. ZF2603. From *Challenger* Sta. 296, SW of Juan Fernandez Island, South Pacific (1825fm.).

Fig. 6 × 35. From *Challenger* Sta. 23, off Sombrero Island, West Indies (450fm.). ZF2605.
Fig. 7 × 120. From *Challenger* Sta. 196B, South Pacific (1825fm.). ZF2606.

Referred by Brady (1884) to *Verneuilina pygmaea* Egger, sp., by Thalmann first (1932) to *V. bradyi* Cushman and later (1937) to *Eggerella bradyi* (Cushman), and also by Barker (1960), Schröder (1986), Zheng (1988), Hermelin (1989), and van Marle (1991) to *Eggerella bradyi* (Cushman). This is the type-species by original designation of the genus *Eggerella* Cushman 1933. Its ultrastructure has been illustrated by Jones (1984c) and Weston (1984).

Figures 8–12 *Verneuilinulla propinqua* (Brady 1884)

[*Verneuilina propinqua* Brady 1884]
Figs 8–9 × 20; Fig. 12 Magnification not given. From *Challenger* Sta. 98, off Africa, Atlantic (1750fm.). ZF2600.
Figs 10–11 × 14. From *Challenger* Sta. 246, North Pacific (2050fm.). ZF2601.

Referred by Brady (1884) and Thalmann (1932) to *Verneuilina*, by Thalmann (1942), Barker (1960) and Schröder (1986) to *Eggerella*, and by Charnock and Jones (1990) to *Verneuilinulla*. *Verneuilinulla* Saidova 1975 differs from *Eggerella* Cushman 1933 in its interio-marginal rather than areal aperture.

Figures 13–14 *Verneuilinulla affixa* (Cushman 1911)

[*Verneuilina affixa* Cushman 1911]
× 20. From *Challenger* Sta. 78, East of the Azores, Atlantic (1000fm.). ZF2602.

Referred by Brady (1884) to *Verneuilina propinqua* (see above), by Thalmann first (1932) to *V. affixa* Cushman and later (1942) to *Eggerella affixa* (Cushman), and by Barker (1960) also to *Eggerella affixa* (Cushman).

Figures 15–17 *Eggerelloides scaber* (Williamson 1858)

[*Bulimina scabra* Williamson 1858]
Fig. 15 × 60. From Lamlash Harbour, the Clyde, Scotland. ZF2598.
Figs 16–17 × 60. From *Challenger* Sta. 354A, Vigo Bay, Spain, East Atlantic (11fm.). ZF2599.

Referred by Brady (1884) to *Verneuilina polystropha* Reuss, sp., by Thalmann (1932) to *V. scabra* (Williamson), by Thalmann (1942) and Barker (1960) to *Eggerella scabra* (Williamson), and by Haynes first (1973a) to *Eggerelloides scab-rum* (Williamson) and later (1973b) to *E. scab-er* (Williamson). This is the type-species by original designation of the genus *Eggerelloides* Haynes 1973. Generic names ending in –*oides* are to be regarded as masculine (International Code of Zoological Nomenclature (Ride *et al.* 1985), Art. 30(b)). Thus, the masculine ending –*er* should be used in preference to the neuter –*rum*.

Figure 18 *Pseudogaudryina atlantica* (Bailey 1851)

[*Textularia atlantica* Bailey 1851]
× 14. From *Challenger* Sta. 24, West Indies (390fm.). ZF2610.

Referred by Brady (1884) and Thalmann (1932) to *Verneuilina triquetra* (Munster), by Barker (1960) to *Gaudryina (Pseudogaudryina)*

atlantica (Bailey), and by van Marle (1991) to *G. atlantica* (Bailey). This is the type-species of the genus *Pseudogaudryina* Cushman 1936.

Figures 19–20 *Pseudogaudryina* sp. nov. (2)

× 20. From *Challenger* Sta. 174C, off Kandavu, Fiji, Pacific (210fm.). 1959.5.11.697.

Referred by Brady (1884) and Thalmann (1932) to *Verneuilina triquetra* (Munster), and by Barker (1960) to *Gaudryina (Pseudogaudryina)* sp. nov.

Figures 21–24 *Triplasia variabilis* (Brady 1884)

[*Verneuilina variabilis* Brady 1884]
Fig. 21 × 14. ZF2611. Figs 22–23 × 14. ZF2612. Fig. 24 × 14. ZF2613. From *Challenger* Sta. 174C, off Kandavu, Fiji, Pacific (210fm.).

Referred by Brady (1884) and Thalmann (1932) to *Verneuilina*, by Thalmann (1942) to *Frankeina*, and by Barker (1960) to *Triplasia*. *Frankeina* Cushman and Alexander 1929 is a junior synonym of *Triplasia* Reuss 1854 (Loeblich and Tappan 1964, 1987).

Plate 48 (XLVIII)

Figures 1–2, ?3, 4–8 *Martinottiella communis* (d'Orbigny 1846)

[*Clavulina communis* d'Orbigny 1846]
Figs 1, 2, 5 × 25. From *Challenger* Sta. 174C, off Kandavu, Fiji, Pacific (210fm.). ZF1277.
Figs 3–4 × 30. From *Challenger* Sta. 24, off Culebra Island, West Indies (390fm.). ZF1278.
Fig. 6 × 25. 1959.5.5.246. Figs 7–8 × 30. From tow net. ZF1280. From *Challenger* Sta. 323, South Atlantic (1900fm.).

Referred by Brady (1884) and Thalmann (1932) to *Clavulina communis* d'Orbigny, by Thalmann (1937) to *Listerella bradyana* Cushman (Figs 1–2, 5) and *Martinottiella communis* (d'Orbigny) (Figs 3–4, 7–8), by Thalmann (1942) to *Schenckiella communis* (d'Orbigny) (Figs 3–4, 7–8), by Barker (1960) to *Martinottiella bradyana* (Cushman) (Figs 1–2, 5) and *M. communis* (d'Orbigny) (Figs 3–4, 6–8), also by Schröder (1986), Hermelin (1989), Charnock and Jones (1990), and van Marle (1991) to *M. communis* (d'Orbigny) (Figs 3–4, 6–8), and by Zheng (1988) to *Martinotiella* (*sic*) *cylindrica* (Cushman) (Fig. 8 only). *Clavulina communis* d'Orbigny 1826 was a *nomen nudum* (van Marle 1991; Tubbs ICZN, pers. comm.). *C. communis* d'Orbigny 1846 is the type-species by original designation of the genus *Martinottiella* Cushman 1933 (which differs from *Clavulina* d'Orbigny 1826 in its initially trochospiral rather than triserial chamber arrangement). *Listerella bradyana* Cushman 1936 is a junior synonym of *C. communis* d'Orbigny 1846. *Schenckiella* Thalmann 1942 (a new name for *Listerella* Cushman 1933, not Jahn 1906) is a junior synonym of *Martinottiella* Cushman 1933).

Figures 9–13 *Multifidella nodulosa* (Cushman 1927)

[*Clavulina communis* var. *nodulosa* Cushman 1927]
Figs 9–11 × 15. ZF1279. Fig. 12 × 15. 1959.5.5.268. Fig. 13 × 100. ZF1281. From *Challenger* Sta. 24, N. of St Thomas, West Indies (390fm.).

Referred by Brady (1884) to *Clavulina communis* d'Orbigny (see above), by Thalmann first (1932) to *C. communis* var. *nodulosa* Cushman (Figs 9–12) and *C. communis* d'Orbigny (Fig. 13), later (1937) to *Listerella nodulosa* (Cushman) (Figs 9–12) and *Martinottiella communis* (d'Orbigny) (Fig. 13), and later still (1942) to *Schenckiella communis* (d'Orbigny) (Fig. 13 only), and by Barker (1960) to *Martinottiella nodulosa* (Cushman). This is in fact the type-species by original designation of the genus *Multifidella* Loeblich and Tappan 1961, which differs from *Martinottiella* Cushman 1933 in possessing a cribrate aperture.

Figures 14–16 *Pseudoclavulina serventyi* Chapman and Parr 1935

× 18. From *Challenger* Sta. 174C, off Kandavu, Fiji, Pacific (210fm.). ZF1284.

Referred by Brady (1884) and Thalmann (1932) to *Clavulina parisiensis* d'Orbigny, and by Barker (1960) and Zheng (1988) to *Pseudoclavulina serventyi* (Chapman and Parr). *Pseudoclavulina* Cushman, 1936 differs from *Clavulina* d'Orbigny 1826 in lacking a 'valvuline' apertural tooth. Its relationship with *Clavulinoides* Cushman 1936 requires investigation.

Figures 17–18 *Clavulina multicamerata* Chapman 1907

× 30. From *Challenger* Sta. 187A, off Booby Island, Torres Strait, Pacific (8fm.). ZF1285.

Referred by Brady (1884) and Thalmann (1932) to *Clavulina parisiensis* d'Orbigny, and by Barker (1960) to *C. mulicamerata* Chapman.

Figures 19–21 *Clavulina humilis* (Brady 1884)

[*Clavulina parisiensis* var. *humulis* Brady 1884]
× 40. From *Challenger* Sta. 209, Philippine Islands (95fm.). ZF1286.

Referred by Brady (1884) to *Clavulina parisiensis* var. *humilis*, by Thalmann first (1932) to *C. humilis* (Brady) and later (1942) to *Pseudoclavulina humilis* (Brady), and by Barker (1960) also to *P. humilis* (Brady). The presence of a 'valvuline' apertural tooth indicates placement in *Clavulina*.

Figures 22–24 *Clavulina pacifica* Cushman 1924

× 35. From *Challenger* Sta. 218A, Nares Harbour, Admiralty Islands, Pacific (15–25fm.). ZF1274.

Referred by Brady (1884) to *Clavulina angularis* d'Orbigny, and by Thalmann (1932), Barker (1960), and Whittaker and Hodgkinson (1979) to *C. pacifica* Cushman.

Figures 25–31 *Clavulina difformis* (Brady 1884)

[*Clavulina angularis* var. *difformis* Brady 1884]
Figs 25–26 × 40. ZF1275. Figs 27–31 × 40. ZF1276. From *Challenger* Sta. 219A, Admiralty Islands, Pacific (17fm.).

Referred by Brady (1884) to *Clavulina angularis* var. *difformis*, and by Thalmann (1932) and later Barker (1960) to *C. difformis* (Brady).

Figures 32–38 *Cylindroclavulina bradyi* (Cushman 1911)

[*Clavulina bradyi* Cushman 1911]
Figs 32–37 × 20. ZF1283. Fig. 38 × 15. ZF1282. From *Challenger* Sta. 174C, Fiji, Pacific (210fm.).

Referred by Brady (1884) to *Clavulina cylindrica* Hantken, by Thalmann first (1932) to *C. bradyi* Cushman and later (1942) to *Liebusella bradyi* (Cushman), and by Barker (1960) and Zheng (1988) to *Cylindroclavulina bradyi* (Cushman). This is the type-species by original designation of the genus *Cylindroclavulina* Bermúdez and Key 1952.

Plate 49 (XLIX)

Figures 1–7 *Tritaxilina caperata* (Brady 1881)

[*Clavulina caperata* Brady 1881]
Fig. 1 × 30. Figs 2, 4–5 × 16. ZF2491. Fig. 6 × 30, Fig. 7 × 120. ZF2493. From *Challenger* Sta. 174C, off Kandavu, Fiji, Pacific (210fm.).
Fig. 3 × 16. From *Challenger* Sta. 23, off Sombrero Island, West Indies (450fm.). ZF2492.

Referred by Brady (1884) to *Tritaxia caperata* Brady, and by Thalmann (1932), Barker (1960), and Zheng (1988) to *Tritaxilina caperata* (Brady) (Figs 1–2, 4–7) and *T. atlantica* (Cushman) (Fig. 3). *Clavulina caperata* Brady 1881 is the type-species by original designation of the genus *Tritaxilina* Cushman 1911. *Tritaxilina atlantica* Cushman 1911 is regarded by the present author as most likely representing nothing more than the megalospheric generation of *T. caperata* (Brady 1881).

Figures 8–9 *Pseudoclavulina tricarinata* (Leroy 1941)

[*Clavulinoides tricarinatus* Leroy 1941]
× 35. From *Challenger* Sta. 185, off Raine Island, Torres Strait, Pacific (155fm.). ZF2496.

Referred by Brady (1884) and Thalmann (1932) to *Tritaxia tricarinata* (Reuss), by Thalmann (1933a) to *Clavulina tricarinata* (Reuss), by Barker (1960) to *Clavulinoides* aff. *instar* Finlay (sp. nov.?), and by van Marle (1991) to *C. tricarinatus* (Leroy). *Clavulinoides* Cushman 1936 is regarded by the present author as a junior synonym of *Pseudoclavulina* Cushman 1936. Note, though, that this requires further investigation.

Figures 10–11 *Latentoverneuilina indiscreta* (Brady 1881)

[*Tritaxia indiscreta* Brady 1881]
× 25. From *Challenger* Sta. 174C, Fiji, Pacific (210fm.). ZF2494.

Referred by Brady (1884) and Thalmann (1932, 1933a) to *Tritaxia*, by Thalmann (1942) and Barker (1960) to *Clavulinoides*, and by Loeblich and Tappan (1985b) to *Latentoverneuilina*. This is the type-species by original designation of the genus *Latentoverneuilina* Loeblich and Tappan 1985.

Figure 12 *Trifarina lepida* (Brady 1881)

[*Tritaxia lepida* Brady 1881]
× 80. From *Challenger* Sta. 45, North Atlantic (1240fm.). ZF2495.

Referred by Brady (1884) and Thalmann (1932, 1933*a*) to *Tritaxia*, and by Thalmann (1942) and Barker (1960) to *Angulogerina* (?). *Angulogerina* Cushman 1927 is regarded by the present author as a junior synonym of *Trifarina* Cushman 1923. The generic identity of the present species is questionable.

Figure 13 *Tritaxis fusca* (Williamson 1858)

[*Rotalina fusca* Williamson 1858]
× 50. From *Challenger* Sta. VIII, off Gomera, Canaries, Atlantic (620fm.). 1959.5.11.679.

Referred by Brady (1884) and Thalmann (1932) to *Valvulina*, and by Barker (1960) and Zheng (1988) to *Tritaxis*. This is the type-species by subsequent designation of the genus *Tritaxis* Schubert 1921. Brönnimann and Whittaker (1984*b*) have reviewed the genus *Tritaxis*, and have designated a neotype for *T. fusca* (Williamson) from material donated to Brady by Williamson.

Figure 14 *Trochamminella* sp.

× 40. From *Challenger* Sta. 166, off New Zealand, South Pacific (275fm.). ZF2596.

Referred by Brady (1884) and Thalmann (1932) to *Valvulina fusca*, and by Barker (1960) and Zheng (1988) to *Tritaxis fusca* (see above). The overall morphology is suggestive of *Trochamminella* Cushman 1943, though the generic placement cannot be confirmed without detailed examination of apertural characteristics (Whittaker, pers. comm.).

Figures 15a, c *Tritaxis australis* Brönnimann and Whittaker 1984

× 50. From *Challenger* Sta. 174C, off Fiji, Pacific (210fm.). ZF2593.

Referred by Brady (1884) Thalmann (1932, 1933*a*), and Barker (1960) to *Valvulina conica* (Parker and Jones), and by the present author tentatively to *Tritaxis australis* Brönnimann and Whittaker 1984.

Figures 15b, 16 *Trochamminella conica* (Parker and Jones 1865)

[*Valvulina triangularis* var. *conica* Parker and Jones 1865]
Fig. 15b × 50, Fig. 16 Magnification not given. From Drobak, Norway. 1886.1.16.90.

Referred by Brady (1884), Thalmann (1932, 1933*a*), and Barker (1960) to *Valvulina conica* (Parker and Jones). These specimens are very similar to that figured by Hodgkinson (1992) (believed to be that figured in the original description by Parker and Jones).

Plate 50 (L)

Figures 1–2 *Bulimina gibba* Fornasini 1901

Fig. 1 × 60; Fig. 2 × 50. From *Porcupine* Sta. 11, West of Ireland, Atlantic (1630fm.). ZF1210.

Referred by Brady (1884) and Thalmann (1932) to *Bulimina elegans* d'Orbigny, by Thalmann (1942) to *B. parkerae*, and by Barker (1960) and Haynes (1973*a*) to *B. gibba* Fornasini.

Figures 3–4 *Bulimina elongata* d'Orbigny 1846

× 50. From *Porcupine* Sta. 10, West of Ireland, Atlantic. ZF1211.

Referred by Brady (1884) and Thalmann (1932) to *Bulimina elegans* d'Orbigny, by Thalmann (1942) to *B. parkerae*, by Barker (1960) to *B. gibba* Fornasini, and by Haynes (1973*a*) to *B. elongata* d'Orbigny. This species has been lectotypified by Papp and Schmid (1985). Certain authors have assigned it to the genus *Caucasina* Khalikov 1951 on account of an initial trochospire, but this has been shown by Nørvang (1968) to be present in the microspheric generation only.

Figures 5–6 *Eubuliminella exilis* (Brady 1884)

[*Bulimina elegans* var. *exilis* Brady 1884]
× 60. From *Porcupine* Sta. 20, NW of Ireland, North Atlantic, (1443fm.). ZF1212.

Referred by Brady (1884) to *Bulimina elegans* var. *exilis*, and by Thalmann (1932), Barker (1960), van Morkhoven *et al.* (1986), and van Marle (1991) to *B. exilis* (Brady) and by Revets (1993*a*) to *Eubuliminella exilis* (Brady). This is the type-species by original designation of the genus *Eubuliminella* Revets 1993, which differs from *Bulimina* d'Orbigny 1826 in its higher seriality.

Figures 7–10 *Globobulimina pacifica* Cushman 1927

Fig. 7 × 40. Fig. 8 × 50. From *Challenger* Sta. 75, off the Azores, Atlantic (450fm.). ZF1222.
Fig. 9 × 50. From *Challenger* Sta. 168, off New Zealand (1100fm.). ZF1223.
Fig. 10 × 50. From *Challenger* Sta. 192, Ki Islands, Central Pacific (129fm.). ZF1224.

Referred by Brady (1884) to *Bulimina pyrula* d'Orbigny, and by Thalmann (1932) and Barker (1960) (albeit tentatively) to *Globobulimina pacifica* Cushman. This is the type-species by original designation of the genus *Globobulimina* Cushman 1927. *G. caribbea* Cushman and Bermúdez, 1945 is arguably synonymous.

Figures 11–12 *Praeglobobulimina spinescens* (Brady 1884)

[*Bulimina pyrula* var. *spinescens* Brady 1884]
× 60. From *Challenger* Sta. 191A, Ki Islands, Central Pacific (580fm.). ZF1225.

Referred by Brady (1884), Thalmann (1932), and Barker (1960) to *Bulimina pyrula* var. *spinescens*, and by Belford (1966) and van Marle (1991) to *Praeglobobulimina spinescens* (Brady). This is the type-species by original designation of the genus *Praeglobobulimina* Hofker 1951.

Figure 13 *Praeglobobulimina ovata* (d'Orbigny 1846)

[*Bulimina ovata* d'Orbigny 1846]
× 50. From *Challenger* Sta. 168, East of New Zealand (110fm.). ZF1220.

Referred by Brady (1884) and Thalmann (1932) to *Bulimina ovata* d'Orbigny, and by Barker (1960) to *B. notovata* Chapman. Papp and Schmid (1985) synonymized *Bulimina ovata* d'Orbigny 1846 with *B. pyrula* d'Orbigny 1846, but the present author has reinstated it on the grounds that its greatest width is above rather than below mid-height.

Figures 14–15 *Praeglobobulimina pupoides* (d'Orbigny 1846)

[*Bulimina pupoides* d'Orbigny 1846]
Fig. 14 × 50. ZF1205. Fig. 15 × 45. ZF1221. From *Challenger* Sta. 306, West of Patagonia, East Pacific (345fm.).

Referred by Brady (1884) and Thalmann (1932, 1933a) to *Bulimina affinis* d'Orbigny (Fig. 14) and *B. pupoides* d'Orbigny (Fig. 15), by Barker (1960) to *B. pupoides* d'Orbigny (Figs 14–15), by Belford (1966) to *Protoglobobulimina pupoides* (d'Orbigny) (Fig. 15), and by van Marle (1991) to *Praeglobobulimina pupoides* (d'Orbigny) s.l. (Fig. 15). Papp and Schmid (1985) synonymized *Bulimina pupoides* d'Orbigny 1846 with *B. pyrula* d'Orbigny 1846, but the present author has reinstated it on the grounds that its greatest width is above rather than below mid-height. *Bulimina pupoides* d'Orbigny 1846 is the type-species by original designation of the genus *Protoglobobulimina* Hofker 1951, regarded by the present author as a junior synonym of *Praeglobobulimina* Hofker 1951.

Figure 16 *Robertina subcylindrica* (Brady 1881)

[*Bulimina subcylindrica* Brady 1881]
× 80. From *Challenger* Sta. 120, off Pernambuco, West Atlantic (675fm.). ZF1228.

Referred by Brady (1884) to *Bulimina*, by Thalmann (1932) to *Buliminella*, by Thalmann (1937) and Barker (1960) to *Robertina*, and by Collins (1958) to *Robertinoides*.

Figure 17 *Robertina tasmanica* Parr 1950

× 60. From *Challenger* Sta. 174B, Fiji, Pacific (610fm.). ZF1229.

Referred by Brady (1884) to *Bulimina subteres* Brady, by Thalmann (1932) to *Buliminella subteres* (Brady), and by Barker (1960) (albeit tentatively) to *Robertina tasmanica* Parr.

Figure 18 *Robertinoides bradyi* (Cushman and Parker, 1936)

[*Robertina bradyi* Cushman and Parker 1936]
× 80. From *Challenger* Sta. 24, West Indies (390fm.). ZF1230.

Referred by Brady (1884) to *Bulimina subteres* Brady, by Thalmann first (1932) to *Buliminella subteres* (Brady), later (1937) to *Robertina bradyi* Cushman and Parker, and later still (1942) to *R. subteres* (Brady), and by Barker (1960) to *Robertinoides bradyi* (Cushman and Parker).

Figure 19 *Robertinoides oceanicus* (Cushman and Parker 1947)

[*Robertina oceanica* Cushman and Parker 1947]
× 70. From *Challenger* Sta. 191A, Ki Islands, Central Pacific (580fm.). ZF1209.

Referred by Brady (1884) to *Bulimina declivis* Reuss, by Thalmann (1932) to *Buliminella declivis* (Reuss), by Barker (1960) to *Robertina oceanica* Cushman and Parker, and by Belford (1966) to *Robertinoides oceanicus* (Cushman and Parker).

Figures 20–22 *Buliminella elegantissima* (d'Orbigny 1839)

[*Bulimina elegantissima* d'Orbigny1839]
× 80. From *Challenger* Sta. 315A, Falkland Islands, South Atlantic (6fm.). ZF1214.

Referred by Brady (1884) to *Bulimina,* and by Thalmann (1932), Barker (1960), and Revets (1990a) to *Buliminella*. This is the type-species by original designation of the genus *Buliminella* Cushman 1911.

Figure 23 *Elongobula* sp. nov.

× 80. From *Challenger* Sta. 162, Bass Strait, Pacific (38–40fm.). ZF1215.

Referred by Brady (1884) to *Bulimina elegantissima* var. *seminuda* Terquem, by Thalmann (1932) to *Buliminella* n. sp., and by Barker (1960) to *B. basicostata* Parr. Regarded by Revets (pers. comm.) as a new species of *Elongobula* Finlay 1939.

Figure 24 *Elongobula arethusae* Revets 1993

× 80. From *Challenger* Sta. 162, Bass Strait, Pacific (38–40fm.). ZF1215.

Referred by Brady (1884) to *Bulimina elegantissima* var. *seminuda* Terquem, by Thalmann (1932) to *Buliminella seminuda* (Terquem), and by Barker (1960) to *B. basicostata* Parr, and by Revets (1993b) to *Elongobula arethusae* sp. nov.

Plate 51 (LI)

Figures 1–2 *Bulimina elongata* d'Orbigny 1846

Fig. 1 × 70; Fig. 2 × 80. From *Porcupine* Sta. 23, off NW Ireland, North Atlantic (630fm.). ZF1216.

Referred by Brady (1884) and Thalmann (1932) to *Bulimina elongata* d'Orbigny (Fig. 1) and to a transient between *B. elongata* d'Orbigny and *B. ovata* d'Orbigny (Fig. 2), by Barker (1960) to *B. elongata* var. *subulata* Cushman and Parker, and by Haynes (1973a) to *B. elongata* var. *lesleyae* Atkinson. *Bulimina elongata* d'Orbigny has been lectotypified by Papp and Schmid (1985). None of the proposed varietal forms of this species seems significantly different from the lectotype.

Figures 3–5 *Bulimina marginata* d'Orbigny 1826

× 80. From *Porcupine* Sta. 11, West of Ireland, North Atlantic (1630fm.). ZF1219.

Referred by Brady (1884), Thalmann (1932), Barker (1960), Belford (1966), Haynes (1973a), and van Marle (1991) (erroneously citing Plate 1, Figs 3–5 as opposed to Plate 51, Figs 3–5) to *Bulimina marginata* d'Orbigny. This is the type-species by original designation of the genus *Bulimina* d'Orbigny 1826. Its internal structure has been well illustrated by Nørvang (1968) and Verhallen (1986).

Figure 6 *Bulimina subornata* Brady 1884

× 80. From *Challenger* Sta. 232, South of Japan, North Pacific (345fm.). ZF1227.

Referred by Brady (1884), Thalmann (1932), Barker (1960), and Belford (1966) to *Bulimina subornata*.

Figures 7–9 _Bulimina aculeata_ d'Orbigny 1826

Fig. 7 ZF1203. Figs 8–9 ZF1204. × 60. From _Challenger_ Sta. 191A, Ki Islands, Central Pacific (580fm.).

Referred by Brady (1884), Thalmann (1932), Barker (1960), Belford (1966), and van Marle (1991) to _Bulimina aculeata_ d'Orbigny.

Figures 10–13 _Bulimina mexicana_ (Cushman 1922)

[_Bulimina inflata_ var. _Mexicana_ Cushman 1922]
Figs 10, 12 × 60. From _Challenger_ Sta. 191A, Ki Islands, Central Pacific (580fm.). ZF1217.
Fig. 11 × 60. From _Challenger_ Sta. 232, South of Japan, North Pacific (345fm.). ZF1218.
Fig. 13 × 60. From _Challenger_ Sta. 279C, off Tahiti, Pacific.

Referred by Brady (1884) to _Bulimina inflata_ Seguenza, by Thalmann (1932) to _B. aculeata_ d'Orbigny (see above), by Barker (1960) to _B. striata_ var. _mexicana_ Cushman? (Figs 10, 12) and _B. costata_ d'Orbigny? (Figs 11, 13), by Hermelin (1989) to _B. mexicana_ (Cushman), and by van Marle (1991) to _B. striata_ d'Orbigny.

Figures 14–15 _Bulimina rostrata_ Brady 1884

× 80. From _Challenger_ Sta. 191A, Ki Islands, Central Pacific (580fm.). ZF1226.

Figures 16–17 _Buliminoides williamsonianus_ (Brady 1881)

[_Bulimina williamsoniana_ Brady 1881]
Fig. 16 × 80. From _Challenger_ Sta. 219A, Admiralty Islands, Pacific (17fm.). ZF1231.
Fig. 17 × 80. From _Challenger_ Sta. 185, Torres Strait, Pacific (155fm.). ZF1232. Lectotype designated by Revets (1990_b_).

Referred by Brady (1884) to _Bulimina,_ and by Thalmann (1932), Barker (1960), and van Marle (1991) to _Buliminoides._ Van Marle (1991) regards _Bulimina williamsoniana_ Brady 1881 as a _nomen nudum,_ but gives no justification. This is the type-species by original designation of the genus _Buliminoides_ Cushman 1911. This genus has been reviewed by Seiglie (1969, 1970) and Revets (1990_b_).

Figures 18–19 _Bulimina_ sp. nov.

× 80. From _Challenger_ Sta. 192, Ki Islands, Central Pacific (129fm.). ZF1206.

Referred by Brady (1884) to _Bulimina buchiana_ d'Orbigny, by Thalmann (1932) (albeit tentatively) to _B. bradyi_ Weinzierl and Applin, and by Barker (1960) to _B._ sp. nov.

Figure 20 _Pleurostomella brevis_ Schwager 1866.

× 40. From _Challenger_ Sta. 192, Ki Islands, Central Pacific (129fm.). ZF2121.

Referred by Brady (1884), Thalmann (1932), Barker (1960), Srinivasan and Sharma (1980), Hermelin (1989), and van Marle (1991) to _Pleurostomella brevis_ Schwager. Srinivasan and Sharma (1980) have neotypified this species.

Figure 21 _Pleurostomella recens_ (Dervieux 1899)

[_Pleurostomella rapa_ var. _recens_ Dervieux 1899]
× 60. From _Challenger_ Sta. 192, Ki Islands, Central Pacific (129fm.). ZF2122.

Referred by Brady (1884) and Thalmann (1932) to _Pleurostomella rapa_ Gumbel, by Thalmann (1933_a_) and Barker (1960) to _P. rapa_ var. _recens,_ and by Hermelin (1989) to _P. recens_ (Dervieux).

Figure 22 _Pleurostomella acuminata_ Cushman 1922

× 70. From _Challenger_ Sta. 192, Ki Islands, Central Pacific (129fm.). ZF2120.

Referred by Brady (1884) to _Pleurostomella alternans_ Schwager, and by Thalmann (1932), Barker (1960), Hermelin (1989), and van Marle (1991) to _P. acuminata_ Cushman.

Figure 23 _Pleurostomella_ sp. nov. (1)

× 60. From _Challenger_ Sta. 192, Ki Islands, Central Pacific (129fm.). ZF2120.

Referred by Brady (1884), Thalmann (1932), and van Marle (1991) to _Pleurostomella alternans_ Schwager, and by Barker (1960) to _P._ sp. nov.

Plate 52 (LII)

Figures 1–3 _Fursenkoina complanata_ (Egger 1893)

[_Virgulina schreibersiana_ var. _complanata_ Egger 1893]
Figs 1, 3 × 70. From _Challenger_ Sta. 276, South Pacific (2350fm.). ZF2618.
Fig. 2 × 70. From _Challenger_ Sta. 332, South Atlantic (2200fm.).

Referred by Brady (1884) and Thalmann (1932, 1933_a_) to _Virgulina schreibersiana_ Czjzek, by Barker (1960) to _V. davisi_ Chapman and Parr (Figs 1, 3) and _Cassidella pacifica_ Hofker? (or 'may be ... _Cassidella complanata_ (Egger)') (Fig. 2), and by van Marle (1991) to _Fursenkoina schreibersiana_ (Czjzek). The generic identity of this species is somewhat questionable. _Virgulina_ d'Orbigny 1826 is a junior homonym of _Virgulina_ Bory de St Vincent 1823.

Figures 4–5 _Fursenkoina pauciloculata_ (Brady 1884)

[_Virgulina pauciloculata_ Brady 1884]
× 80. From _Challenger_ Sta. 217A, Papua, Pacific (37fm.). ZF2617.

Referred by Brady (1884), Thalmann (1932, 1933_a_), Barker (1960) to _Virgulina_ (see above). The generic identity of this species is somewhat questionable.

Figures 6, 14–17 _Fursenkoina texturata_ (Brady 1884)

[_Virgulina texturata_ Brady 1884]
Fig. 6 ZF2622. Figs 14–17 ZF2619. × 40. From _Challenger_ Sta. 296, South Pacific (1825fm.).

Referred by Brady (1884), Thalmann (1932, 1933_a_), and Barker (1960) to _Virgulina texturata_ (Fig. 6) and _V. subdepressa_ (Figs 14–17), and by van Marle (1991) to _Fursenkoina texturata_ (Brady) (Fig. 6 only). _Virgulina subdepressa_ Brady 1884 is regarded by the present author as a junior synonym of *V. texturata* Brady 1884, representing nothing more than the macrospheric as opposed to the microspheric generation. The generic identity of this species is somewhat questionable. *Virgulina* d'Orbigny 1826 is unavailable (see above).

Figures 7–8 *Cassidulinoides tenuis* Phleger and Parker 1951

× 50. From *Challenger* Sta. 279C, off Tahiti (620fm.). ZF2621.

Referred by Brady (1884) to *Virgulina subsquamosa* Egger, by Thalmann (1932) to *V. bradyi* Cushman, by Barker (1960) to *Cassidulinoides* sp. nov., and by the present author to *C. tenuis* Phleger and Parker. Hermelin (1989) referred this species to the genus *Rutherfordoides* McCulloch 1981.

Figure 9 *Fursenkoina bradyi* (Cushman 1922)

[*Virgulina bradyi* Cushman 1922]
× 50. From *Challenger* Sta. 279C, off Tahiti (620fm.). ZF2621.

Referred by Brady (1884) to *Virgulina subsquamosa* Egger, by Thalmann (1932) and Barker (1960) to *V. bradyi* (Cushman), by Hermelin (1989) to *Rutherfordoides bradyi* (Cushman), and by van Marle (1991) to *Fursenkoina bradyi* (Cushman). The generic identity of this species is somewhat questionable.

Figures 10–11 *Fursenkoina rotundata* (Parr 1950)

[*Virgulina rotundata* Parr 1950]
× 70. From *Challenger* Sta. 279C, off Tahiti (620fm.). ZF2620.

Referred by Brady (1884) and Thalmann (1932, 1933a) to *Virgulina subsquamosa* Egger, and by Barker (1960) to *V. rotundata* Parr. The generic identity of this species is somewhat questionable. Note, however, that it bears a close resemblance to *Virgulina mexicana* Cushman 1922, the type-species by original designation of the genus *Hastilina* Nomura 1983 (regarded by Loeblich and Tappan (1987) as a junior synonym of *Rutherfordoides* McCulloch 1981).

Figures 12–13 *Pleurostomella* sp. nov. (2)

Fig. 12 × 50. From *Challenger* Sta. 346, South Atlantic (2350fm.). ZF2124.
Fig. 13 × 50. From *Challenger* Sta. 332, South Atlantic (2200fm.). ZF2123.

Referred by Brady (1884), Thalmann (1932, 1933a), and Hermelin (1989) to *Pleurostomella subnodosa* (Reuss), and by Barker (1960) to *Pleurostomellina* sp. nov. *Pleurostomellina* Schubert 1911 is a junior synonym of *Pleurostomella* Reuss 1860 (Loeblich and Tappan 1964, 1987).

Figures 18–19 *Brizalina earlandi* (Parr 1950)

[*Bolivina earlandi* Parr 1950]
× 60. From *Challenger* Sta. 151, Heard Island, South Pacific (75fm.). ZF1191.

Referred by Brady (1884) to *Bolivina punctata* d'Orbigny, by Thalmann (1932) to *B. beyrichi* Reuss, and by Barker (1960) to *B. earlandi* Parr. Parr's species is transferred by the present author to *Brizalina* as it lacks the retral processes of typical *Bolivina*. It is acknowledged that such a differentiation is at times almost arbitrary.

Revets (pers. comm.) believes that *Bolivina* and *Brizalina* can only be consistently distinguished on differences in the detailed structure of their apertural complexes.

Figures 20–21 *Brizalina spathulata* (Williamson 1858)

[*Textularia variabilis* var. *spathulata* Williamson 1858]
× 70. From *Porcupine* Sta. 18, West of Ireland, East Atlantic (183fm.).

Referred by Brady (1884) to *Bolivina dilatata* Reuss, by Thalmann first (1932) to *B. beyrichi* Reuss and later to *B. spathulata* (Williamson), and by Barker (1960) and van Marle (1991) also to *B. spathulata* (Williamson), and by Haynes (1973a) to *B. (Brizalina) spathulata* (Williamson).

Figure 22 *Parabrizalina porrecta* (Brady 1881)

[*Bolivina porrecta* Brady 1881]
× 50. From *Challenger* Sta. 24, West Indies (390fm.). ZF1190.

Referred by Brady (1884) to *Bolivina*, by Thalmann first (1932) to *Bifarina* and later (1942) to *Loxostomum*, by Barker (1960) first (text) to *Loxostomum* and later (addenda) to *Loxostomoides*, by Zweig-Strykowski and Reiss (1975) to *Brizalina* (*Parabrizalina*), and by Revets (pers. comm.) to *Parabrizalina*. This is the type-species by original designation of *Brizalina* (*Parabrizalina*) Zweig-Strykowski and Reiss 1975. It has also been referred to *Loxostomella* Saidova 1975.

Figure 23 *Brizalina* sp. nov.

× 70. From *Challenger* Sta. 174B, Fiji, Pacific (610fm.). ZF1199.

Referred by Brady (1884) to *Bolivina textilarioides* Reuss, by Thalmann first (1932) to *B. textilarioides* Reuss, later (1933a) to *Bolivina* n. sp., and later still (1942) to *B. subspinescens* Cushman (see below), and by Barker (1960) to *Bolivina* sp. nov. Thalmann 1933.

Figures 24–25 *Brizalina subspinescens* (Cushman 1922)

[*Bolivina subspinescens* Cushman 1922]
× 70. From *Porcupine* Sta. 18, West of Ireland, East Atlantic (183fm.). ZF1200.

Referred by Brady (1884) to *Bolivina textilarioides* Reuss, by Thalmann (1932) to *B. spinescens* Cushman, and by Barker (1960) and van Marle (1991) to *B. subspinescens* Cushman.

Figures 26–28 *Loxostomina limbata* (Brady 1881)

[*Bolivina limbata* Brady 1881]
Fig. 26 × 50. From *Porcupine* Sta. 260A, Honolulu reefs, Pacific (40fm.). ZF1182.
Fig. 27 × 50. From *Challenger* Sta. 218A, Admiralty Islands, Pacific (16–25fm.). ZF1183.
Fig. 28 × 50. From *Challenger* Sta. 219A, Admiralty Islands, Pacific (17fm.). ZF1184.

Referred by Brady (1884) to *Bolivina*, by Thalmann first (1932, 1933a) to *Bolivina* and later (1942) to *Loxostomum*, by Barker first (1960, text) to *Loxostomum* and later (1960, addenda) to *Loxostomoides*, and by Belford (1966) and van Marle (1991) to *Rectobolivina*. Van Marle (1991) regards *Bolivina limbata* Brady 1881 as a *nomen nudum,* but gives no justification. This species was referred by Saidova (1975) to *Loxostomella. Loxostomella* Saidova 1975 is regarded by the present author as a junior synonym of *Loxostomina* Sellier de Civrieux 1969.

Figure 29 *Brizalina subtenuis* (Cushman 1936)

[*Bolivina subtenuis* Cushman 1936]
× 80. From *Challenger* Sta. 174A, Fiji, Pacific (255fm.). ZF1198.

Referred by Brady (1884) to *Bolivina tenuis* Brady, by Thalmann first (1932, 1933*a*) to *B. tenuis* Brady and later (1937) to *B. subtenuis* Cushman, and also by Barker (1960) to *B. subtenuis* Cushman. *Bolivina tenuis* Brady 1881 is a junior homonym of *B. tenuis* Marsson 1878.

Figure 30 *Brizalina nitida* (Brady 1884)

[*Bolivina nitida* Brady 1884]
× 80. From *Challenger* Sta. 185, Torres Strait, Pacific (155fm.). ZF1187.

Referred by Brady (1884), Thalmann (1932, 1933*a*), and Barker (1960) to *Bolivina.*

Figures 31–34 *Sigmavirgulina tortuosa* (Brady 1881)

[*Bolivina tortuosa* Brady 1881]
Figs 31–32 × 70. From *Challenger* Sta. 219A, Admiralty Islands, Pacific (17fm.). ZF1201.
Figs 33–34 × 70. From *Challenger* Sta. 352. Cape Verde Islands, Atlantic (11fm.). ZF1202.

Referred by Brady (1884), Thalmann (1932, 1933*a*) and van Marle (1991) (Figs 31–32 only) to *Bolivina tortuosa,* by Thalmann (1937) to *B. tortuosa* var. *atlantica* Cushman (Figs 33–34 only), by Barker (1960) to *Sigmavirgulina tortuosa* (Brady) (Figs 31–32) and *S.* (?) *tortuosa* var. *atlantica* (Cushman) (Figs 33–34), and by Belford (1966) to *Sigmavirgulina tortuosa* (Brady) (Figs 31–32 only). Van Marle (1991) regards *Bolivina tortuosa* Brady 1881 as a *nomen nudum,* but gives no justification. This is the type-species by original designation of the genus *Sigmavirgulina* Loeblich and Tappan 1957. *Bolivina tortuosa* var. *atlantica* Cushman 1936 appears superfluous.

Plate 53 (LIII)

Figure 1 *Euloxostoma bradyi* (Asano 1938)

[*Bolivina bradyi* Asano 1938]
× 50. From *Challenger* Sta. 305, West of Patagonia, East Pacific (120fm.). ZF1172.

Referred by Brady (1884) to *Bolivina beyrichi* Reuss, by Thalmann first (1932) to *B. beyrichi* Reuss and later (1942) to *B. bradyi* Asano, by Barker first (1960, text) to *B. bradyi* Asano and later (1960, addenda) to *Loxostomoides bradyi* (Asano), and by van Marle (1991) (erroneously citing Pl. 33, Fig. 1 instead of Pl. 53, Fig. 1) to *Brizalina pseudobeyrichi* (Cushman). *Bolivina bradyi* Asano 1938 is the type-species, by synonymy with the originally designated *Loxostoma instabile* Cushman and McCulloch 1942, of the genus *Euloxostoma* McCulloch 1977 (nom. corr.).

Figures 2–4 *Brizalina alata* (Seguenza 1862)

[*Vulvulina alata* Seguenza 1862]
Fig. 2 × 50. From *Challenger* Sta. 191, Arrou Islands, North Pacific (800fm.). ZF1173.
Fig. 3 × 60. From *Challenger* Sta. 209, Philippines, Pacific (95–100fm.). ZF1174.
Fig. 4 × 60.

Referred by Brady (1884) to *Bolivina beyrichi* var. *alata* Seguenza, by Thalmann (1932) and Barker (1960) to *B. alata* Seguenza, and by Belford (1966) and van Marle (1991) to *Brizalina alata* (Seguenza).

Figures 5–6 *Brizalina pygmaea* (Brady 1881)

[*Bulimina (Bolivina) pygmaea* Brady 1881]
× 60. From *Challenger* Sta. 145, Prince Edward Island, South Pacific (50–150fm.). ZF1192.

Referred by Brady (1884), Thalmann (1932), and Barker (1960) to *Bolivina,* and by van Marle (1991) to *Brizalina* (see discussion under *Brizalina earlandi* (Plate 52, Figs 18–19)). Van Marle (1991) regards *Bolivina pygmaea* Brady 1881 as a *nomen nudum,* but gives no justification.

Figures 7–9 *Bolivina robusta* Brady 1881

Fig. 7 × 60. From *Challenger* Sta. 191A, Ki Islands, Central Pacific (580fm.). ZF1194.
Figs 8–9 × 60. From *Challenger* Sta. 174B, Fiji, Pacific (610fm.). ZF1195.

Referred by Brady (1884), Thalmann (1932), Barker (1960), Belford (1966), and van Marle (1991) to *Bolivina robusta* Brady. Van Marle (1991) regards *Bolivina robusta* Brady 1881 as a *nomen nudum,* but gives no justification.

Figures 10–11 *Brizalina subaenariensis* var. *mexicana* (Cushman 1922)

[*Bolivina subaenariensis* var. *mexicana* Cushman 1922]
Figs 10a, 11 × 50. From *Porcupine* Sta. 11, West of Ireland, North Atlantic (1630fm.). ZF1170.
Fig. 10b × 50. From *Challenger* Sta. 17, West of Ireland, North Atlantic (1230fm.). ZF1171.

Referred by Brady (1884) to *Bolivina aenariensis* Costa, sp., by Thalmann (1932) to *B. subaenariensis* Cushman, and by Barker (1960) to *B. subaenariensis* Cushman var.

Figures 12–13 *Bolivina decussata* Brady 1881

× 60. From *Challenger* Sta. 300, North of Juan Fernandez, East Pacific (1375fm.). ZF1176.

Figures 14–15 *Loxostomina mayori* (Cushman 1922)

[*Bolivina mayori* Cushman 1922]
Fig. 14 × 60. From *Challenger* Sta. 219A, Admiralty Islands, Pacific (17fm.). ZF1188.
Fig. 15 × 60. From *Challenger* Sta. 279C, off Tahiti, Pacific (620fm.). ZF1189.

Referred by Brady (1884) and Thalmann (1932) to *Bolivina nobilis* Hantken, by Barker first (1960, text) to *Loxostomum* sp. nov. and later (1960, addenda) to *Loxostomoides* sp. nov., and by Revets (pers. comm.) to *Loxostomina mayori* (Cushman). This is the type-species by original designation of the genus *Loxostomina* Sellier de Civrieux 1969.

Figures 16–18 *Lugdunum hantkenianum* (Brady 1881)

[*Bolivina hantkeniana* Brady 1881]
Fig. 16 × 40. From *Challenger* Sta. 279A, off Tahiti, Pacific (420fm.). ZF1178.
Fig. 17 × 40. From *Challenger* Sta. 279C, off Tahiti, Pacific (620fm.). ZF1179.
Fig. 18 × 60. From *Challenger* Sta. 174C, off Kandavu, Fiji (210fm.). ZF1180.

Referred by Brady (1884), Thalmann (1932), and Barker (1960) to *Bolivina*, and by Revets (pers. comm.) to *Lugdunum*. This is the type-species by original designation of the genus *Lugdunum* Saidova 1975.

Figures 19–21 *Saidovina karreriana* (Brady 1881)

[*Bolivina karreriana* Brady 1881]
× 60. From *Challenger* Sta. 232, South of Japan, North Pacific (345fm.). ZF1181.

Referred by Brady (1884) and Thalmann (1932) to *Bolivina*, by Thalmann (1942) and Barker (1960) to *Loxostomum*, by Belford (1966) and van Marle (1991) to *Brizalina*, and by Revets (pers. comm.) to *Saidovina*. Van Marle (1991) regards *Bolivina karreriana* Brady 1881 as a *nomen nudum*, but gives no justification. This is the type-species by original designation of the genus *Saidovina* Haman 1984, substitute name for *Loxostomina* Saidova 1975, not Sellier de Civrieux 1969.

Figures 22–23 *Pseudobrizalina lobata* (Brady 1881)

[*Bolivina lobata* Brady 1881]
× 80. From *Challenger* Sta. 219A, Admiralty Islands, Pacific (17fm.). ZF1185.

Referred by Brady (1884) and van Marle (1991) to *Bolivina lobata* Brady, by Thalmann first (1932) to *B. lobata* Brady and later (1942) erroneously to *Loxostomum amygdalaeforme* (Brady) (see below), by Barker first (1960, text) to *Loxostomum lobatum* (Brady) and later (1960, addenda) to *Loxostomoides* (?) *lobatum* (Brady), by Zweig-Strykowski and Reiss (1975) to *Brizalina* (*Pseudobrizalina*) *lobata* (Brady), and by Revets (pers. comm.) to *Pseudobrizalina lobata* (Brady). Van Marle (1991) regards *Bolivina lobata* Brady 1881 as a *nomen nudum*, but gives no justification. This is the type-species by original designation of *Brizalina* (*Pseudobrizalina*) Zweig-Strykowski and Reiss 1975, regarded by Loeblich and Tappan (1987) as a junior synonym of *Sagrinella* Saidova 1975, but by Revets (pers. comm.) as distinct.

Figures 24–25 *Lugdunum schwagerianum* (Brady 1881)

[*Bolivina schwageriana* Brady 1881]
× 50. From *Challenger* Sta. 217A, off Papua, Pacific (37fm.). ZF1196.

Referred by Brady (1884), Thalmann (1932), Barker (1960), and

van Marle (1991) to *Bolivina*. Van Marle (1991) regards *Bolivina schwageriana* Brady 1881 as a *nomen nudum*, but gives no justification. The species was referred by Saidova (1975) to *Lugdunum*.

Figures 26–27 *Brizalina semicostata* (Cushman 1911)

[*Bolivina semicostata* Cushman 1911]
× 60. From *Challenger* Sta. 185, Torres Strait, Pacific (155fm.). ZF1175.

Referred by Brady (1884) to *Bolivina costata* d'Orbigny, and by Thalmann (1932) and Barker (1960) to *B. semicostata* Cushman.

Figures 28–29 *Saidovina amygdalaeformis* (Brady 1881)

[*Bolivina amygdalaeformis* Brady 1881]
× 50. From *Challenger* Sta. 209, Philippines, Pacific (95–100fm.). ZF1169.

Referred by Brady (1884) and Thalmann (1932) to *Bolivina*, and by Barker (1960) to *Loxostomum*. The generic identity of this species is somewhat questionable.

Figures 30–31 *Brizalina subreticulata* (Parr 1932)

[*Bolivina subreticulata* Parr 1932]
× 70. From *Challenger* Sta. 177, New Hebrides, South Pacific (125fm.). ZF1193.

Referred by Brady (1884) to *Bolivina reticulata* Hantken, by Thalmann (1932) and Barker (1960) to *B. subreticulata* Parr, by Belford (1966) and van Marle (1991) to *Brizalina subreticulata* (Parr), and by Srinivasan (1966) to *Latibolivina subreticulata* (Parr). *Latibolivina* Srinivasan 1966 is regarded by the present author as a junior synonym of *Brizalina* Costa 1856, but by Revets (pers. comm.) as a junior synonym of *Bolivina* d'Orbigny 1839.

Figures 32–33 *Saidovina subangularis* (Brady 1881)

[*Bolivina subangularis* Brady 1881]
× 60. From *Challenger* Sta. 209, Philippines, North Pacific (95–100fm.). ZF1197.

Referred by Brady (1884) and Thalmann (1932) to *Bolivina*, and by Thalmann (1937), Barker (1960), Belford (1966), and van Marle (1991) to *Bolivinita*. Van Marle (1991) regards *Bolivina subangularis* Brady 1881 as a *nomen nudum*, but gives no justification. The generic identity of this species is somewhat questionable. It was referred by Saidova (1975) to *Loxostomina*. *Loxostomina* Saidova 1975 is a primary junior homonym of *Loxostomina* Sellier de Civrieux 1969. It has recently been renamed *Saidovina* Haman 1984.

Plate 54 (LIV)

Figure 1 *Cassidulina teretis* Tappan 1951

Figs 1a–b ZF1261. Fig. 1c ZF1262. × 75. From *Porcupine* Sta. 11, West of Ireland, North Atlantic (1630fm.).

Referred by Brady (1884) and Thalmann (1932) to *Cassidulina laevigata* d'Orbigny, and by Barker (1960) to *C. teretis* Tappan.

Figures 2–3 Cassidulina laevigata var. carinata Silvestri 1896

Fig. 2 × 75. From *Challenger* Sta. 167, New Zealand (150fm.).
Fig. 3 × 75. From *Porcupine* Sta. 11, West of Ireland, North Atlantic (1630fm.). ZF1263.

Referred by Brady (1884) and Thalmann (1932) to *Cassidulina laevigata* d'Orbigny, by Barker (1960) to *C. carinata* Silvestri?, and also by van Marle (1991) to *C. carinata* Silvestri. The present author regards *Cassidulina laevigata* var. *carinata* Cushman 1922 as both homonymous and synonymous with *C. laevigata* var. *carinata* Silvestri 1896, such that its replacement name *Cassidulina neocarinata* Thalmann 1950 was unnecessary. Rodriguez *et al.* (1980), in contrast, regard Cushman's and Silvestri's forms as distinct.

Figure 4 Cassidulina crassa d'Orbigny 1839

× 60. From *Challenger* Sta. 151, off Heard Island, South Pacific (75fm.). ZF1260.

Referred by Brady (1884) to *Cassidulina crassa* d'Orbigny, by Thalmann (1932) to *C. oblonga* Reuss, by Barker (1960) and van Marle (1991) to *C. crassa* d'Orbigny, and by Belford (1966) to *Globocassidulina crassa* (d'Orbigny). *Cassidulina oblonga* Reuss 1850 is regarded by Marks (1951) as a junior synonym of *C. crassa* d'Orbigny 1839. It is the type-species by original designation of the genus *Cassilongina* Voloshinova 1960, which is regarded by Loeblich and Tappan (1964, 1987) as a junior synonym of *Globocassidulina* Voloshinova 1960.

Figure 5 Cassidulina obtusa Williamson 1858

× 75. From *Porcupine* Sta. 14, West of Ireland, North Atlantic (173fm.). ZF1259.

Referred by Brady (1884), Thalmann (1932), Barker (1960), and van Marle (1991) to *Cassidulina crassa* d'Orbigny (see above). Haynes (1973a) has subsequently suggested that 'many of the North Atlantic references to *C. crassa* such as those of Brady (1884) . . . refer to *C. obtusa*' (which latter species is distinguished by the placement of its aperture parallel with rather than perpendicular to the periphery).

Figures 6–9 Cassidulinoides bradyi (Norman 1881)

[*Cassidulina bradyi* Norman 1881]
× 75. From *Porcupine* Sta. 35, South of Ireland, Atlantic (96fm.). ZF1256.

Referred by Brady (1884) to *Cassidulina*, and by Thalmann (1932), Barker (1960), Belford (1966), and van Marle (1991) to *Cassidulinoides*. Van Marle (1991) regards *Cassidulina bradyi* Norman 1881 as a *nomen nudum* but gives no justification.

Figure 10 Evolvocassidulina orientalis (Cushman 1922)

[*Cassidulina orientalis* Cushman 1922]
× 75. From *Challenger* Sta. 305, West of Patagonia, East Pacific (120fm.). ZF1257.

Referred by Brady (1884) to *Cassidulina bradyi* Norman, by Thalmann (1932) to *Cassidulinoides bradyi* (Norman), by Barker (1960) to *C. orientale* (Cushman), and by Loeblich and Tappan (1985) to *Evolvocassidulina orientalis* (Cushman). *Cassidulina oriental-is* Cush-

man 1922 is the type-species by original designation of the genus *Evolvocassidulina* Eade 1967. It has been lectotypified by Loeblich and Tappan (1985c). The homonymous but not synonymous *C. oriental-e* Cushman 1925 [*Globocassidulina nipponensis* Eade 1969] is the type-species by original designation of the genus *Paracassidulina* Nomura 1983.

Figures 11–16 Cassidulinoides parkerianus (Brady 1881)

[*Cassidulina parkeriana* Brady 1881]
Figs 11–12 × 75. From *Challenger* Sta. 308, West of Patagonia, East Pacific (175fm.). ZF1264.
Figs 13, 15 × 75. From *Challenger* Sta. 305, West of Patagonia, East Pacific (120fm.). ZF1265.
Figs 14, 16 × 75. From *Challenger* Sta. 304, Gulf of Penas, Patagonia, East Pacific (45fm.). ZF1266.

Referred by Brady (1884) to *Cassidulina*, and by Thalmann (1932) and Barker (1960) to *Cassidulinoides*. This is the type-species by original designation of the genus *Cassidulinoides* Cushman 1927.

Figure 17 Globocassidulina subglobosa (Brady 1881)

[*Cassidulina subglobosa* Brady 1881]
× 60. From *Challenger* Sta. 120, off Pernambuco, Atlantic (675fm.). ZF1267.

Referred by Brady (1884), Thalmann (1932), and Barker (1960) to *Cassidulina*, and by Belford (1966), Hermelin (1989), and van Marle (1991) to *Globocassidulina*. Van Marle (1991) regards *Cassidulina subglobosa* Brady 1881 as a *nomen nudum*, but gives no justification. This is the type-species by original designation of the genus *Bradynella* Saidova 1975, regarded by Loeblich and Tappan (1987) as a junior synonym of *Globocassidulina* Voloshinova 1960.

Figure 18 Ceratobulimina jonesiana (Brady 1881)

[*Cassidulina jonesiana* Brady 1881]
× 60. From *Challenger* Sta. 191A, Ki Islands, Central Pacific (580fm.). ZF1207.

Referred by Brady (1884) to *Cassidulina jonesiana* and later (1884) to *Bulimina contraria* Reuss, sp., by Thalmann (1932) to *Ceratobulimina contraria* (Reuss), and by Barker (1960), Belford (1966), and van Marle (1991) to *C. pacifica* Cushman and Harris. *Ceratobulimina pacifica* Cushman and Harris 1927 is a junior synonym of *C. jonesiana* (Brady 1881).

Plate 55 (LV)

Figure 1 Ehrenbergina pupa (d'Orbigny 1839)

[*Cassidulina pupa* d'Orbigny 1839]
× 75. From *Challenger* Sta. 75, off the Azores, Atlantic (450fm.). ZF1436.

Referred by Brady (1884), Thalmann (1932), Barker (1960), and van Marle (1991) to *Ehrenbergina pupa* (d'Orbigny) (see also Plate 113, Fig. 10).

Figures 2–3, 5 *Ehrenbergina trigona* (Goes 1896)

[*Ehrenbergina serrata* var. *trigona* Goes 1896]
Fig. 2 × 60. From *Challenger* Sta. 296, SW of Juan Fernandez, Pacific (1825fm.). ZF1437.
Fig. 3 × 75. From *Challenger* Sta. 283, South Pacific (2075fm.). ZF1438.
Fig. 5 × 75. From *Challenger* Sta. 192, Ki Islands, Central Pacific (129fm.).

Referred by Brady (1884) and Leroy (1944b) to *Ehrenbergina serrata* Reuss, by Thalmann (1932) to *E. bradyi* Cushman, and by Barker (1960) (Figs 2–3 and questionably Fig. 5) to *E. serrata* var. *trigona* Goes. *Ehrenbergina bradyi* Cushman appears synonymous (Phleger *et al.* 1953).

Figures 4, 6–7 *Ehrenbergina pacifica* Cushman 1927

Fig. 4 × 75; Figs 6–7 × 60. From *Challenger* Sta. 192, Ki Islands, Central Pacific (129fm.). ZF1434.

Referred by Brady (1884) and Leroy (1944b) to *Ehrenbergina serrata* Reuss, by Thalmann (1932), and Barker (1960) (with some doubt about Fig. 4), and van Marle (1991) to *E. pacifica* Cushman, and by Hermelin (1989) to *E. undulata* Parker.

Figures 8–11 *Ehrenbergina hystrix* Brady 1881

× 60. From *Challenger* Sta. 280, South Pacific (1940fm.). ZF1434.

Referred by Brady (1884), Thalmann (1932), Barker (1960), and van Marle (1991) to *Ehrenbergina*. Van Marle (1991) regards *Ehrenbergina hystrix* Brady 1881 as a *nomen nudum*, but gives no justification. This is the type-species by original designation of the genus *Reissia* Loeblich and Tappan, 1964, regarded by the present author as a junior synonym of *Ehrenbergina* Reuss 1850.

Figures 12–14, 17–18 *Chilostomella oolina* Schwager 1878

Figs 12, 14, 17 × 50. From *Challenger* Sta. 191A, Ki Islands, Central Pacific (580fm.). ZF1269.
Fig. 13 × 50. From *Challenger* Sta. 232, South of Japan, North Pacific (345fm.).
Fig. 18 × 50. From *Challenger* Sta. 209, Philippines, North Pacific (95–100fm.). ZF1272.

Referred by Brady (1884) and Thalmann (1932) to *Chilostomella ovoidea* Reuss (see below), and by Thalmann (1933a), Barker (1960), Hermelin (1989), and van Marle (1991) to *C. oolina* Schwager.

Figures 15–16, 19–23 *Chilostomella ovoidea* Reuss 1850

Figs 19, 23 × 50. From *Challenger* Sta. 191A, Ki Islands, Central Pacific (580fm.). ZF1269.
Figs 15–16, 20 × 50. From *Challenger* Sta. 33, off Bermuda, Atlantic (435fm.). ZF1271.
Figs 21–22 × 70. From *Challenger* Sta. 232, South of Japan, North Pacific (345fm.). ZF1270.

This is the type-species by subsequent designation of the genus *Chilostomella* Reuss 1849.

Figures 24–26 *Allomorphina pacifica* Hofker 1951

Fig. 24 × 60. From *Challenger* Sta. 232, South of Japan, North Pacific (345fm.). ZF1045.
Fig. 25 × 60. 1959.5.5.51. Fig. 26 × 100. ZF1044. From *Challenger* Sta. 279C, off Tahiti, Pacific (620fm.).

Referred by Brady (1884) and Thalmann (1932) to *Allomorphina trigona* Reuss, by Leroy (1944a) to *A. trigonula* (sic) Reuss, and by Barker (1960), Belford (1966), Hermelin (1989), and van Marle (1991) to *A. pacifica* Hofker.

Plate 56 (LVI)

Figures 1–2, ?3 *Oolina globosa* (Montagu 1803)

[*Vermiculum globosum* Montagu 1803]
Fig. 1 × 60–75. From *Challenger* Sta. 224, North Pacific (1850fm.). ZF1664.
Fig. 2 × 60–75. From *Challenger* Sta. 162, Bass Strait, Pacific (38fm.). ZF1665.
Fig. 3 × 60–75. From *Challenger* Sta. 300, North Pacific (1375fm.). ZF1666.

Referred by Brady (1884) and Thalmann (1932) to *Lagena*, and by Barker (1960), Hofker (1976), Albani and Yassini (1989), and van Marle (1991) to *Oolina* (see also Text-figure 11). This is the type-species by original designation of the genus *Entolagena* Silvestri 1900. Loeblich and Tappan treated this genus first (1964) as a junior synonym of *Oolina* d'Orbigny 1839, and later (1987) as of uncertain status.

Figure 4 *Parafissurina felsinea* (Fornasini 1894)

[*Lagena felsinea* Fornasini 1894]
× 75. From *Challenger* Sta. 276, South Pacific (2350fm.). ZF1619.

Referred by Brady (1884) and Thalmann (1932) to *Lagena apiculata* (Reuss), and by Barker (1960) to *Oolina felsinea* (Fornasini).

Figure 5 *Oolina ovum* (Ehrenberg 1843)

[*Miliola ovum* Ehrenberg 1843]
× 75. From *Challenger* Sta. 241. North Pacific (2300fm.). ZF1731.

Referred by Brady (1884) and Thalmann (1932) to *Lagena*, and by Barker (1960) to *Oolina*.

Figure 6 *Parafissurina botelliformis* (Brady 1881)

[*Lagena botelliformis* Brady 1881]
× 75. From *Challenger* Sta. 352A, Cape Verde Islands (11fm.). ZF1632.

Referred by Brady (1884) and Thalmann (1932) to *Lagena*, and by Barker (1960) to *Oolina*.

Figure 7 *Lagena laevis* (Montagu 1803)

[*Vermiculum laeve* Montagu 1803]
× 75. From *Challenger* Sta. 315A, Falkland Islands, South Atlantic (6fm.). ZF1696.

Figures 8?–9 *Procerolagena clavata* (d'Orbigny 1846)

[*Oolina clavata* d'Orbigny 1846]

Fig. 8 × 75. From *Challenger* Sta. 315A, Falkland Islands, South Atlantic (6fm.). ZF1696.

Fig. 9 × 75. From *Challenger* Sta. 185, Torres Strait, Pacific (155fm.). ZF1697.

Referred by Brady (1884), Thalmann (1932), and Barker (1960) to *Lagena laevis* (Montagu) (see above), and by the present author to *Procerolagena clavata* (d'Orbigny). This species has been lectotypified by Papp and Schmid (1985). Well preserved specimens possess the phialine lips characteristic of the genus *Procerolagena* Puri 1954 (see, for instance, Whittaker and Hodgkinson 1979).

Figures 10–11, ?13 *Lagena hispidula* Cushman 1913

Fig. 10 × 75. From *Challenger* Sta. 160, Southern Ocean (2600fm.). ZF1698.

Fig. 11 × 75. From *Challenger* Sta. 332, South Atlantic (2200fm.). ZF1699.

Fig. 13 × 75. From *Challenger* Sta. 5, SW of the Canaries, Atlantic (2740fm.). ZF1701.

Referred by Brady (1884) to *Lagena laevis* Montagu, sp. (see above), by Thalmann (1932) (Figs 10–11, ?13) to *L. laevis* var. *nebulosa* Cushman (see below), and by Thalmann (1933a) (Figs 10–11 only), Barker (1960) (Figs 10–11, ?13), and Harmelin (1989) (Figs 10–11 only) to *L. hispidula* Cushman. This is the type-species by original designation of the genus *Pygmaeoseistron* Patterson and Richardson, in Loeblich and Tappan 1987 [*Pygmaeoseistron* Patterson and Richardson 1988], regarded by the present author as a junior synonym of *Lagena* Walker and Jacob 1798.

Figure 12 *Lagena nebulosa* (Cushman 1923)

[*Lagena laevis* var. *nebulosa* Cushman 1923]

× 75. From *Challenger* Sta. 279C, Tahiti, Pacific (2740fm.). ZF1700.

Referred by Brady (1884) to *Lagena laevis* Montagu, sp. (see above), Thalmann (1932) to *L. laevis* var. *nebulosa* Cushman, and by Barker (1960) to *L. nebulosa* Cushman. This species was referred by Patterson and Richardson (1988) to *Pygmaeoseistron* (see above).

Figure 14 *Lagena* sp. nov. (1)

× 75. From *Challenger* Sta. 224, North Pacific (1850fm.). ZF1702.

Referred by Brady (1884) to *Lagena laevis* Montagu, sp., by Thalmann (1932) tentatively to *L. laevis* var. *nebulosa* Cushman, and by Barker (1960) to *L.* aff. *laevis* (Montagu).

Figures 15–16 *Oolina globosa* (Montagu 1803) var.

× 75. From *Porcupine* Sta. 67–68, East of the Shetlands, Scotland (64–75fm.). ZF1616.

Referred by Brady (1884) to *Lagena apiculata* Reuss, sp., by Thalmann (1932) to *Ellipsolagena apiculata* (Reuss), and by Barker (1960) to *Oolina apiculata* Reuss. This is clearly erroneous as Reuss's species possesses a radiate aperture and lacks an internal tube (and is incidentally the type by original designation of *Lagena* (*Reussoolina*) Colom 1956). The present author regards Brady's specimens as no

more than varietally (i.e. infra-subspecifically) distinct from *Oolina globosa* (Montagu) (compare with Fig. 1 above).

Figures 17–18 *Parafissurina lateralis* (Cushman 1913)

[*Lagena lateralis* Cushman 1913]

Fig. 17 × 75. ZF1617. From *Challenger* Sta. 151, off Heard Island (75fm.).

Fig. 18 × 75. ZF1618. From *Porcupine* Sta. 67–68, East of Shetland (64–75fm.).

Referred by Brady (1884) to *Lagena apiculata* Reuss, sp., by Thalmann (1932) to *Ellipsolagena apiculata* (Reuss), and by Barker (1960) and van Marle (1991) to *Parafissurina lateralis* (Cushman).

Figures 19–22, 24–29 *Procerolagena gracillima* (Seguenza 1862)

[*Amphorina gracillima* Seguenza 1862]

Fig. 19 × 60. From *Challenger* Sta. 151, Heard Island, South Pacific (75fm.).

Fig. 20 × 60. From *Challenger* Sta. 149E, Kerguelen Island, South Pacific (28fm.). ZF1667.

Figs 21–22 × 60. From *Challenger* Sta. 168, NE of New Zealand (1100fm.). ZF1668.

Figs 24, 26 × 60. From *Challenger* Sta. 149I, Kerguelen Island, South Pacific (120fm.). ZF1669.

Fig. 25 × 60. From *Challenger* Sta. 189, S. of New Guinea (28fm.). ZF1671.

Fig. 27 × 60. From *Challenger* Sta. 217A, Humboldt Bay, Papua, Pacific (37fm.). ZF1680.

Fig. 28 × 60. From *Challenger* Sta. 160, Southern Ocean (2600fm.). ZF1670.

Fig. 29 × 60. From *Challenger* Sta. 218, North of Papua, Pacific (1070fm.).

Referred by Brady (1884) to *Lagena gracillima* Seguenza, sp. (Figs 19–22, 24–28) and *L. elongata* Ehrenberg, sp. (Fig. 29), by Thalmann (1932) and Barker (1960) to *L. gracillima* (Seguenza) (Figs 19–22, 24–26) and *L. elongata* (Ehrenberg) (Figs 27–29), by Hofker (1976) to *L. elongata* (Ehrenberg) (Fig. 29 only), by Whittaker and Hodgkinson (1979) to *L. gracillima* (Seguenza) (Figs 25–26) and *L. elongata* (Ehrenberg) (Fig. 29), by Albani and Yassini (1989) to *Procerolagena gracillima* (Seguenza) (Figs 19–22, 24–26), and by van Marle (1991) to *Lagena elongata* (Ehrenberg) (Figs 27–28). *Miliola elongata* Ehrenberg 1844 is the type-species by original designation of the genus *Phialinea* Jones 1984, regarded by Loeblich and Tappan (1987) as unrecognizable, and by Patterson and Richardson (1987) as a junior synonym of *Procerolagena* Puri 1954. Forms previously erroneously assigned to *Phialinea* Jones 1984 can certainly be accommodated in *Procerolagena* Puri 1954. Forms previously erroneously assigned to the unrecognizable *Miliola elongata* Ehrenberg 1844 can be accommodated in *Amphorina gracillima* Seguenza 1862. This is the type-species, by synonymy with the originally designated *Hyalinonetrion sahulense* Patterson and Richardson, in Loeblich and Tappan 1987 [*H. sahulense* Patterson and Richardson 1988] of the genus *Hyalinonetrion* Patterson and Richardson, in Loeblich and Tappan 1987 [*Hyalinonetrion* Patterson and Richardson 1988]. *Hyalinonetrion* Patterson and Richardson 1987 is regarded by the present author as a junior synonym of *Procerolagena* Puri 1954.

Figure 23 *Procerolagena amphora* (Williamson 1848)

[*Lagena amphora* Williamson 1848]
× 60. From *Challenger* Sta. 279C, Tahiti, Pacific (620fm.). ZF1792.

Referred by Brady (1884), Thalmann (1932), and Barker (1960) to *Lagena gracillima* (Seguenza), and by Albani and Yassini (1989) to *Procerolagena gracillima* (Seguenza) (see above).

Figure 30 *Procerolagena clavata* var. *setigera* (Millett 1901)

[*Lagena clavata* var. *setigera* Millett 1901]
× 75. From *Challenger* Sta. 205A, Hong Kong, North Pacific (7fm.). ZF1703.

Referred by Brady (1884), Thalmann (1932), and Barker (1960) to *Lagena laevis* (Montagu) var., and by Whittaker and Hodgkinson (1979) to *L. clavata* var. *setigera* Millett. *Lagena clavata* var. *setigera* Millett 1901 has recently been lectotypified by Jones (1984*b*)

Figures 31–32 *Oolina truncata* (Brady 1884)

[*Lagena truncata* Brady 1884]
Fig. 31 × 75. From *Challenger* Sta. 283, South Pacific (2075fm.). ZF1799.
Fig. 32 × 75. ZF1798. From *Challenger* Sta. 332, South Atlantic (2200fm.).

Referred by Brady (1884), Thalmann (1932), and Barker (1960) to *Lagena*. This species is clearly entosolenian and belongs in *Oolina*.

Figures 33–35, ?36 *Oolina globosa* var. *setosa* (Earland 1934)

[*Lagena globosa* var. *setosa* Earland 1934]
Fig. 33 × 75. From *Challenger* Sta. 5, SW of the Canaries, Atlantic (2740fm.). ZF1713.
Fig. 34 × 75. From *Challenger* Sta. 218, North of Papua, Pacific (1070fm.). ZF1715.
Fig. 35 × 75. From *Challenger* Sta. 160, Southern Ocean (2600fm.). ZF1714.
Fig. 36 × 50. From *Challenger* Sta. 146, Southern Ocean (1375fm.). ZF1716.

Referred by Brady (1884) to *Lagena longispina* Brady, 'globular form', by Thalmann first (1932) to *L. longispina* Brady and later (1937) to *L. globosa* var. *setosa* Earland (Figs 33–35) and *L. longispina* Brady (Fig. 36), by Barker (1960) to *Oolina globosa* var. *setosa* (Earland) (Figs 33–35) and *O. longispina* (Brady), 'globular form' (Fig. 36), by Hermelin (1989) to *Oolina setosa* (Earland) (Figs 33–35), and by van Marle (1991) to *Oolina longispina* (Brady) (Fig. 36). *Lagena globosa* var. *setosa* Earland 1934 has recently been lectotypified by Jones (1948*b*). *L. longispina* Brady 1881 is regarded by van Marle (1991) as a *nomen nudum* (for no apparent reason).

Plate 57 (LVII)

Figures 1–2 *Lagena hispida* Reuss 1858

Fig. 1 × 60. From *Challenger* Sta. 323, South Atlantic (1900fm.). ZF1687.
Fig. 2 × 60. From *Challenger* Sta. 232, South of Japan, North Pacific (345fm.). ZF1688.

Referred by Brady (1884), Thalmann (1932), Barker (1960), Hermelin (1989), and van Marle (1991) to *Lagena hispida* Reuss.

Figure 3 *Lagena aspera* var. *crispata* (Matthes 1939)

[*Lagena hispida* var. *crispata* Matthes 1939]
× 60. From *Challenger* Sta. 185, Torres Strait, Pacific (155fm.). ZF1691.

Referred by Brady (1884), Thalmann (1932), Barker (1960, text), Hermelin (1989), and van Marle (1991) to *Lagena hispida* Reuss, and by Barker (1960, addenda) to *L. hispida* var. *crispata* Matthes.

Figures 4, 6–7 *Lagena aspera* Reuss 1861

Fig. 4 × 60. From *Challenger* Sta. 232, South of Japan, North Pacific (345fm.). ZF1692.
Fig. 6 × 75. ZF1621. Fig. 7 × 75. ZF1620. From *Challenger* Sta. 185, Torres Strait, Pacific (155fm.).

Referred by Brady (1884), Thalmann (1932), and Barker (1960) to *Lagena hispida* Reuss (Fig. 4) and *L. aspera* Reuss (Figs 6–7), and by Hermelin (1989) and van Marle (1991) to *L. hispida* Reuss (Fig. 4 only). *Lagena aspera* Reuss differs from *L. hispida* Reuss (see above) in its less elongate test.

Figures 5, ?12 *Oolina ampulladistoma* (Rymer Jones 1874)

[*Lagena ampulladistoma* Rymer Jones 1874]
Fig. 5 × 60. From *Challenger* Sta. 185, Torres Strait, Pacific (155fm.). ZF1615.
Fig. 12 × 60. From *Challenger* Sta. 279C, Tahiti, Pacific (620fm.). ZF1626.

Referred by Brady (1884), Thalmann (1932), and Barker (1960) to *Lagena ampulladistoma* Rymer Jones (Fig. 5) and *L. aspera* Reuss (Fig. 12).

Figures 8–9, ?10 *Lagena gibbera* Buchner 1940

Fig. 8 × 60. From *Challenger* Sta. 323, South Atlantic (1900fm.). ZF1622.
Fig. 9 × 60. From *Challenger* Sta. 300, North of Juan Fernandez, East Pacific (1375fm.). ZF1623.
Fig. 10 × 45. From *Challenger* Sta. 218, North of Papua, Pacific (1075fm.). ZF1624.

Referred by Brady (1884) and Thalmann (1932) to *Lagena aspera* Reuss (see above), and by Barker (1960) to *L. gibbera* Buchner (Figs 8–9) and *L. aspera* Reuss (Fig. 10).

Figure 11 *Lagena* sp. nov. (2)

× 45. From *Challenger* Sta. 241, North Pacific (2300fm.). ZF1625.

Referred by Brady (1884), Thalmann (1932), and Barker (1960) to *Lagena aspera* Reuss (see above).

Figure 13 *Entosolenia lineata* Williamson 1848

× 60. From *Challenger* Sta. 149D, Kerguelen Island, South Pacific (20–50fm.). ZF1712.

Referred by Brady (1884) and Thalmann (1932) to *Lagena*, and by

Barker (1960) to *Oolina*. This is the type-species by subsequent designation of the genus *Entosolenia* Williamson 1858. *Entosolenia* Williamson 1858 was synonymized by Loeblich and Tappan (1964, 1987) with *Oolina* d'Orbigny 1839. It is herein reinstated as it possesses not only an 'endosolen' but also an 'exosolen' (Knight 1986).

Figures 14, 16 *Lagena semistriata* (Williamson 1848)

[*Lagena striata* var. *semistriata* Williamson 1848]
Fig. 14 ×75. From *Challenger* Sta. 162, Bass Strait, Pacific (38–40fm.). ZF1754.
Fig. 16 ×75. From *Challenger* Sta. 217A, off Papua, Pacific (37fm.). ZF1755.

Referred by Brady (1884) and Thalmann (1932) to *Lagena semistriata* (Williamson), and by Barker (1960) and van Marle (1991) to *L. laevis* (Montagu).

Figures 15, 21 *Lagena crenata* Parker and Jones, 1865

Fig. 15 ×75. From *Challenger* Sta. 162, Bass Strait, Pacific (38–40fm.). ZF1640.
Fig. 21 ×75. From *Porcupine* Sta. 11, West of Ireland, North Atlantic (1630fm.). ZF1639.

This species has been lectotypified by Hodgkinson (1992).

Figures 17, 35–36 *Cushmanina stelligera* (Brady 1881)

[*Lagena stelligera* Brady 1881]
Fig. 17 ×75. From *Porcupine* Sta. 67–68, East of the Shetlands, Scotland, North Atlantic (64–75fm.). ZF1756.
Fig. 35 ×75. From *Challenger* Stas. 276 (Tahiti, 2350fm.). and 271 (South Pacific, 2425fm.). ZF1771. Lectotype designated by Jones (1984*b*).
Fig. 36 ×75. From *Challenger* Sta. 149D, Kerguelen Island, South Pacific (20–60fm.). ZF1772.

Figure 17 was referred by Brady (1884) and Thalmann (1932) to *Lagena semistriata* (Williamson), and by Barker (1960) to *L. laevis* (Montagu). Figures 35–36 were referred by Brady (1884), Thalmann (1932), Barker (1960), and Hermelin (1989) to *Lagena stelligera* Brady, and by Jones (1984*a*) to *Oolina stelligera* (Brady). All three figures are referred herein to *Cushmanina stelligera* (Brady). *Cushmanina* Jones 1984*a* differs from *Lagena* Walker and Jacob 1798 in possessing an entosolenian tube ('endosolen' of Knight 1986), and from *Oolina* d'Orbigny 1839 in possessing an apertural neck.

Figure 18 *Oolina* sp.

×75. From *Challenger* Sta. 185, Torres Strait, Pacific (155fm.). ZF1757.

Referred by Brady (1884) and Thalmann (1932) tentatively to *Lagena semistriata* (Williamson), and by Barker (1960) (albeit tentatively) and van Marle (1991) to *L. laevis* (Montagu). This specimen is probably entosolenian and therefore referable to the genus *Oolina*.

Figure 19 *Lagena substriata* Williamson 1848

×60. From *Challenger* Sta. 185, Torres Strait, Pacific (155fm.). ZF1774.

Referred by Brady (1884), Thalmann (1932), and Barker (1960) to *Lagena striata* (d'Orbigny).

Figure 20 *Oolina sidebottomi* (Earland 1934)

[*Lagena sidebottomi* Earland 1934]
×75. From *Challenger* Sta. 276, North of Tahiti, Pacific (2350fm.). ZF1791.

Referred by Brady (1884) and Thalmann (1932) to a transient between *Lagena crenata* (q.v.) and *L. semistriata* (q.v.), and by Thalmann (1937) and Barker (1960) to *L. sidebottomi* Earland (a new name for *L. intermedia* Sidebottom 1912 not *L. striata* var. *intermedia* Rzehak 1886). This is the type-species by original designation of the genus *Exsculptina* Patterson and Richardson, in Loeblich and Tappan 1987 [*Exsculptina* Patterson and Richardson 1988], regarded by the present author as a junior synonym of *Oolina* d'Orbigny 1839.

Figures 22, 24 *Lagena striata* (d'Orbigny 1839)

[*Oolina striata* d'Orbigny 1839]
Fig. 22 ×60. From *Challenger* Sta. 217A, Papua, Pacific (37fm.). ZF1773.
Fig. 24 ×60. From *Challenger* Sta. 232, South of Japan, North Pacific (345fm.). ZF1775.

Referred by Brady (1884), Thalmann (1932), Barker (1960), Hermelin (1989), and van Marle (1991) to *Lagena striata* (d'Orbigny).

Figures 23, 25–27, 33–34 *Lagena sulcata* (Walker and Jacob 1798)

[*Serpula* (*Lagena*) *sulcata* Walker and Jacob 1798]
Figs 23, 25 ×60. From *Challenger* Sta. 149I, Kerguelen Island, South Pacific (120fm.).
Fig. 26 ×60. From *Challenger* Sta. 191A, Ki Islands, Central Pacific (580fm.). ZF1785.
Fig. 27 ×60. From *Challenger* Sta. 218, North of Papua, Pacific (1070fm.). ZF1790.
Fig. 33 ×60. From *Challenger* Sta. 271, South Pacific (2425fm.). ZF1786.
Fig. 34 ×60. From *Challenger* Sta. 149D, Kerguelen Island, South Pacific (20–60fm.). ZF1787.

Referred by Brady (1884) and Thalmann (1932) to *Lagena sulcata* (Walker and Jacob) (Figs 23, 26, 33–34) and *L. sulcata* var. *interrupta* Williamson (Figs 25, 27), by Leroy (1944*a*) to *L. semistriata* Williamson (Figs 23, 26, 33–34) and *L. sulcata* var. *interrupta* Williamson (Figs 25, 27), by Barker (1960) to *L. laevis* (Montagu) (Figs 23, 25–27) and *L. sulcata* (Walker and Jacob) (Figs 33–34), and by van Marle (1991) to *L. sulcata* (Walker and Jacob) (Figs 33–34 only). No formal distinction is maintained herein for any of the proposed varietal forms of *Lagena sulcata* (Walker and Jacob) as there appears to be a complete intergradation from one to the next. This is the type-species by subsequent designation of the genus *Lagena* Walker and Jacob 1798.

Figure 28 *Lagena* sp.

×60. From *Challenger* Sta. 285, S Pacific (2375fm.). ZF1776.

Referred by Brady (1884), Thalmann (1932), and Barker (1960) (albeit tentatively) to *Lagena striata* (d'Orbigny).

Figures 29–30 *Bifarilaminella advena* (Cushman 1923)

[*Lagena advena* Cushman 1923]
Fig. 29 ×60. From *Challenger* Sta. 338, South Atlantic (1990fm.). ZF1777.
Fig. 30 ×60. From *Challenger* Sta. 346, South Atlantic (2350fm.). ZF1778.

Referred by Brady (1884) to *Lagena striata* d'Orbigny, sp., and by Thalmann (1932), and Barker (1960), and Hermelin (1989) (Fig. 30 only) to *L. advena* Cushman. This is the type-species by original designation of the genus *Bifarilaminella* Patterson and Richardson, in Loeblich and Tappan 1987 [*Bifarilamellina* Patterson and Richardson 1988], which differs from *Lagena* Walker and Jacob, 1798 in its compound wall.

Figure 31 *Oolina* sp. nov. (1)

×60–75. From *Challenger* Sta. 151, off Heard Island, South Pacific (75fm.). ZF1607.

Referred by Brady (1884) and Thalmann (1932) to *Lagena acuticosta* Reuss, and by Barker (1960) to *L.* sp. nov. (1).

Figure 32 *Oolina apiopleura* (Loeblich and Tappan 1953)

[*Lagena apiopleura* Loeblich and Tappan 1953]
×60–75. From *Challenger* Sta. 276, South Pacific (2350fm.). ZF1608.

Referred by Brady (1884) and Thalmann (1932) to *Lagena acuticosta* Reuss, and by Barker (1960) to *L. elegantissima* Bornemann.

Plate 58 (LVIII)

Figure 1 *Oolina exsculpta* (Brady 1881)

[*Lagena exsculpta* Brady 1881]
×75. From *Challenger* Sta. 160, Southern Ocean (2600fm.). ZF1648.

Referred by Brady (1884) and Thalmann (1932) to *Lagena*, and by Barker (1960) to *Fissurina*. This species was referred by Patterson and Richardson (1988) to *Exsculptina*. *Exsculptina* Patterson and Richardson in Loeblich and Tappan 1987 [*Exsculptina* Patterson and Richardson 1988] is regarded by the present author as a junior synonym of *Oolina* d'Orbigny, 1839.

Figures 2–3, 7–8, 22–24 *Lagena multilatera* McCulloch 1977

Fig. 2 ×75. ZF1672. Figs 8, 22–23 ×75. ZF1674. From *Challenger* Sta. 276, North of Tahiti, Pacific (2350fm.).
Fig. 3 ×75. From *Challenger* Sta. 160, Southern Ocean (2600fm.). ZF1673.
Fig. 7 ×75. From *Challenger* Sta. 283, South Pacific (2075fm.). ZF1675.
Fig. 24 ×75. From *Challenger* Sta. 271, South Pacific (2425fm.). ZF1679.

Referred by Brady (1884), Thalmann (1932), Barker (1960), and van Marle (1991) (Figs 22–24 only) to *Lagena gracilis* Williamson (see below). Referred by the present author to *Lagena multilatera* McCulloch, which species is interpreted as varying in the number of ribs that it possesses and the extent to which they are raised, and also in the degree of development of the apical spine.

Figures 4?, 17 *Lagena semilineata* var. *spinigera* Earland 1934

Fig. 4 ×74. Fig. 17 ×75. ZF1782. From *Challenger* Sta. 276, North of Tahiti, Pacific (2350fm.).

Referred by Brady (1884) to *Lagena sulcata*, 'apiculate form', by Thalmann (1932) (Fig. 4 only) to *L. sulcata* var. *apiculata* Cushman, by Barker (1960) to *L. sulcata* var. *spicata* Cushman and McCulloch, and by the present author to *L. semilineata* var. *spinigera* Earland. This species was referred by Patterson and Richardson (1988) to *Exsculptina*. *Exsculptina* Patterson and Richardson 1988 is regarded by the present author as a junior synonym of *Oolina* d'Orbigny 1839.

Figures 5–6, 10?, 18 *Lagena sulcata* (Walker and Jacob 1798)

[*Serpula* (*Lagena*) *sulcata* Walker and Jacob 1798]
Figs 5–6 ×75. From *Challenger* Sta. 276, North of Tahiti, Pacific (2350fm.).
Fig. 10 ×75. From *Challenger* Sta. 192, Ki Islands, Central Pacific (129fm.).
Fig. 18 ×75. From *Challenger* Sta. 232, South of Japan, North Pacific (345fm.). ZF1783.

Figures 5–6 and 18 were referred by Brady (1884) and Thalmann (1932) to *Lagena sulcata* var. *interrupta* Williamson, 'apiculate forms', and by Barker (1960) to *L. sulcata* var. *spicata* Cushman and McCulloch. Figure 10 was referred by Brady (1884), Thalmann (1932), and Barker (1960) to *Lagena gracilis* Williamson. All are referred herein to *L. sulcata* (Walker and Jacob) (see also Plate 57, Figs 23, 25–27, 33–34).

Figures 9, ?11–15 *Procerolagena gracilis* (Williamson 1848)

[*Lagena gracilis* Williamson 1848]
Fig. 9 ×75. From *Challenger* Sta. 306, west coast of Patagonia, East Pacific (345fm.). ZF1676.
Fig. 11 ×45–60. From *Challenger* Sta. 149I, Kerguelen Islands, South Pacific (120fm.). ZF1643.
Figs 12–15 ×45–60. From *Porcupine* Sta. 67–68, East of the Shetlands, North Atlantic (64–75fm.). ZF1644.

Figure 9 was referred by Brady (1884), Thalmann (1932), and Barker (1960) to *Lagena gracilis* Williamson. Figures 11–15 were referred by Brady (1884), Thalmann (1932), Barker (1960), Popescu (1983), and van Marle (1991) to *L. distoma* Parker and Jones. *Lagena distoma* Parker and Jones, 1865 (lectotypified by Hodgkinson 1992) is regarded herein as falling within the range of variability exhibited by *L. gracilis* Williamson 1848 (as also incidentally are *L. distoma* var. *ingens* Buchner 1940 and *L. gracillima* var. *mollis* Cushman 1944). *Lagena gracilis* Williamson 1848 is the type-species by original designation of the genus *Procerolagena* Puri 1954. In an earlier paper (Jones, 1984*a*), I referred it to my new genus *Phialinea*. I now regard *Phialinea* Jones 1984 as a junior synonym of *Procerolagena* Puri 1954 (see also Patterson and Richardson 1987). In another paper (Jones 1984*b*), I designated a lectotype for *gracilis*.

Figure 16 *Procerolagena distomamargaritifera* (Rymer Jones 1872)

[*Lagena vulgaris* var. *distoma-margaritifera* Rymer Jones 1872]
×40. From *Challenger* Sta. 162, Bass Strait, Pacific (38–40fm.). ZF1645.

Referred by Brady (1884), Thalmann (1932), and Barker (1960) to *Lagena*. This species was referred by Albani and Yassini (1989) to *Procerolagena*.

Figure 19 *Lagena meridionalis* (Wiesner 1931)

[*Lagena gracilis* var. *meridionalis* Wiesner 1931]
× 75. From *Challenger* Sta. 145, Prince Edward Island, South Pacific (50–150fm.). ZF1678.

Referred by Brady (1884) and Thalmann (1932) to *Lagena gracilis* Williamson, by Thalmann (1937) to *L. gracilis* var. *meridionalis* Wiesner, and by Barker (1960) and Hermelin (1989) to *L. meridionalis* (Wiesner).

Figure 20 *Pseudofissurina* sp. nov.

× 75. From *Challenger* Sta. 241, North Pacific (2300fm.). ZF1609.

Referred by Brady (1884) and Thalmann (1932) to *Lagena acuticosta* Reuss, and by Barker (1960) to *L.* sp. nov. (2). Examination of Brady's figured specimen reveals that it possesses a parafissurine aperture and a compressed test, and is referable to *Pseudofissurina* Jones 1984.

Figure 21 *Oolina apiopleura* (Loeblich and Tappan 1953)

[*Lagena apiopleura* Loeblich and Tappan 1953]
× 75. From *Challenger* Sta. 271, South Pacific (2425fm.). ZF1610.

Referred by Brady (1884), Thalmann (1932), and Leroy (1944a,b) to *Lagena acuticosta* Reuss, by Loeblich and Tappan (1953) to *L. apiopleura*, and by Barker (1960) to *L.* sp. nov.?

Figures 25, 27 *Cushmanina plumigera* (Brady 1881)

[*Lagena plumigera* Brady 1881]
Fig. 25 × 75. ZF1732. Fig. 27 × 75. ZF1734. From *Challenger* Sta. 271, South Pacific (2425fm.).

Referred by Brady (1884), Thalmann (1932), and Barker (1960) to *Lagena*.

Figure 26 *Lagena flexa* Cushman and Gray 1946

× 75. From *Challenger* Sta. 276, North of Tahiti, Pacific (2350fm.). ZF1733.

Referred by Brady (1884) and later Thalmann (1932) and Barker (1960) to *Lagena plumigera* Brady (see above).

Figures 28–32 *Oolina squamosa* (Montagu 1803)

[*Vermiculum squamosum* Montagu 1803]
Fig. 28 × 75. From *Challenger* Sta. 151, Heard Island, South Pacific (75fm.). ZF1761.
Fig. 29 × 75. From *Challenger* Sta. 149D, Kerguelen Island, South Pacific (20–50fm.). ZF1762.
Fig. 30 × 75.
Fig. 31 × 75. From *Challenger* Sta. 145, Prince Edward Island, South Pacific (50–150fm.). ZF1763.
Fig. 32 × 75. From *Challenger* Sta. 260A, off Honolulu reefs, Pacific (40fm.). ZF1685.

Referred by Brady (1884) to *Lagena squamosa* Montagu, sp. (Figs 28–31) and *L. hexagona* Williamson, sp. (Fig. 32), by Thalmann (1932) to *L. squamosa* (Montagu) (Figs 28–30), *L. montagui* Silvestri (Fig. 31) and *L. hexagona* (Williamson) (Fig. 32), by Barker (1960), Hermelin (1989), and van Marle (1991) to *Oolina melo* d'Orbigny (Figs 28–31) and *O. hexagona* (Williamson) (Fig. 32), and by Haynes (1973a) to *O. hexagona* (Williamson) (Fig. 32). Knight (1986) has subsequently referred all *Oolina* species with vertically-aligned ornament to the *O. squamosa* (Montagu) plexus. Patterson and Richardson (1988) referred such species to *Favulina*, herein regarded as a junior synonym of *Oolina* d'Orbigny 1839.

Figure 33 *Oolina hexagona* (Williamson 1848)

[*Entosolenia squamosa* var. *hexagona* Williamson 1848]
× 75. From *Challenger* Sta. 271, South Pacific (2425fm.).

Referred by Brady (1884) and Thalmann (1932) to *Lagena hexagona* (Williamson), and by Barker (1960), Haynes (1973a), and Hermelin (1989) to *Oolina hexagona* (Williamson). This is the type-species by original designation of the genus *Favulina* Patterson and Richardson, in Loeblich and Tappan 1987 [*Favulina* Patterson and Richardson 1988], herein regarded as a junior synonym of *Oolina* d'Orbigny 1839. In an earlier paper (Jones, 1984b), I designated a lectotype for this species.

Figure 34 *Oolina seminuda* (Brady 1884)

[*Lagena seminuda* Brady 1884]
× 64. From *Challenger* Sta. 302. South Pacific (1450fm.). ZF1751.

Referred by Brady (1884) and Thalmann (1932) to *Lagena*, and by Barker (1960) to *Oolina*.

Figure 35 *Oolina favosopunctata* (Brady 1881)

[*Lagena favosopunctata* Brady 1881]
× 75. From *Challenger* Sta. 185, Torres Strait, Pacific (155fm.). ZF1650.

Referred by Brady (1884), Thalmann (1932), and Barker (1960) to *Lagena*.

Figure 36 *Sipholagena hertwigiana* (Brady 1881)

[*Lagena hertwigiana* Brady 1881]
× 75, From *Challenger* Sta. 185, off Raine Island, Torres Strait, Pacific (155fm.). ZF1681.

Referred by Brady (1884), Thalmann (1932), and Barker (1960) (albeit tentatively) to *Lagena* (see also Text-figure 12). This species was referred by Moncharmont-Zei and Sgarrella (1977) to their new genus *Buchneria*. The appropriate generic assignment of the present species is to *Sipholagena* Moncharmont-Zei and Sgarrella 1980 (the replacement name for *Buchneria* Moncharmont-Zei and Sgarrella 1977, not Borner 1952).

Figures 37, 40 *Cushmanina striatopunctata* (Parker and Jones 1865)

[*Lagena sulcata* var. *striatopunctata* Parker and Jones 1865]
Fig. 37 × 75. From *Challenger* Sta. 241, North Pacific (2300fm.). ZF1780.

Fig. 40 ×75. From *Challenger* Sta. 185, Torres Strait, Pacific (155fm.). ZF1781.

Referred by Brady (1884) and Thalmann (1932) to *Lagena,* and by Barker (1960) and van Marle (1991) (Fig. 40 only) to *Oolina. Lagena sulcata* var. *striatopunctata* Parker and Jones 1865 has recently been lectotypified (more correctly, neotypified (Hodgkinson 1992)) by Jones (1984*b*). The presence of both an internal tube and an external apertural neck indicates placement in *Cushmanina* Jones 1984.

Figures 38–39 *Cushmanina feildeniana* (Brady 1878)

[*Lagena feildeniana* Brady 1878]
Fig. 38 ×75. From *Challenger* Sta. 241, North Pacific (2300fm.). ZF1780.
Fig. 39 ×75. From *Challenger* Sta. 332, South Atlantic (2200fm.). ZF1652.

Referred by Brady (1884), Thalmann (1932), and Barker (1960) to *Lagena.*

Figure 41 *Cushmanina torquata* (Brady 1878)

[*Lagena torquata* Brady 1878]
×75. From *Challenger* Sta. 300, North of Juan Fernandez, West Pacific (1375fm.). ZF1793.

Referred by Brady (1884) and Thalmann (1932) to *Lagena,* and by Barker (1960) (albeit tentatively) to *Oolina.*

Figures 42–43 *Cushmanina desmophora* (Rymer Jones 1872)

[*Lagena vulgaris* var. *desmophora* Rymer Jones 1872]
Fig. 42 ×75. From *Challenger* Sta. 120, off Pernambuco, Atlantic (675fm.). ZF1641.
Fig. 43 ×75. From *Challenger* Sta. 224, North Pacific (1850fm.). ZF1642.

Referred by Brady (1884) and Thalmann (1932) to *Lagena,* and by Barker (1960) and Hermelin (1989) to *Oolina.* This is the type-species by original designation of the genus *Cushmanina* Jones 1984, which differs from *Lagena* Walker and Jacob 1798 in possessing an entosolenian tube ('endosolen' of Knight 1986), and from *Oolina* d'Orbigny 1839 in possessing an apertural neck.

Plate 59 (LIX)

Figure 1 *Fissurina seguenziana* (Fornasini 1887)

[*Lagena seguenziana* Fornasini 1887]
×75. From *Challenger* Sta. VIII, off the Canaries, Atlantic (620fm.). ZF1725.

Referred by Brady (1884) to *Lagena orbignyana* Seguenza, sp., by Thalmann first (1932) to *L. orbignyana* (Seguenza) and later (1937) to *L. seguenziana* Fornasini, and by Barker (1960) (albeit tentatively) to *Fissurina seguenziana* (Fornasini).

Figure 2 *Fissurina* sp. nov. (1)

×75. From *Challenger* Sta. 24, West Indies (390fm.). ZF1693.

Referred by Brady (1884) and Thalmann (1932) to *Lagena hispida*

Reuss, 'compressed var.', and by Barker (1960) to *Fissurina hispida* (Reuss), 'compressed var.'.

Figure 3 *Fissurina quadrata* (Williamson 1858)

[*Entosolenia marginata* var. *quadrata* Williamson 1858]
×75. From *Porcupine* Sta. 67–68, East of the Shetlands, Scotland (64–75fm.). ZF1737.

Referred by Brady (1884) and Thalmann (1932) to *Lagena quadrata* (Williamson), and by Barker (1960) to *Fissurina quadrata* (Williamson).

Figure 4 *Fissurina* sp. nov. (2)

×75, From *Challenger* Sta. 219A, Admiralty Islands, Pacific (17fm.). ZF1651.

Referred by Brady (1884), Thalmann (1932), and Barker (1960) to *Lagena favoso-punctata* Brady, 'compressed var.'.

Figure 5 *Fissurina* sp. nov. (3)

×75. From *Challenger* Sta. 241, North Pacific (2300fm.). ZF1694.

Referred by Brady (1884) and Thalmann (1932) to *Lagena hispida* Reuss, 'compressed var.', and by Barker (1960) to *Fissurina hispida* (Reuss), 'compressed var.'.

Figure 6 *Fissurina incomposita* (Patterson and Pettis 1986)

[*Lagenosolenia incomposita* Patterson and Pettis 1986]
×60. From *Challenger* Sta. 219A, Admiralty Islands, Pacific (17fm.).

Referred by Brady (1884) to *Lagena acuta* Reuss, by Thalmann first (1932) to *L. sacculus* Fornasini, later (1937) to *L. marginata* var. *spinifera* Earland, and later still (1942) to *L. crebra* Matthes, and by Barker (1960) and Hermelin (1989) to *Fissurina crebra* (Matthes). *Lagena marginata* var. *spinifera* Earland 1934 is a primary junior homonym of *L. aspera* var. *spinifera* Chapman 1895. It has recently been renamed *Lagenosolenia incomposita* by Patterson and Pettis (1986). *Lagenosolenia* McCulloch 1977 is regarded by the present author as a junior synonym of *Fissurina* Reuss 1850.

Figures 7, 15 *Fissurina annectens* (Burrows and Holland 1895)

[*Lagena annectens* Burrows and Holland 1895]
Fig. 7 ×60. From *Challenger* Sta. 332, South Atlantic (2200fm.). ZF1742.
Fig. 15 ×75. From *Challenger* Sta. 149D, Kerguelen Island, South Pacific (20–50fm.). ZF1741.

Referred by Brady (1884) to *Lagena quadricostulata* Reuss, by Thalmann (1932) to *L. quadricostulata* Reuss (Fig. 7) and *L. fasciata* (Egger) (Fig. 15), and by Barker (1960) (Figs 7?, 15) and Hermelin (1989) (Fig. 15 only) to *Fissurina annectens* (Burrows and Holland).

Figures 8–11 *Fissurina staphyllearia* Schwager 1866

Fig. 8 ×45. From *Challenger* Sta. 151, Heard Island, South Pacific (75fm.). ZF1767.
Figs 9–10 ×75. From *Challenger* Sta. 149D, Kerguelen Island, South Pacific (20–50fm.). ZF1768.

Fig. 11 ×75. From *Challenger* Sta. 160, Southern Ocean (2600fm.). ZF1769.

Referred by Brady (1884) and Thalmann (1932) to *Lagena staphyllearia* Schwager, and by Barker (1960) and Hermelin (1989) to *Fissurina kerguelenensis* Parr. *F. kerguelensis* Parr 1850 is regarded by the present author as a junior synonym of *F. staphyllearia* Schwager 1866.

Figure 12 *Fissurina unguiculata* (Brady 1881)

[*Lagena unguicalata* Brady 1881]
×60. From *Challenger* Sta. 332, South Atlantic (2200fm.). ZF1801.

Referred by Brady (1884) and Thalmann (1932) to *Lagena*, and by Barker (1960) to *Fissurina*.

Figures 13–14 *Fissurina longispina* (Brady 1881)

[*Lagena longispina* Brady 1881]
×30. From *Challenger* Sta. 98, off West Africa (1750fm.). ZF1717.

Referred by Brady (1884) and Thalmann (1932) to *Lagena longispina* Brady, 'compressed form', and by Barker (1960) to *Oolina longispina* (Brady), 'compressed form'. Barker (1960) noted that 'this variant approaches a *Fissurina*'.

Figure 16 *Fissurina quadrata* var. *carinata* (Chapman 1909)

[*Lagena quadrata* var. *carinata* Chapman 1909]
×75. From *Challenger* Sta. 135, Tristan d'Acunha, Atlantic (100–150fm.). ZF1738.

Referred by Brady (1884) and Thalmann (1932) to *Lagena quadrata* Williamson, 'partially carinate specimen', by Thalmann (1937) to *L. quadrata* var. *carinata* Chapman, and by Barker (1960) to *Fissurina quadrata* var. *carinata* (Chapman).

Figures 17, 19 *Fissurina semimarginata* (Reuss 1870)

[*Lagena marginata* var. *semimarginata* Reuss 1870]
Fig. 17 ×60–75. From *Challenger* Sta. 145, Prince Edward Island, South Pacific (50–150fm.). ZF1721.
Fig. 19 ×60–75. From *Challenger* Sta. 332, South Atlantic (2200fm.). ZF1722.

Referred by Brady (1884) and Thalmann (1932) to *Lagena marginata* var. *semimarginata* Reuss, and by Barker (1960) to *Fissurina semimarginata* (Reuss).

Figure 18 *Fissurina orbignyana* Seguenza 1862
×75–90. From *Challenger* Sta. 145, Prince Edward Island, South Pacific (50–150fm.). ZF1726.

Referred by Brady (1884) and Thalmann (1932) to *Lagena*, and by Barker (1960) to *Fissurina*.

Figure 20 *Fissurina baccata* (Heron-Allen and Earland 1922)

[*Lagena orbignyana* var. *baccata* Heron-Allen and Earland 1922]
×75–90.

Referred by Brady (1884) and Thalmann (1932) to *Lagena orbignyana* (Seguenza), and by Barker (1960) to *Fissurina orbignyana* Seguenza.

Figures 21–22 *Fissurina submarginata* Boomgaart 1949

Fig. 21 ×45. ZF1718. Fig. 2 ×40. ZF1719. From *Challenger* Sta. 300, North of Juan Fernandez, East Pacific (1375fm.).

Referred by Brady (1884) and Thalmann (1932) to *Lagena marginata* (Walker and Boys), by Barker (1960) and van Marle (1991) to *Fissurina submarginata* Boomgaart, and by Hermelin (1989) to *F. marginata* (Montagu).

Figure 23 *Pseudosolenina wiesneri* (Barker 1960)

[*Fissurina wiesneri* Barker 1960]
×25. From *Challenger* Sta. 168, NE of New Zealand (1100fm.). ZF1720.

Referred by Brady (1884) and Thalmann (1932) to *Lagena marginata* (Walker and Boys), by Barker (1960) to *Fissurina wiesneri* (a new name for *Lagena marginata* var. *carinata* Wiesner 1931, not *L. quadrata* var. *carinata* Chapman 1909), and also by Hermelin (1989) to *F. wiesneri* Barker. This species is transferred herein to the genus *Pseudosolenina* Jones 1984 on account of its subterminal rather than terminal aperture.

Figure 24 *Fissurina bradii* Silvestri 1902

×45. From *Challenger* Sta. 300, North of Juan Fernandez, East Pacific (50–150fm.). ZF1727.

Referred by Brady (1884) to *Lagena orbignyana* Seguenza, sp., by Thalmann (1932) to *L. orbignyana* var. *caribaea* Cushman, and by Barker (1960) to *Fissurina bradii* Silvestri.

Figure 25 *Fissurina cucullata* (Silvestri 1902)

[*Lagena cucullata* Silvestri 1902]
×30. From *Challenger* Sta. 296, SW of Juan Fernandez, East Pacific (1825fm.). ZF1729.

Referred by Brady (1884) and Thalmann (1932) to *Lagena orbignyana* (Seguenza), and by Barker (1960) to *Fissurina cucullata* Silvestri.

Figure 26 *Fissurina bradyiformata* (McCulloch 1977)

[*Lagenosolenia bradyiformata* McCulloch 1977]
×60. From *Challenger* Sta. 271, South Pacific (2425fm.). ZF1730.

Referred by Brady (1884) and Thalmann (1932) to *Lagena orbignyana* (Seguenza), and by Barker (1960) to *Fissurina orbignyana* Seguenza.

Figure 27 *Fissurina siliqua* (Rymer Jones 1874)

[*Lagena vulgaris* var. *siliqua* Rymer Jones 1874]
×45. From *Challenger* Sta. 346, South Atlantic (2350fm.). ZF1759.

Referred by Brady (1884) and Thalmann (1932) to *Lagena siliqua*, and by Barker (1960) to *Fissurina siliqua*.

Figures 28–30 *Fissurina seminiformis* (Schwager 1866)

[*Lagena seminiformis* Schwager 1866]
Fig. 28 ×45–60. From *Challenger* Sta. 224, North Pacific (1850fm.). ZF1747.

Fig. 29 × 45–60. From *Challenger* Sta. 296, South Pacific (1825fm.). ZF1749.

Fig. 30 × 45–60. From *Challenger* Sta. 283, South Pacific (2075fm.). ZF1750.

Referred by Brady (1884) and Thalmann (1932) to *Lagena*, and by Barker (1960) (albeit tentatively), Srinivasan and Sharma (1980), and Hermelin (1989) to *Fissurina*. Srinivasan and Sharma (1980) have neotypified and reillustrated this species.

Plate 60 (LX)

Figures 1–2 *Fissurina lacunata* (Burrows and Holland 1895)

[*Lagena lacunata* Burrows and Holland 1895]
Fig. 1 × 75. From *Challenger* Sta. 162, Bass Strait, Pacific (38–40fm.).
Fig. 2 × 75.

Referred by Brady (1884) to *Lagena castrensis* Schwager, by Thalmann (1932) to *L. orbignyana* var. *lacunata* (Burrows and Holland), and by Barker (1960) and van Marle (1991) to *Fissurina lacunata* (Burrows and Holland). This species was referred by Patterson (1986) to the new genus *Cerebrina*, which is regarded by the present author as a junior synonym of *Fissurina* Reuss 1850. It was referred by Alabani and Yassini (1989) to *Palliolatella* Patterson and Richardson 1987, which is also regarded as a junior synonym of *Fissurina* Reuss 1850.

Figure 3 *Fissurina contusa* Parr 1945

× 75. From *Challenger* Sta. 185, Torres Strait, Pacific (38–40fm.). ZF1635.

Referred by Brady (1884) to *Lagena castrensis* Schwager, by Thalmann (1932) (albeit tentatively) to *L. orbignyana* var. *lacunata* (Burrows and Holland) (see above), and by Barker (1960) to *Fissurina contusa* Parr. This species was referred by Patterson (1986) to *Cerebrina* (see above).

Figure 4 *Fissurina clathrata* (Brady 1884)

[*Lagena clathrata* Brady 1884]
× 75. From *Challenger* Sta. 191, Arrou Islands (800fm.) or 191A, Ki Islands, Central Pacific (580fm.). ZF1636.

Referred by Brady (1884) to *Lagena clathrata*, by Thalmann (1932) to *L. orbignyana* var. *clathrata* Brady, and by Barker (1960) and Hermelin (1989) to *Fissurina clathrata* (Brady).

Figure 5 *Fissurina rizzae* Seguenza 1862

× 75. From *Challenger* Sta. 162, Bass Strait, Pacific (38fm.). ZF1740.

Referred by Brady (1884) to *Lagena quadrata* var., by Thalmann (1932) to *L. quadrata* var. *rizzae* Seguenza, and by Barker (1960) to *Fissurina quadrata* var. *rizzae* (Seguenza).

Figures 6–7, 9 *Fissurina lagenoides* (Williamson 1858)

[*Entosolenia marginata* var. *lagenoides* Williamson 1858]
Fig. 6 × 60–75. From *Challenger* Sta. 164A, off Sydney, Australia (410fm.). ZF1704.

Fig. 7 × 60–75. From *Challenger* Sta. 120, off Pernambuco, Atlantic (675fm.). ZF1708.

Fig. 9 × 60–75. From *Challenger* Sta. 5, SW of the Canaries, Atlantic (2740fm.). ZF1705.

Referred by Brady (1884) and later Thalmann (1932) to *Lagena lagenoides* (Williamson), and by Barker (1960) to *Fissurina lagenoides* (Williamson).

Figure 8 (18 on plate) *Fissurina pacifica* Parr 1950

× 75. From *Challenger* Sta. 162, Bass Strait, Pacific (38–40fm.). ZF1658.

Referred by Brady (1884) and Thalmann (1932) to *Lagena formosa* Schwager, and by Barker (1960) to *Fissurina pacifica* Parr.

Figure 10 *Fissurina* sp. nov. (4)

× 60. From *Challenger* Sta. 145, Prince Edward Island, South Pacific (50–150fm.).

Referred by Brady (1884) and Thalmann (1932) to *Lagena formosa* Schwager, and by Barker (1960) to *Fissurina* sp. nov.

Figure 11 *Fissurina tenuistriatiformis* (McCulloch 1977)

[*Lagenosolenia* (?) *tenuistriatiformis* McCulloch 1977]
× 60–75. From *Challenger* Sta. 338, South Atlantic (1900fm.). ZF1710.

Referred by Brady (1884) and Thalmann (1932) to *Lagena lagenoides* var. *tenuistriata* (Brady), by Barker (1960) to *Fissurina lagenoides* var. *tenuistriata* (Brady), and by Jones (1984a) to *Solenina tenuistriatiformis* (McCulloch). This is the type-species by original designation of the genus *Solenina* Jones 1984, regarded herein as a junior synonym of *Fissurina* Reuss 1850. *Lagena tubulifera* var. *tenuistriata* Brady 1881 is a primary junior homonym of *L. tenuistriata* Stache 1865.

Figure 12 *Fissurina* sp. nov. (5)

× 60–75. From *Challenger* Sta. 185, Torres Strait, Pacific (155fm.). ZF1706.

Referred by Brady (1884) and Thalmann (1932) to *Lagena lagenoides* (Williamson), and by Barker (1960) to *Fissurina lagenoides* (Williamson).

Figures 13–14 *Fissurina radiata* Seguenza 1862

Fig. 13 × 60–75. From *Challenger* Sta. 185, Torres Strait, Pacific (155fm.). ZF1706.

Fig. 14 × 60–75. From *Challenger* Sta. 85, off the Canaries, Atlantic (1125fm.). ZF1707.

Referred by Brady (1884) to *Lagena lagenoides* Williamson, sp., by Thalmann (1932) to *L. sublagenoides* Cushman, and by Barker (1960) and van Marle (1991) to *Fissurina radiata* Seguenza. This is the type-species by original designation of the genus *Lagnea* Popescu 1983, regarded by the present author as a junior synonym of *Fissurina* Reuss 1850.

Figures 15–16 *Fissurina radiata* var. *striatula* (Cushman 1913)

[*Lagena sublagenoides* var. *striatula* Cushman 1913]
Fig. 15 × 60–75. From *Challenger* Sta. 338, South Atlantic (1990fm.). ZF1709.
Fig. 16 × 60–75. From *Porcupine* Sta. 23, off NW Ireland, Atlantic (630fm.). ZF1711.

Referred by Brady (1884) to *Lagena lagenoides* var. *tenuistriata* Brady, by Thalmann (1932) to *L. sublagenoides* var. *striatula* Cushman, and by Barker (1960) to *Fissurina radiata* var. *striatula* (Cushman).

Figure 17 *Fissurina* sp. nov. (6)

× 90. From *Challenger* Sta. 352, Cape Verde Islands, Atlantic (11fm.). ZF1659.

Referred by Brady (1884) and Thalmann (1932) to *Lagena formosa* Schwager, and by Barker (1960) to *Fissurina* sp. indet.

Figures 18–19 *Fissurina formosa* (Schwager 1866)

[*Lagena formosa* Schwager 1866]
Fig. 18 × 50. From *Challenger* Sta. 283, South Pacific (2075fm.). ZF1660.
Fig. 19 × 50. From *Challenger* Sta. 218, North of Papua, Pacific (1070fm.). ZF1661.

Referred by Brady (1884) and Thalmann (1932) to *Lagena formosa* Schwager, by Barker (1960) to *Fissurina* sp. nov., and by the present author to *F. formosa* (Schwager). Srinivasan and Sharma (1980) have neotypified and reillustrated this species.

Figure 20 *Fissurina foliformis* (Buchner 1940)

[*Lagena foliformis* Buchner 1940]
× 60. From *Challenger* Sta. 283, South Pacific (2075fm.). ZF1660.

Referred by Brady (1884) and Thalmann (1932) to *Lagena formosa* Schwager (see above), by Barker (1960) to *Fissurina foliformis* (Buchner), and by Jones (1984a) to *Solenina foliformis* (Buchner). *Solenina* Jones 1984 is herein regarded as a junior synonym of *Fissurina* Reuss 1850 (see *F. tenuistriatiformis* above).

Figure 21 *Fissurina formosa* var. *favosa* (Brady 1884)

[*Lagena formosa* var. *favosa* Brady 1884]
× 60. From *Challenger* Sta. 224, North Pacific (1850fm.). ZF1663.

Referred by Brady (1884) and Thalmann (1932) to *Lagena*, and by Barker (1960) to *Fissurina*.

Figure 22 *Fissurina formosa* var. *comata* (Brady 1884)

[*Lagena formosa* var. *comata* Brady 1884]
× 75. From *Challenger* Sta. 224, North Pacific (1850fm.). ZF1663.

Referred by Brady (1884) and Thalmann (1932) to *Lagena*, and by Barker (1960) to *Fissurina*.

Figure 23 *Fissurina squamosoalata* (Brady 1881)

[*Lagena squamosoalata* Brady 1881]
× 60. From *Porcupine* Sta. 23, off NW Ireland, Atlantic (650fm.). ZF1765.

Referred by Brady (1884) and Thalmann (1932) to *Lagena*, and by Barker (1960) to *Fissurina*.

Figure 24 *Fissurina squamosomarginata* (Parker and Jones 1865)

[*Lagena sulcata* var. *marginata* subvar. *squamosomarginata* Parker and Jones 1865]
× 60. From *Challenger* Sta. 168, off New Zealand (1100fm.). ZF1766.

Referred by Brady (1884) and Thalmann (1932) to *Lagena squamosomarginata*, and by Barker (1960) to *Fissurina squamosomarginata*. This species was referred by Patterson (1986) to the new genus *Cerebrina*, regarded by the present author as a junior synonym of *Fissurina* Reuss 1850. It has been lectotypified by Hodgkinson (1992).

Figure 25 *Fissurina alveolata* var. *caudigera* (Brady 1884)

[*Lagena alveolata* var. *caudigera* Brady 1884]
× 40. From *Challenger* Sta. 296, South Pacific (1825fm.). ZF1613.

Referred by Brady (1884) and Thalmann (1932) to *Lagena*, and by Barker (1960) to *Fissurina*.

Figures 26–27 *Fissurina fimbriata* (Brady 1881)

[*Lagena fimbriata* Brady 1881]
Fig. 26 × 60. From *Challenger* Sta. 271, South Pacific (2425fm.). ZF1654. Lectotype designated by Jones (1984b).
Fig. 27 × 60. From *Challenger* Sta. 276, North of Tahiti, Pacific (2350fm.). ZF1655.

Referred by Brady (1884) and Thalmann (1932) to *Lagena*, and by Barker (1960), Popescu (1983), and Hermelin (1989) to *Fissurina*.

Figure 28 *Fissurina* sp. nov. (7)

× 60. From *Challenger* Sta. 191A, Ki Islands, Pacific (580fm.). ZF1656.

Referred by Brady (1884) and Thalmann (1932) to *Lagena fimbriata* Brady, and by Barker (1960) and Hermelin (1989) to *Fissurina fimbriata* (Brady) (see above).

Figure 29 *Fissurina auriculata* (Brady 1881)

[*Lagena auriculata* Brady 1881]
× 60. From *Challenger* Sta. 276, South Pacific (2350fm.). ZF1628.

Referred by Brady (1884) and Thalmann (1932) to *Lagena*, and by Barker (1960), Hermelin (1989), and van Marle (1991) to *Fissurina*. Van Marle (1991) regards *Lagena alveolata* Brady 1881 as a *nomen nudum*, but gives no justification.

Figures 30–32 *Fissurina alveolata* (Brady 1884)

[*Lagena alveolata* Brady 1884]
Fig. 30 × 60. From *Challenger* Sta. 64, North Atlantic (2750fm.). ZF1612. Lectotype designated by Jones (1984b).
Fig. 32 × 60. From *Challenger* Sta. 332, South Atlantic (2200fm.). ZF1611. Paralectotype designated by Jones (1984b).

Referred by Brady (1884) and Thalmann (1932) to *Lagena*, and by Barker (1960) and van Marle (1991) to *Fissurina*.

Text-figures in the 'Challenger Report' and their explanations

Text-figure 1 (p. 141) *Pyrgoella sphaera* (d'Orbigny 1839)

[*Biloculina sphaera* d'Orbigny 1839]
× 40. From *Challenger* Sta. 218, N. of Papua (1070fm.). British Museum (Natural History) (BM(NH)) Slide Register Number ZF1166

Referred by Brady (1884) to *Biloculina*, and by Barker (1960) to *Pyrgoella* (see also Plate 2). This is the type-species by original designation of the genus *Pyrgoella* Cushman and White 1936.

Text-figures 2–3 (pp. 158–9) *Quinqueloculina seminulum* (Linné 1758)

[*Serpula seminulum* Linné 1758]
2a. From Plancus. No magnification given. 2b. From Gualtieri. No magnification given. 2c. From Martini. No magnification given.
3a. Recent, from Williamson. No magnification given. 3b. Recent, from Parker and Jones. × 16. 3c. Fossil, from the Crag. × 16.

Referred by Brady (1884) to *Serpula* (Fig. 2) and *Miliolina* (Fig. 3), and by Barker (1960) to *Quinqueloculina* (see also Plate 5). This is the type-species by subsequent designation of the genus *Quinqueloculina* d'Orbigny 1826.

Text-figure 4 (p. 179) *Edentostomina rupertiana* (Brady 1881)

[*Miliolina rupertiana* Brady 1881]
a–d × 40; e × 200. From Tamatavé, Madagascar (shore sand). ZF1885.

Referred by Brady (1884) to *Miliolina*, and by Barker (1960) to *Triloculina* (see also Plate 7). This is the type-species by original designation of the genus *Rupertianella* Loeblich and Tappan 1985, regarded by the present author as a junior synonym of *Edentostomina* Collins 1958.

Text-figure 5a (p. 194) *Nummoloculina contraria* (d'Orbigny 1846)

[*Biloculina contraria* d'Orbigny 1846]
× 40.

Referred by Brady (1884) to *Planispirina*, and by Barker (1960) to *Nummoloculina* (see also Plate 11). This is the type-species by original designation of the genus *Nummoloculina* Steinmann 1881.

Text-figure 5b (p. 194) *Planispirinella exigua* (Brady 1879)

[*Planispirina exigua* Brady 1879]
× 80. ZF2108 (locality not known).

Referred by Brady (1884) and later Barker (1960) to *Planispirina* (see also Plate 112). This is the type-species by original designation of the genus *Planispirinella* Wiesner 1931.

Text-figure 5c (p. 194) *Sigmoilina sigmoidea* (Brady 1884)

[*Planispirina sigmoidea* Brady 1884]
× 40. From *Challenger* Sta. 120. ZF2113.

Referred by Brady (1884) to *Planispirina sigmoidea* n. sp., and by Barker (1960) to *Nummoloculina contraria* (d'Orbigny) (see also Plate 2). This is the type-species by original designation of the genus *Sigmoilina* Schlumberger 1887.

Text-figure 6 (p. 211) 'Diagrammatic representation of the progressive development of the simple type of *Orbitolites* to the most complex'

After Carpenter. No magnification given.

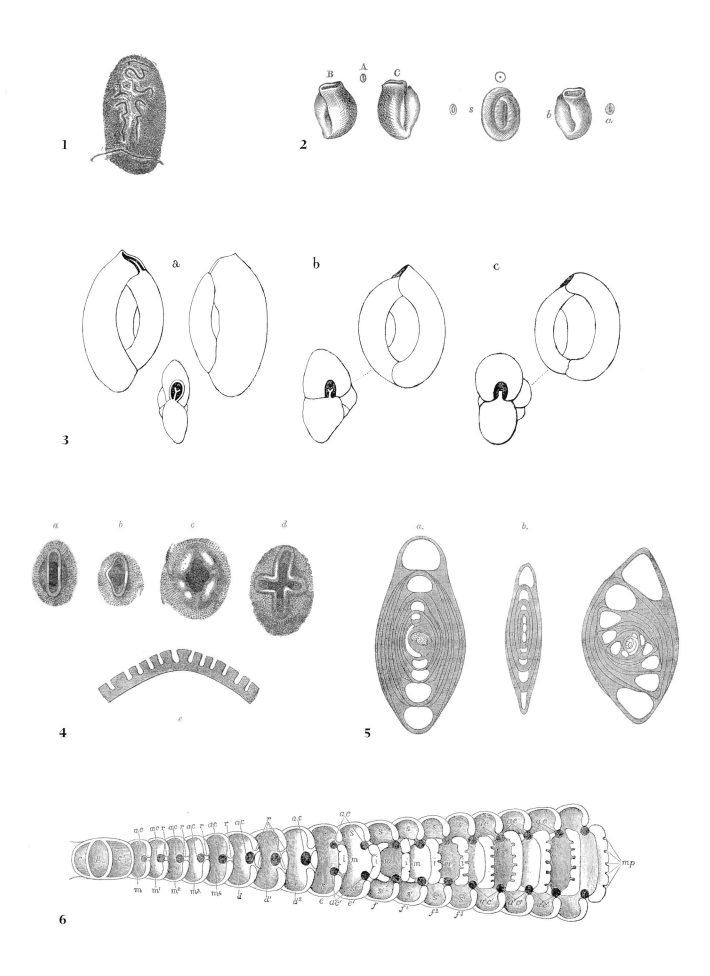

Text-figure 7 (p. 217) *Amphisorus hemprichii* Ehrenberg 1839

a–b × 12; c × 50.

Referred by Brady (1884) to *Orbitolites duplex* Carpenter, and by Barker (1960) to *Amphisorus hemprichii* Ehrenberg (see also Plate 16). This is the type-species by original designation of the genus *Amphsorus* Ehrenberg 1839.

Text-figure 8 (p. 225) *Keramosphara murrayi* Brady 1882

a × 1; b × 25; c × 50; d × 100.

This is the type-species by original monotypy of the genus *Keramosphaera* Brady 1882.

Text-figure 9 (p. 242) *Syringammina fragilissima* Brady 1883

a–b × 1; c × 8.

This is the type-species by original designation of the genus *Syringammina* Brady 1883 (regarded by Loeblich and Tappan (1987) as xenophyophorean).

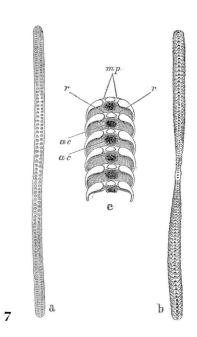

7
mp.
r
r
a c
a c
c
a
b

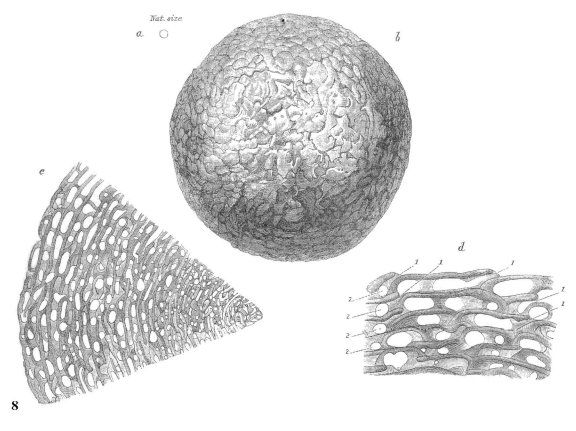

Nat. size
a
b
c
d
1
1
1
1
1
2
2
2
2
2
8

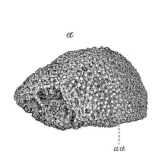
a
a a

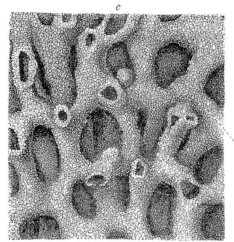
c
c c

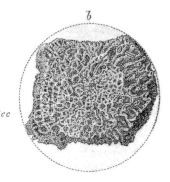
b

9

Text-figure 10 (p. 263) *Dendrophrya arborescens* **(Norman 1881)**

[*Psammatodendron arborescens* Norman 1881]
Dredged off Cumbrae. × 20.

Referred by Brady (1884) to *Hyperammina*, and by Barker (1960) to *Psammatodendron* (see also Plate 28). This is the type-species by original monotypy of the genus *Psammatodendron* Norman 1881, regarded by the present author as a junior synonym of *Dendrophrya* T.S. Wright 1861.

Text-figures 11a–b, h, k–l (p. 441) *Oolina globosa* **(Montagu 1803)**

[*Vermiculum globosum* Montagu 1803]
× 120.

Referred by Brady to *Lagena globosa* Montagu, sp., and by Barker (1960) to *Oolina globosa* (Montagu) (see also Plate 56). This is the type-species by original designation of the genus *Entolagena* Silvestri 1900, regarded by the present author as a junior synonym of *Oolina* d'Orbigny 1839. Text-figures 11k–l are regarded as representing an elongate variety with fistulose overgrowth.

Text-figures 11c–d, i (p. 441) *Pseudoolina fissurinea* **Jones 1984**

× 120.

Referred by Brady (1884) to *Lagena globosa* Montagu, sp., by Barker (1960) to *Oolina globosa* (Montagu), and by the present author tentatively to *Pseudoolina fissurinea* Jones. This is the type-species by original designation of the genus *Pseudoolina* Jones 1984.

Text-figure 11e (p. 441) *Galwayella* **sp.**

× 120.

Referred by Brady to *Lagena globosa* Montagu, sp., by Barker (1960) to *Oolina globosa* (Montagu), and by the present author tentatively to *Galwayella* sp.

Text-figures 11f–g, j (p. 441) *Anturina haynesi* **Jones 1984**

× 120.

Referred by Brady to *Lagena globosa* Montagu, sp., by Barker (1960) to *Oolina globosa* (Montagu), and by the present author tentatively to *Anturina haynesi* Jones. This is the type-species by original designation of the genus *Anturina* Jones 1984.

Text-figure 11m (p. 441) *Oolina globosa* **var.** *caudigera* **(Wiesner 1931)**

[*Lagena* (*Entosolenia*) *globosa* var. *caudigera* Wiesner 1931]
× 120.

Referred by Brady to *Lagena globosa* Montagu, sp., by Barker (1960) to *Oolina globosa* (Montagu), and by the present author to *O. globosa* var. *caudigera* Wiesner.

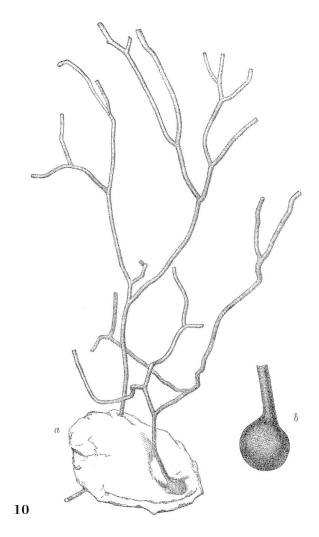

10

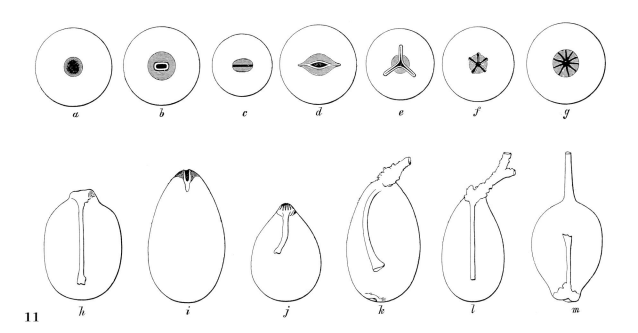

11

Text-figure 12 (p. 470) *Sipholagena hertwigiana* (Brady 1884)

[*Lagena hertwigiana* Brady 1884]
× 200. From *Challenger* Sta. 185, Raine Island (155fm.). ZF1683.

Referred by Brady (1884) and later Barker (1960) to *Lagena* (see also Plate 58).

Text-figures 13a–b (p. 499) *Dentalina farcimen* (Soldani 1798)

[*Orthoceras farcimen* Soldani 1798]
a. Recent, from Soldani. No magnification given. B. Fossil, from the Crag of Antwerp (Reuss). No magnification given.

Referred by Brady (1884) to *Nodosaria farcimen* Soldani, sp., and by Barker (1960) to *N. (Dentalina) farcimen* (Soldani).

Text-figure 13c (p. 499) *Dentalina sp.*

Fossil, from the Cretaceous. No magnification given.

Referred by Brady (1884) to *Nodosaria farcimen* Soldani, sp., and by Barker (1960) to *N. (Dentalina) farcimen* (Soldani).

Text-figure 14a (p. 501) *Dentalina elegans* d'Orbigny 1846

From d'Orbigny (Miocene). No magnification given.

Referred by Brady (1884) to *Nodosaria pauperata* d'Orbigny, sp., and by Barker (1960) to *Stilostomella pauperata* (d'Orbigny). Papp and Schmid (1985) regard *Dentalina pauperata* d'Orbigny 1846 as a junior synonym of *D. elegans* d'Orbigny 1846. The generic identity of this species is somewhat questionable.

Text-figure 14b (p. 501) *Dentalina baggi* Galloway and Wissler 1927

From the Recent. No magnification given.

Referred by Brady (1884) to *Nodosaria pauperata* d'Orbigny, sp., and by Barker (1960) tentatively to *Dentalina baggi* Galloway and Wissler.

Text-figure 14c (p. 501) *Dentalina sp.*

From the Liassic. No magnification given.

Referred by Brady (1884) to *Nodosaria pauperata* d'Orbigny, sp., and by Barker (1960) tentatively to *Stilostomella* sp. The identity of this form, is questionable.

Text-figure 15 (p. 516) *Amphicoryna intercellularis* (Brady 1881)

[*Nodosaria interceullaris* Brady 1881]
a × 20; b × 40. From *Challenger* Sta. 33, Bermuda (435fm.). ZF1592.

Referred by Brady (1884) to *Nodosaria*, and by Barker (1960) to *Amphicoryna* (see also Plate 65). The generic identity of this species is somewhat questionable in that it appears to possess a compound wall.

Text-figure 16 (p. 524) *Frondicularia millettii* Brady 1884

× 120.

Referred by Brady (1884) to *Frondicularia*, and by Barker (1960) to *Plectofrondicularia*.

Text-figure 17 (p. 598) *Globorotalia sp.*

Surface specimen taken in the tow-net off New Guinea. From a sketch by Mr Wild. No magnification given.

Referred by Brady (1884) to *Globigerina marginata* Reuss, sp., and by Barker (1960) to *Globorotalia* sp. nov.

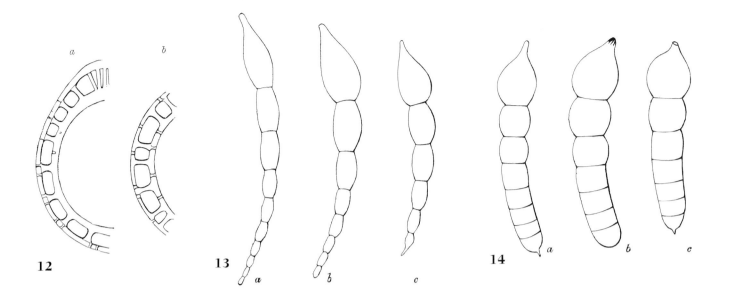

12   13   *a*   *b*   *c*   14   *a*   *b*   *c*

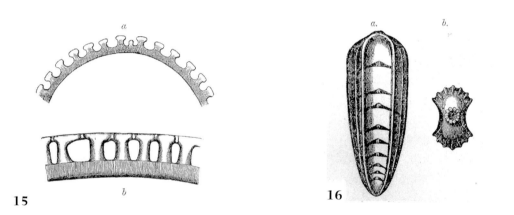

a

15   *b*   16   *a*   *b*

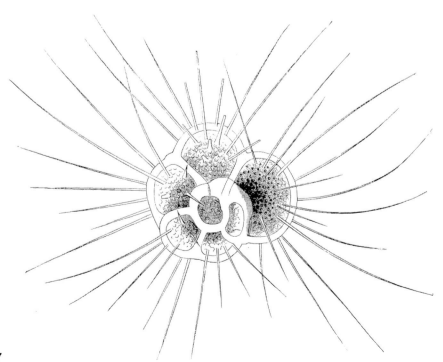

17

Text-figure 18 (p. 616) *Pullenia bulloides* (d'Orbigny 1846)

[*Nonionina bulloides* d'Orbigny 1846]
× 100. From *Knight Errant* Sta. 8 (540fm.). ZF2200.

Referred by Brady (1884) to *Pullenia sphaeroides* d'Orbigny, sp., and by Barker (1960) to *P. bulloides* (d'Orbigny) (see also Plate 84). *Nonionina bulloides* d'Orbigny 1846 is the type-species of the genus *Pullenia* Parker and Jones 1862 (*N. bulloides* d'Orbigny 1826 and *N. sphaeroides* d'Orbigny 1826 being *nomina nuda*). It has recently been lectotypified by Papp and Schmid (1985).

Text-figure 19 (p. 635) *Alanwoodia campanaeformis* (Brady 1884)

[*Patellina campanaeformis* Brady 1884]
× 200. From *Challenger* Sta. 185, off Raine Island, Torres Strait (155fm.). ZF2065.

Referred by Brady (1884) to *Patellina*, and by Barker (1960) to *Alanwoodia*. This is the type-species by original designation of the genus *Alanwoodia* Loeblich and Tappan, 1955.

Text-figure 20 (p. 639) *Cymbaloporetta plana* (Cushman 1924)

[*Tretomphalus bulloides* var. *plana* Cushman 1924]
× 60.

Referred by Brady (1884) to *Cymbalopora* (*Tretomphalus*) *bulloides* d'Orbigny, and by Barker (1960) to *Tretomphalus planus* Cushman (see also Plate 102). This is the type-species by original designation of the genus *Pseudotretomphalus* Hofker 1979, regarded by Loeblich and Tappan (1987) as junior synonym of *Cymbaloporetta* Cushman 1928.

Text-figure 21 (p. 700) *Hoeglundina elegans* (d'Orbigny 1826)

[*Rotalia* (*Turbinuline*) *elegans* d'Orbigny 1826]
× 40. ZF2235 or ZF2236.

Referred by Brady (1884) to *Pulvinulina partschiana* d'Orbigny, sp., and by Barker (1960) to *Hoglundina* (sic) *elegans* (d'Orbigny) (see also Plate 105). This is the type-species by original designation of the genus *Hoeglundina* Brotzen 1948.

Text-figure 22 (p. 749) *Operculinella cumingii* (Carpenter 1860)

[*Amphistegina cumingii* Carpenter 1860]
× 20.

Referred by Brady (1884) to *Nummulites cumingi* Carpenter, sp., by Barker (1960) to *Operculina ammonoides* (Gronovius), and by the present author to *Operculinella cumingii* (Carpenter) (see also Plate 112). This is the type-species by original designation of the genus *Operculinella* Yabe 1908.

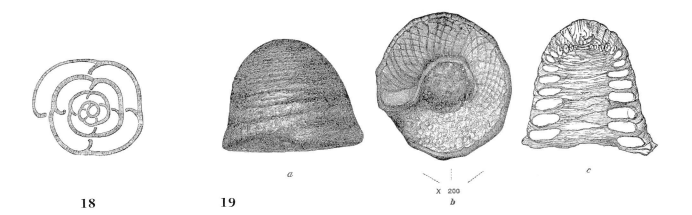

18   19

a

X 200
b

c

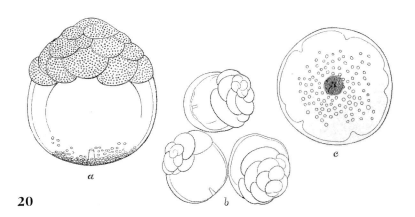

a

20

b

c

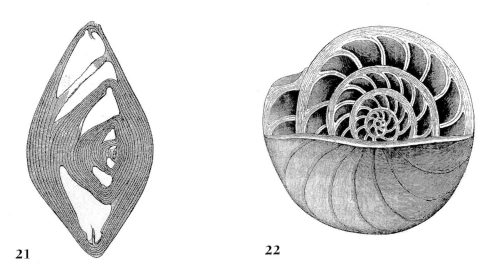

21   22

Plates in the
'*Challenger* Report'

Plate 1

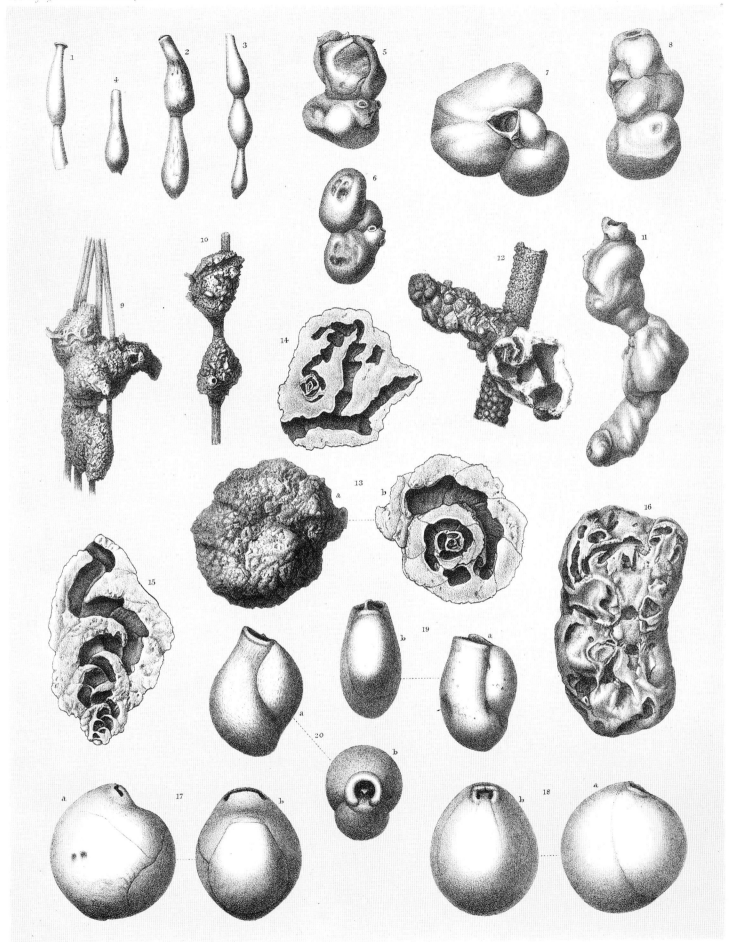

A.T.Hollick ad nat. del. et lith.

Hanhart imp

NUBECULARIA_MILIOLA , (Biloculina.) .

Plate 2

MILIOLA. (Biloculina.)

Plate 3

MILIOLA (Biloculina and Miliolina.)

Plate 4

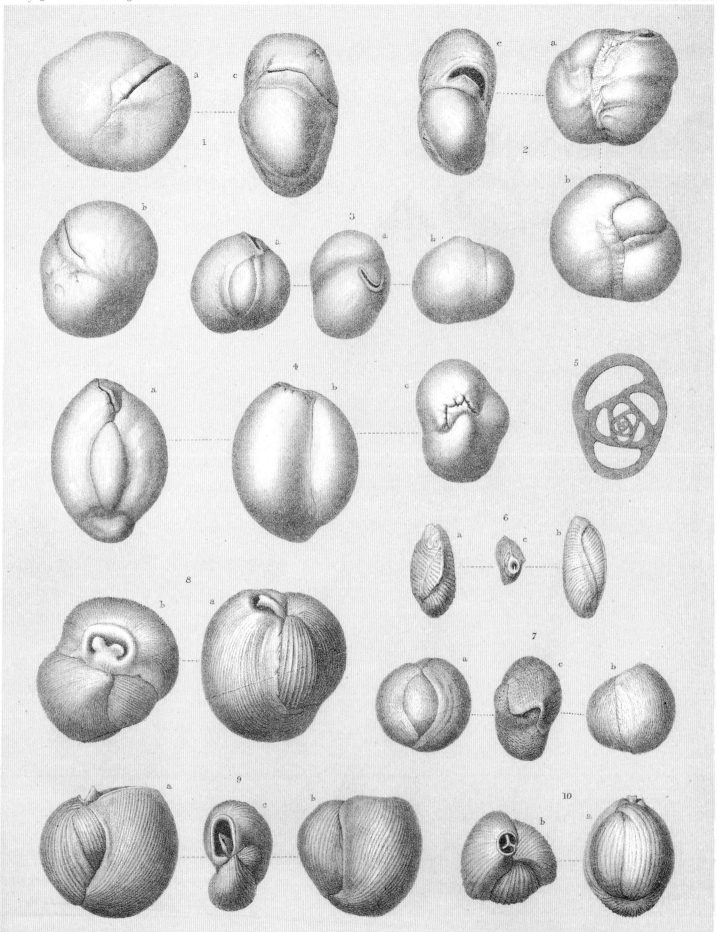

A.T.Hollick ad nat del. et lith.

MILIOLA. (Miliolina.)

Hanhart imp.

Plate 5

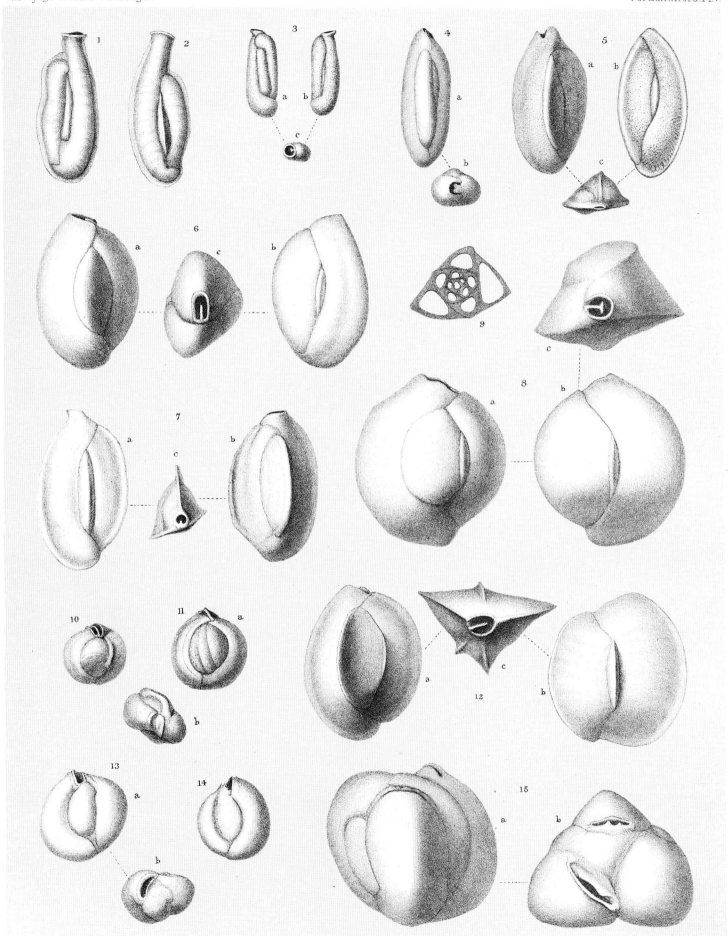

MILIOLA, (Miliolina).

Plate 6

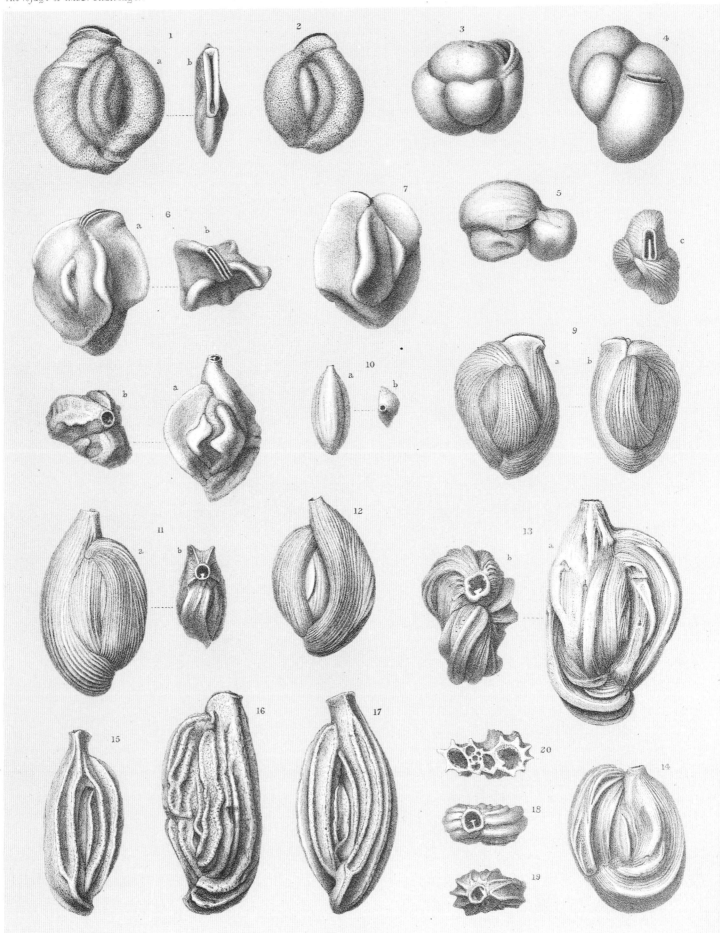

MILIOLA, (Miliolina).

Plate 7

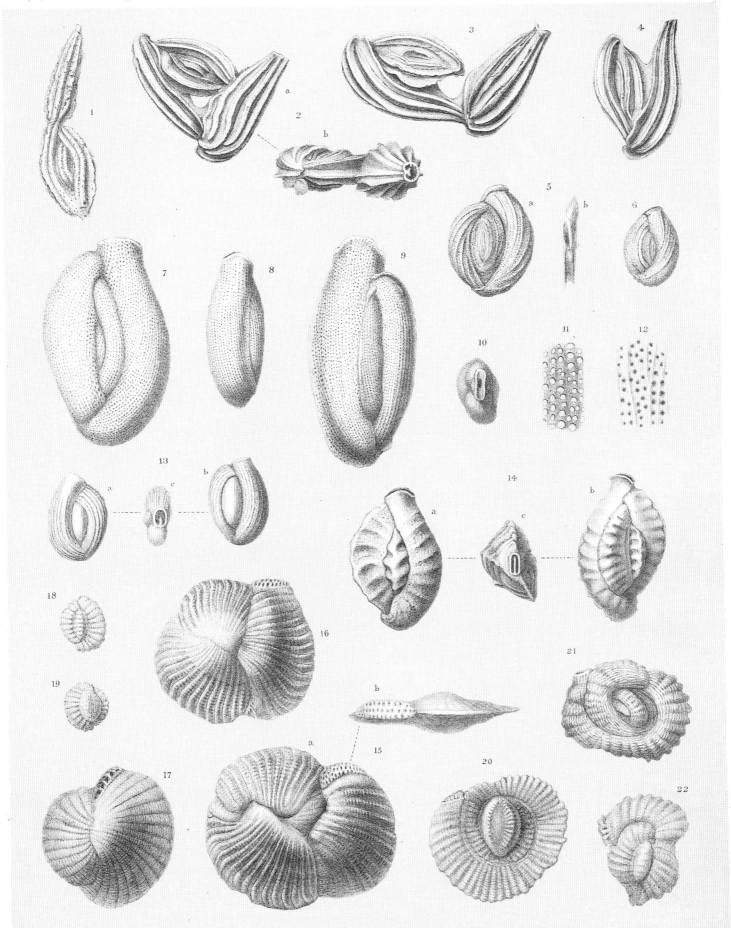

MILIOLA (Miliolina).

Plate 8

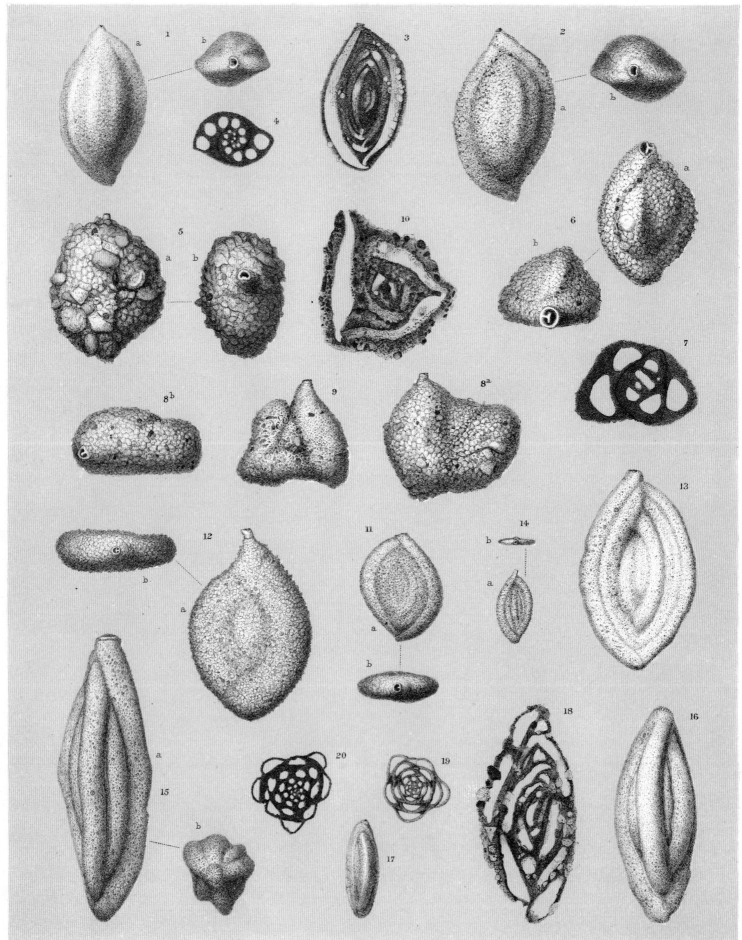

MILIOLA (Miliolina_Spiroloculina).

Plate 9

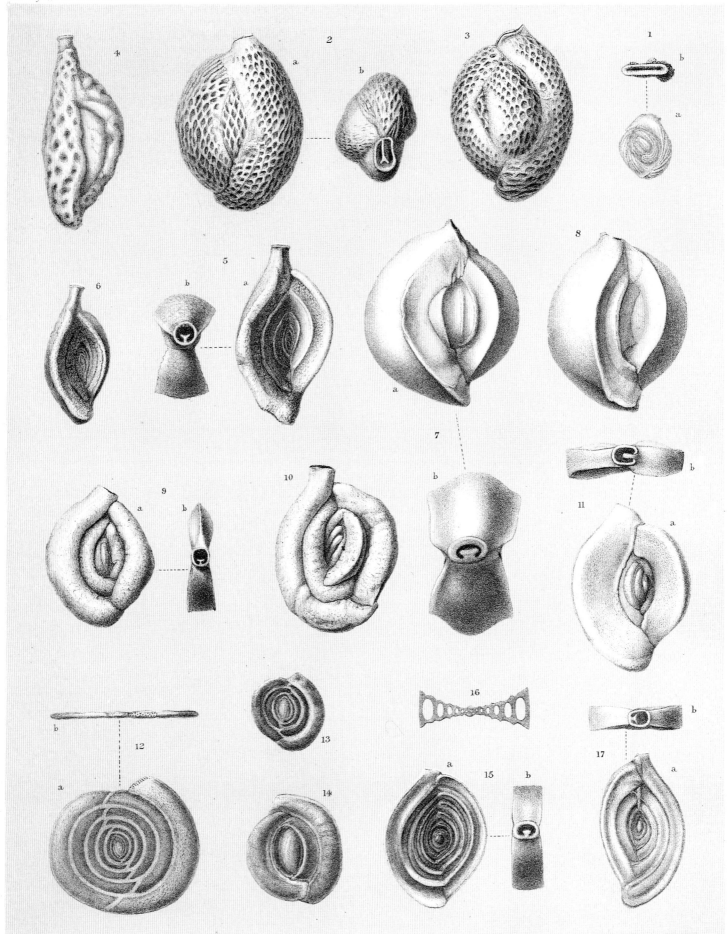

T.Hollick del.et lith.

Hanhart imp.

MILIOLA (Miliolina & Spiroloculina.).

Plate 10

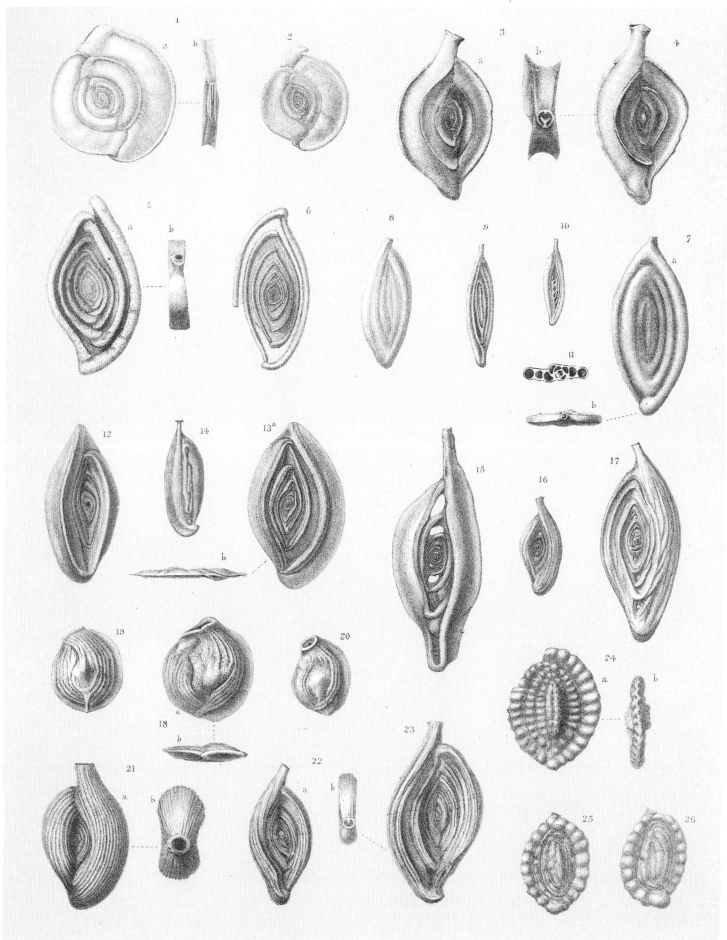

MILIOLA (Spiroloculina.).

Plate 11

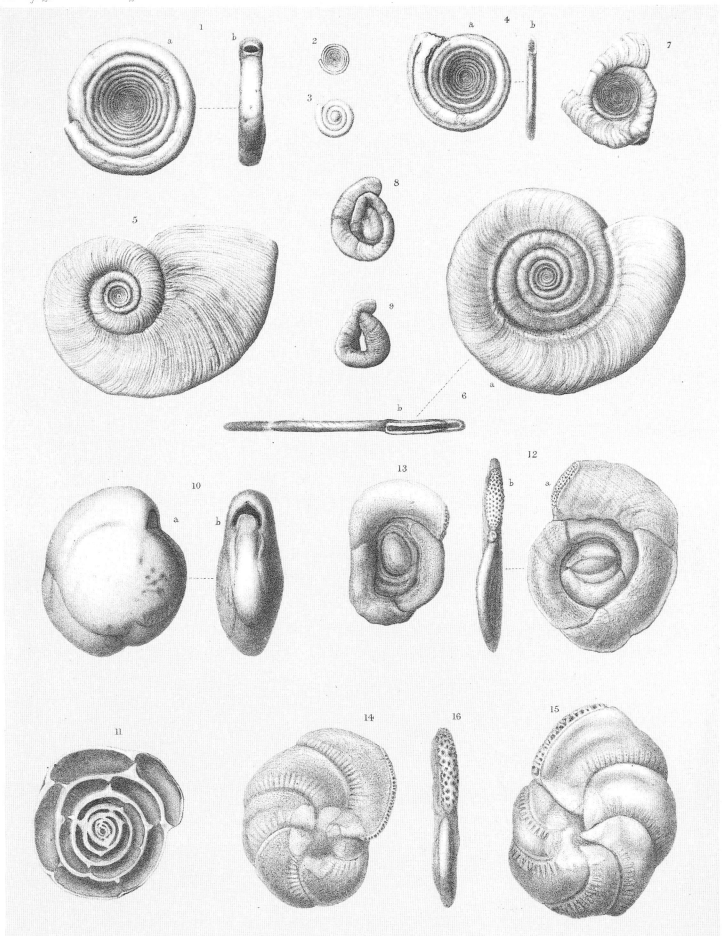

CORNUSPIRA_HAUERINA.

Plate 12

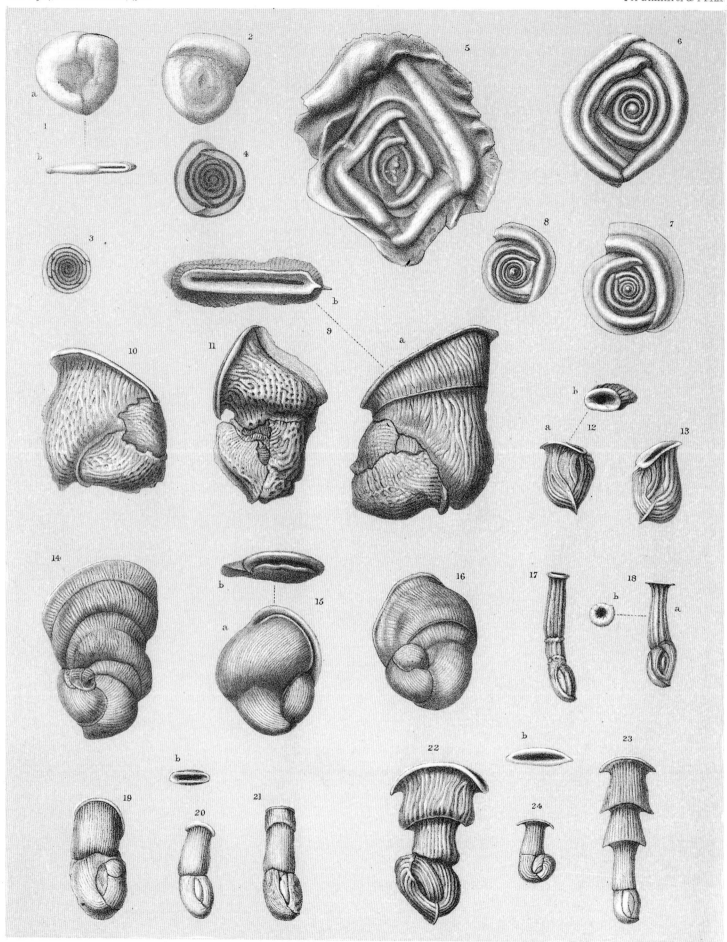

HAUERINA_VERTEBRALINA_ARTICULINA.

Plate 13

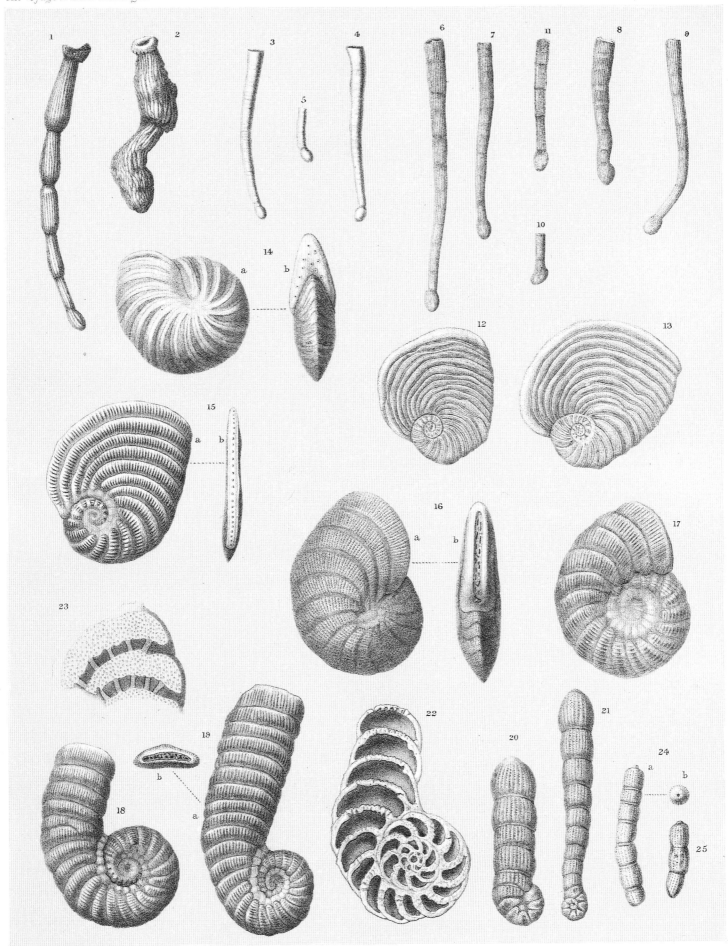

ARTICULINA _ PENEROPLIS.

Plate 14

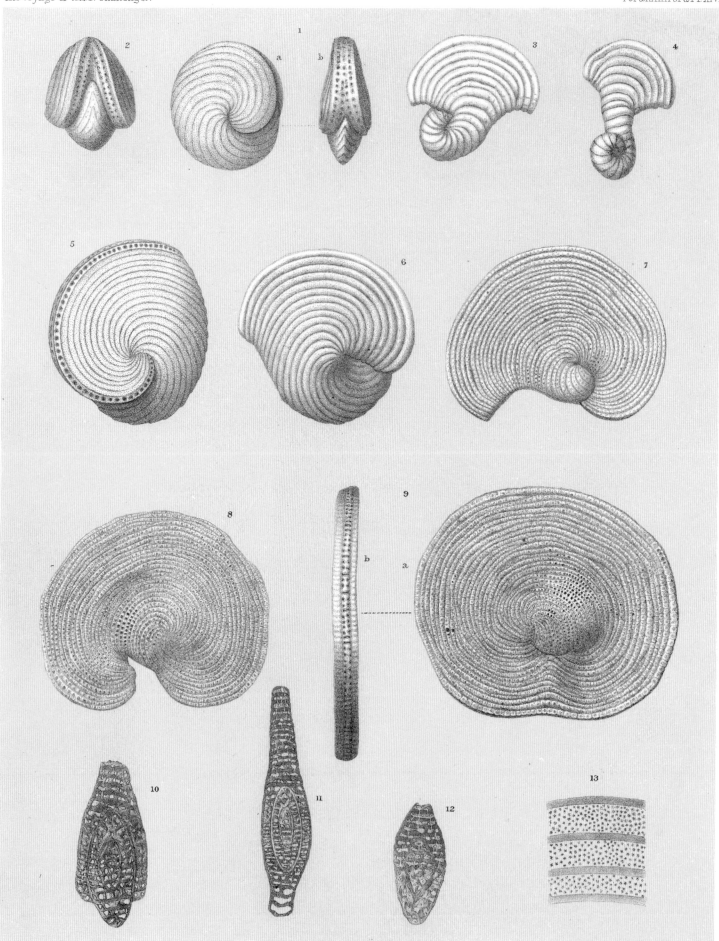

ORBICULINA.

Plate 15

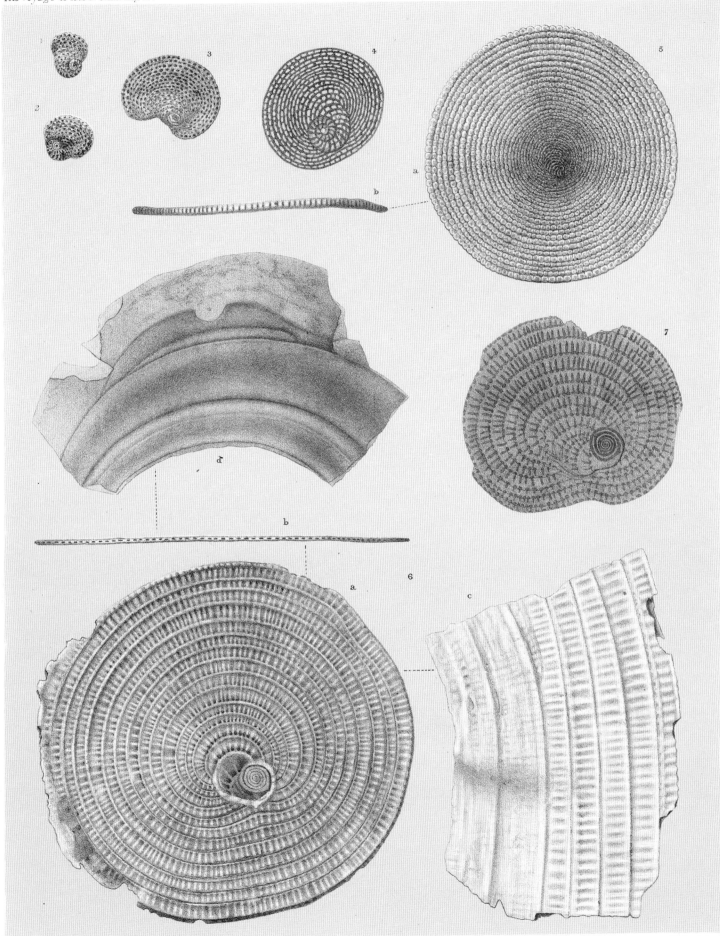

ORBITOLITES.

Plate 16

ORBITOLITES.

Plate 17

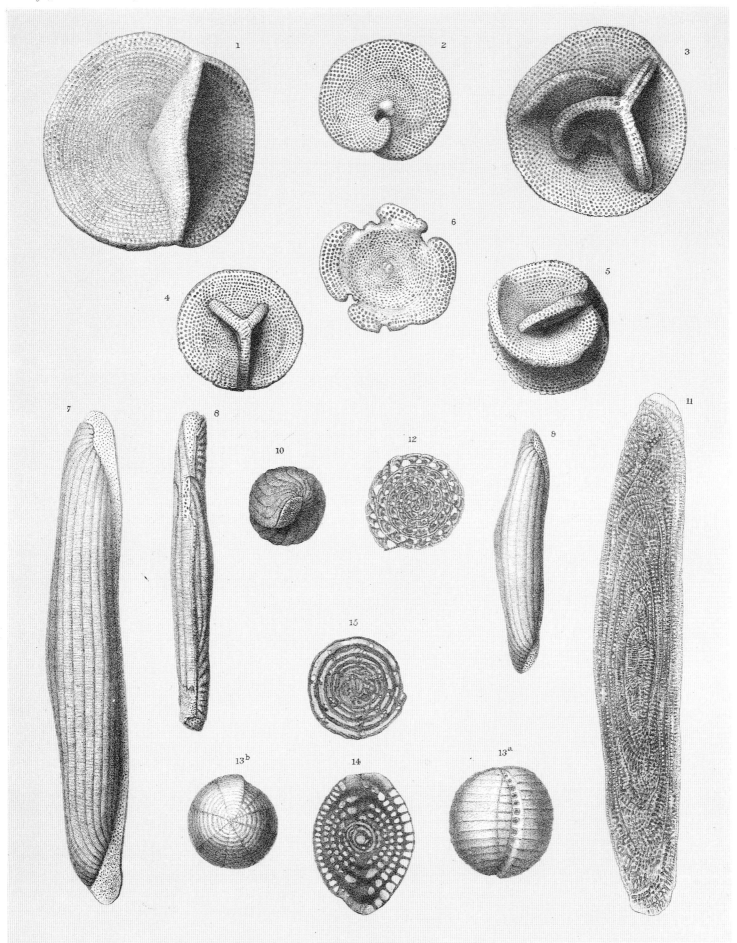

ORBITOLITES_ALVEOLINA.

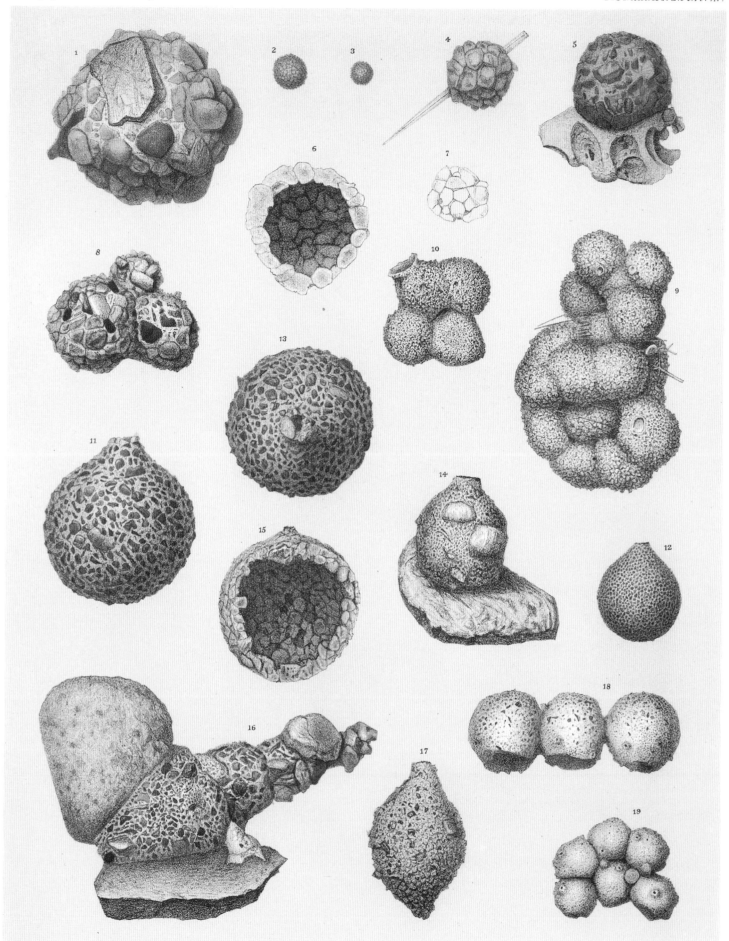

A.T.Hollick ad.nat.del.et lith.

Hanhart imp.

PSAMMOSPHÆRA-SOROSPHÆRA-SACCAMMINA.

Plate 19

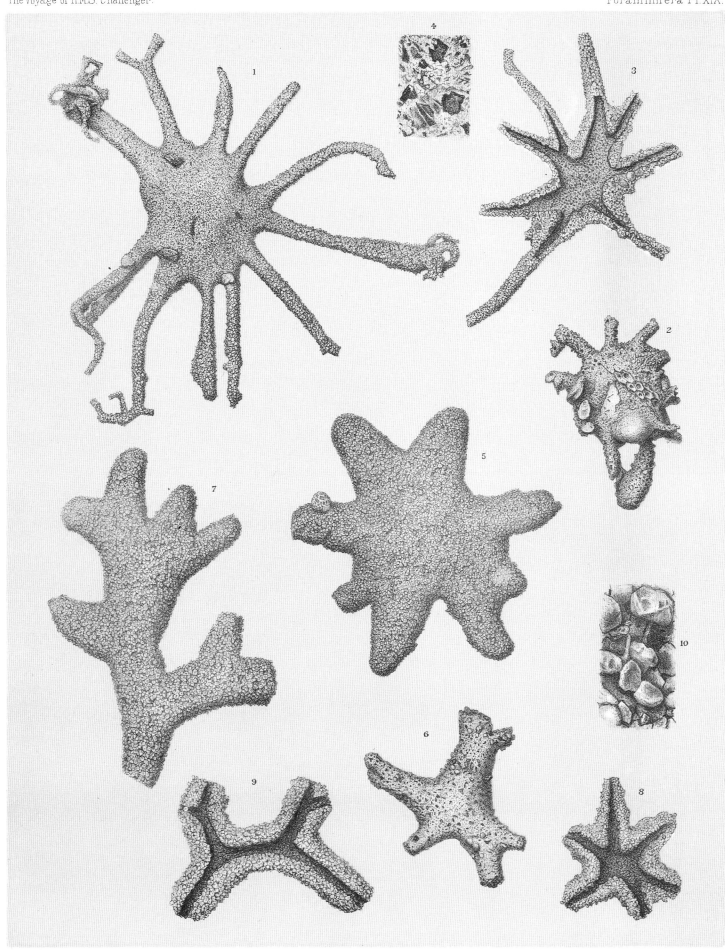

A.T.Hollick ad nat.del.et lith.

ASTRORHIZA.

Hanhart imp.

Plate 20

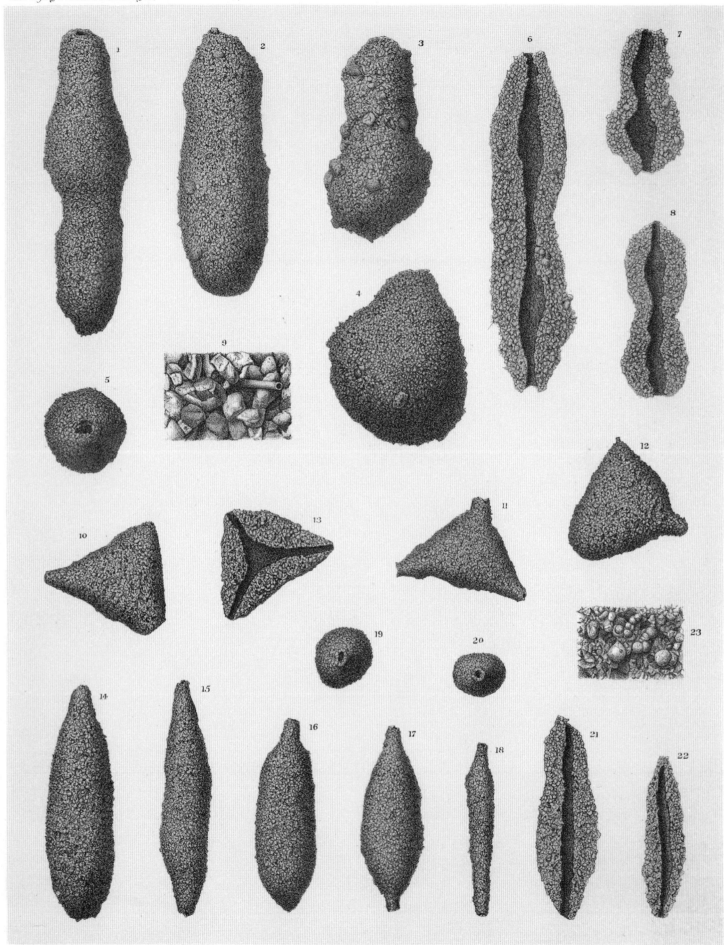

A.T.Hollick ad nat.del.et lith. Hanhart imp.

ASTRORHIZA .

Plate 21

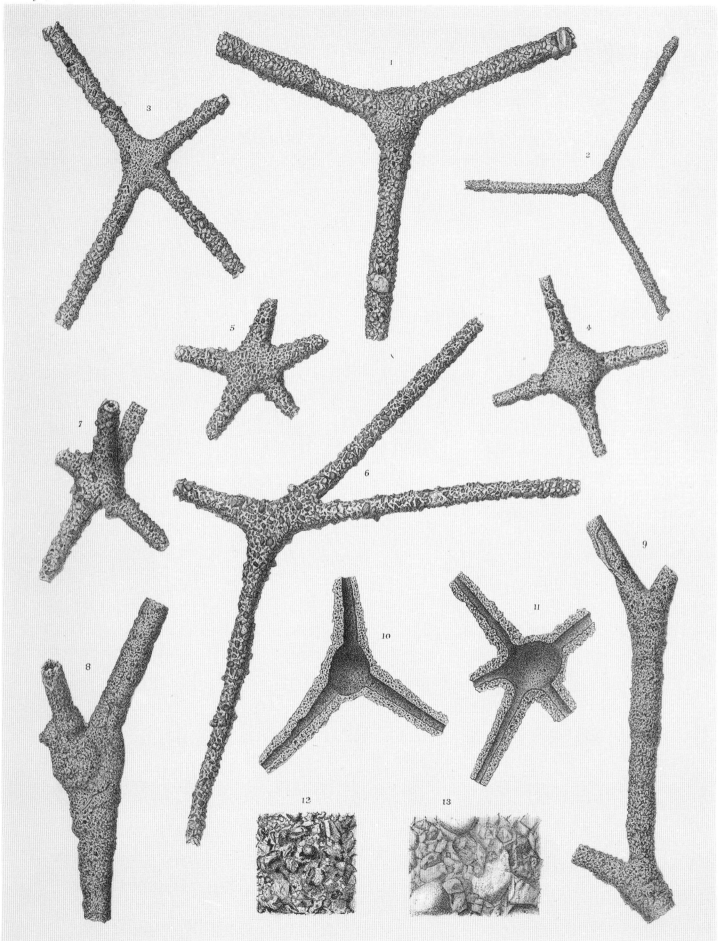

A.T.Hollick ad nat.del.et lith.

Hanhart imp.

RHABDAMMINA.

Plate 22

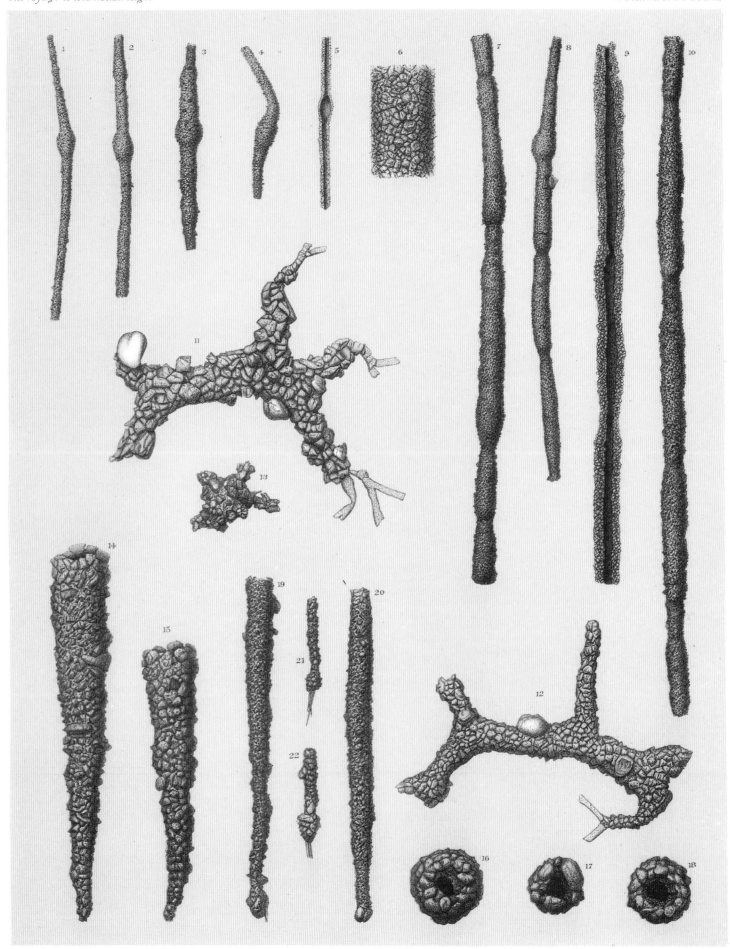

A.T.Hollick ad nat.del.et.lith

Hanhart imp.

RHABDAMMINA_JACULELLA.

Plate 23

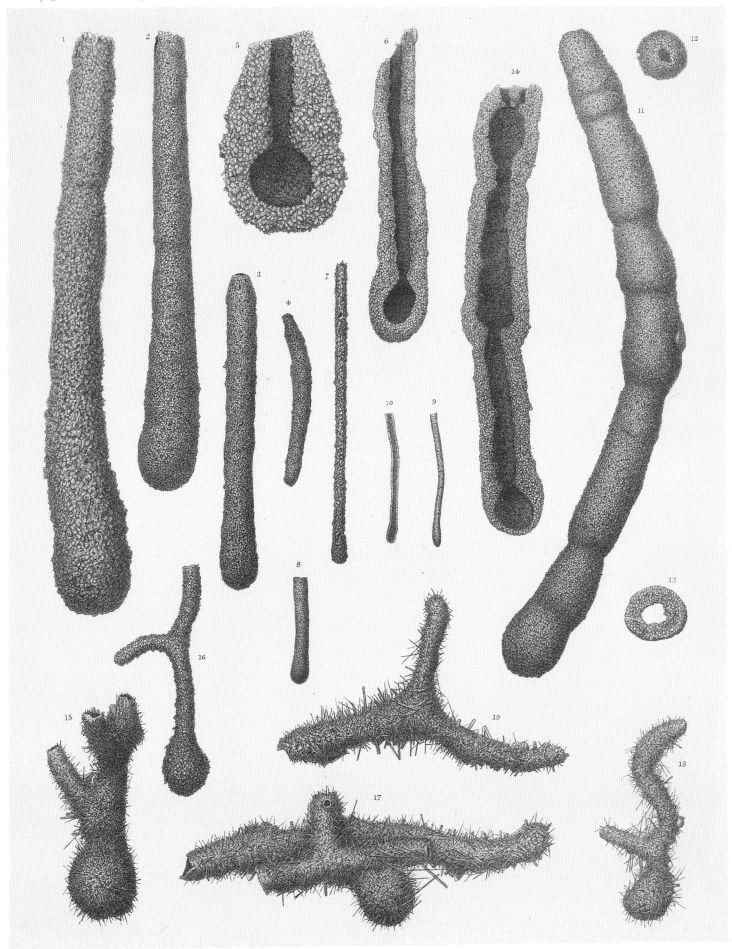

A.T.Hollick ad nat.del.et lith.

Hanhart imp.

HYPERAMMINA.

Plate 24

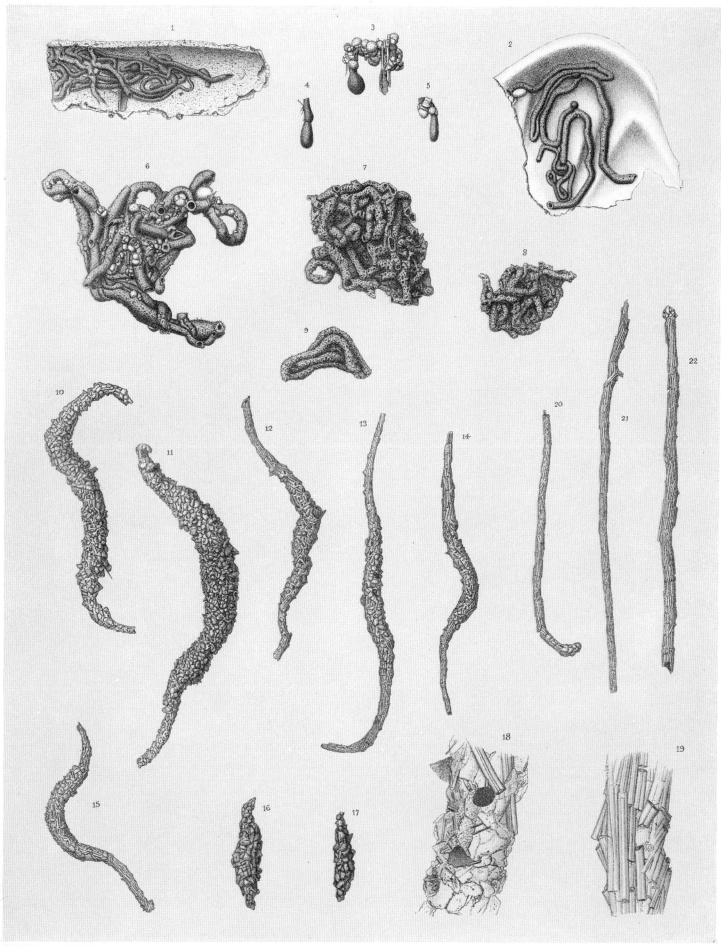

HYPERAMMINA _ MARSIPELLA

Plate 25

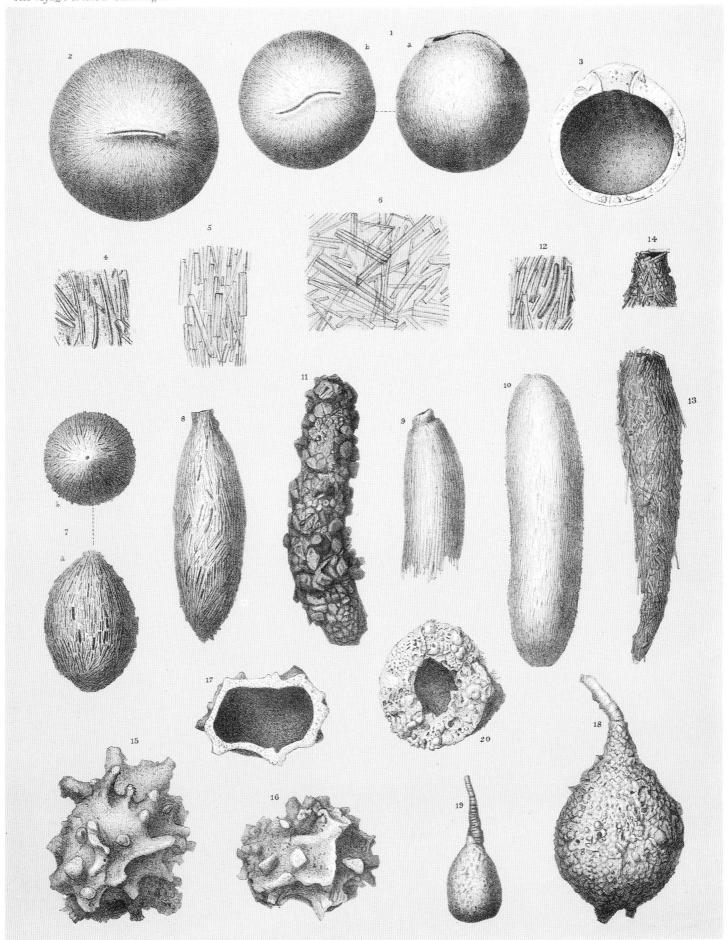

A.T.Hollick ad nat.del.et lith.

Hanhart imp.

PILULINA–TECHNITELLA–STORTHOSPHÆRA–PELOSINA.

Plate 26

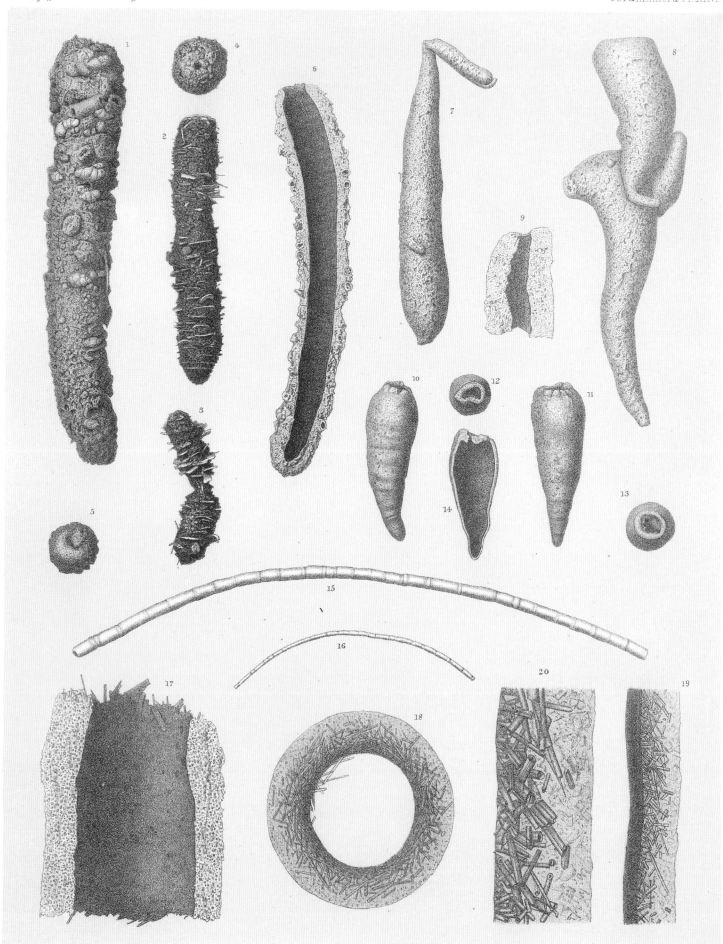

PELOSINA_HIPPOCREPINA_BATHYSIPHON.

Plate 27

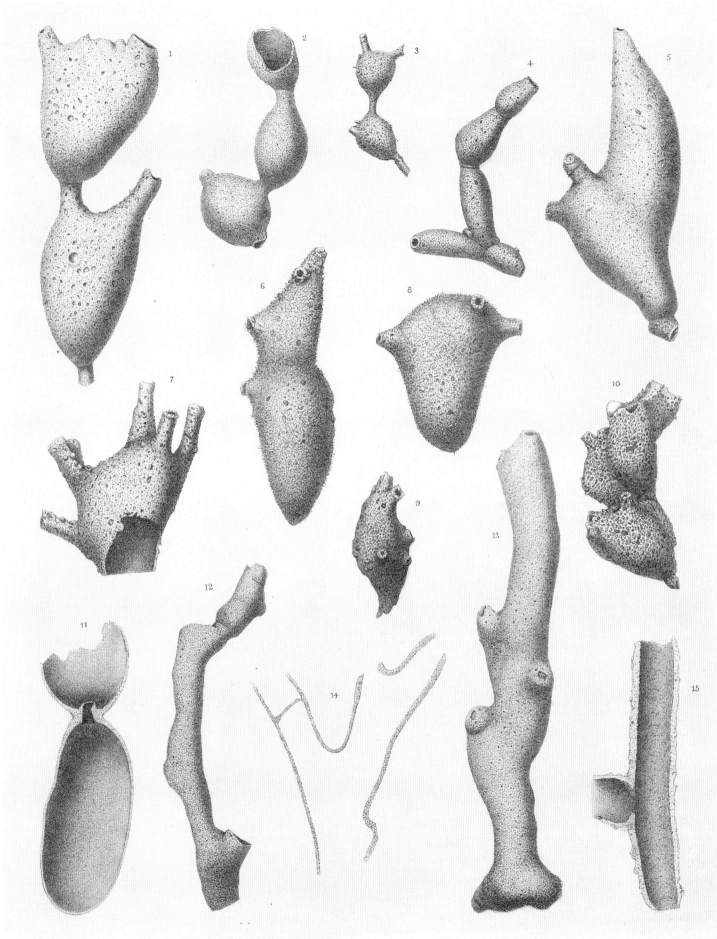

A.T.Hollick ad nat del et lith.

ASCHEMONELLA.

Hanhart imp.

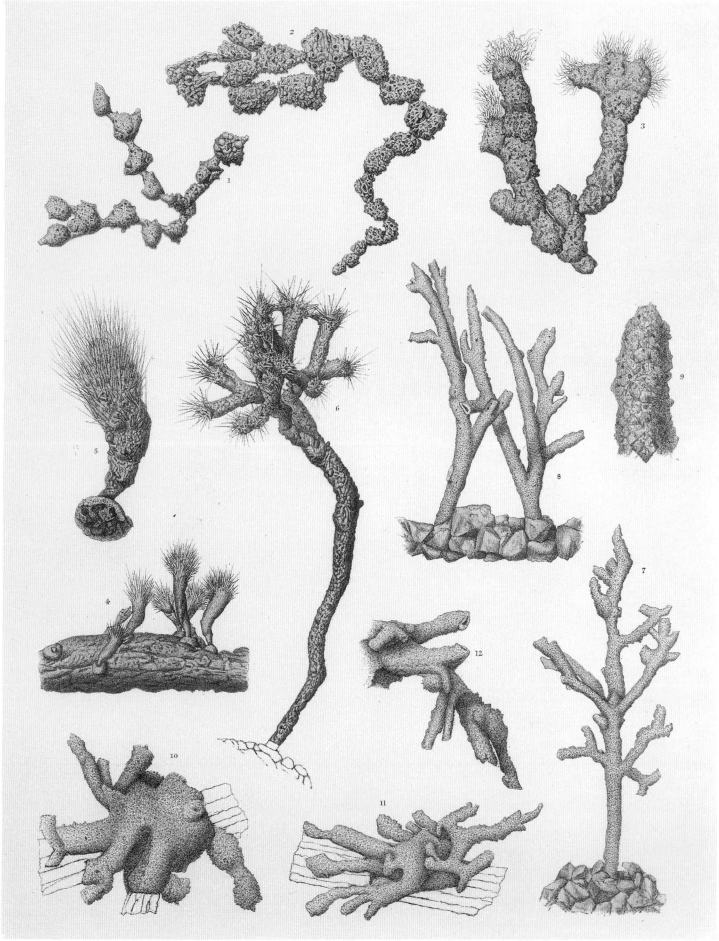

ASCHEMONELLA _ HALIPHYSEMA _ DENDROPHRYA .

Plate 28

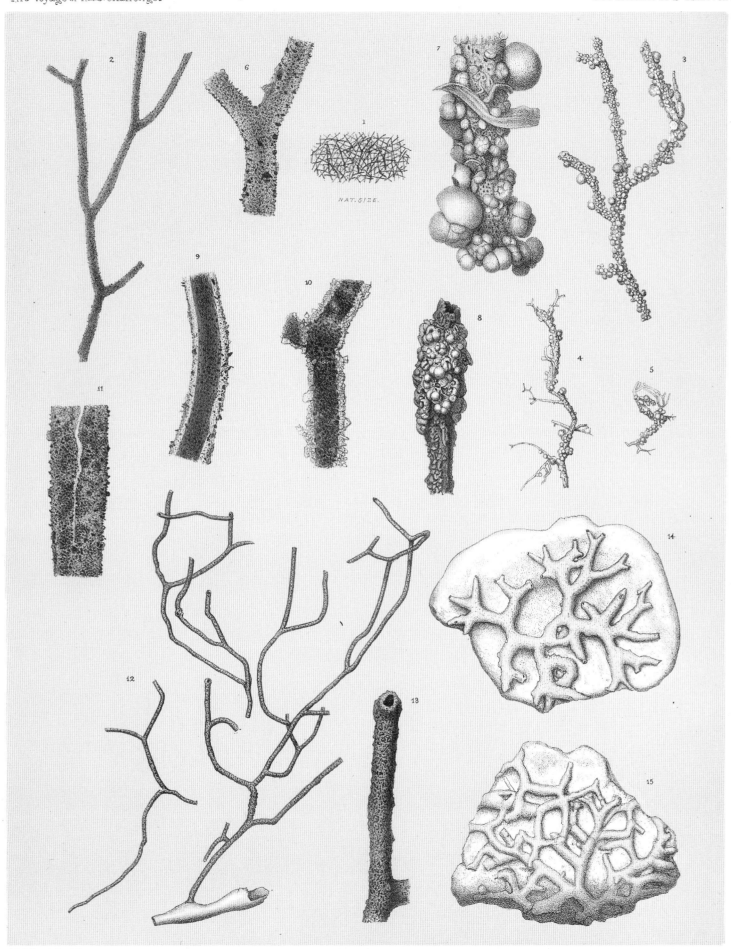

NAT.SIZE.

A.T.Hollick ad.nat.del.et.lith.

Hanhart imp.

RHIZAMMINA_HYPERAMMINA_SAGENELLA.

Plate 29

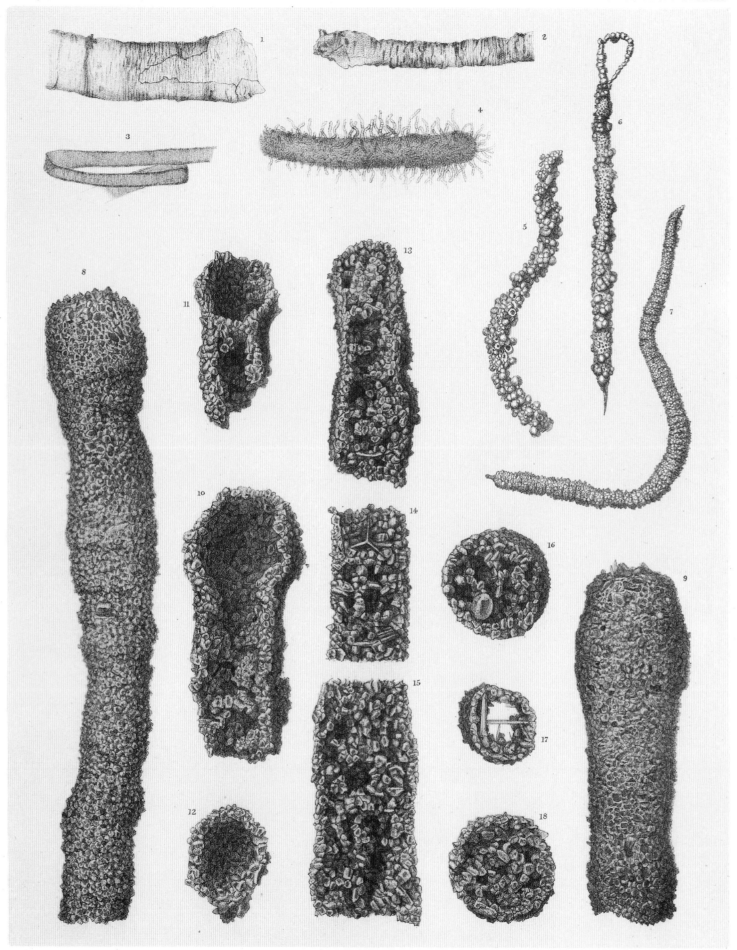

A.T.Hollick ad nat.del.et lith.

Hanhart imp.

RHIZAMMINA_BOTELLINA.

Plate 30

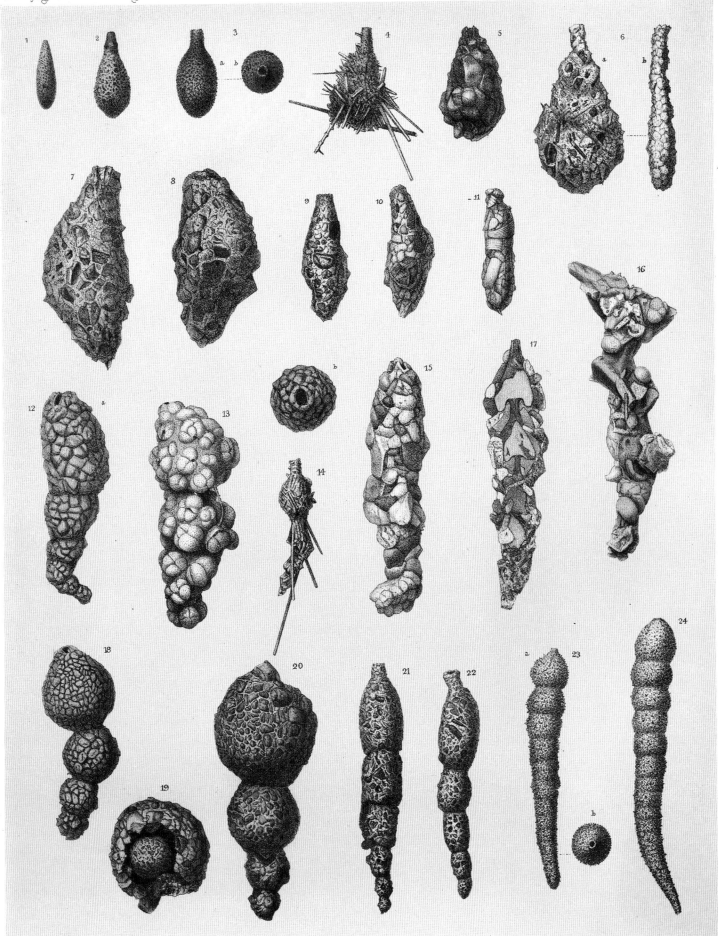

LITUOLA (Reophax)

Plate 31

The Voyage of H.M.S.Challenger

Foraminifera. Pl.XXXI.

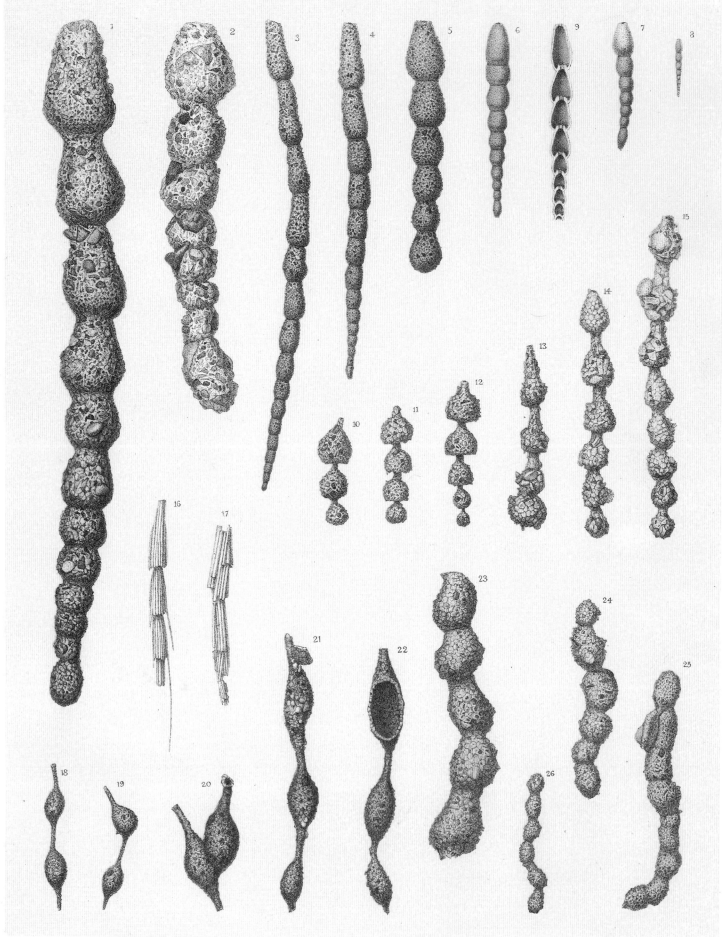

Plate 32

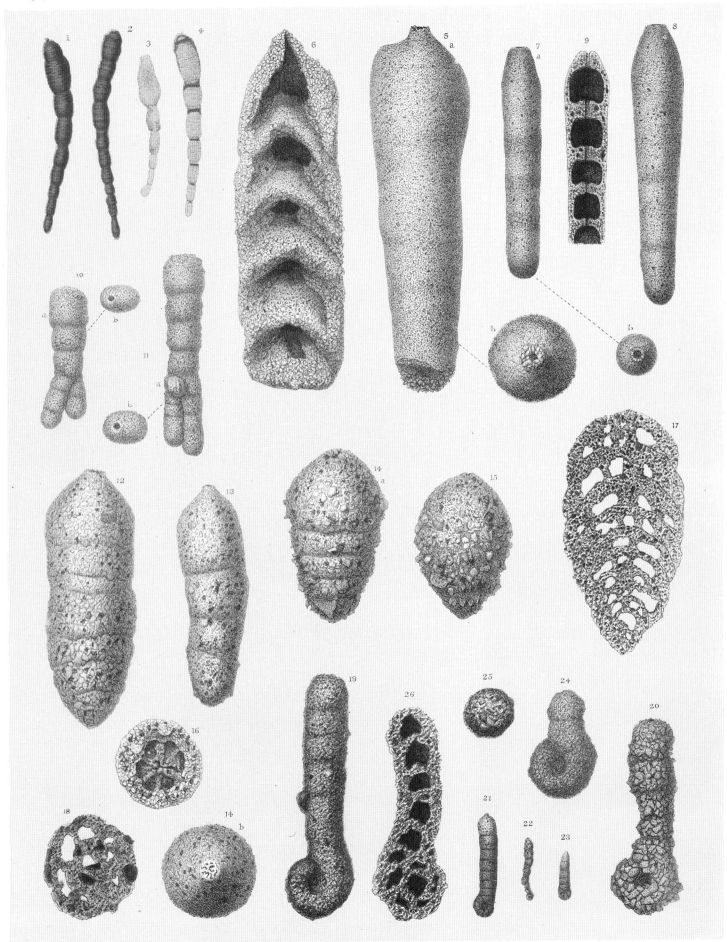

LITUOLA.(Reophax.Haplostiche. Haplophragmium.)

Plate 33

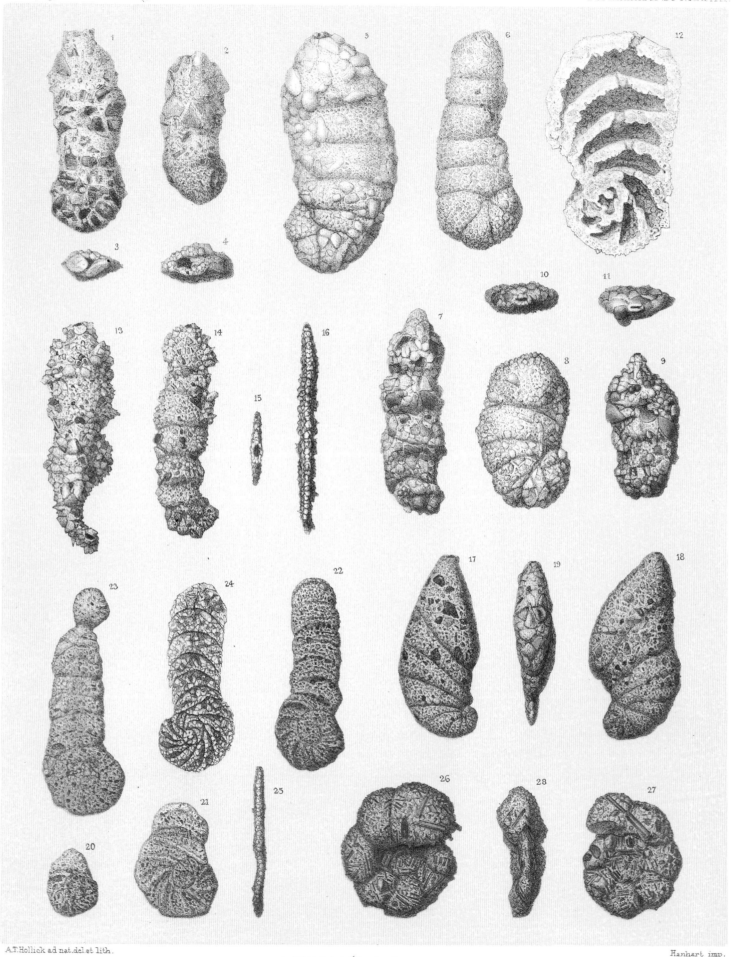

A.T.Hollick ad nat.del.et lith.

Hanhart imp.

LITUOLA (Haplophragmium)

Plate 34

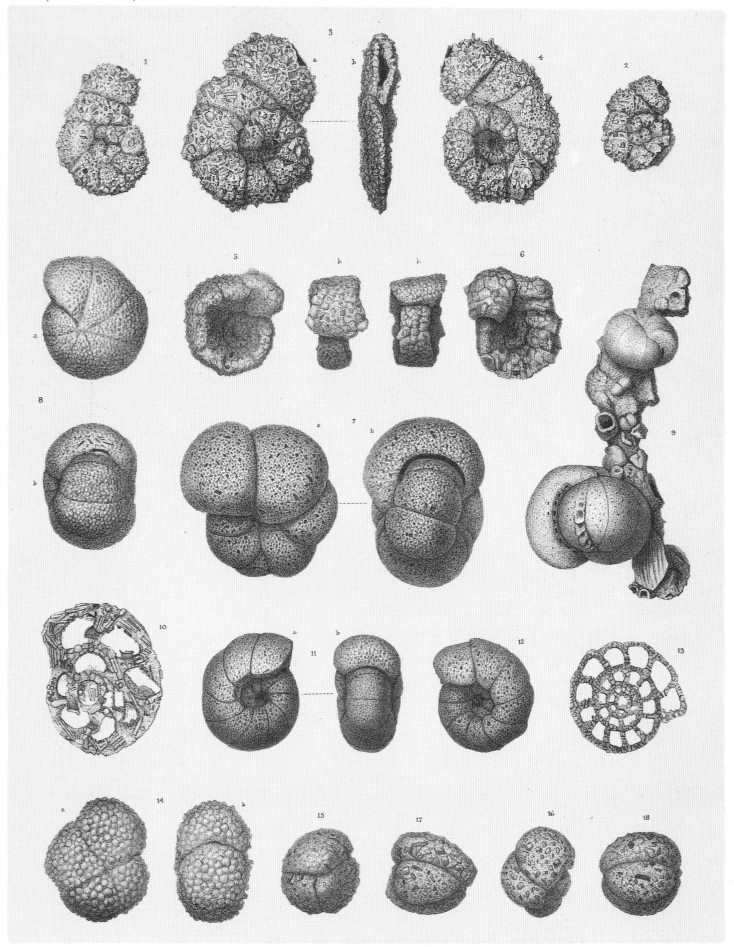

A.T.Hollick ad nat del et lith.

Hanhart imp.

LITUOLA, (Haplophragmium.)

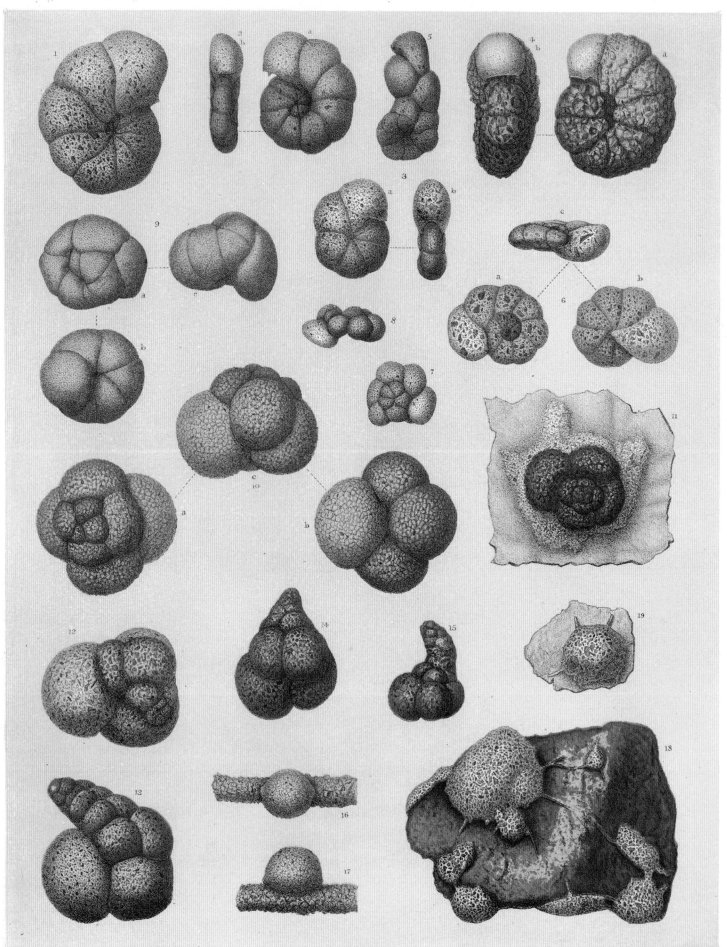

A.T.Hollick ad.nat.del.et.lith. Hanhart imp.

LITUOLA.(Haplophragmium_Placopsilina.)

Plate 36

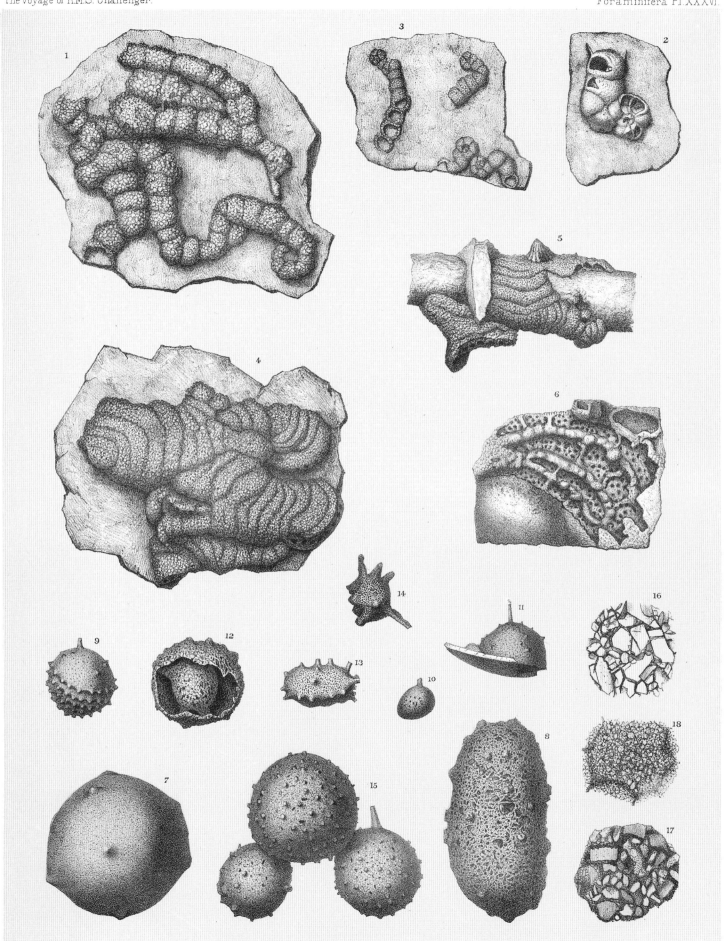

LITUOLA (Placopsilina-Bdelloidina)-THURAMMINA.

Plate 37

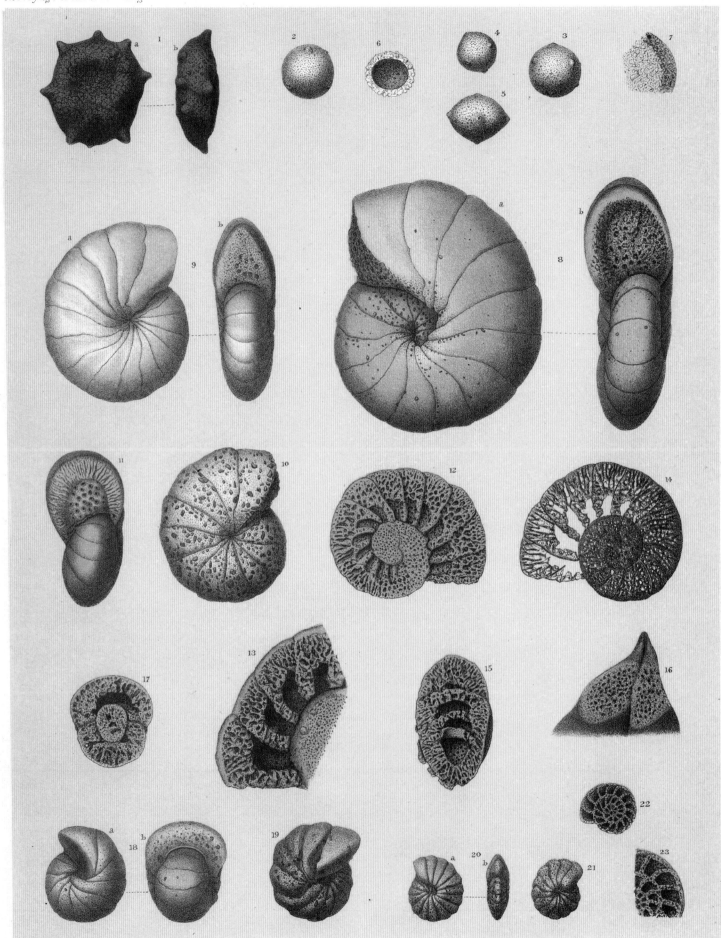

A.T.Hollick ad.nat.del et lith.

Hanhart imp.

THURAMMINA _ CYCLAMMINA.

Plate 38

The Voyage of H.M.S. "Challenger"

Foraminifera Pl.XXXVIII.

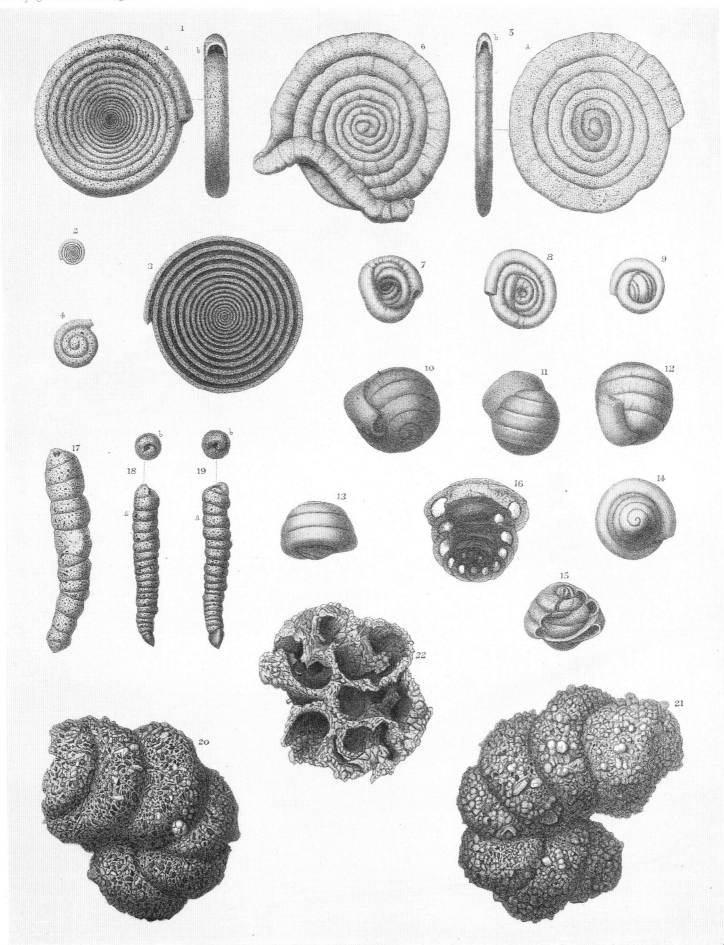

A.T. Hollick ad nat. del. et lith.

Hanhart imp.

TROCHAMMINA.(Ammodiscus)

Plate 39

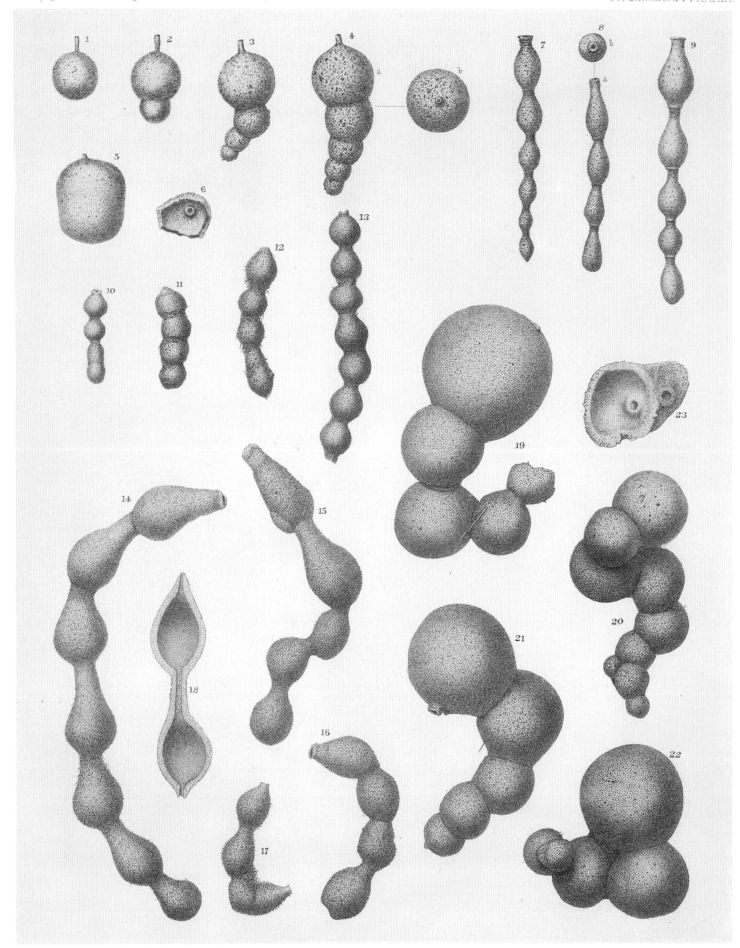

TROCHAMMINA. (Hormosina.)

Plate 40

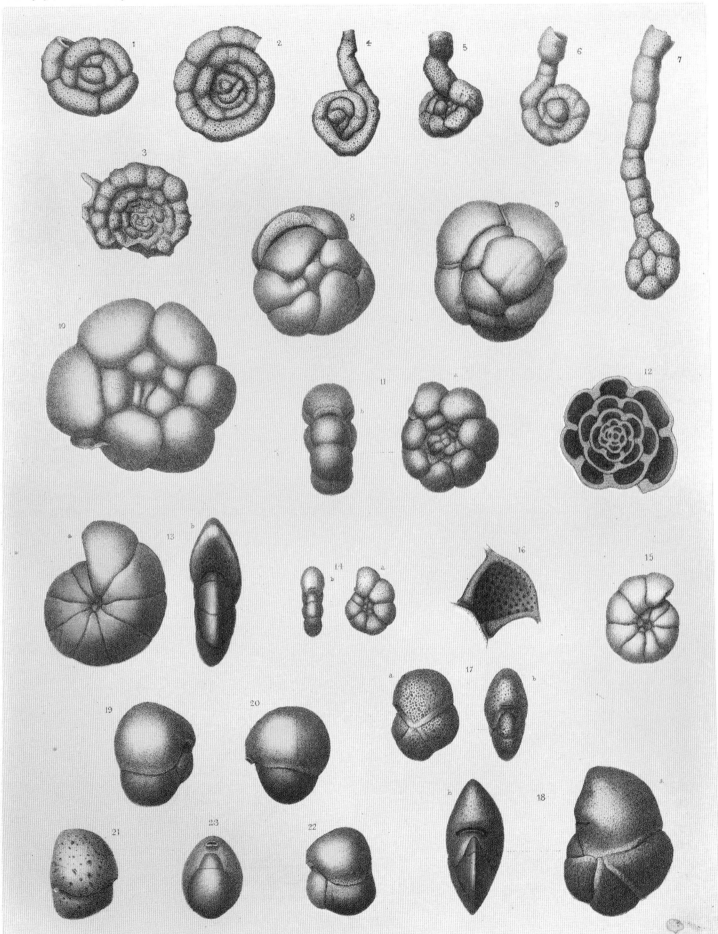

TROCHAMMINA.

Plate 41

The Voyage of H.M.S."Challenger".

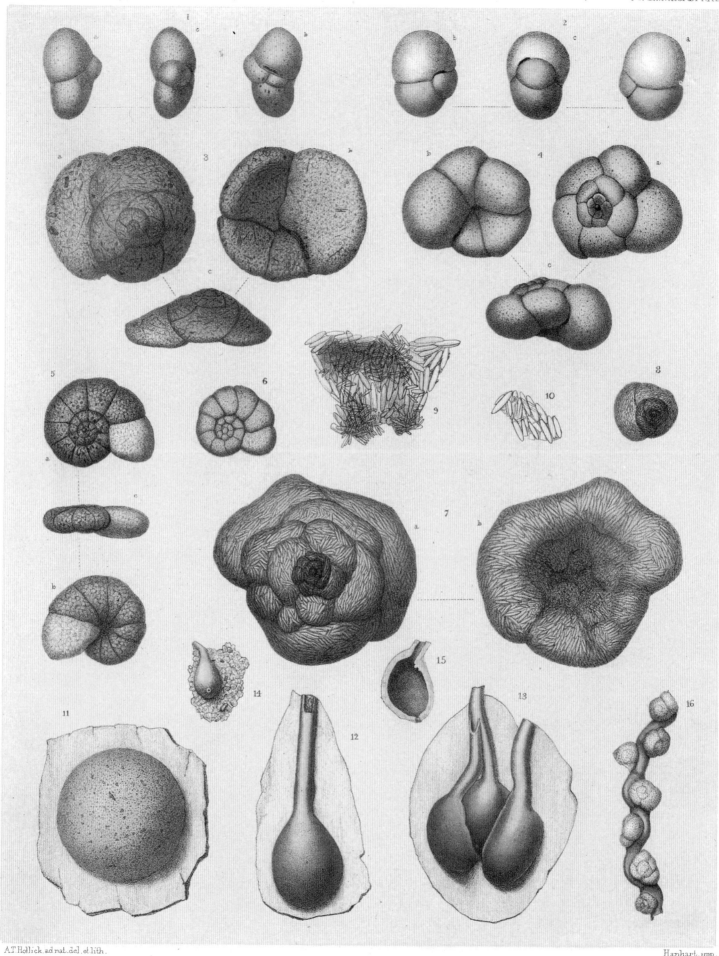

A.T.Hollick.ad nat.del.et lith.

Hanhart.imp.

TROCHAMMINA. (Trochammina_Carterina_Webbina.)

Plate 42

TEXTULARIA.

Plate 43

The Voyage of H.M.S. "Challenger"

Foraminifera Pl. XLIII

Plate 44

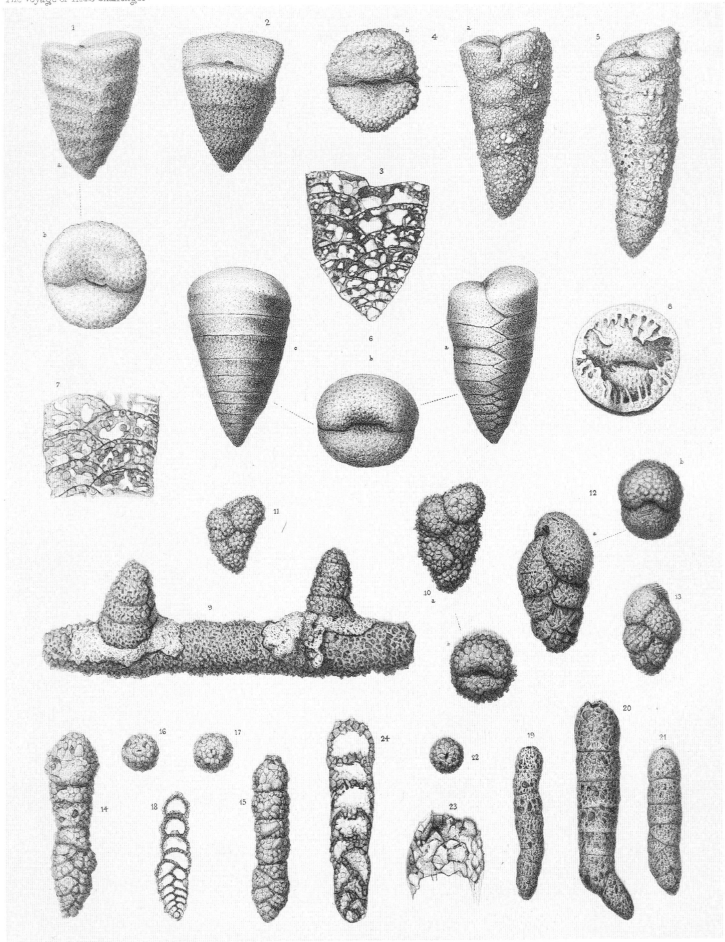

A.T Hollick ad.nat del et lith.

Hanhart imp

TEXTULARIA. (Textularia Bigenerina.)

Plate 45

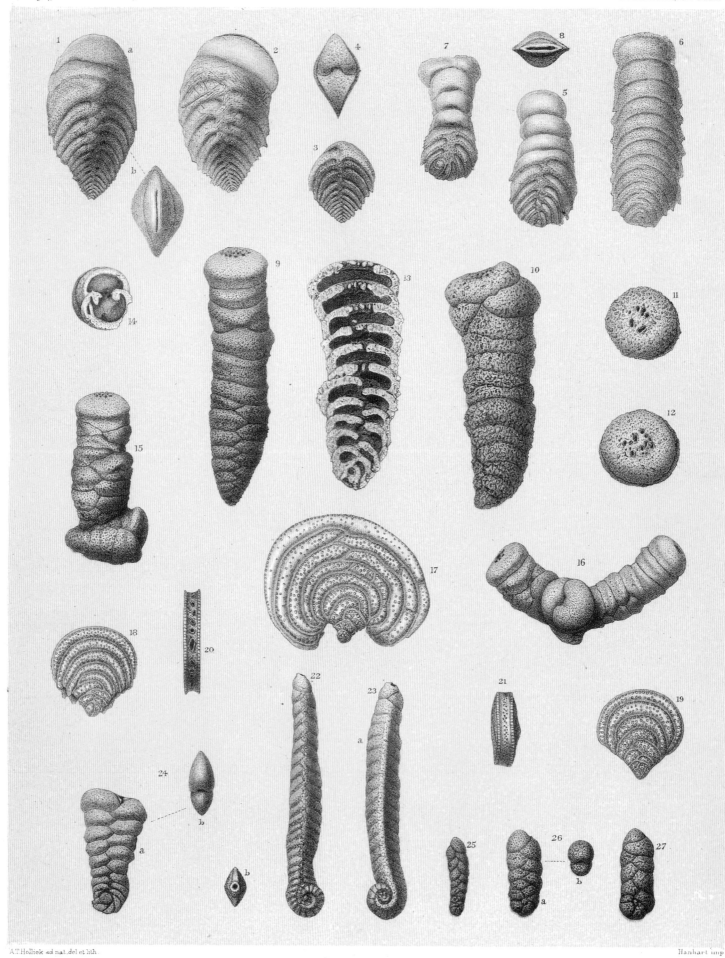

A.T.Hollick ad nat.del et lith.

Hanhart imp

TEXTULARIA (Bigenerina_Pavonina_Spiroplecta.)

Plate 46

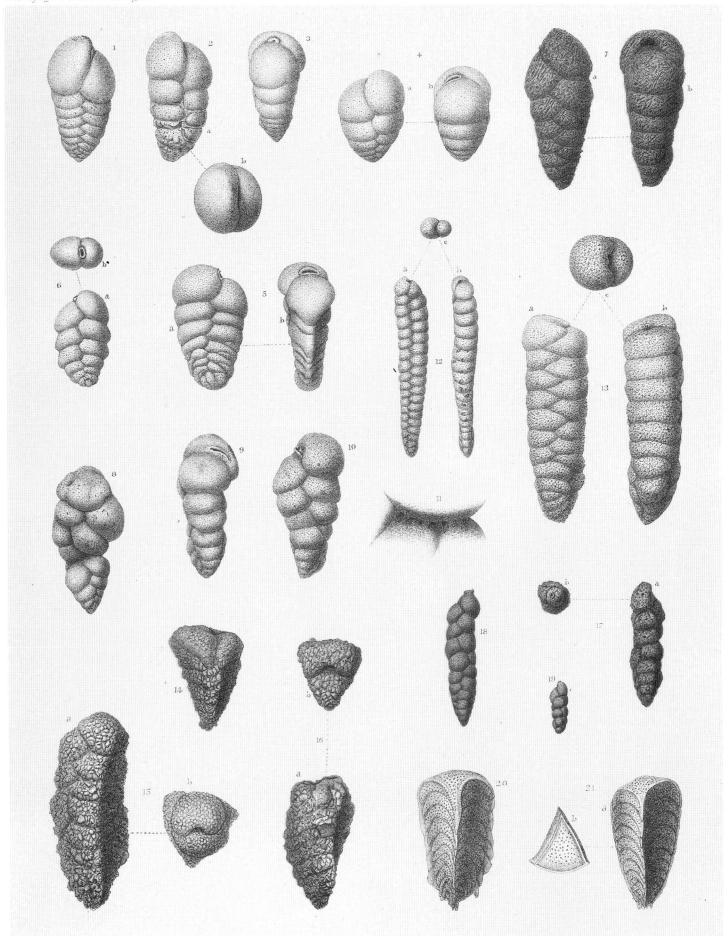

A.T.Hollick ad nat.del.et lith.

Hanhart imp.

TEXTULARIA, (Gaudryina _ Chrysalidina.)

Plate 47

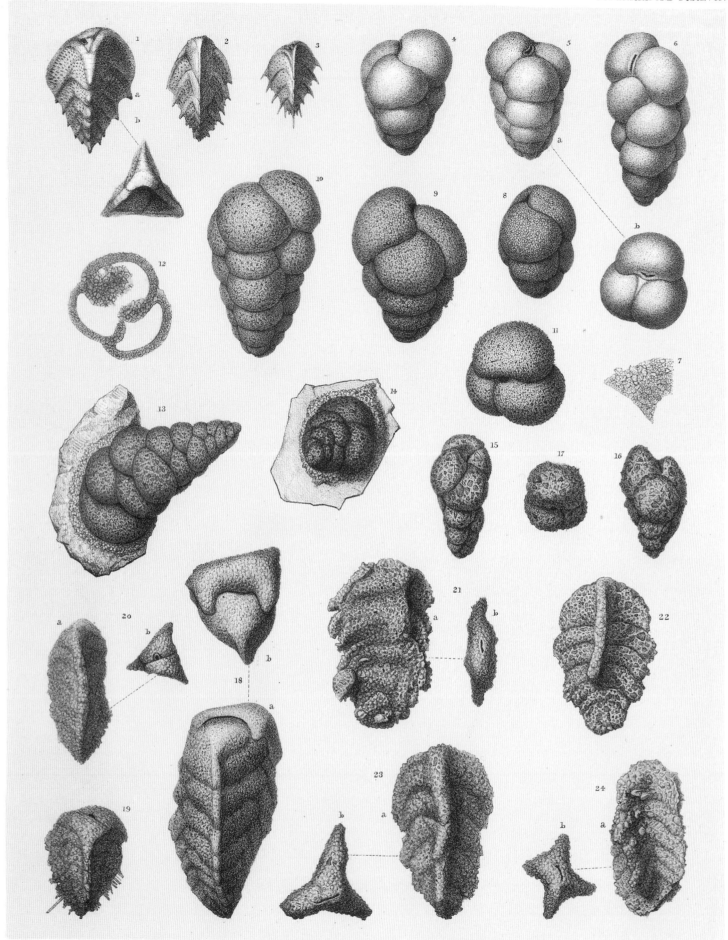

A.T.Hollick ad.nat.del.et lith.

Hanhart imp.

TEXTULARIA (Verneuilina.)

Plate 48

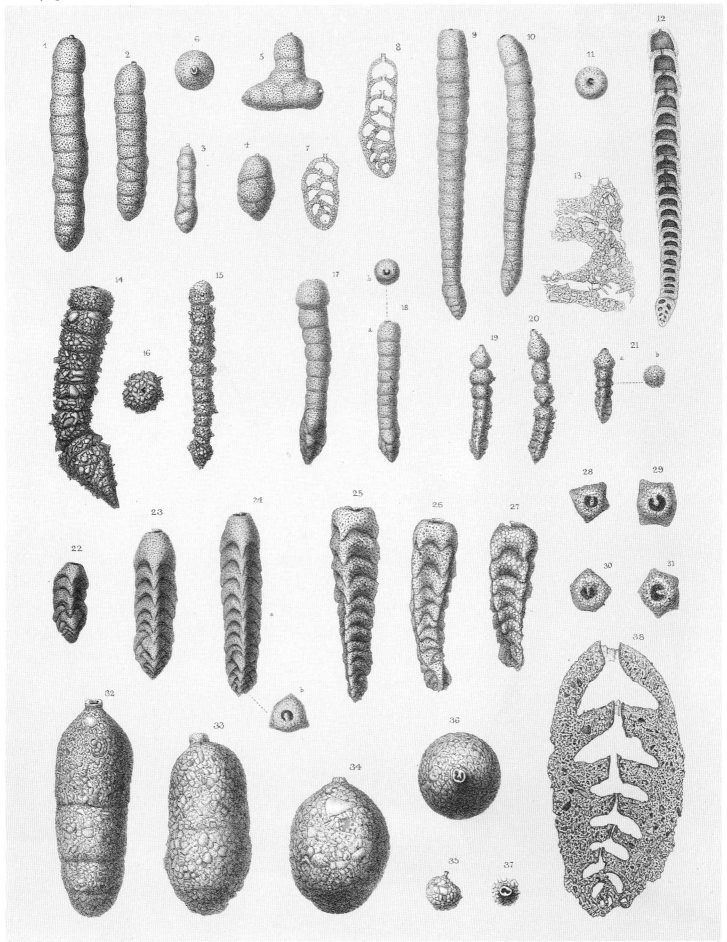

VALVULINA (Clavulina)

Plate 49

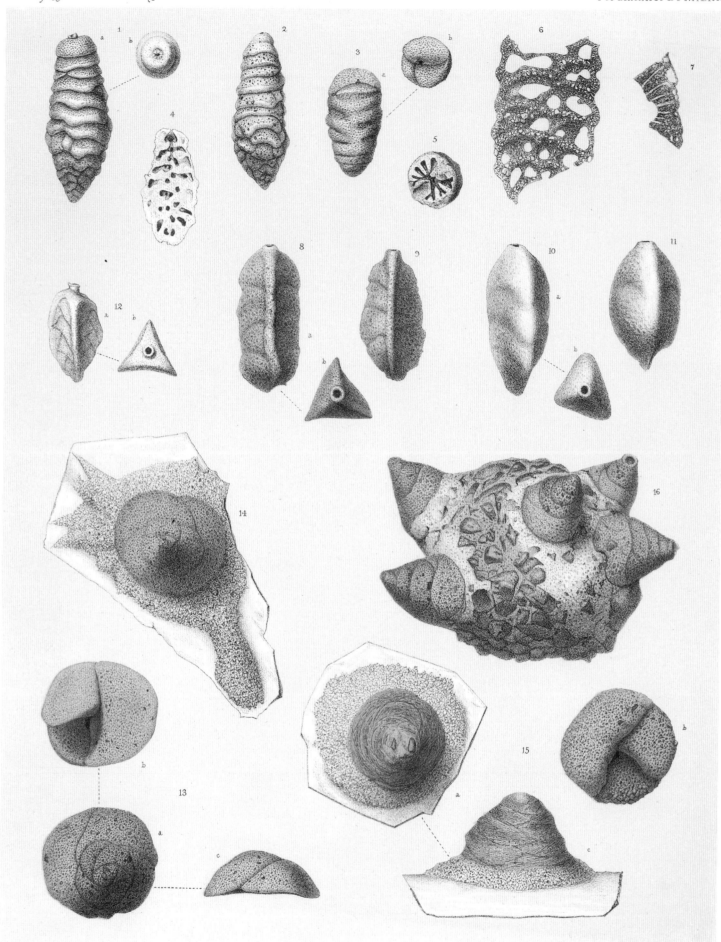

TEXTULARIA (Tritaxia)_VALVULINA.

Plate 50

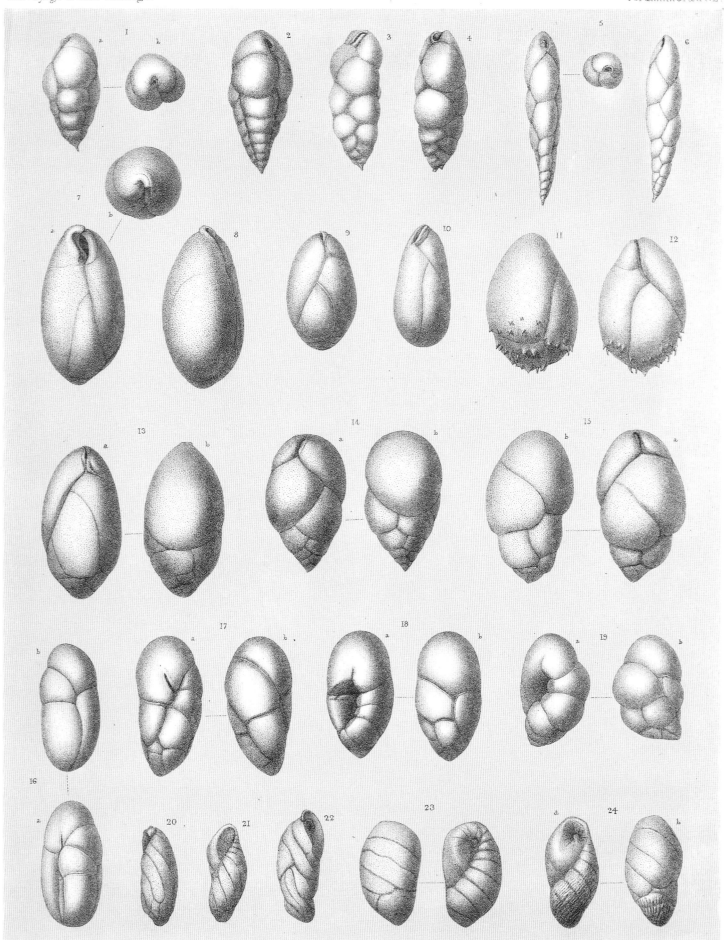

A.J. Hollick ad nat. del et lith.

BULIMINA.

Hanhart imp.

Plate 51

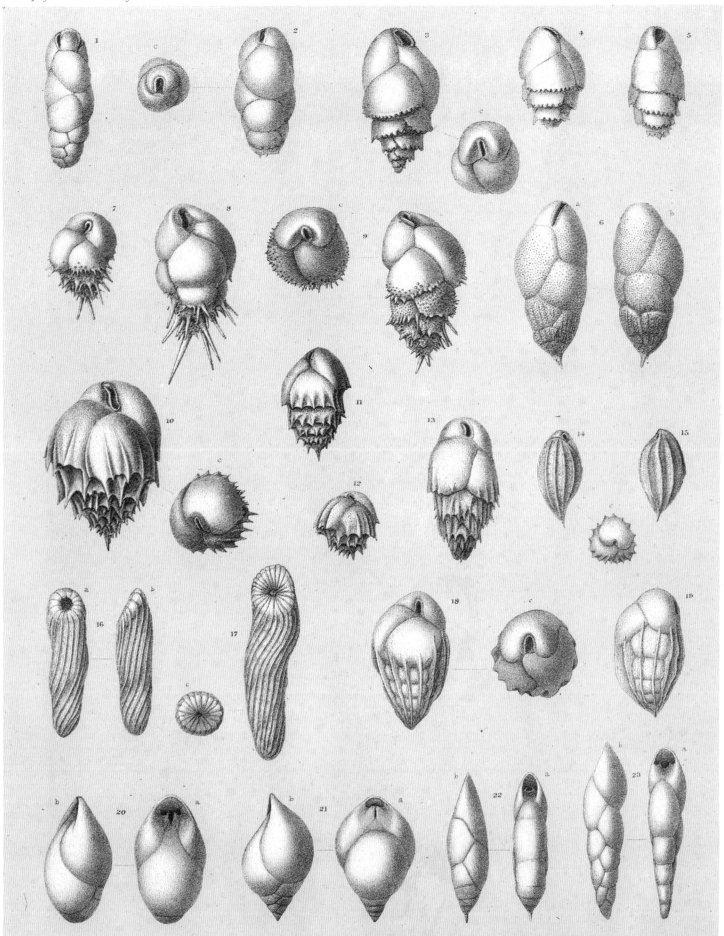

BULIMINA, (Bulimina_Pleurostomella).

Plate 52

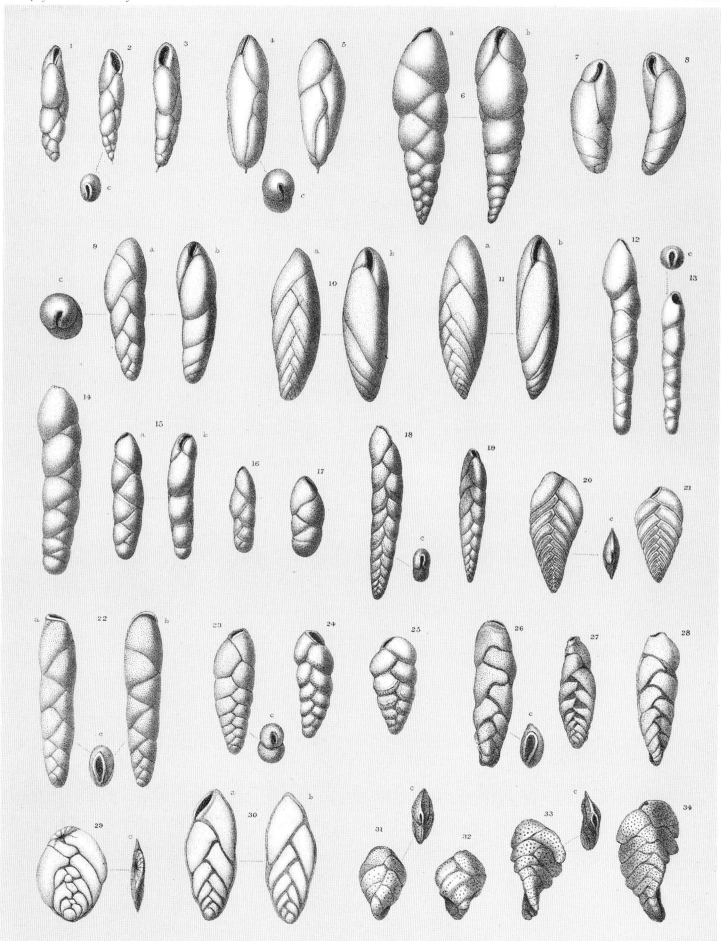

A.T.Hollick ad nat.del.et lith.

BULIMINA, (Virgulina_Bolivina).

Hanhart imp.

Plate 53

A.T.Hollick ad nat.del.et lith.

Hanhart imp.

BULIMINA, (Bolivina).

Plate 54

The Voyage of H.M.S. "Challenger"

Foraminifera Pl.LIV.

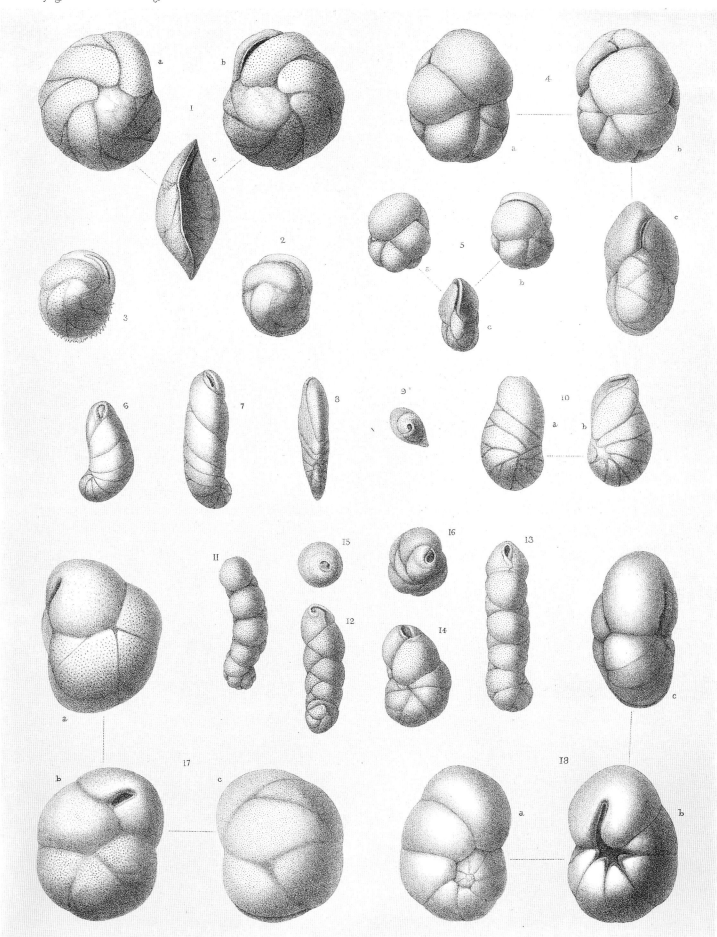

Plate 55

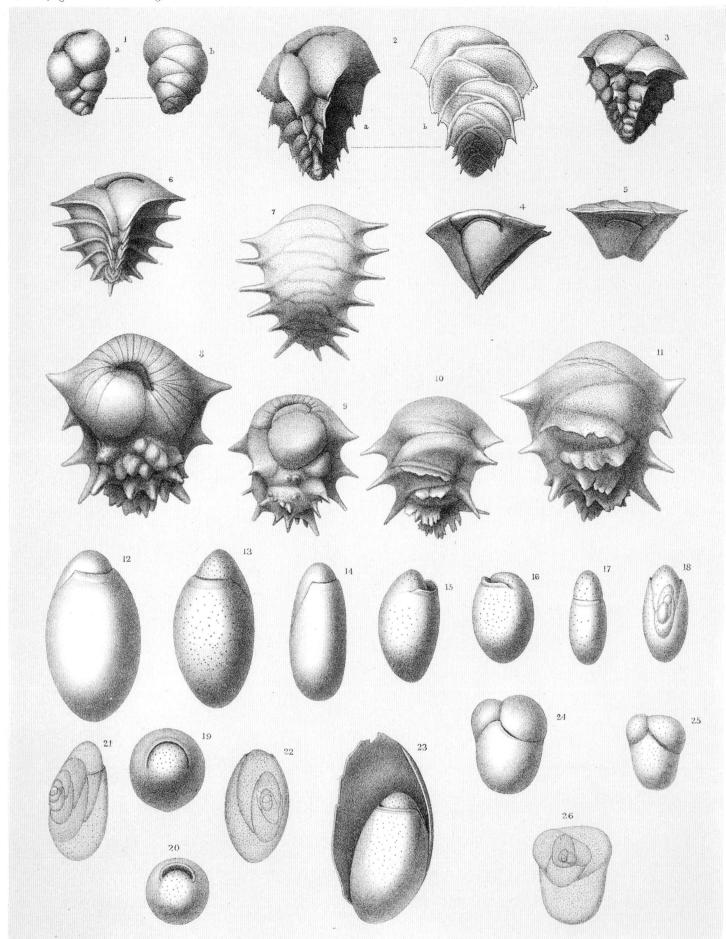

A.T.Hollick, ad nat del et lith.

Hanhart imp

EHRENBERGINA— CHILOSTOMELLA—ALLOMORPHINA.

Plate 56

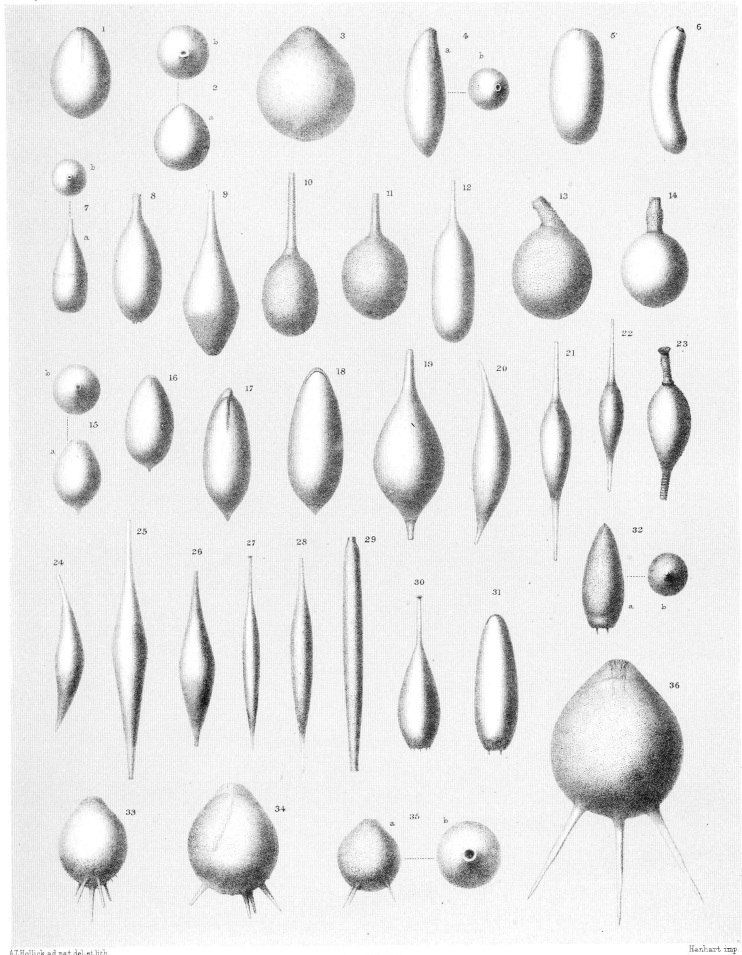

LAGENA.

Plate 57

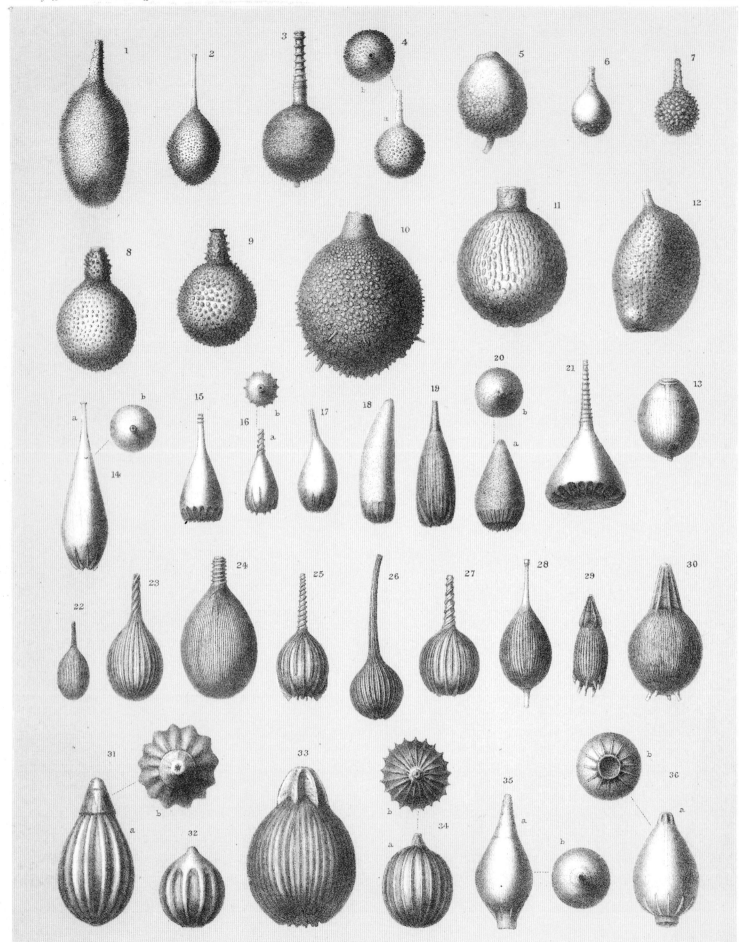

LAGENA.

Plate 58

LAGENA.

Plate 59

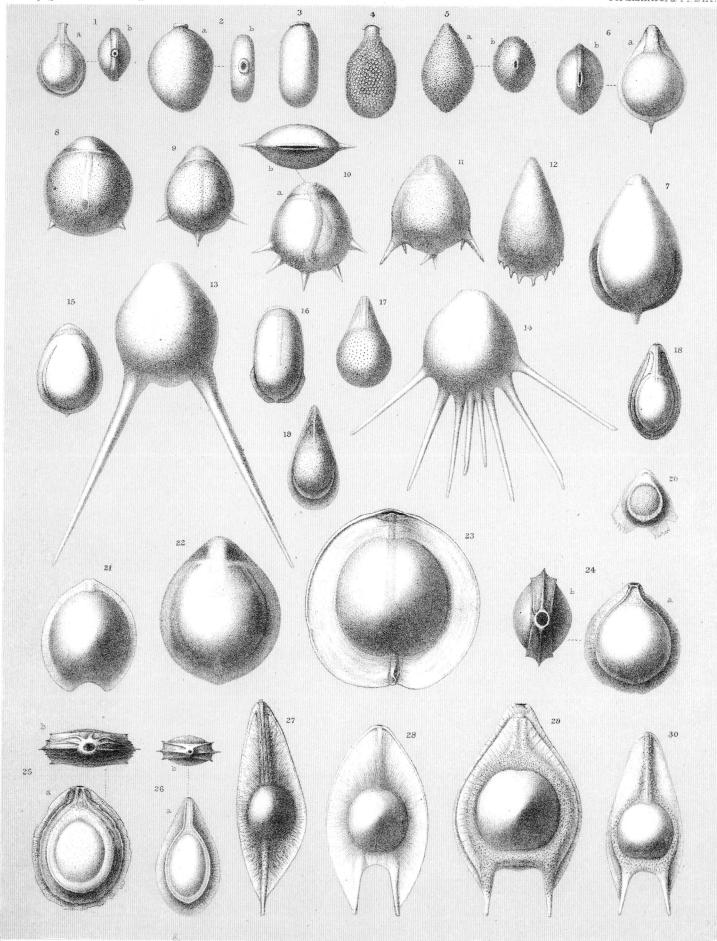

LAGENA.

Plate 60

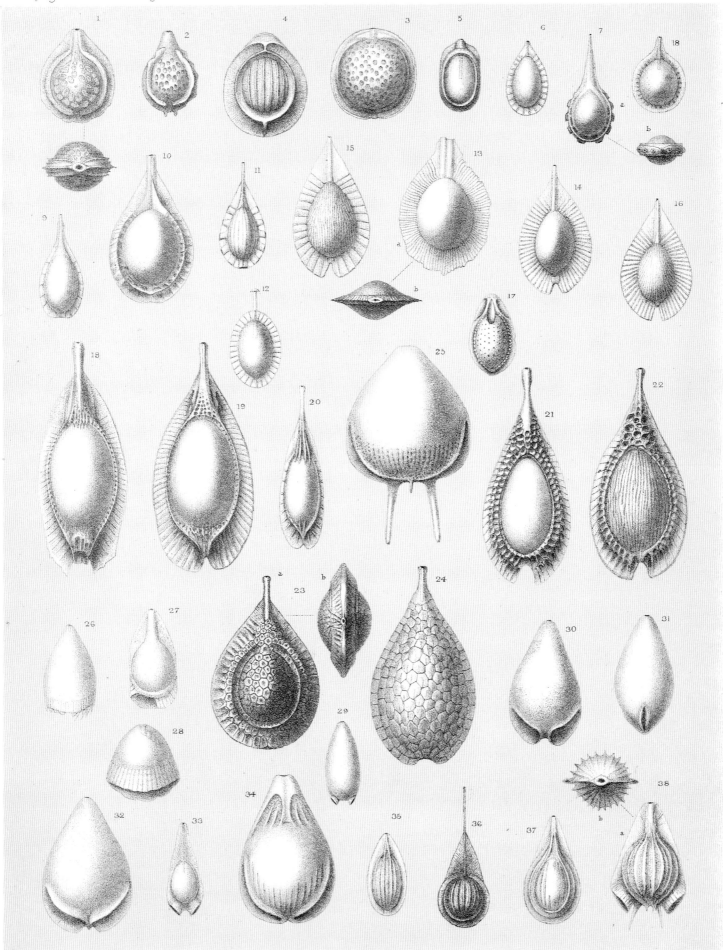

A.T.Hollick ad nat. del.et lith. Hanhart imp.

LAGENA.

Plate 61

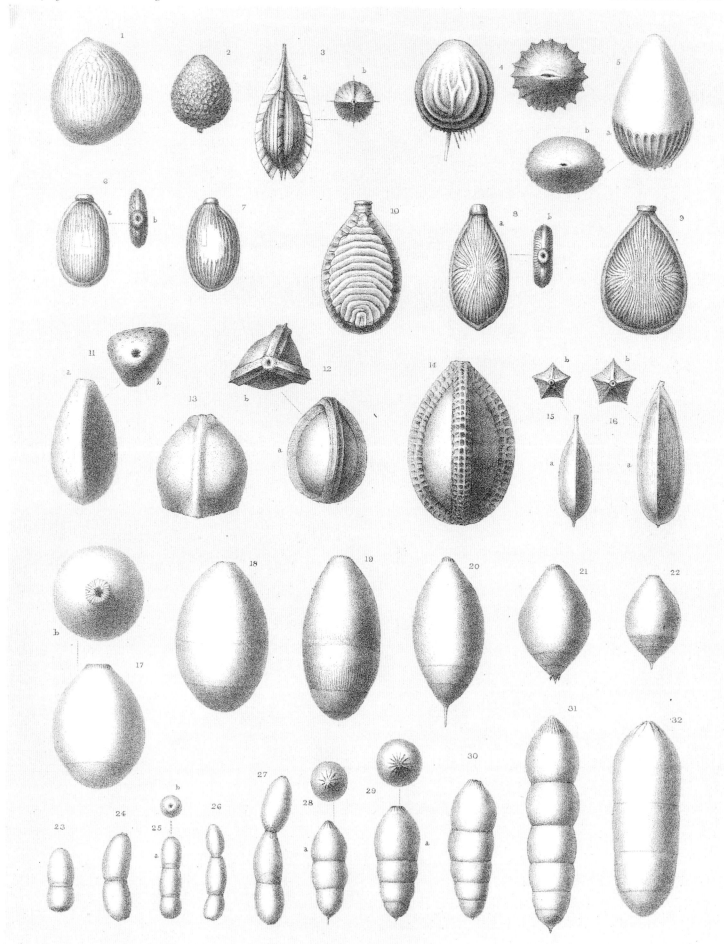

LAGENA _ NODOSARIA.

Plate 62

The Voyage of H.M.S."Challenger."

Foraminifera Pl. LXII.

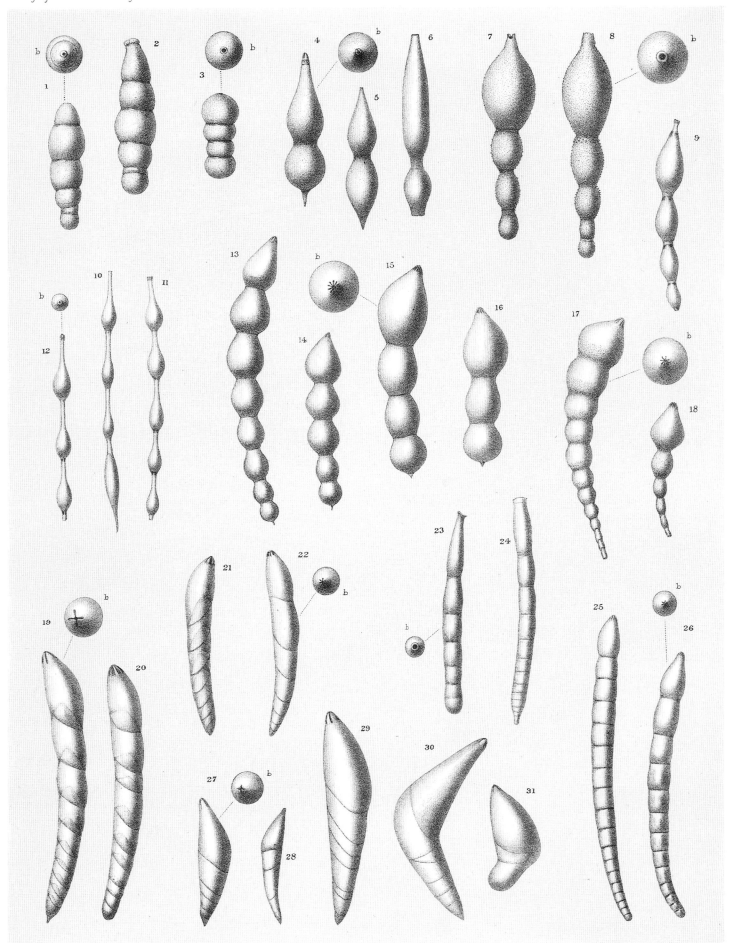

A.T.Hollick ad nat.del.et lith.

Hanhart imp.

NODOSARIA.

Plate 63

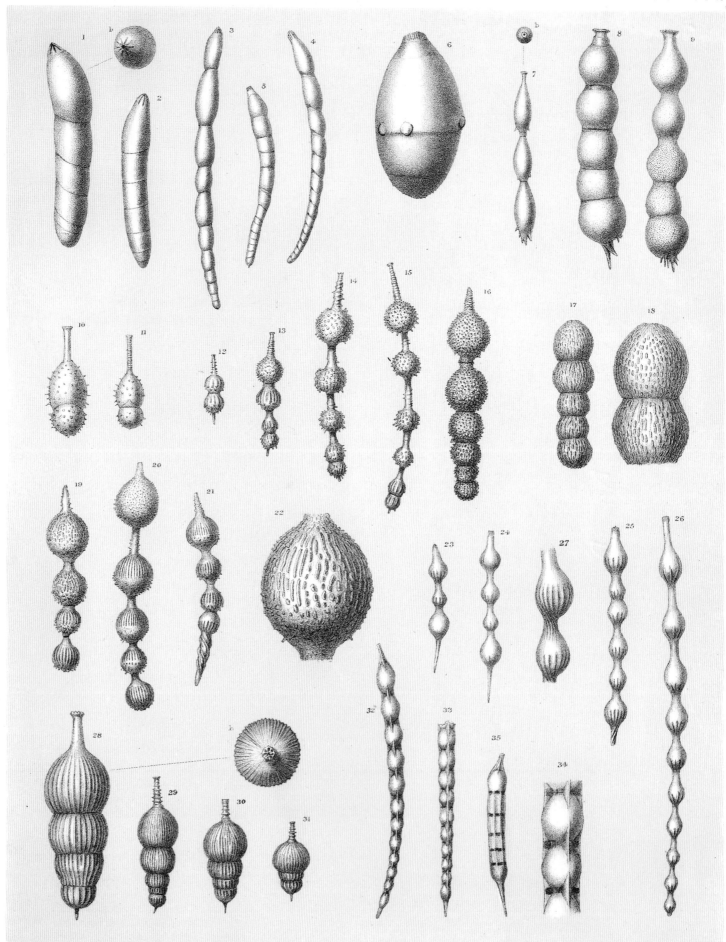

A.T.Hollick ad nat. del. et lith.

Hanhart imp.

NODOSARIA.

Plate 64

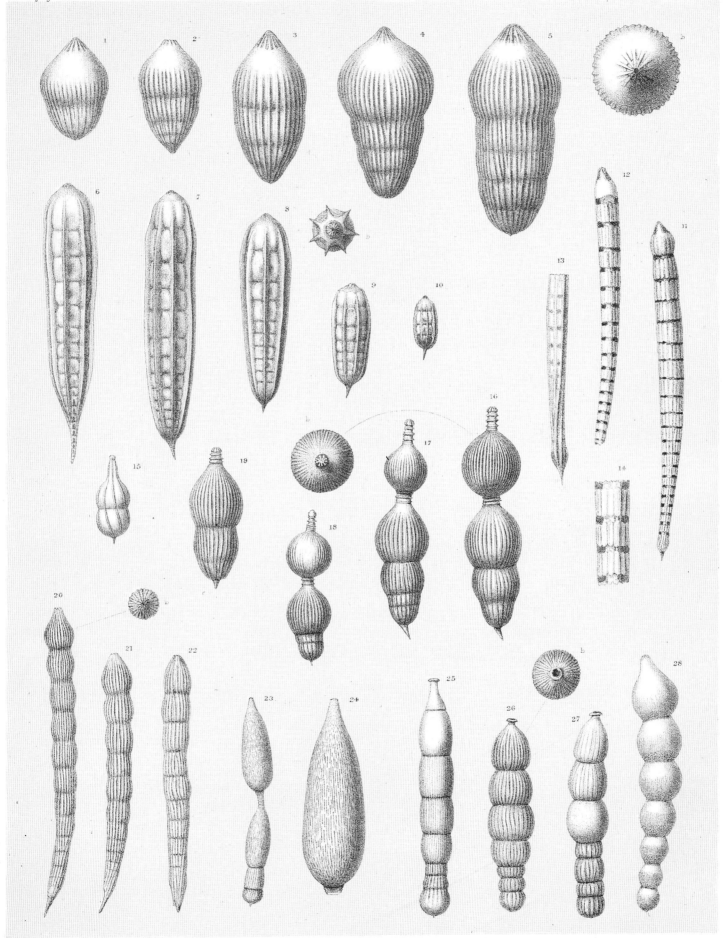

NODOSARIA.

Plate 65

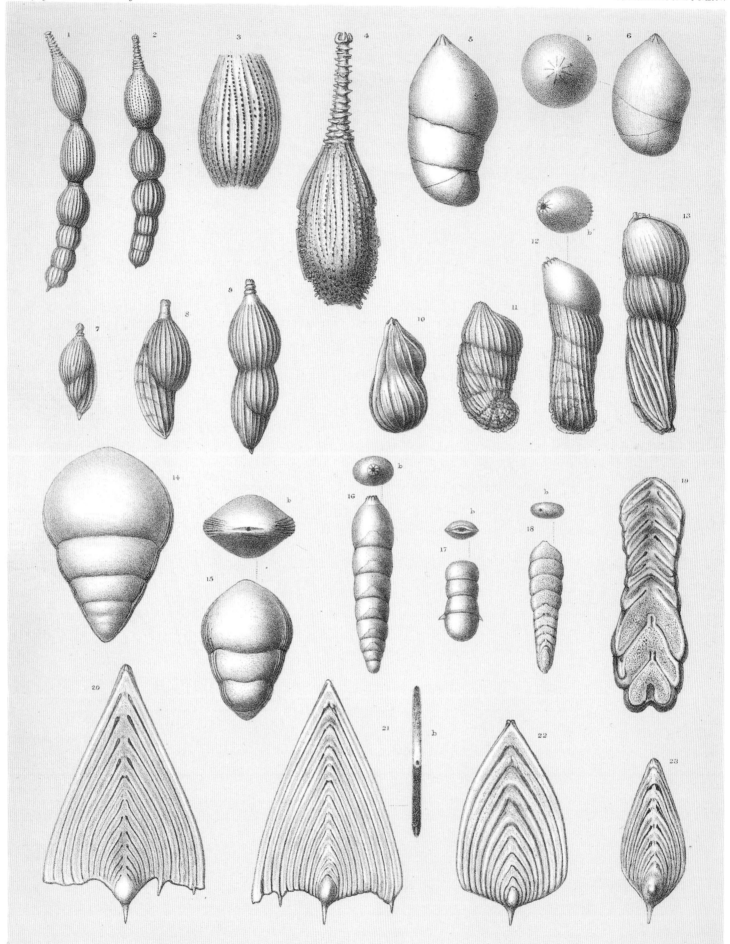

NODOSARIA _ MARGINULINA _ LINGULINA _ FRONDICULARIA.

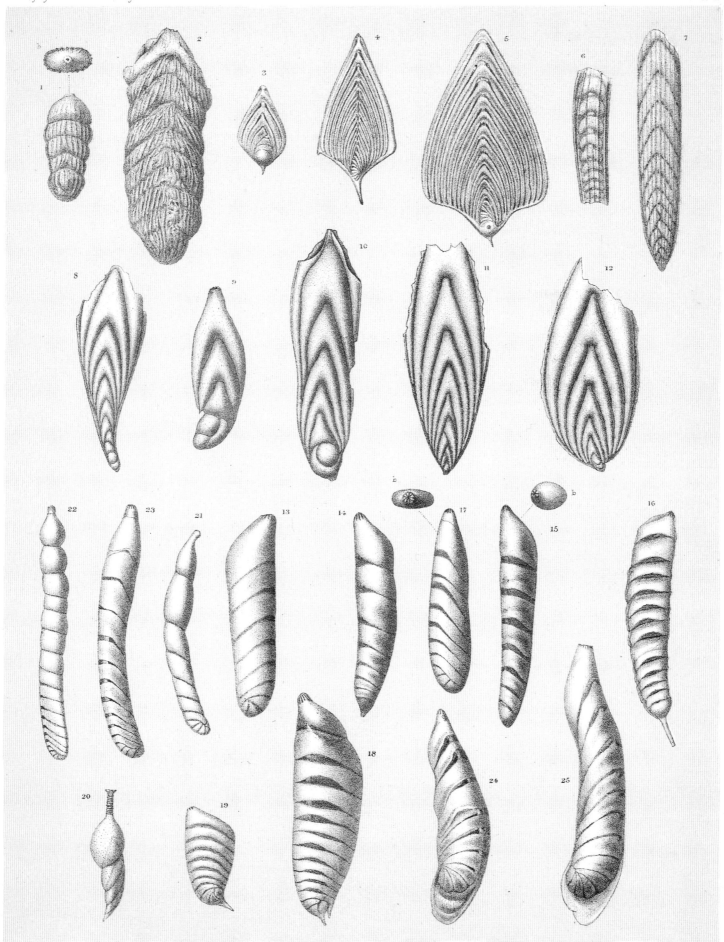

A.T.Hollick ad nat.del.et lith.

Hanhart imp.

FRONDICULARIA, (Flabellina)._VAGINULINA_CRISTELLARIA.

Plate 67

RHABDOGONIUM—VACINULINA—CRISTELLARIA.

CRISTELLARIA.

Plate 69

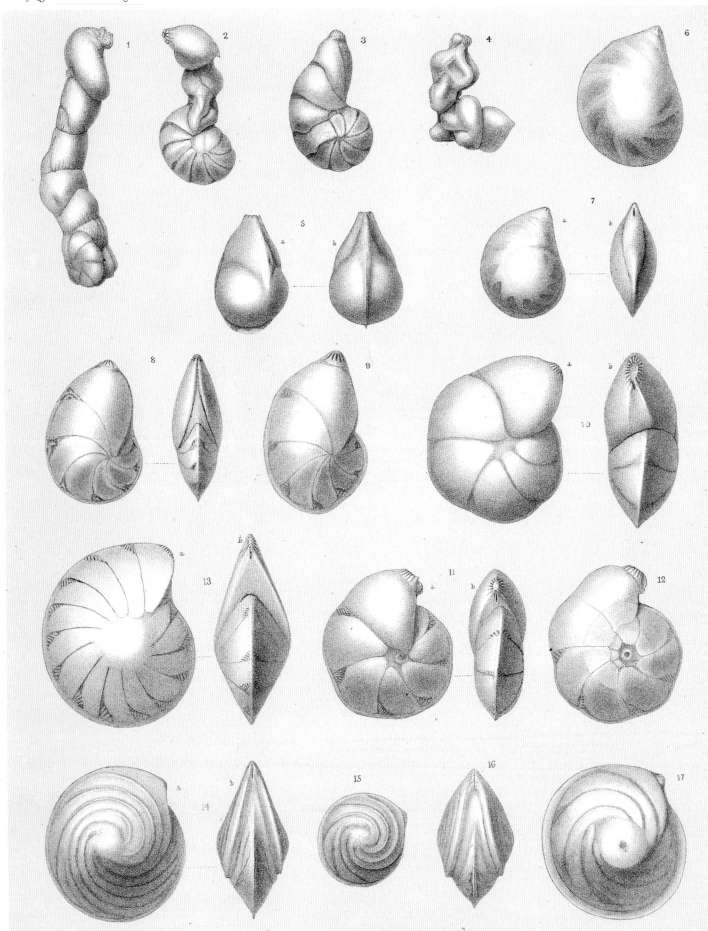

CRISTELLARIA

Plate 70

CRISTELLARIA.

Plate 71

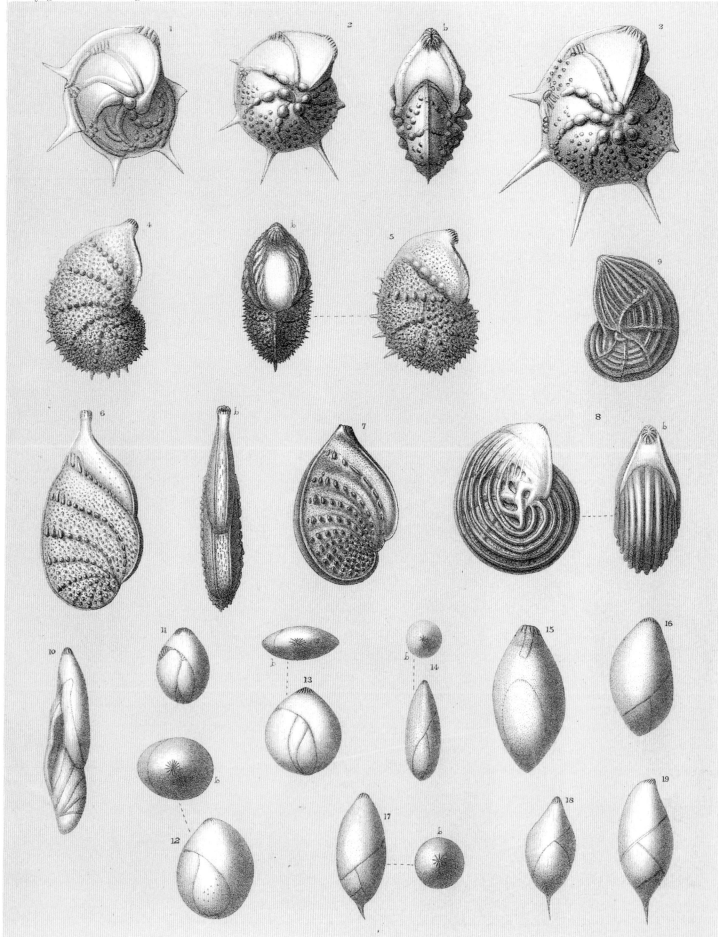

A.T.Hollick ad.nat.del.et lith.

Hanhart imp.

CRISTELLARIA-POLYMORPIHNA

Plate 72

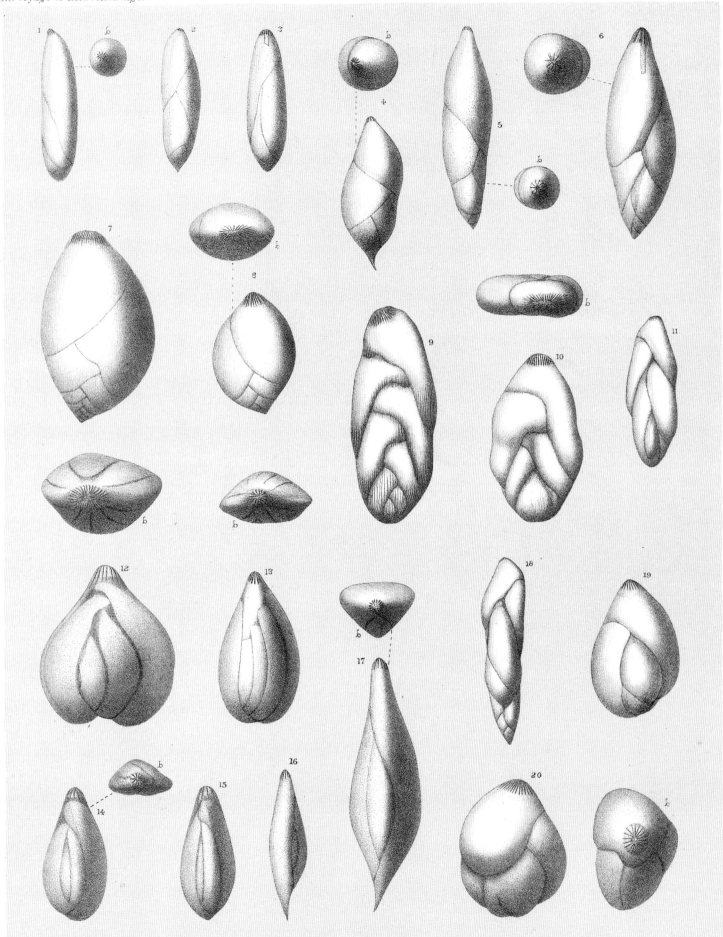

POLYMORPHINA

Plate 73

The Voyage of H.M.S. "Challenger"

Foraminifera Pl. LXXIII

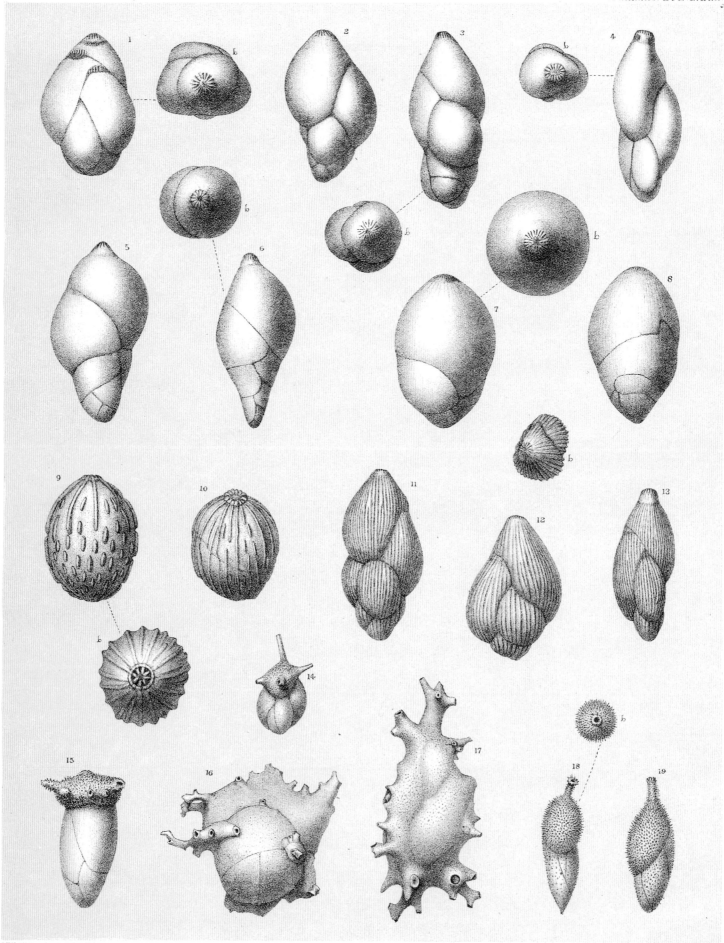

POLYMORPHINA

Plate 74

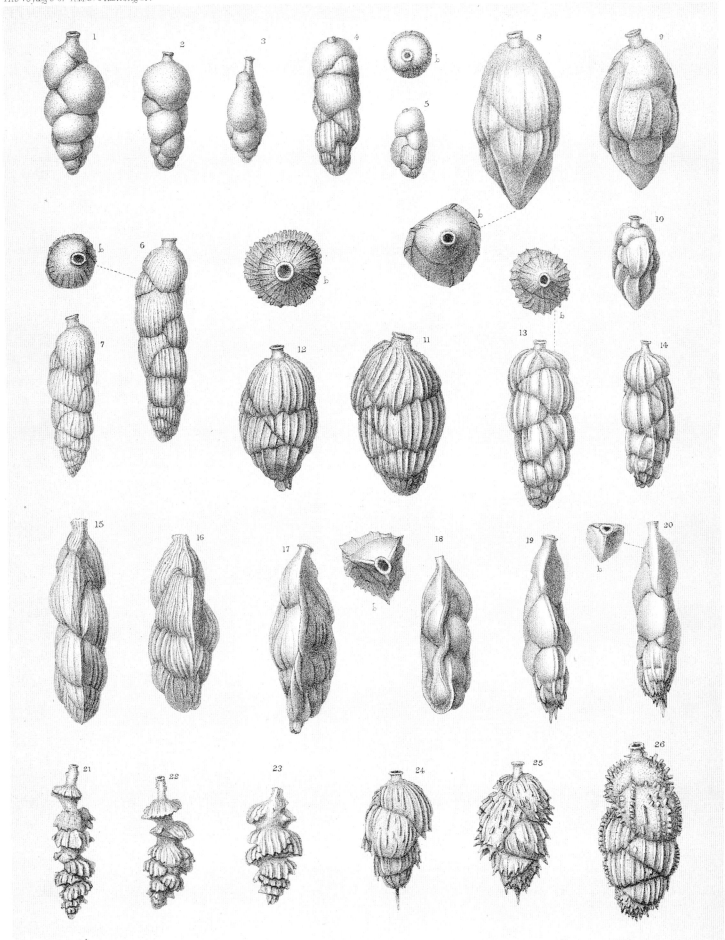

UVIGERINA

Plate 75

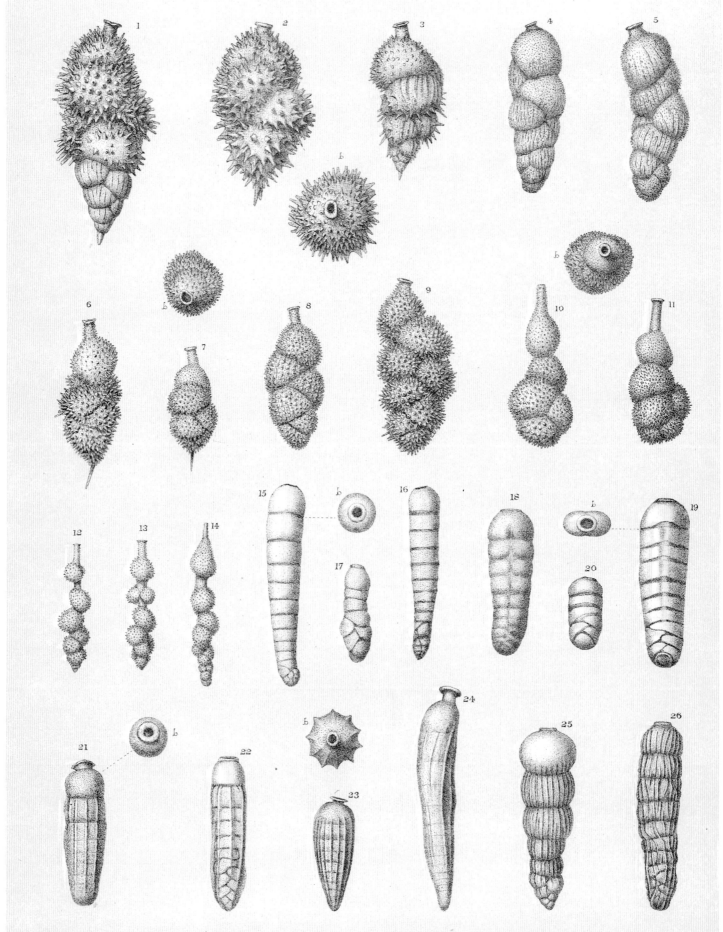

UVIGERINA (Uvigerina - Sagrina.)

Plate 76

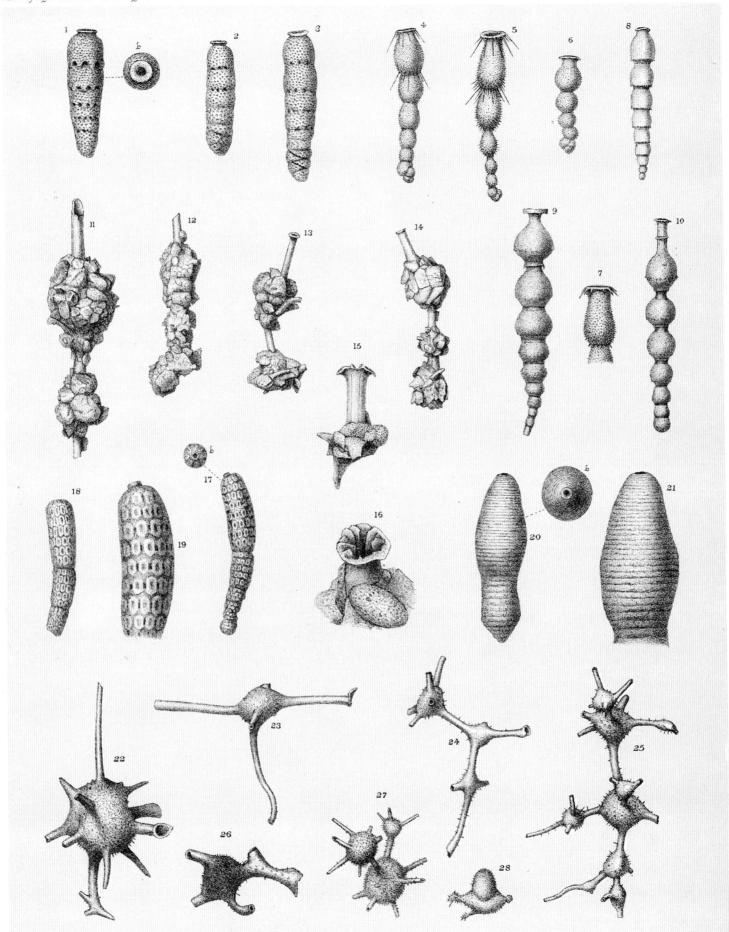

A.T.Hollick ad nat.del et lith.

Hanhart imp.

UVIGERINA (Sagrina) - RAMULINA. etc.

Plate 77

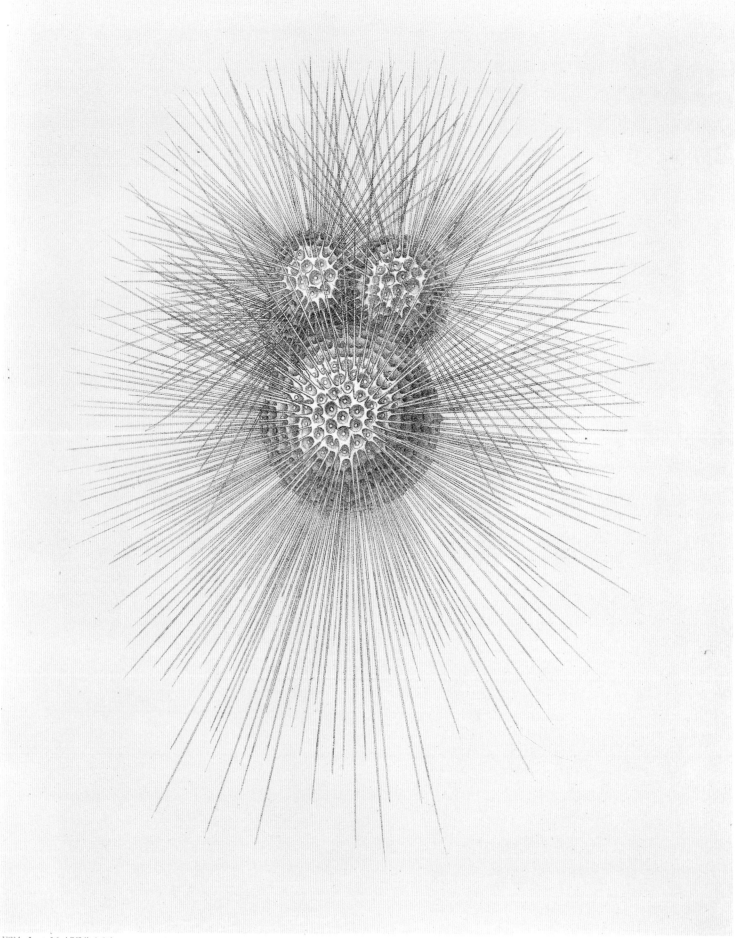

GLOBIGERINA.

Plate 78

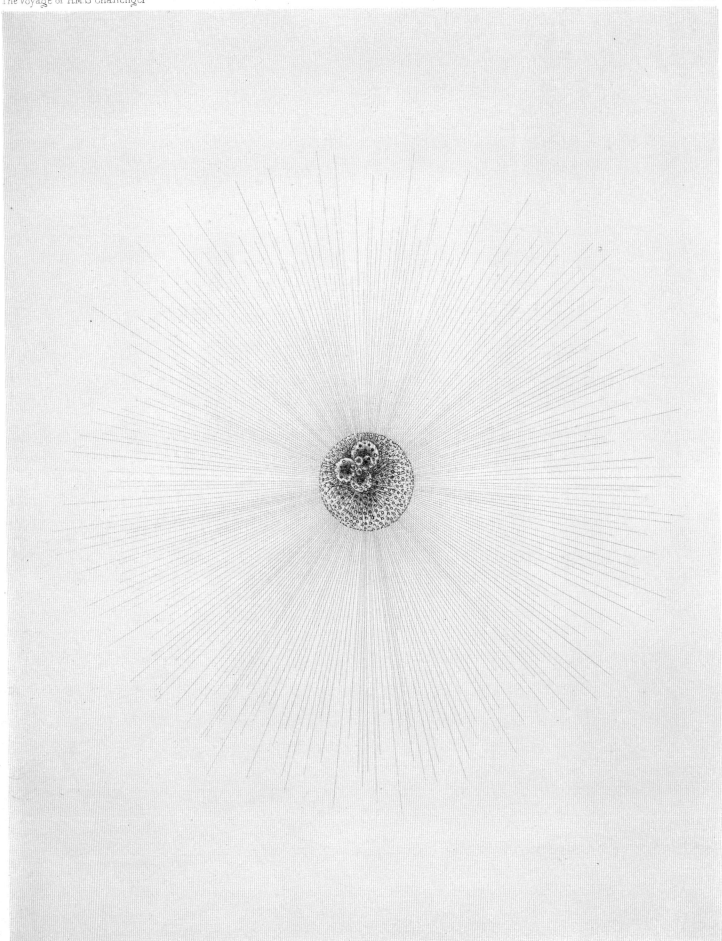

J.J.Wild ad nat. del. A.T.Hollick lith.

Hanhart imp.

GLOBIGERINA (Orbulina.)

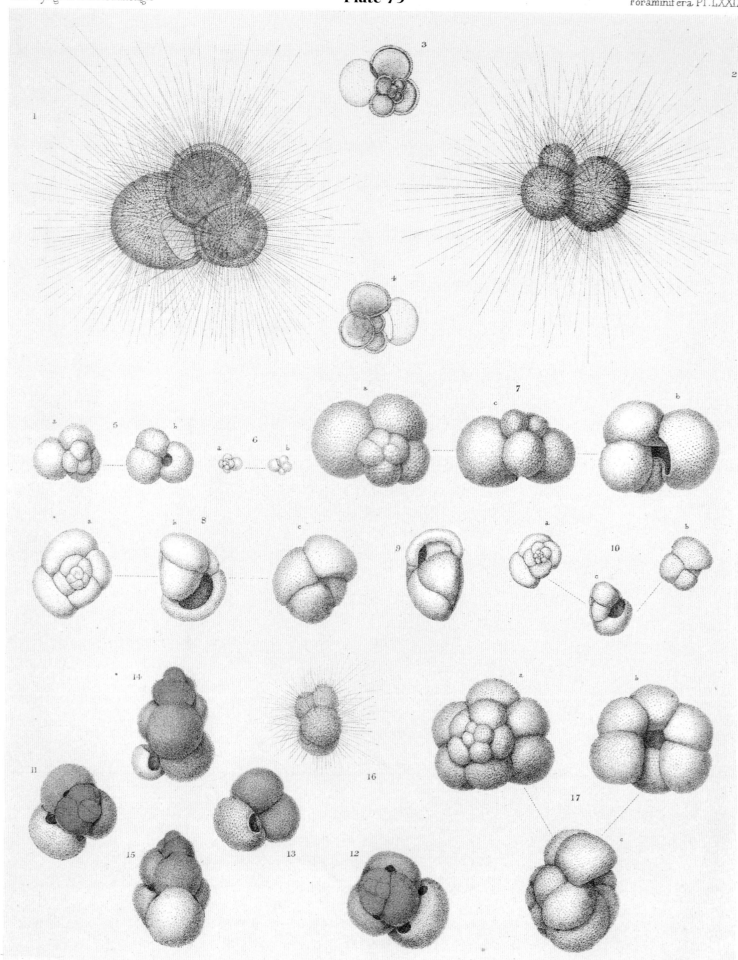

A.T.Hollick ad.nat.del.et lith.

Hanhart imp.

GLOBIGERINA.

Plate 80

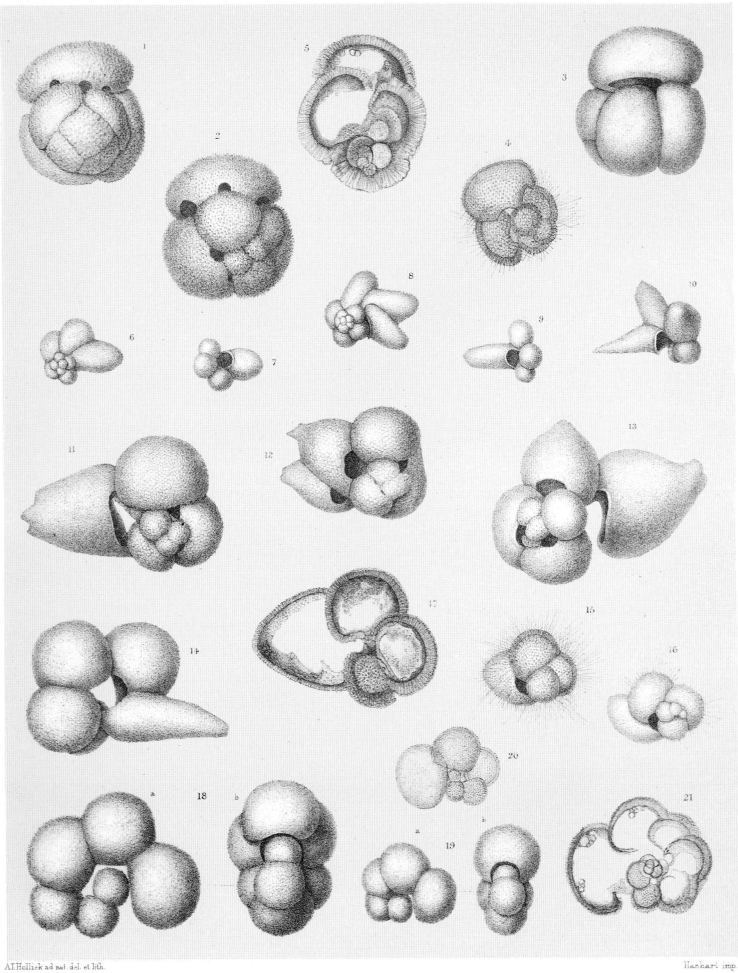

A.T.Hollick ad nat.del.et lith.

Hanhart imp.

GLOBIGERINA.

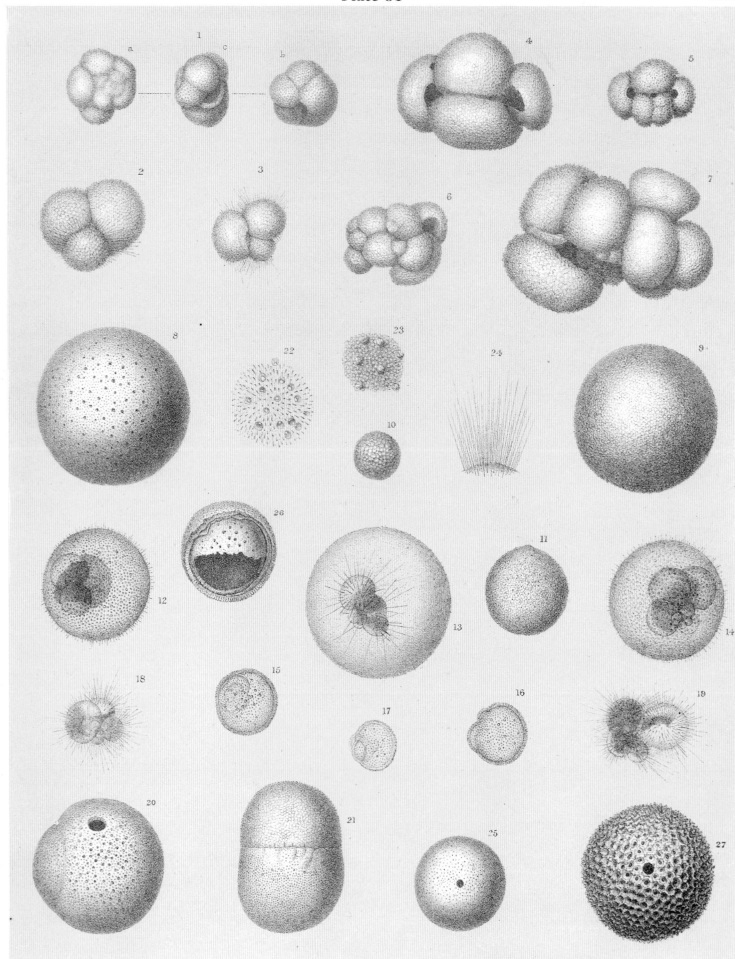

A.T.Hollick ad nat.del.et lith.

Hanhart imp.

GLOBIGERINA (Globigerina_Orbulina).

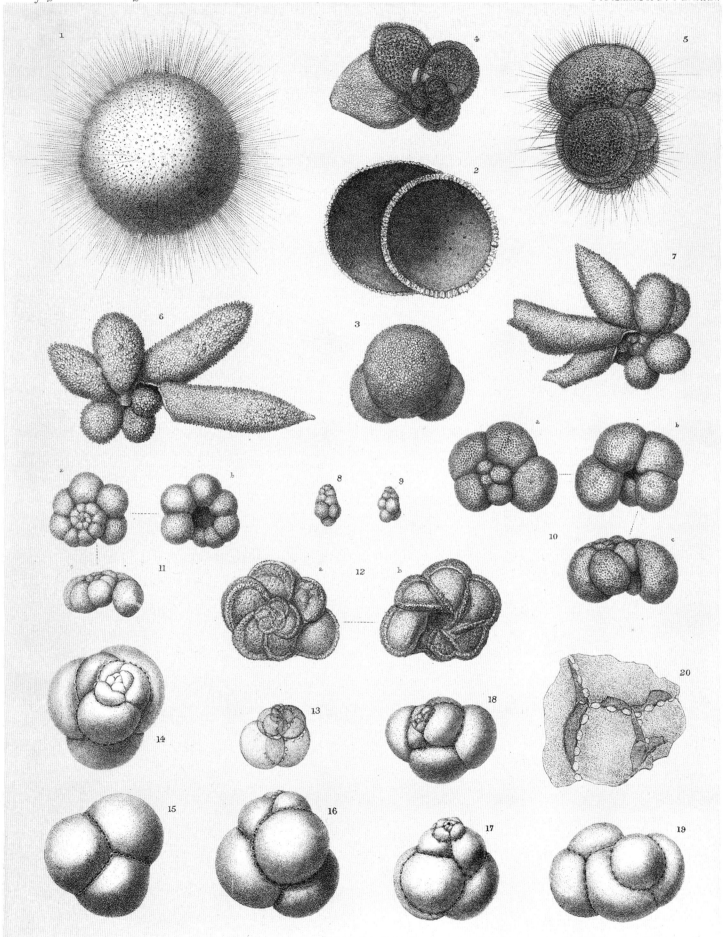

A.T.Hollick ad nat.del.et lith.

Hanhart imp.

GLOBIGERINA (Globigerina-Orbulina.) CANDEINA.

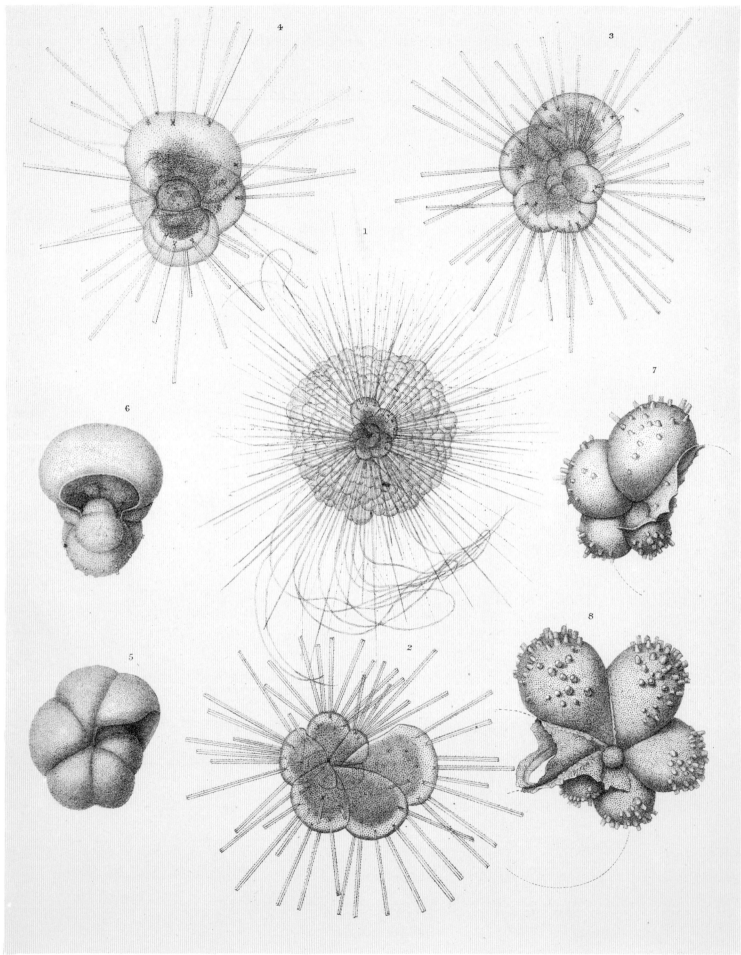

J.J.Wild & A.T.Hollick, ad nat.del. A.T.Hollick lith.   Hanhart imp.

HASTIGERINA.

Plate 84

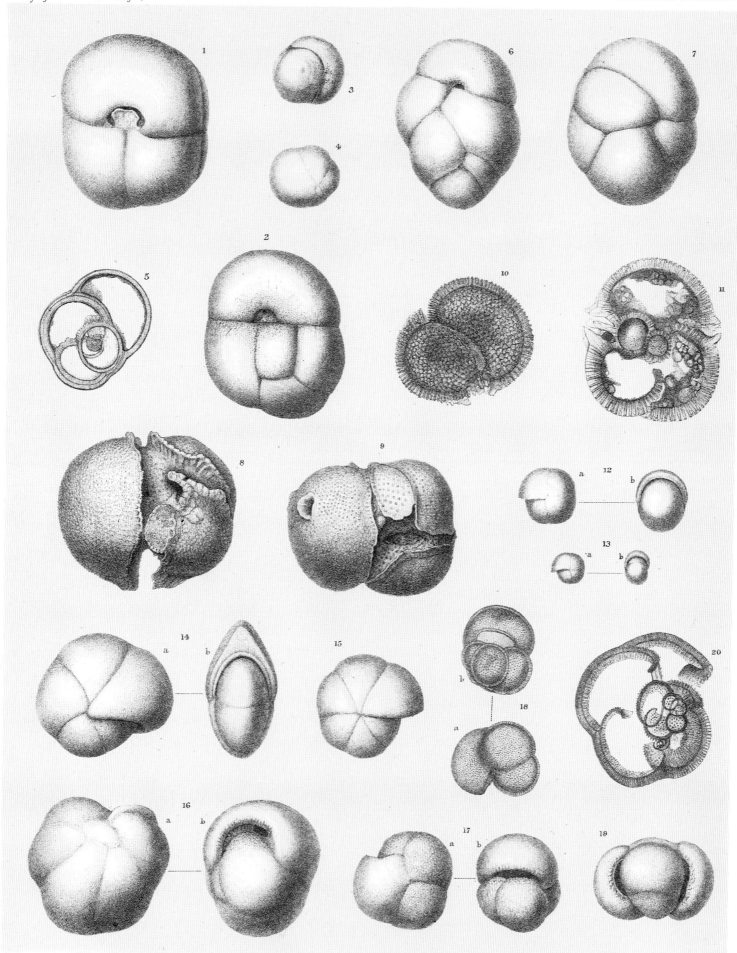

A.T.Hollick ad nat. del. et lith.

Hanhart imp.

SPHÆROIDINA — PULLENIA.

Plate 85

The Voyage of H.M.S "Challenger".

Foraminifera. Pl. LXXXV.

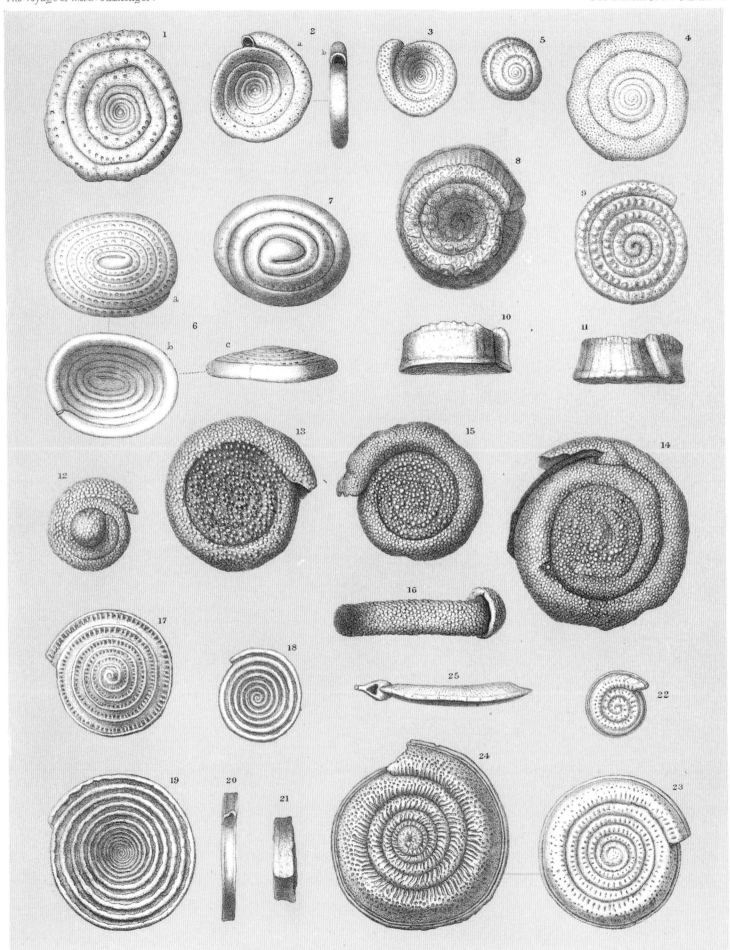

Plate 86

A.T.Hollick, ad nat. del. et lith.

Hanhart imp.

PATELLINA — DISCORBINA.

Plate 87

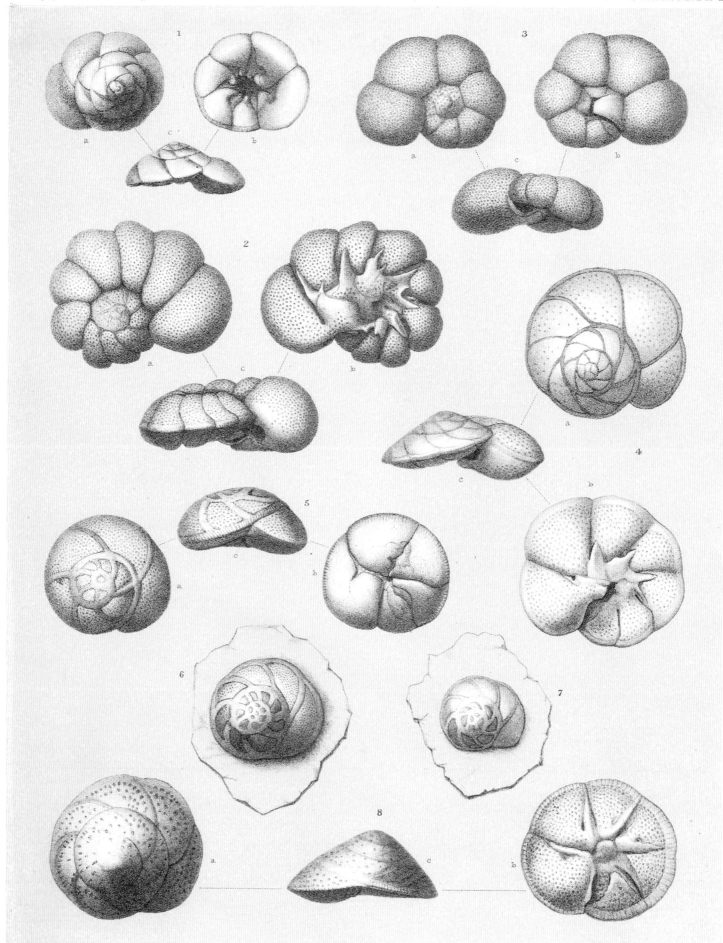

DISCORBINA.

Plate 88

A.T.Hollick ad nat.del.et lith.

DISCORBINA.

Hanhart imp.

Plate 89

A.T.Hollick ad nat.del.et lith.

DISCORBINA.

Hanhart imp.

Plate 90

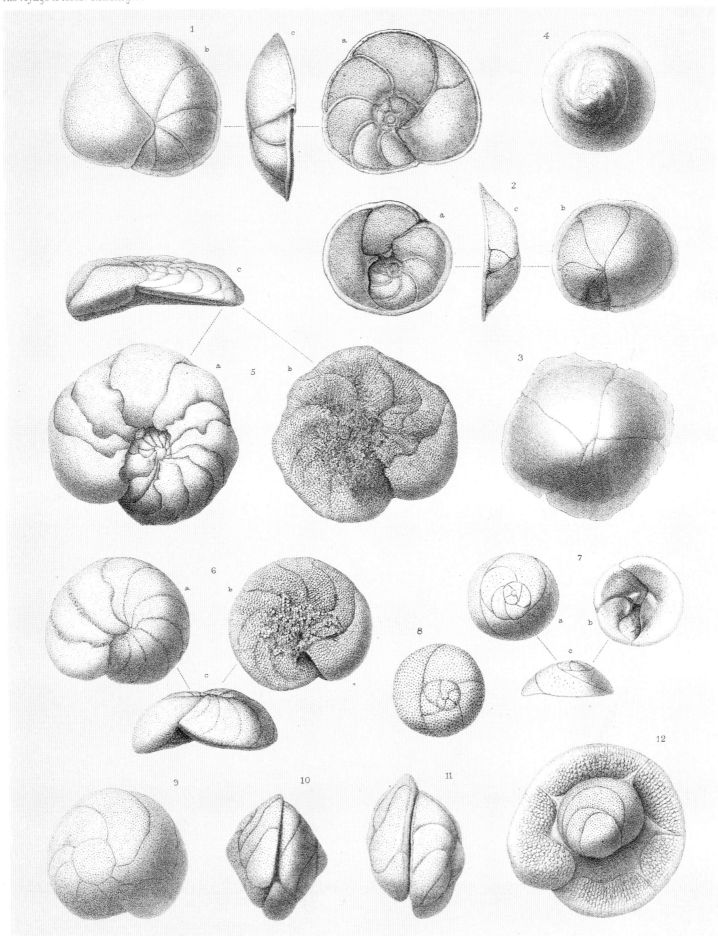

A.T.Hollick ad nat,del.et lith.

Hanhart imp.

DISCORBINA.

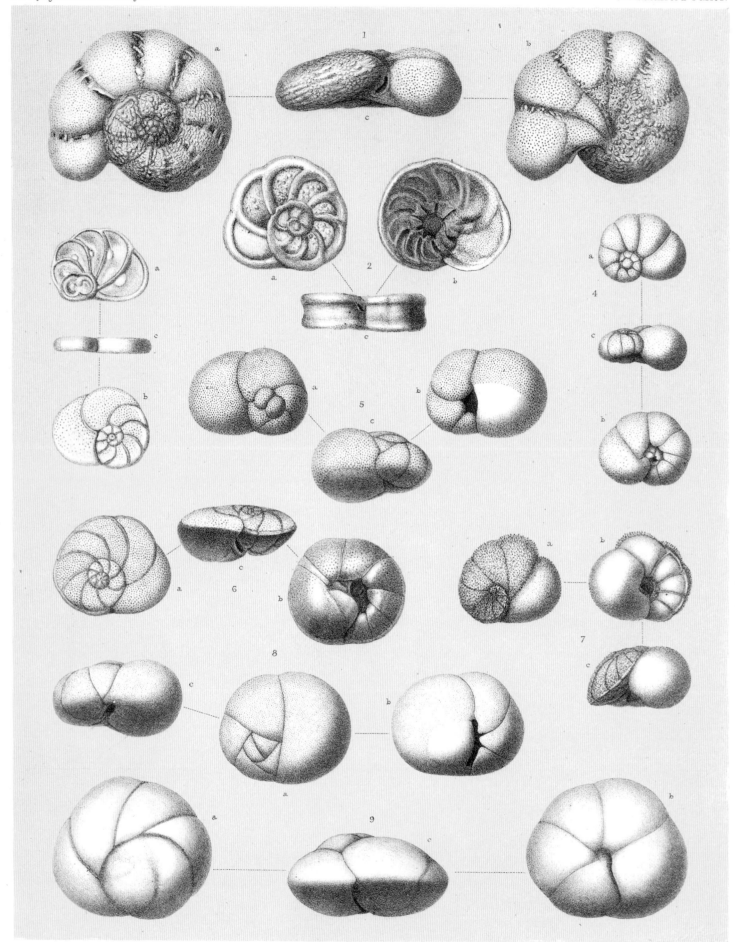

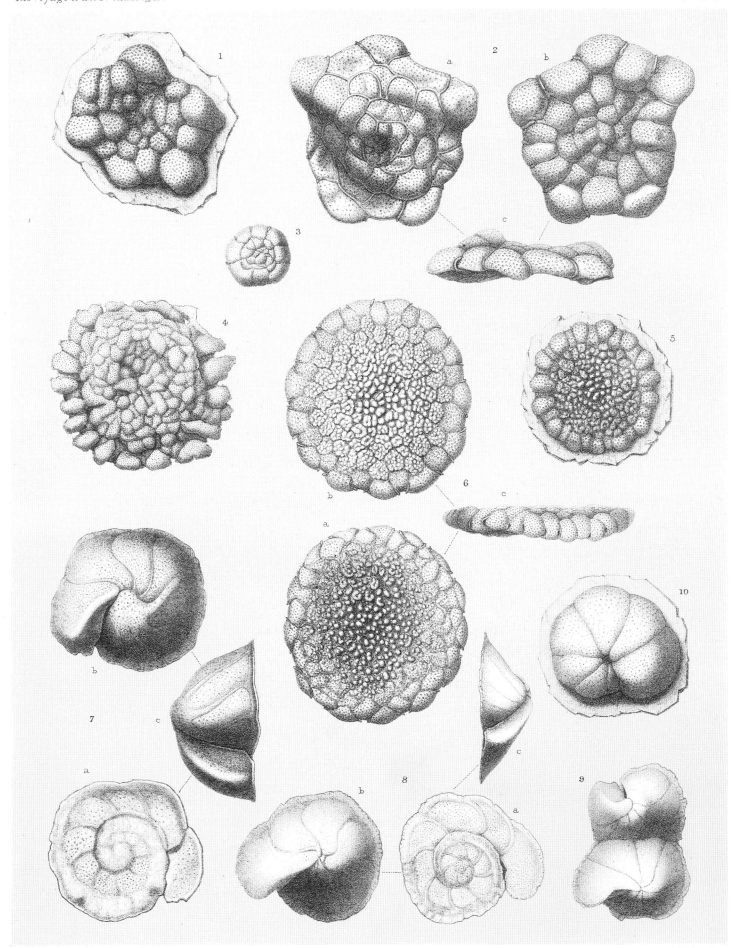

PLANORBULINA (Planorbulina_Truncatulina).

Plate 93

The Voyage of H.M.S. "Challenger."

Foraminifera. Pl. XCIII.

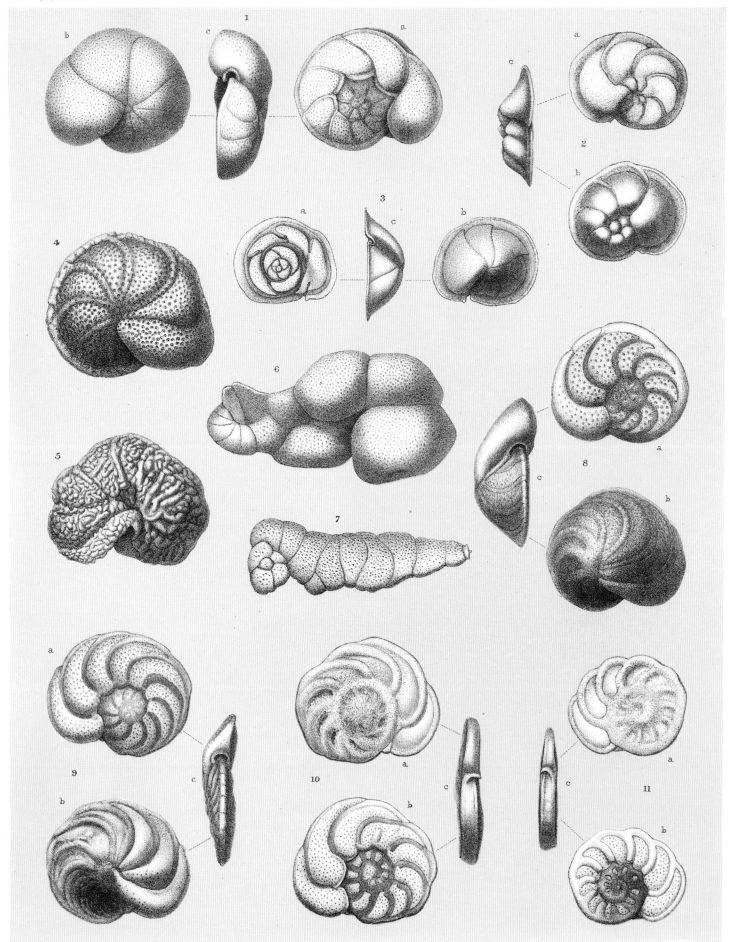

PLANORBULINA (Truncatulina).

Plate 94

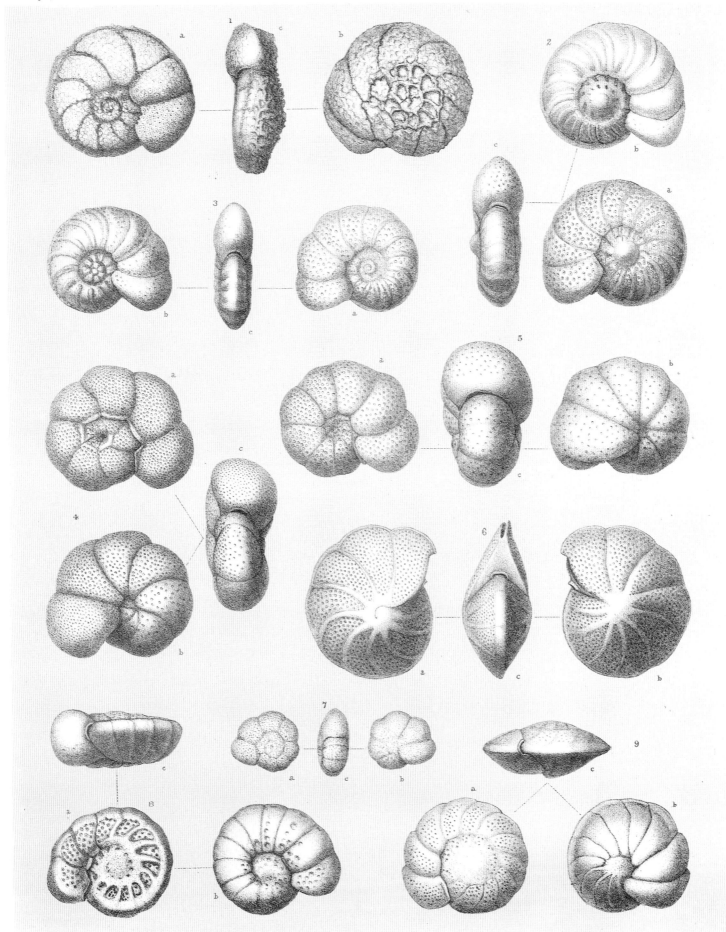

PLANORBULINA (Truncatulina_Anomalina.)

Plate 95

PLANORBULINA (Truncatulina.)

Plate 96

PLANORBULINA (Truncatulina).

Plate 97

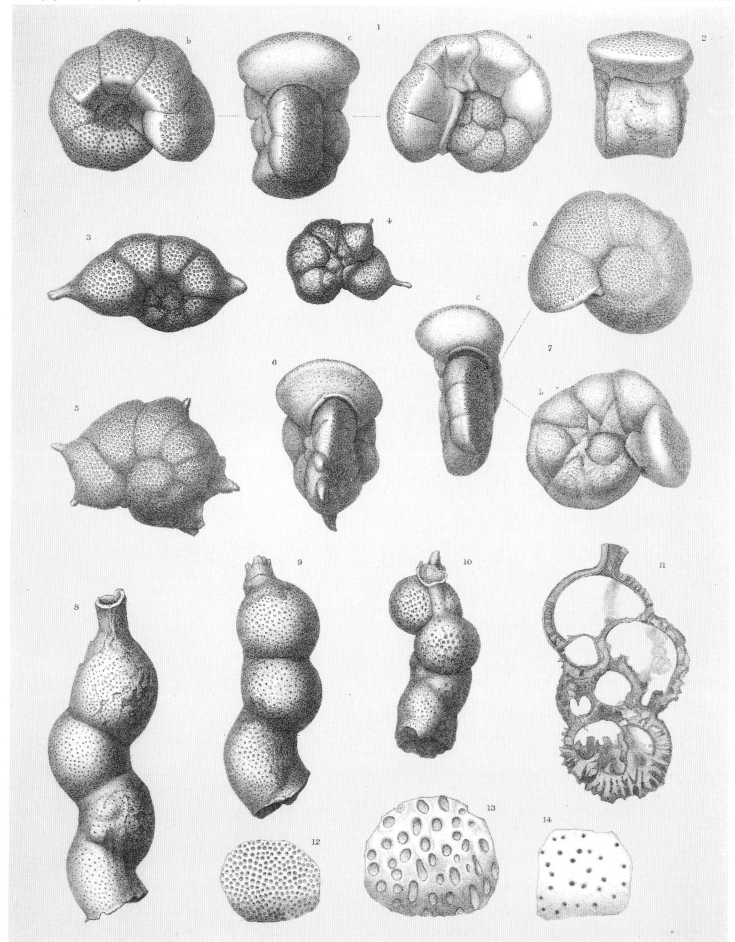

A.T. Hollick ad nat. del. et lith.

Hanhart imp.

PLANORBULINA (Anomalina).—RUPERTIA.

Plate 98

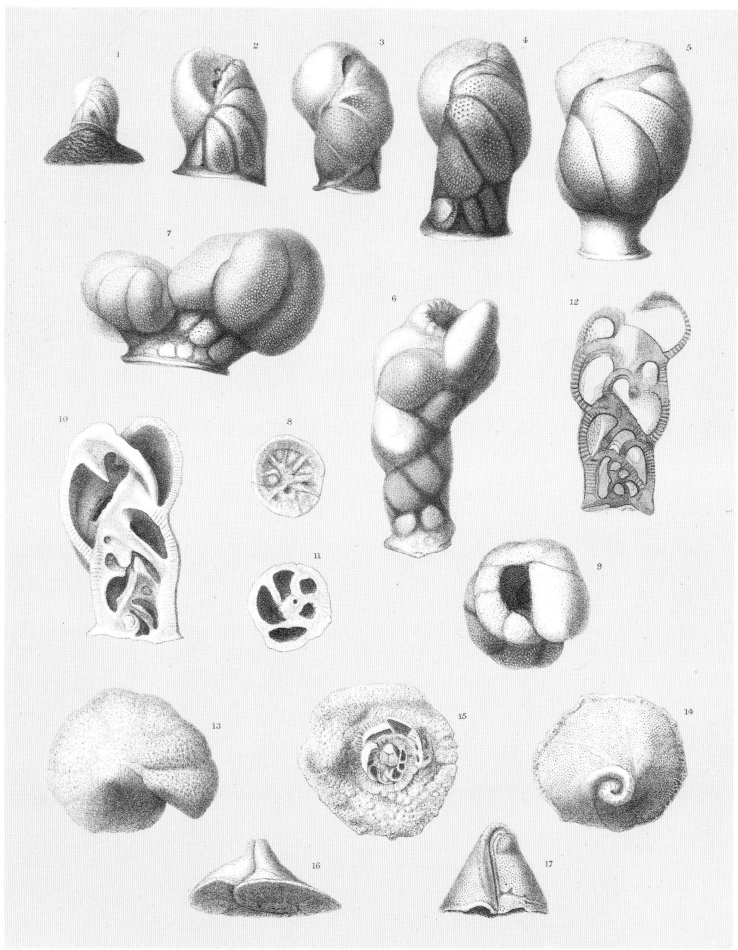

RUPERTIA ___ CARPENTERIA.

Plate 99

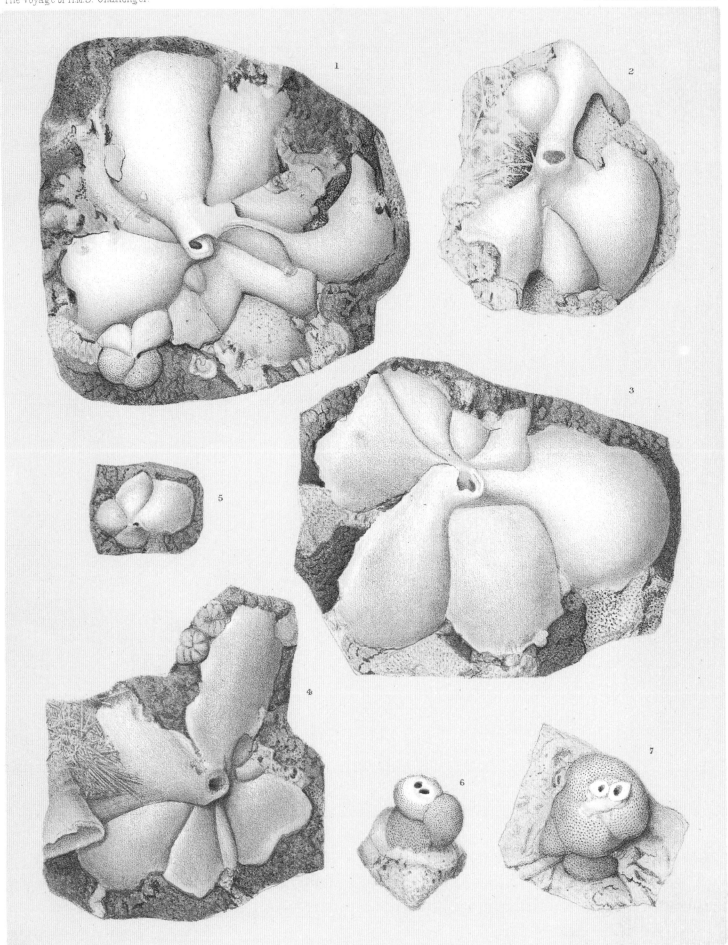

CARPENTERIA.

Plate 100

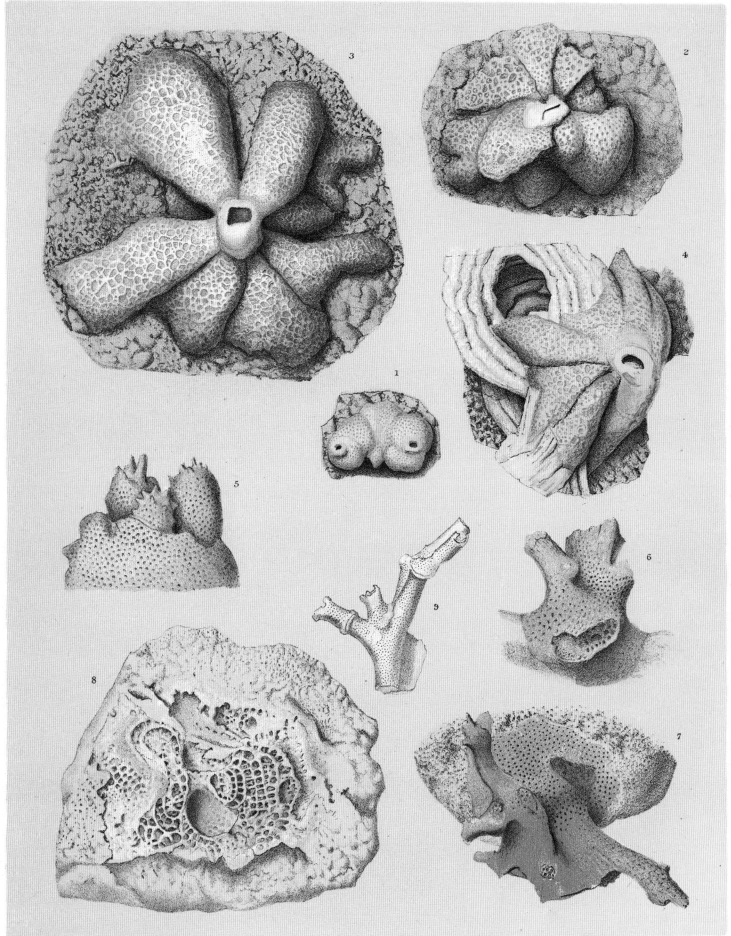

CARPENTERIA ___ POLYTREMA.

Plate 101

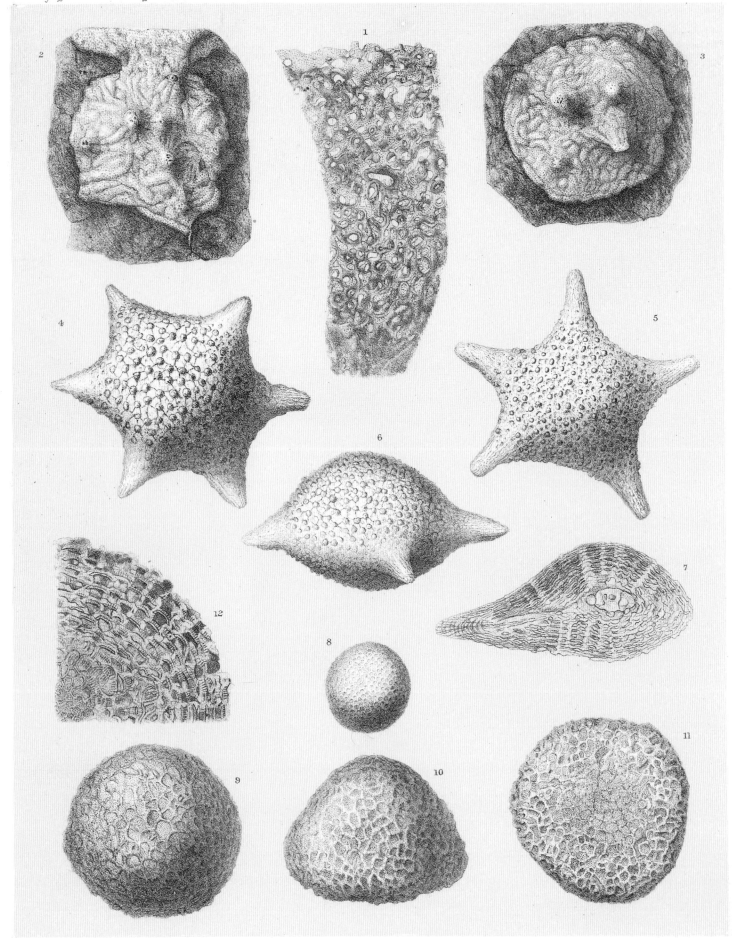

A.T.Hollick ad nat.del.et lith.

Hanhart imp.

POLYTREMA_TINOPORUS. (Tinoporus_Gypsina.)

Plate 102

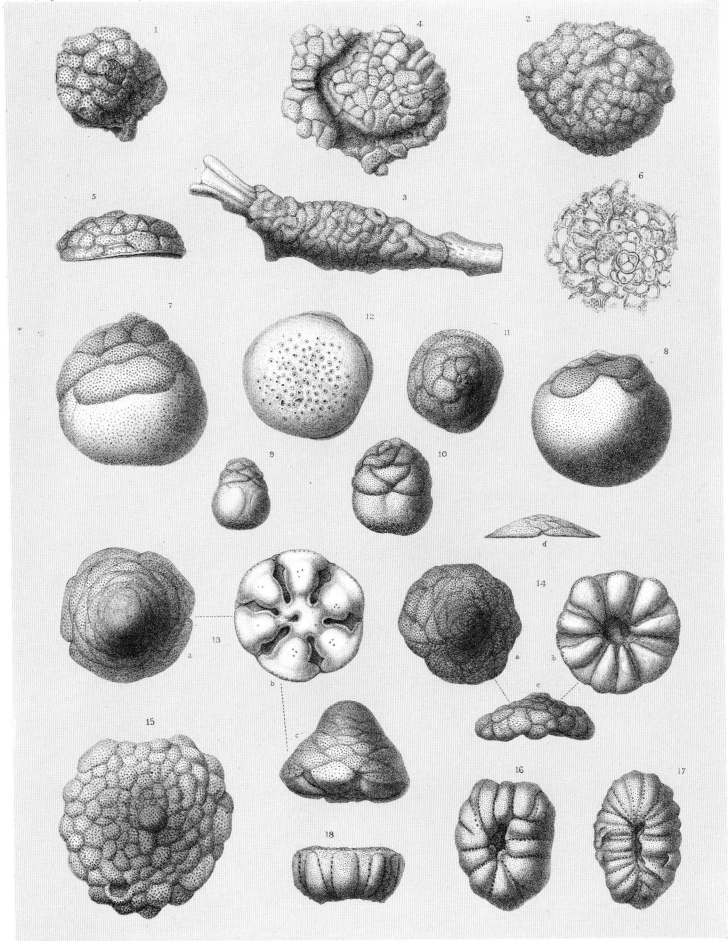

A.T.Hollick,ad nat.del et lith.

Hanhart imp

TINOPORUS (Gypsina) CYMBALOPORA.

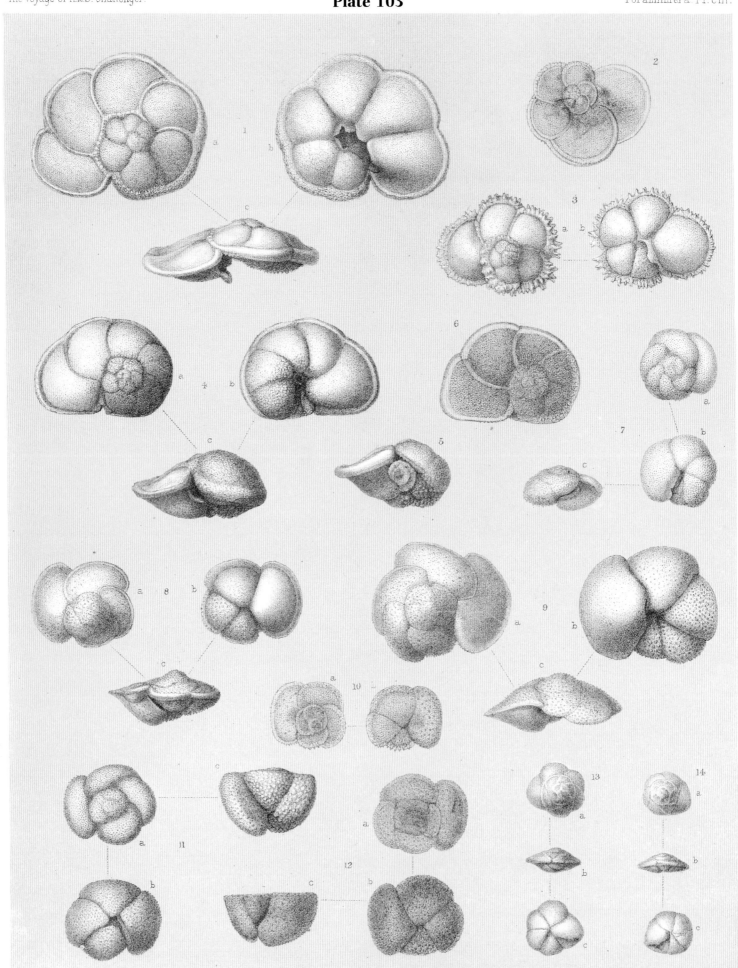

PULVINULINA.

Plate 104

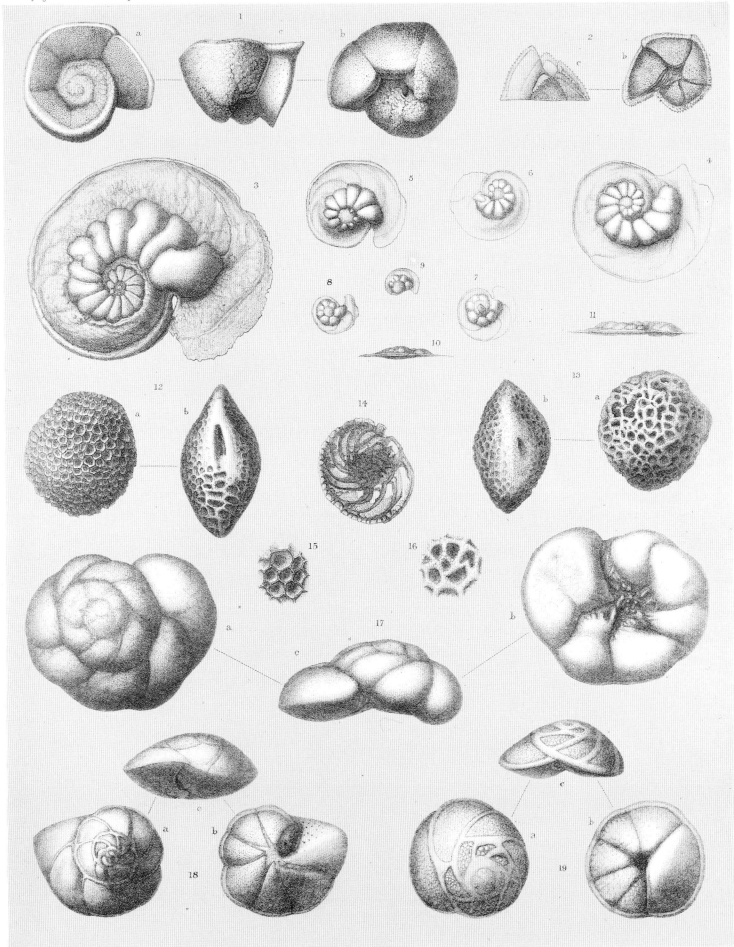

PULVINULINA.

Plate 105

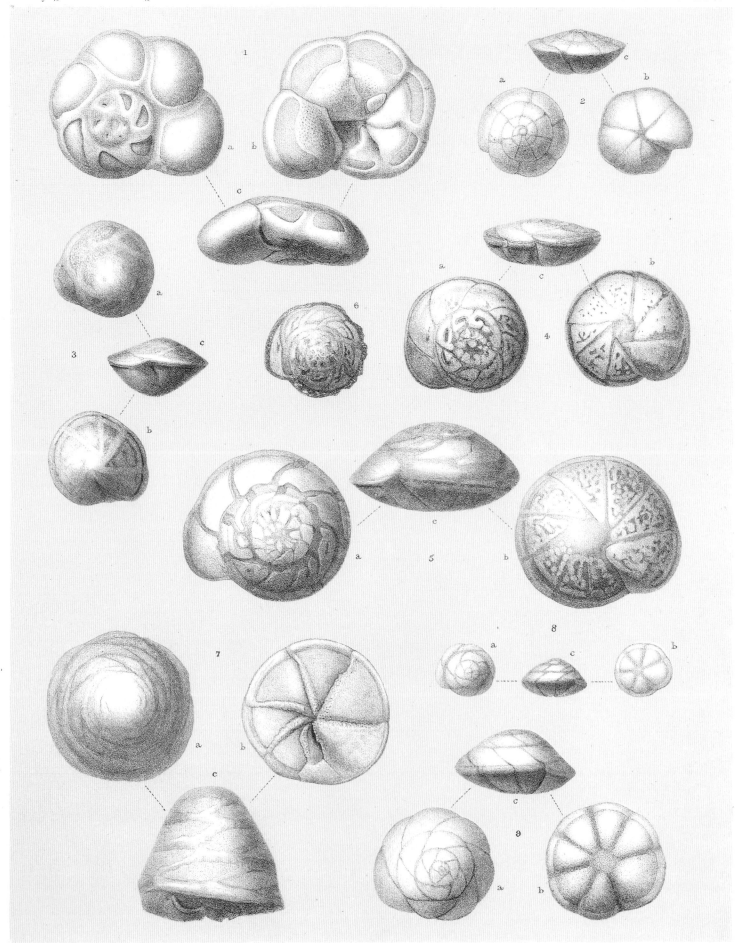

PULVINULINA.

Plate 106

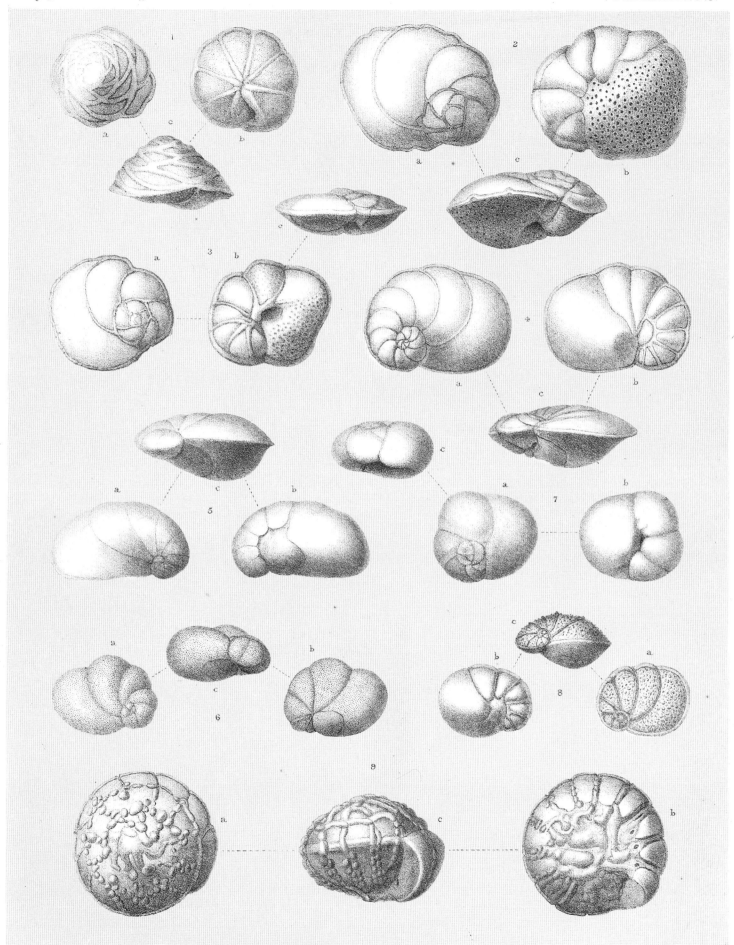

A.T.Hollick ad nat. del. et lith. Hanhart imp.

PULVINULINA_ROTALIA.

Plate 107

The Voyage of H.M.S."Challenger". Foraminifera Pl. CVII.

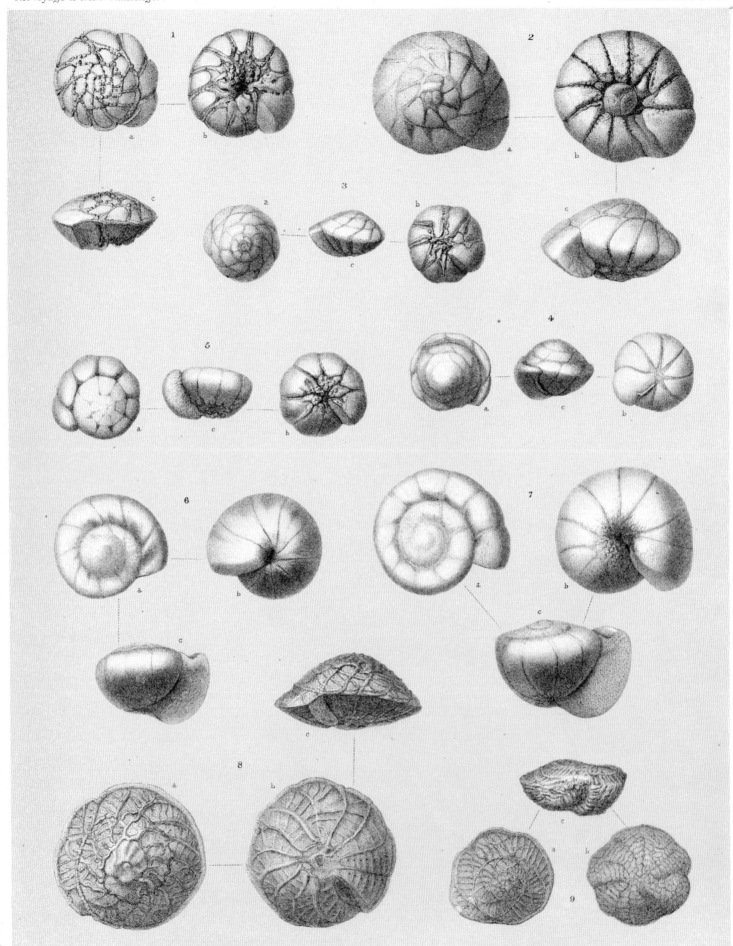

A.T.Hollick ad nat.del et lith. Hanhart imp.

ROTALIA.

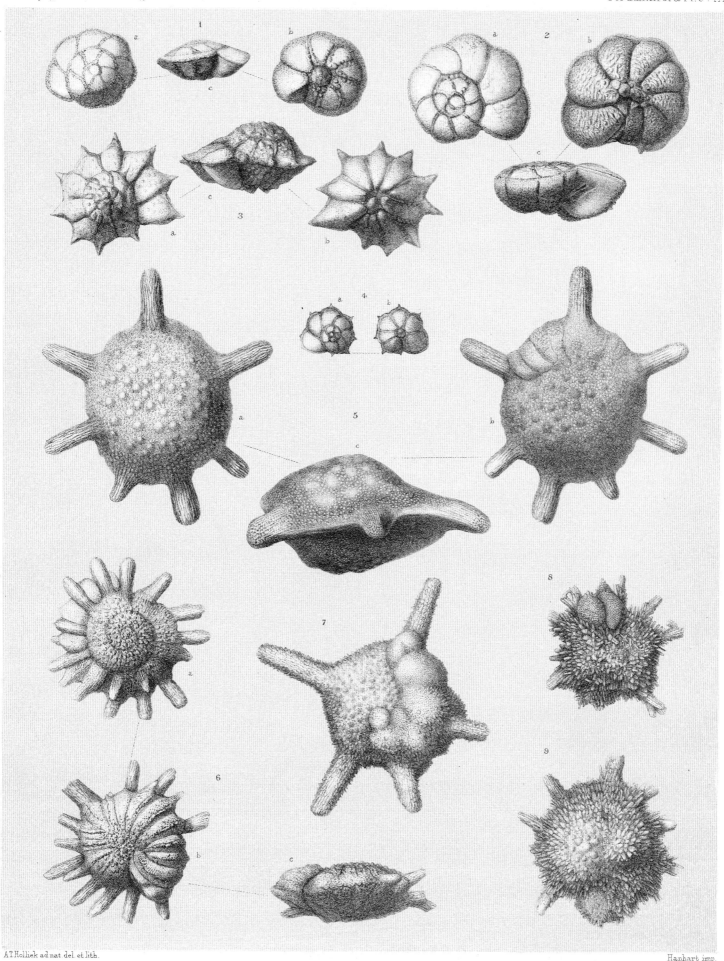

A.T.Holliek ad.nat. del. et.lith.

Hanhart imp.

ROTALIA _ CALCARINA.

Plate 109

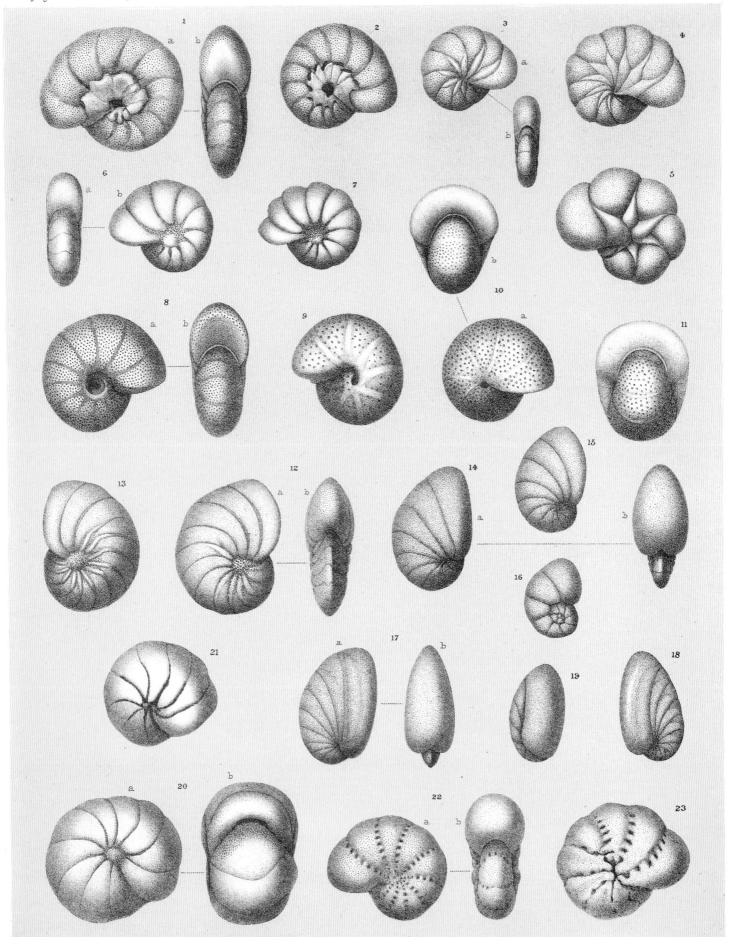

A.T.Hollick ad nat. del. et lith.

Hanhart imp.

NONIONINA ___ POLYSTOMELLA.

Plate 110

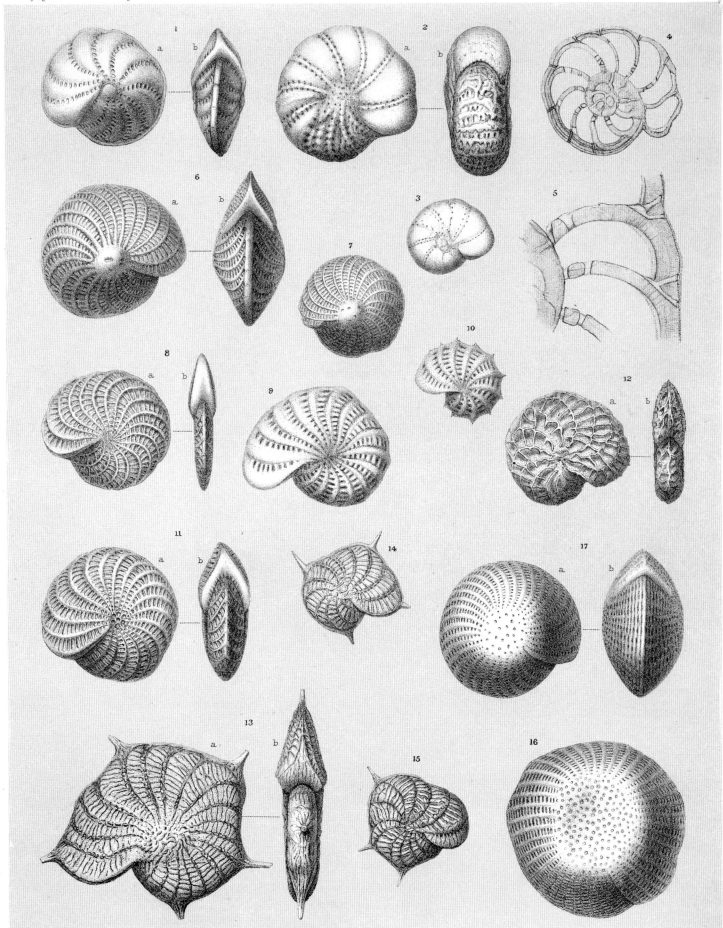

A.T.Hollick ad nat.del.et lith.

POLYSTOMELLA.

Hanhart imp.

Plate 111

A.T.Hollick ad nat. del. et lith.

Hanhart imp.

AMPHISTEGINA ‗ CYCLOCLYPEUS.

Plate 112

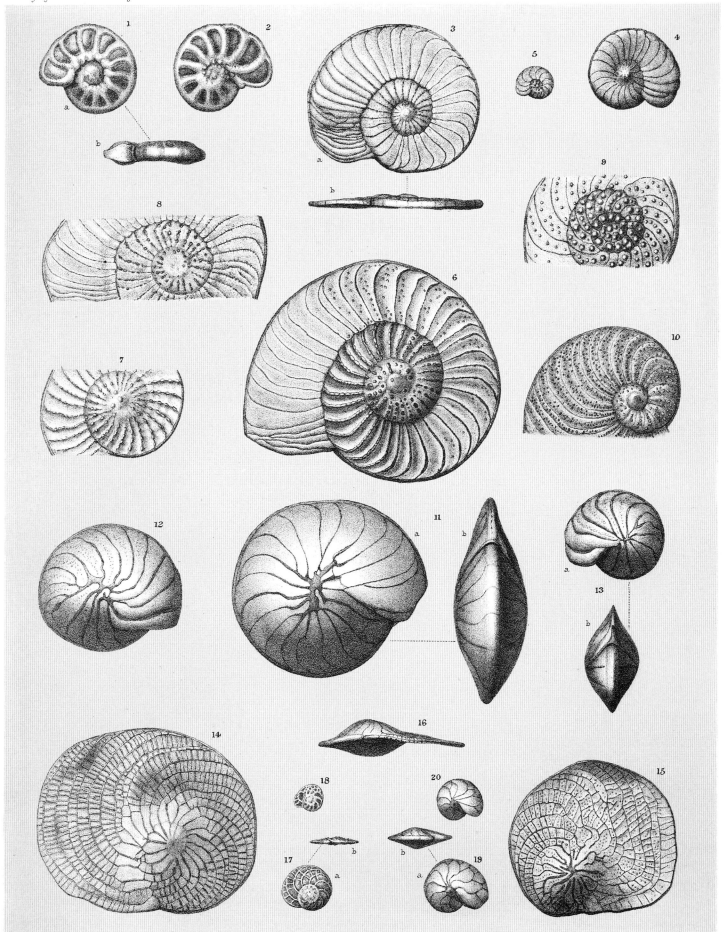

OPERCULINA _ NUMMULINA _ HETEROSTEGINA.

Plate 113

The Voyage of H.M.S "Challenger".

Foraminifera Pl. CXIII.

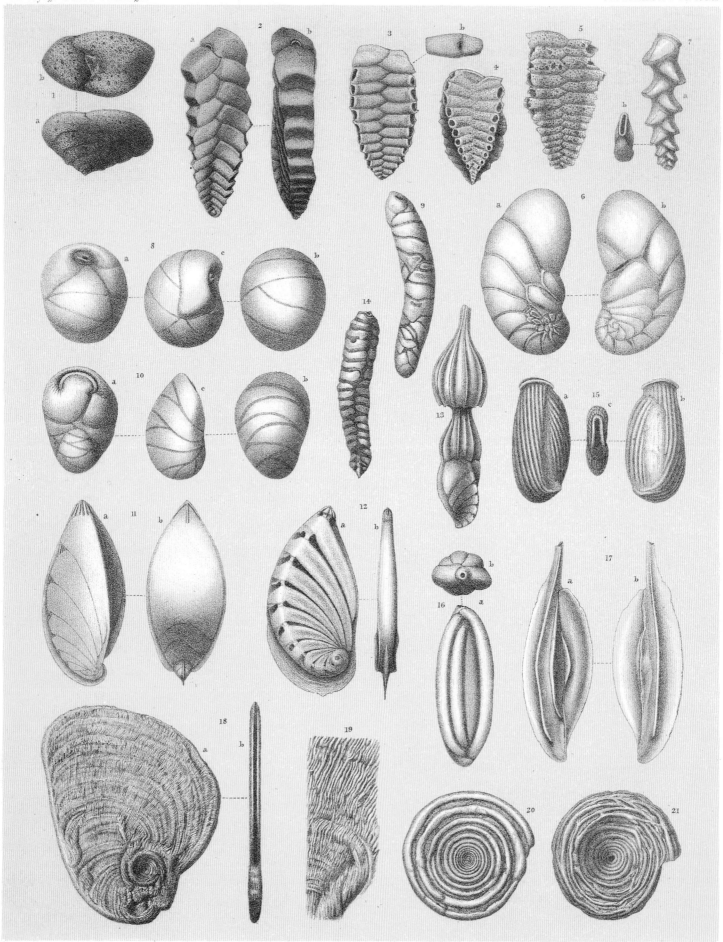

TEXTULARIA_CASSIDULINA, etc.

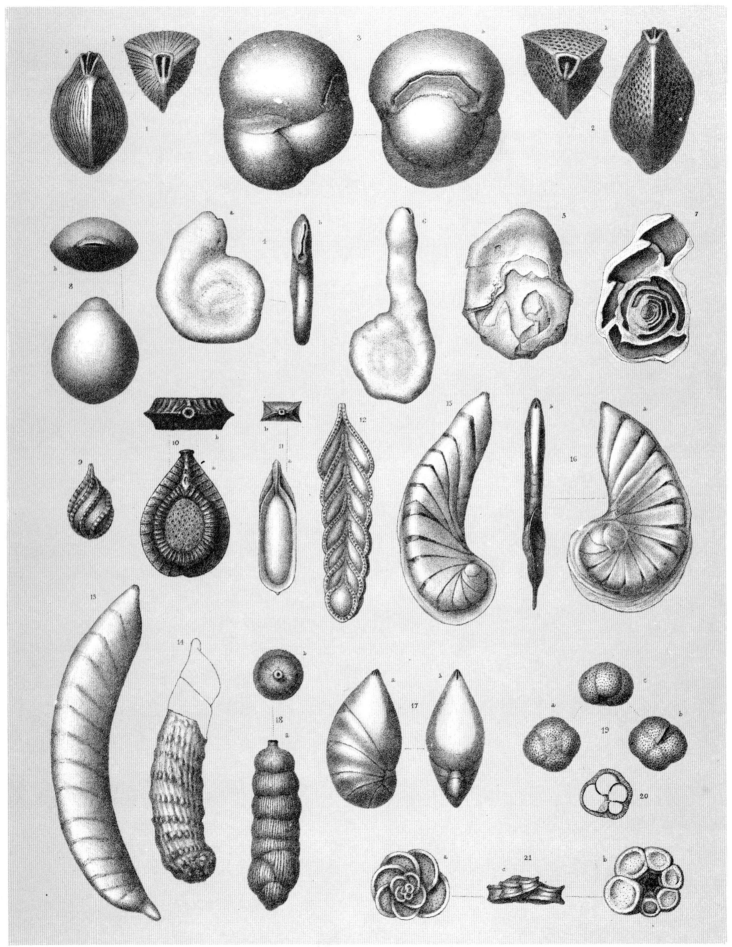

MILIOLINA_LAGENA, etc.

Plate 115

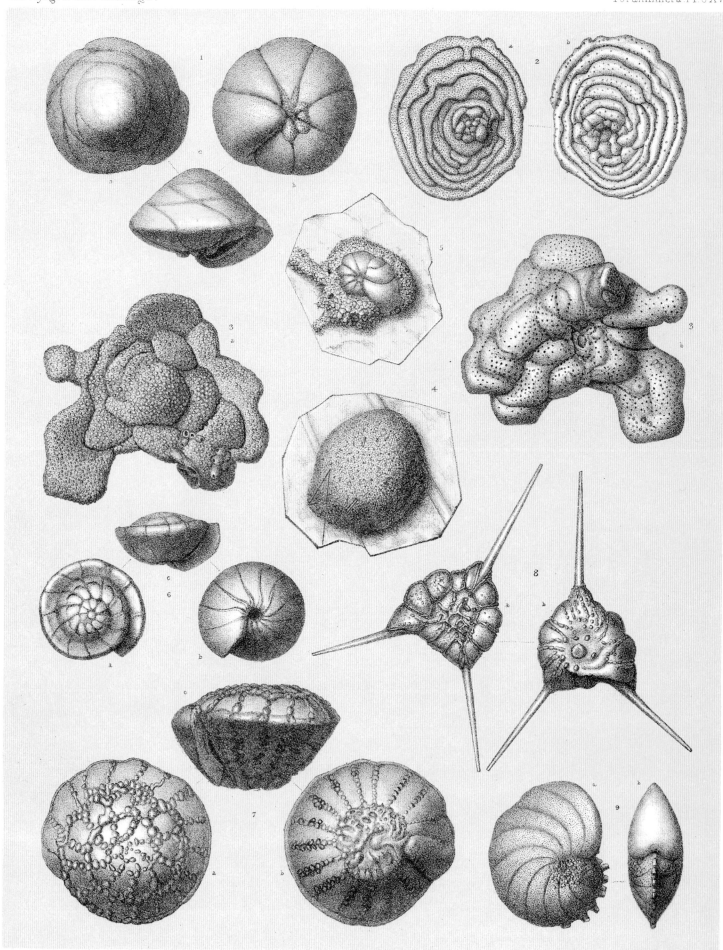

A.T.Hollick, ad. nat. del. et lith.

Hanhart imp.

PULVINULINA ROTALIA, etc.

Supplementary plates
and their explanations

Plate 1

Figures 1–2 *Dentalina albatrossi* **(Cushman 1923)**
From *Challenger* Sta. 209, Philippines, Pacific (95–100fm.). British Museum (Natural History) register number ZF1989. Figure 1 apertural view; Fig. 2 general view, image capture system (scale bar = 100μ).
Compare Brady's Plate 63, Fig. 35 and Plate 64, Figs 11–14.

Figures 3–4, 14 *Dentalina catenulata* **(Brady 1884)**
From *Challenger* Sta. 209, Philippines, Pacific (95–100fm.). ZF1931. Figure 3 apertural view; Fig. 4 general view, image capture system (scale bar = 100μ); Fig. 14 apertural view, scanning electron microscopy (scale bar = 10μ).
Compare Plate 63, Figs 32–4. Note atypical rounded aperture.

Figures 5, 15 *Dentalina inflexa* **(Reuss 1866)**
From *Challenger* Sta. 192, Ki Islands, Pacific (129fm.). ZF1962. Figure 5 general view, image capture system (scale bar = 100μ); Fig. 15 apertural view, scanning electron microscopy (scale bar = 10μ).
Compare Plate 62, Fig. 9.

Figures 6–7 *Glandulonodosaria ambigua* **(Neugeboren 1856)**
From *Challenger* Sta. 192, Ki Islands, Pacific (129fm.). ZF1971. Figure 6 apertural view; Fig. 7 general view, image capture system (scale bar = 100μ).
Compare Plate 62, Fig. 3.

Figures 8–9 *Glandulonodosaria annulata* **(Brady 1884)**
From *Challenger* Sta. 260A, off Honolulu Reefs, Pacific (40fm.). ZF2344. Figure 8 apertural view; Fig. 9 general view, image capture system (scale bar = 100μ).
Compare Plate 76, Figs 20–1.

Figures 10–11 *Glandulonodosaria calomorpha* **(Reuss 1865)**
From *Challenger* Sta. 314A, South Atlantic (6fm.). ZF1929. Figure 10 apertural view; Fig. 11 general view, image capture system (scale bar = 100μ).
Compare Plate 61, Figs 23–6, ?27.

Figures 12–13 *Glandulonodosaria* **sp. nov.**
From *Challenger* Sta. 192, Ki Islands, Pacific (129fm.). ZF1986. Figure 12 apertural view; Fig. 13 general view, image capture system (scale bar = 100μ).
Compare Plate 63, Figs 17–18.

Figure 16 *Orthomorphina jedlitschkai* **(Thalmann 1937)**
From *Challenger* Sta. 192, Ki Islands, Pacific (129fm.). ZF1962. Apertural view, scanning electron microscopy (scale bar = 10μ).
Compare Plate 62, Figs 1–2. Note rounded aperture.

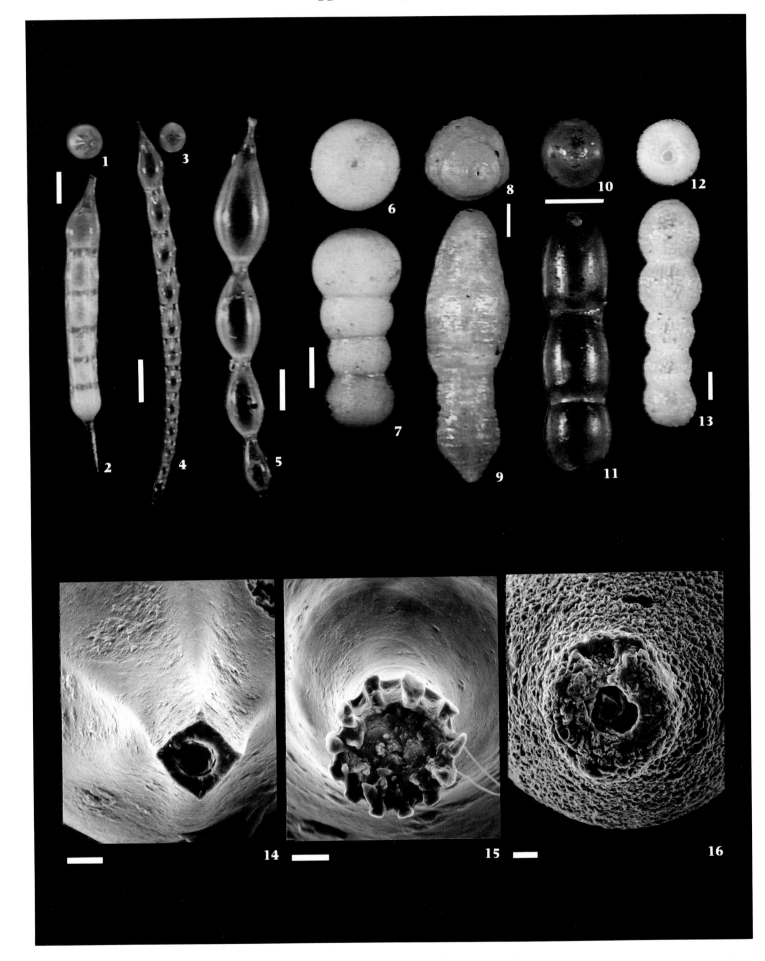

Plate 2

Figure 1 *Nodogenerina antillea* (Cushman 1923)
From *Challenger* Sta. 192, Ki Islands, Pacific (129fm.). British Museum (Natural History) register number 1959.5.11.369. General view, image capture system (scale bar = 100μ).
Compare Plate 76, Figs 9–10.

Figures 2–3, 15 *Nodogenerina virgula* (Brady 1879)
From *Challenger* Sta. 120, off Pernambuco, Atlantic (675fm.). ZF2363. Figure apertural view; Fig. 3 general view, image capture system (scale bar = 100μ); Fig. 15 apertural view, scanning electron microscopy (scale bar = 10μ).
Compare Plate 76, Fig. 8. Note lack of apertural dentition.

Figures 4–5, 16 *Orthomorphina challengeriana* (Thalmann 1937)
From *Challenger* Sta. 192, Ki Islands, Pacific (129fm.). ZF1972. Figure 4 apertural view; Fig. 5 general view, image capture system (scale bar = 100μ); Fig. 16 apertural view, scanning electron microscopy (scale bar = 100μ).
Compare Plate 64, Figs 25–7. Note rounded aperture.

Figures 6–7 *Orthomorphina jedlitschkai* (Thalmann 1937)
From *Challenger* Sta. 192, Ki Islands, Pacific (129fm.). ZF1972. Figure 6 apertural view; Fig. 7 general view, image capture system (scale bar = 100μ).
Compare Plate 62, Figs 1–2. See also Supplementary Plate 1, Fig. 16. Note rounded aperture.

Figures 8–9 *Stilostomella abyssorum* (Brady 1884)
From *Challenger* Sta. 296, South Pacific (1825fm.). ZF3649 [Ex ZF1962]. Figure 8 apertural view; Fig. 9 general view, image capture system. Length = 2.5mm.
Compare Plate 63, Fig. 8. Note apertural dentition.

Figures 10–11, 14 *Stilostomella consobrina* (d'Orbigny 1846)
From *Challenger* Sta. 300, North of Juan Fernandez, East Pacific (1375fm.). Figure 10 apertural view; Fig. 11 general view, image capture system (scale bar = 100μ); Fig. 14 apertural view, scanning electron microscopy (scale bar = 10μ).
Compare Plate 62, Figs 23–4. Note apertural dentition.

Figures 12–13 *Stilostomella fistuca* (Schwager 1866)
From *Challenger* Sta. 192, Ki Islands (129fm.). ZF1980. Figure 12 apertural view; Fig. 13 general view, image capture system (scale bar = 100μ).
Compare Plate 62, Figs 7–8. Note crescentic aperture.

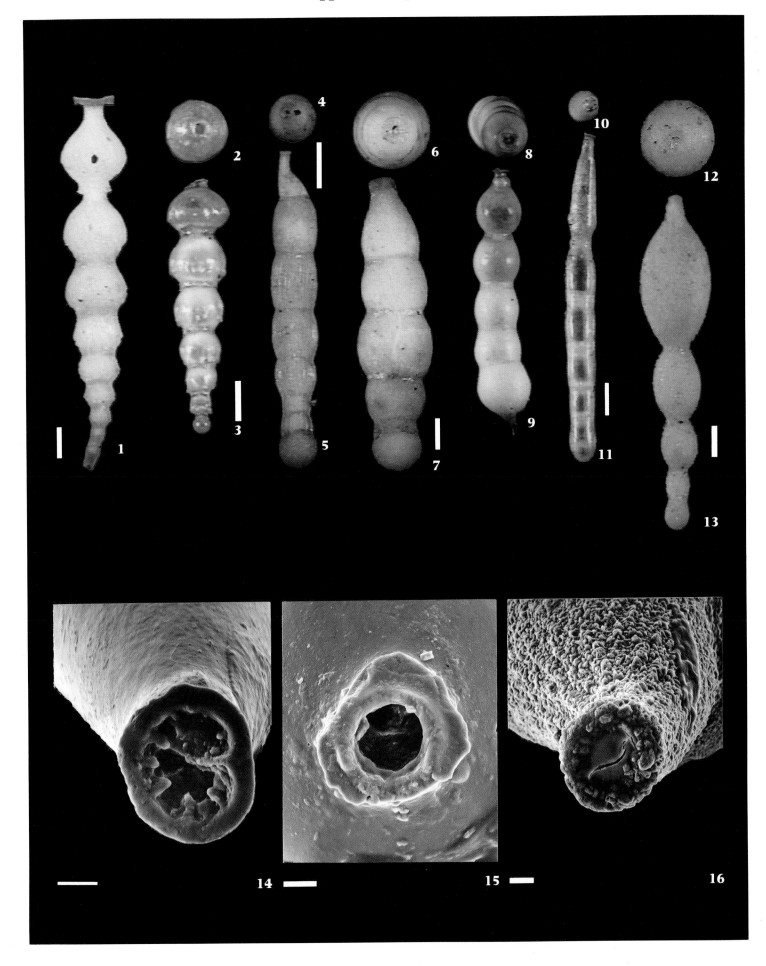

Figure 31 *Fisurina auriculata* var. *linearituba* (Cushman 1913)

[*Lagena auriculata* var. *linearituba* Cushman 1913]
× 60. From *Challenger* Sta. 346, South Atlantic (2350fm.). ZF1630.

Referred by Brady (1884) to *Lagena auriculata* Brady, by Thalmann (1932) to *L. auriculata* var. *linearituba* Cushman, and by Barker (1960) to *Fissurina alveolata* var. *linearituba* (Cushman).

Figure 33 *Fissurina auriculata* var. *duplicata* (Sidebottom 1912)

[*Lagena auriculata* var. *duplicata* Sidebottom 1912]
× 60. From *Challenger* Sta. 5, SW of the Canaries, Atlantic (2740fm.). ZF1629.

Referred by Brady (1884) and Thalmann (1932) to *Lagena auriculata* Brady, by Thalmann (1937) to *L. auriculata* var. *duplicata* Sidebottom, and by Barker (1960) to *Fissurina auriculata* var. *duplicata* (Sidebottom).

Figure 34 *Fissurina alveolata* var. *semisculpta* Parr 1950

× 60. From *Challenger* Sta. 146, Southern Ocean (1375fm.). ZF1614.

Referred by Brady (1884) and Thalmann (1932) to *Lagena alveolata* var. *substriata* Brady, and by Barker (1960) to *Fissurina alveolata* var. *semisculpta* Parr. *Lagena alveolata* var. *substriata* Brady 1881 is a primary junior homonym of *L. substriata* Williamson 1848.

Figures 35, 37 *Fissurina striolata* (Sidebottom 1912)

[*Lagena marginata* var. *striolata* Sidebottom 1912]
Fig. 35 × 75. From *Challenger* Sta. 276, North of Tahiti, Pacific (2350fm.).
Fig. 36 × 75. From *Challenger* Sta. 224, North Pacific (1850fm.). ZF1789.

Referred by Brady (1884) and Thalmann (1932) to *Lagena sulcata* (Walker and Jacob), 'winged variety', and by Barker (1960) to *Fissurina sulcata* (Walker and Jacob), 'winged variety'.

Figure 36 *Fissurina* sp. nov. (8)

× 75. From *Challenger* Sta. 5, SW of the Canaries, Atlantic (2740fm.). ZF1788.

Referred by Brady (1884) and Thalmann (1932) to *Lagena sulcata* (Walker and Jacob), 'winged variety', and by Barker (1960) to *Fissurina sulcata* (Walker and Jacob), 'winged variety'.

Figure 38 *Fissurina auriculata* var. *costata* (Brady 1881)

[*Lagena auriculata* var. *costata* Brady 1881]
× 60. From *Challenger* Sta. 276, South Pacific (2350fm.). ZF1631. Lectotype designated by Jones (1984b).

Referred by Brady (1884) and Thalmann (1932) to *Lagena auriculata* var. *costata* Brady, by Barker (1960) to *Fissurina auriculata* var. *costata* (Brady), and by Jones (1984a) to *Solenina costata* (Brady). *Solenina* Jones 1984 is herein regarded as a junior synonym of *Fissurina* Reuss 1850 (see *F. tenuistriatiformis*, Fig. 11, above).

Plate 61 (LXI)

Figure 1 *Oolina variata* (Brady 1881)

[*Lagena variata* Brady 1881]
× 75. From *Challenger* Sta. 162, Bass Strait, Pacific (38–40fm.). ZF1801.

Referred by Brady (1884) and Thalmann (1932) to *Lagena*, and by Barker (1960) and Poignant (1984) to *Oolina*.

Figure 2 *Oolina ampulladistoma* (Rymer Jones 1874)

[*Lagena ampulladistoma* Rymer Jones 1874]
× 100. From *Challenger* Sta. 219A, Admiralty Islands, Pacific (17fm.). ZF1649.

Referred by Brady (1884) and Thalmann (1932) to *Lagena favosopunctata* Brady, and by Thalmann (1937) and Barker (1960) to *L. ampulladistoma* Rymer Jones.

Figure 3 *Cushmanina quadralata* (Brady 1881)

[*Lagena quadralata* Brady 1881]
× 100. From *Challenger* Sta. 160, Southern Ocean (2600fm.). ZF1735.

Referred by Brady (1884), Thalmann (1932) and Barker (1960) to *Lagena*, and by the present author to *Cushmanina* (see also discussion under *C. desmophora* on Plate 58).

Figure 4 *Pseudoolina multicosta* (Karrer 1877)

[*Fissurina multicosta* Karrer 1877]
× 75. From *Challenger* Sta. 346, South Atlantic (2350fm.). ZF1724.

Referred by Brady (1884) and Thalmann (1932) to *Lagena multicosta* (Karrer), by Barker (1960) to *Oolina multicostata* (sic) (Karrer), and by Hermelin (1989) to *Globofissurella* sp. 1. *Globofissurella* Patterson 1986 is herein regarded as a junior synonym of *Pseudoolina* Jones 1984.

Figure 5 *Oolina exsculpta* (Brady 1881)

[*Lagena exsculpta* Brady 1881]
× 75. From *Challenger* Sta. 168, NE of New Zealand (1100fm.). ZF1647.

Referred by Brady (1884) and Thalmann (1932) to *Lagena*, and by Barker (1960) to *Fissurina*. This species was referred by Patterson and Richardson (1988) to *Exsculptina*. *Exsculptina* Patterson and Richardson, in Loeblich and Tappan 1987 [*Exsculptina* Patterson and Richardson 1988] is regarded by the present author as a junior synonym of *Oolina* d'Orbigny 1839. Brady's specimen is figured as having a 'fissurine' aperture, but close physical inspection reveals that the aperture is in fact typically 'ooline' (though the test is atypically semi-compressed).

Figures 6–7 *Fissurina wrightiana* (Brady 1881)

[*Lagena wrightiana* Brady 1881]
× 90. From *Challenger* Sta. 219A, Admiralty Islands, Pacific (17fm.). ZF1803.

Referred by Brady (1884) and Thalmann (1932) to *Lagena*, and by

Barker (1960) and van Marle (1991) to *Fissurina*. The internal tube of this form is well illustrated by Poignant (1984), who refers it to *Fissurina* sp. H). Van Marle (1991) regards *Lagena wrightiana* Brady 1881 as a *nomen nudum,* but gives no justification.

Figures 8–9 *Fissurina radiatomarginata* (Parker and Jones 1865)

[*Lagena radiatomarginata* Parker and Jones 1865]
Fig. 8 ×90.
Fig. 9 ×90. From *Challenger* Sta. 185, Torres Strait, Pacific (155fm.). ZF1745.

Referred by Brady (1884) and Thalmann (1932) to *Lagena,* and by Barker (1960) to *Fissurina.*

Figure 10 *Fissurina schulzeana* (Brady 1881)

[*Lagena schulzeana* Brady 1881]
×90. From *Challenger* Sta. 185, Torres Strait, Pacific (155fm.). ZF1746.

Referred by Brady (1884) and Thalmann (1932) to *Lagena,* and by Barker (1960) (albeit tentatively) to *Fissurina.* This species was referred by Patterson (1986) to the new genus *Cerebrina,* regarded by the present author as a junior synonym of *Fissurina* Reuss 1850.

Figure 11 *Galwayella oblonga* (Seguenza 1862)

[*Trigonulina oblonga* Seguenza 1862]
×75. From *Challenger* Sta. 279C, off Tahiti, Pacific (620fm.). ZF1797.

Referred by Brady (1884) and Thalmann (1932) to *Lagena trigono-oblonga* Seguenza and Siddall (*sic*), and by Barker (1960) to *Fissurina trigono-oblonga* (Seguenza and Siddall) (*sic*). The fact that *Trigonulina oblonga* Seguenza 1862 was perhaps an inappropriate name for this species did not constitute sufficient grounds for its redescription as *Lagena (trigono)oblonga* Siddall 1879 (International Code on Zoological Nomenclature, Ride *et al.* 1985, Articles 18 and 23(m)).

Figures 12–13 *Galwayella trigonomarginata* (Parker and Jones 1865)

[*Lagena sulcata* var. *trigono-marginata* Parker and Jones 1865]
Fig. 12a ×100. ZF1794. Fig. 12b ×100. ZF1795. From *Porcupine* Sta. 6, W. of Ireland, North Atlantic (90fm.).
Fig. 13 ×100. From *Challenger* Sta. 241, North Pacific. ZF1796.

Referred by Brady (1884) and Thalmann (1932) to *Lagena,* and by Barker (1960) to *Fissurina.*

Figure 14 *Galwayella trigonoornata* (Brady 1881)

[*Lagena trigono-ornata* Brady 1881]
×60. From *Challenger* Sta. 218, North of Papua, Pacific (1070fm.). Specimen lost.

Referred by Brady (1884) and Thalmann (1932) to *Lagena,* and by Barker (1960) to *Fissurina.* This species was referred by Patterson and Pettis (1986) to the new genus *Galwayella.*

Figures 15–16 *Lagena quinquelaterata* Brady 1881

Fig. 15 ×100. From *Challenger* Sta. 276, South Pacific (2350fm.). ZF1743.
Fig. 16 ×100. From *Challenger* Sta. 145, off Prince Edward Island, South Pacific (50–150fm.). ZF1744.

Referred by Brady (1884) and Thalmann (1932) to *Lagena,* and by Barker (1960) to *Fissurina.*

Figures 17–22 *Glandulina ovula* d'Orbigny 1846

Figs 17–19 ×30. From *Porcupine* Sta. 36, off SW Ireland, Atlantic (725fm.). ZF1974.
Figs 20–22 ×30. From *Challenger* Sta. 24, West Indies (390fm.). ZF1955.

Figures 17–19 were referred by Brady (1884) to *Nodosaria (Glandulina) rotundata* Reuss, by Thalmann (1932) to *Glandulina rotundata* Reuss, by Leroy (1944a) to *G. laevigata* d'Orbigny, and by Barker (1960) to *Rectoglandulina rotundata* (Reuss). Figures 20–22 were referred by Brady (1884) to *Nodosaria (Glandulina) laevigata* d'Orbigny, by Thalmann (1932), Leroy (1944a,b), and van Marle (1991) to *Glandulina laevigata* (d'Orbigny), and by Barker (1960) to *Rectoglandulina torrida* (Cushman). All are herein referred to *Glandulina ovula* d'Orbigny 1846 (Figs 17–19 being interpreted as representing megalospheric individuals and Figs 20–22 microspheric). This species has been lectotypified by Papp and Schmid (1985). *Glandulina laevigata* d'Orbigny 1846 (? not *Glandulina laevigata* d'Orbigny 1839) appears conspecific. Problems in glandulinid taxonomy are discussed by Nyholm (1973) and Taylor *et al.* (1985).

Figures 23–26 ?27 *Glandulonodosaria calomorpha* (Reuss 1865)

[*Nodosaria calomorpha* Reuss 1865]
Figs 23–26 ×50. From *Challenger* Sta. 315A, South Atlantic (6fm.). ZF1929.
Fig. 27 ×50. From *Challenger* Sta. 332, South Atlantic (2200fm.). ZF1930.

Referred by Brady (1884), Thalmann (1932), and Barker (1960) to *Nodosaria.* The specimen figured by Brady as Figure 25 is herein refigured on Supplementary Plate 1.

Figures 28–31 *Pseudoglandulina glanduliniformis* (Dervieux 1893)

[*Nodosaria radicula* var. *glanduliniformis* Dervieux 1893]
Fig. 28 ×30. ZF1968. Fig. 31 ×30. ZF1970. From *Challenger* Sta. 166, W. of New Zealand (275fm.).
Figs 29–30 ×30. From *Challenger* Sta. 174C, off Fiji, Pacific (210fm.). ZF1969.

Referred by Brady (1884) to *Nodosaria radicula* Linne, sp., by Thalmann first (1932) to *Glandulina radicula* (Linne) and later (1937) to *Glandulonodosaria radicula* (Linne), by Leroy (1944a,b) to *Nodosaria radicula* (Linne), by Barker (1960) to *N. radicula* var. *glanduliniformis* Dervieux, and by van Marle (1991) to *Pseudonodosaria radicula* (Linnaeus). *Nautilus radicula* Linne 1767 is the type-species by subsequent designation of the genus *Nodosaria* Lamarck 1812. *Pseudoglandulina* Cushman 1929 differs from *Nodosaria* Lamarck 1812 in the strongly embracive nature of its chambers. *Glandulonodo-*

saria Silvestri 1900 differs from both genera in possessing a rounded rather than radiate aperture.

Figure 32 *Pseudoglandulina aequalis* (Reuss 1863)

[*Glandulina aequalis* Reuss 1863]
× 30. From *Challenger* Sta. 24, West Indies (390fm.). ZF1927.

Referred by Brady (1884) to *Nodosaria* (*Glandulina*) *aequalis* Reuss, by Thalmann (1932) to *Glandulina aequalis* Reuss, by Barker (1960) to *Rectoglandulina* (?) *aequalis* (Reuss), and by van Marle (1991) to *Pseudonodosaria aequalis* (Reuss). *Pseudonodosaria* Boomgaart 1949 and *Rectoglandulina* Loeblich and Tappan 1955 are regarded by the present author as junior synonyms of *Pseudoglandulina* Cushman 1929.

Plate 62 (LXII)

Figures 1–2 *Orthomorphina jedlitschkai* (Thalmann 1937)

[*Nodogenerina jedlitschkai* Thalmann 1937]
× 50. From *Challenger* Sta. 192, Ki Islands, Pacific (129fm.). ZF1972.

Referred by Brady (1884) to *Nodosaria radicula* var. *annulata* Terquem and Berthelin, by Thalmann first (1932) to *N. annulata* (Terquem and Berthelin) and later (1937) to *Nodogenerina jedlitschkai* 'nom. nov.', and by Barker (1960) and Hermelin (1989) to *Orthomorphina jedlitschkai* (Thalmann). The specimen figured by Brady as Figure 2 is herein refigured on Supplementary Plate 2.

Figure 3 *Glandulonodosaria ambigua* (Neugeboren 1856)

[*Nodosaria ambigua* Neugeboren 1856]
× 50. From *Challenger* Sta. 192, Ki Islands, Pacific (129fm.). ZF1971.

Referred by Brady (1884) to *Nodosaria radicula* var. *ambigua* Neugeboren, by Thalmann first (1932) to *N. ambigua* Neugeboren and later (1942) to *Ellipsoglandulina rotunda* (d'Orbigny), and by Barker (1960) to *Orthomorphina ambigua* (Neugeboren). This is the type-species by original monotypy of the genus *Glandulonodosaria* Silvestri 1900. Brady's figured specimen is herein refigured on Supplementary Plate 1.

Figures 4–5 *Nodosaria simplex* Silvestri 1872

× 35–40. From *Challenger* Sta. 192, Ki Islands, Pacific (129fm.). ZF1980.

Referred by Brady (1884), Thalmann (1932), Barker (1960), and van Marle (1991) to *Nodosaria simplex* Silvestri. The generic placement of this species is somewhat questionable.

Figures 6?, 10–12 *Grigelis* sp. nov.

Fig. 6 × 35–40. From *Challenger* Sta. 166, West coast of New Zealand (275fm.). ZF1981.
Fig. 10 × 30–40. From *Challenger* Sta. 191A, Ki Islands, Pacific (580fm.). ZF1966.
Figs 11–12 × 30–40. From *Challenger* Sta. 209, Philippines, Pacific (95–100fm.). ZF1967.

Referred by Brady (1884), Thalmann (1932), and van Marle (1991) to *Nodosaria simplex* Silvestri (Fig. 6) and *N. pyrula* d'Orbigny (Figs

10–12), by Barker (1960) tentatively to *N. simplex* Silvestri (Fig. 6) and *Dentalina guttifera* d'Orbigny (Figs 10–12), and by Hofker (1976) to *N. pyrula* d'Orbigny (Figs 10–12 only). Brady's specimens appear distinct from both *Nodosaria pyrula* d'Orbigny 1826 (a junior synonym of *Orthoceras monile* Soldani 1798, which is the type-species by original designation of the genus *Grigelis* Mikhalevich 1981) and *Dentalina guttifera* d'Orbigny 1846, as lectotypified by Papp and Schmid 1985.

Figures 7–8 *Stilostomella fistuca* (Schwager 1866)

[*Nodosaria fistuca* Schwager 1866]
× 50. From *Challenger* Sta. 192, Ki Islands, Pacific (129fm.). ZF1980.

Referred by Brady (1884) to *Nodosaria subtertenuata* Schwager, by Thalmann (1932) to *N. fistuca* Schwager, and by Barker (1960) to *Stilostomella fistuca* (Schwager). Srinivasan and Sharma (1980) have recently designated a neotype for this species, which they also refer to *Stilostomella*. The specimen figured by Brady as Figure 8 is herein refigured on Supplementary Plate 2.

Figure 9 *Dentalina inflexa* (Reuss 1866)

[*Nodosaria* (*Dentalina*) *inflexa* Reuss 1866]
× 30. From *Challenger* Sta. 174C, off Fiji, Pacific (210fm.). ZF1951.

Referred by Brady (1884), Barker (1960), and van Marle (1991) to *Nodosaria,* and by Thalmann (1932) to *Dentalina.* Brady's figured specimen is herein refigured on Supplementary Plate 1.

Figures 13–16 *Dentalina subsoluta* (Cushman 1923)

[*Nodosaria subsoluta* Cushman 1923]
Figs 13–14 × 12. From *Challenger* Sta. 24, West Indies (390fm.). ZF1982.
Figs 15–16 × 12. From *Porcupine* Sta. 23, off NW Ireland, North Atlantic (630fm.). ZF1983.

Referred by Brady (1884) to *Nodosaria* (*Dentalina*) *soluta* Reuss, by Thalmann (1932) to *Dentalina soluta* (Reuss), by Barker (1960) and van Marle (1991) to *D. subsoluta* (Cushman), and by Hofker (1976) to *Nodosaria subsoluta* Cushman.

Figures 17–18 *Dentalina* sp. nov.

× 15, From *Challenger* Sta. 24, West Indies (390fm.). ZF1942.

Referred by Brady (1884) to *Nodosaria* (*Dentalina*) *farcimen* Soldani, sp. 'not typical', and by Thalmann (1932) and Barker (1960) to *Dentalina farcimen* (Soldani).

Figures 19–20 *Dentalina bradyensis* (Dervieux 1894)

[*Nodosaria inornata* var. *bradyensis* Dervieux 1894]
Fig. 19 × 35–40. From *Challenger* Sta. 24, West Indies (390fm.). ZF1934.
Fig. 20 × 35–40. From *Challenger* Sta. 33, off Bermuda (435fm.). ZF1935.

Referred by Brady (1884) to *Nodosaria* (*Dentalina*) *communis* d'Orbigny, by Thalmann first (1932) to *Dentalina communis* d'Orbigny and later (1942) to *Nodosaria inornata* var. *bradyensis* Dervieux, by Leroy (1944a) to *Dentalina* cf. *communis* d'Orbigny, by Barker (1960) to *D.*

inornata var. *bradyensis* (Dervieux), and by Hofker (1976) to *D. bradyensis* (Dervieux).

Figures 21–22 *Dentalina aphelis* (Loeblich and Tappan 1986)

[*Laevidentalina aphelis* Loeblich and Tappan 1986]
× 35–40. From *Challenger* Sta. 174A, off Fiji, Pacific (255fm.). ZF1936.

Referred by Brady (1884) to *Nodosaria* (*Dentalina*) *communis* d'Orbigny, by Thalmann (1932), Barker (1960), Hofker (1976), Hermelin (1989), and van Marle (1991) to *Dentalina communis* (d'Orbigny), and by Leroy (1944a) to *D.* cf. *communis* d'Orbigny. Loeblich and Tappan (1986) have recently pointed out that *Nodosaria* (*Dentalina*) *communis* d'Orbigny 1826 is an invalid species since d'Orbigny (1826) cited a valid prior species (*Nautilus recta* Montagu 1803) as a synonym. Loeblich and Tappan (1986) proposed *Laevidentalina aphelis* as a replacement name for forms previously erroneously referred to *Dentalina communis* (d'Orbigny) *L. aphelis* Loeblich and Tappan 1986 is the type-species by original designation of the genus *Laevidentalina* Loeblich and Tappan 1986, regarded herein as a junior synonym of *Dentalina* Risso 1826.

Figures 23–24 *Stilostomella consobrina* (d'Orbigny 1846)

[*Dentalina consobrina* d'Orbigny 1846]
Fig. 23 × 50. From *Challenger* Sta. 191A, Ki Islands, Pacific (580fm.). ZF1938.
Fig. 24 × 50. From *Challenger* Sta. 300, North of Juan Fernandez, East Pacific (1375fm.). ZF1937.

Referred by Brady (1884) to *Nodosaria* (*Dentalina*), by Thalmann (1932) to *Dentalina*, by Barker (1960) and van Marle (1991) to *Stilostomella*, and by Hermelin (1989) to *Siphonodosaria*. Papp and Schmid (1985) have recently lectotypified this species and transferred it to *Stilostomella*. The specimen figured by Brady as Figure 24 is herein refigured on Supplementary Plate 2. It clearly exhibits the characteristic apertural tooth.

Figures 25–26 *Dentalina subemaciata* Parr 1950

× 15. From *Challenger* Sta. 24, West Indies (390fm.). ZF1939.

Referred by Brady (1884) to *Nodosaria consobrina* var. *emaciata* Reuss, by Thalmann (1932) to *Dentalina emaciata* Reuss, and by Barker (1960) (albeit tentatively) to *D. subemaciata* Parr.

Figures 27–31 *Dentalina ariena* Patterson and Pettis 1986

Fig. 27 × 50. From *Challenger* Sta. 300. North of Juan Fernandez, East Pacific (1375fm.). ZF1956.
Figs 28–29 × 50; Fig. 31 × 40. From *Challenger* Sta. 332, South Atlantic (2200fm.). ZF1957.
Fig. 30 × 40. From *Challenger* Sta. 272, South Pacific (2600fm.). ZF1958.

Referred by Brady (1884) to *Nodosaria* (*Dentalina*) *mucronata* Neugeboren, by Thalmann (1932) to *Dentalina mucronata* Neugeboren, by Barker (1960) and Hermelin (1989) to *D. intorta* (Dervieux), and by Patterson and Pettis (1986) to *D. ariena. Dentalina intorta* (Dervieux 1894) is a secondary (subjective) junior homonym of *D. intorta* Terquem 1870.

Plate 63 (LXIII)

Figure 1 *Dentalina advena* (Cushman 1923)

[*Nodosaria advena* Cushman 1923]
× 20. From *Challenger* Sta. 24, West Indies (390fm.). ZF1975.

Referred by Brady (1884) to *Nodosaria* (*Dentalina*) *roemeri* Neugeboren, by Thalmann first (1932) to *N. advena* Cushman and later (1933a) to *Dentalina advena* (Cushman), and also by Barker (1960) and van Marle (1991) to *D. advena* (Cushman).

Figure 2 *Dentalina plebeia* Reuss 1855

× 20. From *Challenger* Sta. 33, off Bermuda (435fm.). ZF1964.

Referred by Brady (1884) to *Nodosaria* (*Dentalina*), and by Thalmann (1932) and Barker (1960) to *Dentalina*.

Figures 3–5 *Dentalina filiformis* (d'Orbigny 1826)

[*Nodosaria* (*Nodosaire*) *filiformis* d'Orbigny 1826]
Fig. 3 × 10. ZF1943. Fig. 4 × 20. ZF1935. From *Challenger* Sta. 33, off Bermuda (435fm.).
Fig. 5 × 20. From *Challenger* Sta. 145, Prince Edward Island, South Pacific. ZF1944.

Referred by Brady (1884) to *Nodosaria* (*Dentalina*), and by Thalmann (1932), Barker (1960), and van Marle (1991) to *Dentalina*.

Figure 6 *Glandulina ovula* d'Orbigny 1846

× 50. From *Porcupine* Sta. 36, SW of Ireland, Atlantic (725fm.). ZF1928.

Referred by Brady (1884) to *Nodosaria* (*Glandulina*) *armata* Reuss, sp., by Thalmann (1932) to *Glandulina armata* Reuss, by Barker (1960) to *G.* (?) *armata* Reuss, and by the present author to *G. ovula* d'Orbigny (see also Plate 61, Figs 17–22)

Figure 7 *Stilostomella retrorsa* (Reuss 1863)

[*Nodosaria retrorsa* Reuss 1863]
× 50. From *Challenger* Sta. 191A, Ki Islands, Pacific (580fm.). ZF1938.

Referred by Brady (1884) to *Nodosaria* (*Dentalina*), by Thalmann (1932) to *Dentalina*, and by Barker (1960) to *Orthomorphina*. The generic placement of this species is somewhat questionable.

Figures 8–9 *Stilostomella abyssorum* (Brady 1881)

[*Nodosaria* (?) *abyssorum* Brady 1881]
× 20. From *Challenger* Sta. 296, South Pacific (1825fm.). ZF1926. Lectotype designated by Loeblich and Tappan (1964) (ZF3649) ex this slide.

Referred by Brady (1884) to *Nodosaria* (?) *abyssorum* Brady, by Thalmann first (1932) to *Sagrinnodosaria abyssorum* (Brady) and later (1937) to *Siphonodosaria pauperata* (d'Orbigny), by Barker (1960) and van Marle (1991) to *Stilostomella abyssorum* (Brady), and by Hermelin (1989) to *Siphonodosaria abyssorum* (Brady). *Nodosaria abyssorum* Brady 1881, regarded by van Marle (1991) as a *nomen nudum* (for no apparent reason), is the type-species by subsequent designation of the genus *Siphonodosaria* Silvestri 1924, regarded by the present author as a junior synonym of *Stilostomella* Guppy 1894. The specimen figured by Brady as Figure 8 is herein refigured on Supplementary Plate 2.

Figures 10–11 *Amphicoryna papillosa* (Silvestri 1872)

[*Nodosaria papillosa* Silvestri 1872]
Fig. 10 ×75. From *Challenger* Sta. 174C, Fiji, Pacific (210fm.). ZF1945.
Fig. 11 ×75. From *Challenger* Sta. 217A, off Papua, Pacific (37fm.). ZF1946.

Referred by Brady (1884) and Thalmann (1932) to 'young or arrested' specimens of *Nodosaria hispida* d'Orbigny or *N. setosa* Schwager (*N. papillosa* Silvestri?), by Thalmann (1937) to *Nodogenerina hispida* (d'Orbigny), and by Barker (1960) to *Amphicoryna hirsuta* (d'Orbigny).

Figures 12–15 *Amphicoryna hirsuta* (d'Orbigny 1826)

[*Nodosaria (Nodosaire) hirsuta* d'Orbigny 1826]
Fig. 12 ×45; Figs 13–14 ×30. From *Challenger* Sta. 167, West coast of New Zealand (150fm.). ZF1947.
Fig. 15 ×25. From *Challenger* Sta. 166, West coast of New Zealand (275fm.). ZF1948.

Referred by Brady (1884) to *Nodosaria hispida* d'Orbigny, by Thalmann first (1932) to *N. hirsuta* d'Orbigny, later (1933a) to *Nodogenerina hirsuta* (d'Orbigny), and later still (1951) to *Amphicoryna hirsuta* (d'Orbigny), and also by Barker (1960) to *A. hirsuta* (d'Orbigny).

Figure 16 *Amphicoryna* sp. nov. (1)

×25. From *Challenger* Sta. 24, West Indies (390fm.). ZF1949.

Referred by Brady (1884) to *Nodosaria hispida* d'Orbigny, by Thalmann (1932) to *N. hirsuta* d'Orbigny, and by Barker (1960) to *Amphicoryna hirsuta* (d'Orbigny) (see above).

Figures 17–18 *Glandulonodosaria* sp. nov.

Fig. 17 ×50; Fig. 18 ×100. From *Challenger* Sta. 192, Ki Islands, Pacific (129fm.). ZF1986.

Referred by Brady (1884) and Thalmann (1932) to *Nodosaria verruculosa,* and by Barker (1960) to *Orthomorphina* (?) sp. nov. Brady's figured specimen is herein refigured on Supplementary Plate 1.

Figures 19–22 *Amphicoryna sublineata* (Brady 1884)

[*Nodosaria hispida* var. *sublineata* Brady 1884]
Figs 19–21 ×35; Fig. 22 ×100. From *Challenger* Sta. 33, off Bermuda, Atlantic (435fm.). ZF1950.

Referred by Brady (1884) to *Nodosaria hispida* var. *sublineata,* by Thalmann first (1932) to *Nodosaria sublineata* (Brady) and later

(1937) to *Nodogenerina sublineata* (Brady), and by Barker (1960) to *Amphicoryna* (?) *sublineata* (Brady).

Figures 23–27 *Grigelis semirugosa* (d'Orbigny 1846)

[*Nodosaria semirugosa* d'Orbigny 1846]
Figs 23, 25–26 ×15; Fig. 27 ×30. From *Challenger* Sta. 33, off Bermuda (435fm.). ZF1941.
Fig. 24 ×15. From *Challenger* Sta. 24, West Indies (390fm.). ZF1940.

Referred by Brady (1884) to *Nodosaria costulata* Reuss, by Thalmann (1937) to *N. pyrula* var. *semirugosa* (d'Orbigny), and by Barker (1960) to *Dentalina guttifera* var. *semirugosa* (d'Orbigny). Papp and Schmid (1985) synonymized *Nodosaria semirugosa* d'Orbigny 1846 with *N. pyrula* d'Orbigny 1826, but are not followed here. Loeblich and Tappan (1987) figure specimens of *Grigelis semirugosa* (d'Orbigny) from Puerto Rico (pl. 441, fig. 1) that are practically identical to Brady's.

Figures 28–31 *Amphicoryna scalaris* (Batsch 1791)

[*Nautilus (Orthoceras) scalaris* Batsch 1791]
Fig. 28 ×45–50. From *Porcupine* Sta. 10, West of Ireland, North Atlantic (85fm.). ZF1977.
Figs 29–31 ×45–50. From *Challenger* Sta. 209, Philippines, Pacific (95fm.). ZF1976.

Referred by Brady (1884) and Thalmann (1932) to *Nodosaria,* by Thalmann (1937) to *Nodogenerina,* by Leroy (1944a,b) to *Lagenonodosaria,* and by Barker (1960) and van Marle (1991) to *Amphicoryna.* This is the type-species, by synonymy with *Nodosaria (Marginulina) raphanus* var. *falx* Jones and Parker 1860 (the erstwhile type-species fixed by subsequent designation) of the genus *Amphicoryna* Schlumberger 1881.

Figures 32–34 *Dentalina catenulata* (Brady 1884)

[*Nodosaria catenulata* Brady 1884]
Figs 32–33 ×40; Fig. 34 ×100. From *Challenger* Sta. 209, Philippines, Pacific (95–100fm.). ZF1931.

Referred by Brady (1884), Thalmann (1932), Barker (1960), and van Marle (1991) to *Nodosaria.* The specimen figured by Brady as Figure 32 is herein refigured on Supplementary Plate 1. The rounded aperture of this species is atypical of the genus.

Figure 35 *Dentalina albatrossi* (Cushman 1923)

[*Nodosaria vertebralis* var. *albatrossi* Cushman 1923]
×40. From *Challenger* Sta. 209, Philippines, Pacific (95–100fm.). ZF1989.

Referred by Brady (1884) to *Nodosaria vertebralis* Batsch, sp., by Thalmann first (1932) to *N. vertebralis* var. *albatrossi* Cushman and later (1937) to *Nodogenerina albatrossi* (Cushman), by Leroy (1944a) to *Nodosaria* aff. *vertebralis* (Batsch), and by Barker (1960) (apparently on the understanding that Brady's specimens were distinct from Cushman's) to *N. vertebralis* subsp. nov. Brady's figured specimen is herein refigured on Supplementary Plate 1. The rounded aperture of this species is atypical of the genus.

Plate 64 (LXIV)

Figures 1–5 *Pseudoglandulina comatula* (Cushman 1923)

[*Nodosaria comatula* Cushman 1923]
Figs 1–2, 4–5 × 50. From *Challenger* Sta. 33, off Bermuda, Atlantic
(435fm.). ZF1932.
Fig. 3 × 50. From *Challenger* Sta. 23, West Indies (450fm.). ZF1933.

Referred by Brady (1884) to *Nodosaria comata* Batsch, sp., by
Thalmann first (1932) to *Glandulina comatula* (Cushman) and later
(1942) to *Pseudoglandulina comatula* (Cushman), by Barker (1960) to
Rectoglandulina comatula (Cushman), and by van Marle (1991) to
Pseudonodosaria comatula (Cushman). *Pseudonodosaria* Boomgaart
1949 and *Rectoglandulina* Loeblich and Tappan 1955 are regarded by
the present author as junior synonyms of *Pseudoglandulina* Cushman
1929. Note, however, that *Rectoglandulina* Loeblich and Tappan
1955 is regarded by Loeblich and Tappan (1987) as a junior synonym
of *Pyramidulina* Fornasini 1894.

Figures 6–10 *Nodosaria lamnulifera* Thalmann 1950

Figs 6–7, 9–10 × 15. Fig. 8 × 15. ZF1973. From *Challenger* Sta.
174C, off Fiji, Pacific (210fm.).

Referred by Brady (1884) and Thalmann (1932) to *Nodosaria
raphanus* (Linné), and by Thalmann (1950) and Barker (1960) to *N.
lamnulifera*, a new name for *N. bradyi* Boomgaart 1949, a secondary
(subjective) junior homonym of *N. bradyi* (Spandel 1901).

Figures 11–14 *Dentalina albatrossi* (Cushman 1923)

[*Nodosaria vertebralis* var. *albatrossi* Cushman 1923]
Figs 11–12 × 15–18; Fig. 14 × 30. ZF1987. Fig. 13 × 15–18.
ZF1988. From *Challenger* Sta. 33, off Bermuda (435fm.).

Referred by Brady (1884) to *Nodosaria vertebralis* Batsch, sp., by
Thalmann first (1932) to *N. vertebralis* var. *albatrossi* Cushman and
later (1937) to *Nodogenerina albatrossi* (Cushman), and by Barker
(1960) to *Nodosaria albatrossi* Cushman (see also Plate 63, Fig. 35).

Figure 15 *Amphicoryna proxima* (Silvestri 1872)

[*Nodosaria proxima* Silvestri 1872]
× 60. From *Challenger* Sta. 209, Philippines, Pacific (95–100fm.).
ZF1965.

Referred by Brady (1884) and Thalmann (1932) to *Nodosaria*, by
Thalmann (1937) to *Nodogenerina*, and by Barker (1960) to
Amphicoryna.

Figures 16–19 *Amphicoryna separans* (Brady 1884)

[*Nodosaria scalaris* var. *separans* Brady 1884]
Figs 16, 19 × 40. Specimens lost (1948). Figs 17–18 × 40. ZF1979.
From *Challenger* Sta. 166, West coast of New Zealand, South Pacific
(275fm.).

Referred by Brady (1884) to *Nodosaria scalaris* var. *separans*, by
Thalmann first (1932) to *Lagenonodosaria separans* (Brady) and later
(1937) to *Nodogenerina separans* (Brady), and by Barker (1960) to
Amphicoryna separans (Brady). This is the type-species by subsequent

designation of the genus *Lagenonodosaria* Silvestri 1900, regarded
herein as a junior synonym of *Amphicoryna* Schlumberger 1881.

Figures 20–22 *Dentalina flintii* (Cushman 1923)

[*Nodosaria flintii* Cushman 1923]
Figs 20–21 × 16. From *Porcupine* Sta. 51, Faroe Channel, Scotland,
North Atlantic (440fm.). ZF1960.
Fig. 22 × 16. From *Porcupine* Sta. 23, off NW Ireland, North Atlantic
(630fm.). ZF1961.

Referred by Brady (1884) to *Nodosaria obliqua* Linné, sp., by
Thalmann first (1932) to *N. flintii* Cushman and later (1933a) to *N.
vertebralis* (Batsch), and by Barker (1960) and Hofker (1976) to *N.
flintii* Cushman.

Figures 23–24 *Amphicoryna substriatula* (Cushman 1917)

[*Nodosaria substriatula* Cushman 1917]
Fig. 23 × 40; Fig. 24 × 80. From *Challenger* Sta. 297A, off Tahiti,
Central Pacific (420fm.). ZF1985.

Referred by Brady (1884) to *Nodosaria* (*Dentalina*) *subcanaliculata*
Neugeboren, var., by Thalmann (1932) and Barker (1960) to *N.
substriatula*, and by Hofker (1976) (evidently in error) to *N. flintii*
Cushman (Fig. 23 only). The generic identity is questionable.

Figures 25–27 *Orthomorphina challengeriana* (Thalmann 1937)

[*Nodogenerina challengeriana* Thalmann 1937]
× 50–60. From *Challenger* Sta. 192, Ki Islands, Pacific (129fm.).
ZF1962.

Referred by Brady (1884) to *Nodosaria perversa* Schwager, also by
Thalmann first (1932) to *N. perversa* Schwager and later (1937) to
Nodogenerina challengeriana, and by Barker (1960), Hermelin (1989),
and van Marle (1991) to *Orthomorphina challengeriana* (Thalmann).
The specimen figured by Brady as Figure 25 is herein refigured on
Supplementary Plate 2.

Figure 28 *Dentalina subsoluta* (Cushman 1923), var.

× 15. From *Challenger* Sta. 23, West Indies (450fm.). ZF1984.

Referred by Brady (1884) and Thalmann (1932) to *Nodosaria soluta*,
'striate variety', and by Barker (1960) to *Dentalina subsoluta*
(Cushman).

Plate 65 (LXV)

Figures 1–4 *Amphicoryna intercellularis* (Brady 1881)

[*Nodosaria intercellularis* Brady 1881]
Figs 1–2 × 40; Figs 3–4 × 100. From *Challenger* Sta. 33, off Bermuda,
Atlantic (435fm.). ZF1954.

Referred by Brady (1884) to *Nodosaria* (*Dentalina*), by Thalmann
(1932) to *Dentalina*, by Thalmann (1937) to *Nodogenerina*, and by
Barker (1960) to *Amphicoryna* (see also Text-figure 15). The generic
identity of this species is somewhat questionable in that it appears to
possess a compound wall.

Figures 5–6 *Marginulina obesa* (Cushman 1923)

[*Marginulina glabra* var. *obesa* Cushman 1923]
Fig. 5 × 30. ZF1809. Fig. 6 × 30. ZF1810. From *Challenger* Sta. 246, North Pacific (2050fm.).

Referred by Brady (1884), Thalmann (1932), and van Marle (1991) to *Marginulina glabra* d'Orbigny, and by Barker (1960) to *M. obesa* (Cushman).

Figures 7–9 *Amphicoryna scalaris* (Batsch 1798)

[*Nautilus* (*Orthoceras*) *scalaris* Batsch 1791]
Fig. 7 × 50–60. ZF1066. Figs 8–9 × 50–60. ZF1065. From *Challenger* Sta. 166, West coast of New Zealand (275fm.).

Referred by Brady (1884) and Thalmann (1932) to *Amphicoryne* (sic) *falx* (Jones and Parker), and by Barker (1960) to *Amphicoryna scalaris* (Batsch). This is the type-species of the genus *Amphicoryna* Schlumberger 1881 (see also Plate 63, Figs 28–31).

Figure 10 Nodosariid juvenarium

× 40. From *Challenger* Sta. 346, South Atlantic (2350fm.).

Referred by Brady (1884) and Thalmann (1932) to *Marginulina costata* (Batsch) (see below), and by Barker (1960) (albeit tentatively) to *Marginulinopsis bradyi* (Goes) (see below).

Figures 11–12 *Vaginulinopsis bradyi* (Goes 1894)

[*Cristellaria bradyi* Goes 1894]
Fig. 11 × 30. From *Challenger* Sta. 26, West Indies. ZF1808. Specimen figured by Loeblich and Tappan (1964; fig. 403.9). Fig. 12 × 20. From *Challenger* Sta. 33, off Bermuda, Atlantic (435fm.).

Referred by Brady (1884) to *Marginulina costata* Batsch, sp., also by Thalmann first (1932) to *M. costata* (Batsch) and later (1937) to *Marginulinopsis densicostata* nov., and by Barker (1960) to *M. bradyi* (Goes). This is the type-species, by synonymy with *Marginulinopsis densicostata* Thalmann 1937 (the erstwhile type-species fixed by subsequent designation), of the genus *Marginulinopsis* Silvestri 1904. *Marginulinopsis* Silvestri 1904 is regarded by the present author as a junior synonym of *Vaginulinopsis* Silvestri 1904.

Figure 13 *Marginulina costata* (Batsch 1791)

[*Nautilus* (*Orthoceras*) *costatus* Batsch 1791]
× 20. From *Challenger* Sta. 323, South Atlantic (1900fm.). ZF1807.

Referred by Brady (1884) and Thalmann (1932) to *Marginulina costata* (Batsch), and by Barker (1960) to *Marginulinopsis bradyi* (Goes) (see above).

Figures 14–15 *Lingulina seminuda* (Hantken 1875)

[*Lingulina costata* var. *seminuda* Hantken 1875]
Fig. 14 × 12. 1959.5.5.617. Fig. 15 × 12. From *Challenger* Sta. 24, off Culebra Island, West Indies (390fm.).

Referred by Brady (1884) to *Lingulina carinata* var. *seminuda* Hantken, and by Thalmann (1932), Barker (1960), and Hofker (1976) to *L. seminuda* Hantken.

Figure 16 *Frondicularia bradii* (Silvestri 1903)

[*Lingulonodosaria bradii* Silvestri 1903]
× 90. From *Challenger* Sta. 191A, Ki Islands, Pacific (580fm.). ZF1805.

Referred by Brady (1884) to *Lingulina carinata* d'Orbigny, also by Thalmann first (1932) to *L. carinata* d'Orbigny, later (1937) (erroneously citing Fig. 6 rather than Fig. 16) to *Lingulonodosaria bradyi* (sic) Silvestri, and later still (1942) to *L. bradii* Silvestri, and by Barker (1960) to *Frondicularia bradii* (Silvestri).

Figure 17 *Lingulina* sp. nov.

× 90. From *Challenger* Sta. 135, Tristan d'Acunha, Atlantic (100–150fm.). ZF1804.

Referred by Brady (1884) and Thalmann (1932) to *Lingulina carinata* d'Orbigny, and by Barker (1960) to *L.* sp. nov.

Figure 18 *Frondicularia kiensis* Barker 1960

× 60. From *Challenger* Sta. 192, Ki Islands, Pacific (129fm.). ZF1447.

Referred by Brady (1884) to *Frondicularia spathulata*, by Thalmann (1932) to *F. bradyi* Cushman, and by Barker (1960) to *F. kiensis* nom. nov. *Frondicularia spathulata* Brady 1884 is a primary junior homonym of *F. sphathulata* Williamson 1858. *F. bradyi* Cushman 1913 is a secondary (subjective) junior homonym of *F. bradii* (Silvestri 1903)

Figure 19 *Frondicularia compta* Brady 1879

× 60. From *Challenger* Sta. 162, Bass Strait, Pacific (38–40fm.). Holotype. ZF1443.

Figures 20–23 *Frondicularia sagittula* (van den Broeck 1876)

[*Frondicularia alata* var. *sagittula* van den Broeck 1876]
Fig. 20 × 15. From *Challenger* Sta. 24, off Culebra Island, West Indies (390fm.). 1959.5.5.341.
Fig. 21 × 15. ZF1439. Fig. 22 × 15; Fig. 23 × 20. ZF1440. From *Challenger* Sta. 33, off Bermuda (435fm.).

Referred by Brady (1884) and Thalmann (1932) to *Frondicularia alata* d'Orbigny, and by Barker (1960) and Hofker (1976) to *F. sagittula* (van den Broeck).

Plate 66 (LXVI)

Figures 1–2 *Frondicularia robusta* Brady 1884

Fig. 1 × 75. From *Challenger* Sta. 192, Ki Islands, Pacific (129fm.). ZF1446.
Fig. 2 × 17. From *Challenger* Sta. 260A, off Honolulu (40fm.). 1959.5.5.345.

Referred by Brady (1884), Thalmann (1932), Barker (1960), Hofker (1976), and van Marle (1991) to *Frondicularia robusta*.

Figures 3–5 *Frondicularia sagittula* (van den Broeck 1896)

[*Frondicularia alata* var. *sagittula* van den Broeck 1896]
Figs 3, 5 × 15–20. From *Challenger* Sta. 33, off Bermuda (435fm.).
ZF1440.
Fig. 4 × 15–20. From *Challenger* Sta. 24, West Indies (390fm.).
ZF1441.

Referred by Brady (1884) to *Frondicularia alata* d'Orbigny, by
Thalmann (1932) to *F. sagittula* var. *lanceolata* (van den Broeck), by
Barker (1960) to *F. sagittula* (van den Broeck), and by van Marle
(1991) to *F. lanceolata* van den Broeck. Hofker (1976) synonymized
Frondicularia alata var. *lanceolata* van den Broeck 1876 with *F. alata*
var. *sagittula* van den Broeck, 1876 (see also Plate 65, Figs 20–23).

Figures 6?–7 *Plectofrondicularia helenae* (Chapman 1941)

[*Parafrondicularia helenae* Chapman 1941]
× 60. From *Challenger* Sta. 192, Ki Islands, Pacific (129fm.). ZF1447.

Referred by Brady (1884) to *Frondicularia interrupta* Karrer, also by
Thalmann first (1932) to *F. interrupta* Karrer and later (1942) to
Plectofrondicularia interrupta (Karrer), by Barker (1960) to *Parafrondicularia helenae* Chapman, and by van Marle (1991) to *Plectofrondicularia helenae* (Chapman). *Frondicularia interrupta* Karrer 1877 is a
primary junior homonym of *F. interrupta* Costa 1855. *Plectofrondicularia messinae* Souaya 1965, the new name for *Frondicularia
interrupta* Karrer 1877, not Costa 1855, is a probable junior synonym
of *Parafrondicularia helenae* Chapman 1941. *Parafrondicularia* Asano
1938 is regarded by the present author as a junior synonym of
Plectofrondicularia Liebus 1902.

Figures 8–12 *Plectofrondicularia advena* (Cushman 1923)

[*Frondicularia advena* Cushman 1923]
Figs 8–10 × 60. From *Challenger* Sta. 191A, Ki Islands, Pacific
(580fm.). ZF1444.
Fig. 11 × 60; Fig. 12 × 45. From *Challenger* Sta. 192, Ki Islands,
Pacific (129fm.). ZF1445.

Referred by Brady (1884) to *Frondicularia inaequalis* Costa, by
Thalmann (1923) to *F. advena* Cushman, and by Barker (1960) and
van Marle (1991) to *Parafrondicularia advena* (Cushman). This is the
type-species by original designation of the genus *Proxifrons* Vella
1963, regarded by the present author as a junior synonym of
Plectofrondicularia Liebus 1902.

Figure 13 *Vaginulinopsis sublegumen* Parr 1950

× 30. From Mounts Bay, Cornwall, England. ZF2587.

Referred by Brady (1884) and Thalmann (1932) to *Vaginulina
legumen* (Linné), and by Barker (1960) and van Marle (1991) to
Vaginulinopsis sublegumen Parr.

Figures 14–15 *Vaginulina subelegans* Parr 1950

Fig. 14 × 30. From Mounts Bay, Cornwall, England. ZF2587.
Fig. 15 × 30. From *Challenger* Sta. 279C, off Tahiti, Pacific (620fm.).
ZF2586.

Referred by Brady (1884) and Thalmann (1932) to *Vaginulina
legumen* (Linné), by Barker (1960) to *Vaginulinopsis sublegumen* Parr

(Fig. 14) and *Vaginulina subelegans* Parr? (Fig. 15), and by van Marle
(1991) to *Vaginulinopsis sublegumen* Parr (Fig. 14 only).

Figure 16 *Vaginulina margaritifera* (Batsch 1791)

[*Nautilus (Orthoceras) margaritifera* Batsch 1791]
× 13. From *Porcupine* Sta. 27, Rockall Bank, East Atlantic. ZF2590.

Figures 17?, 21–23 *Vaginulinopsis tenuis* (Bornemann 1855)

[*Marginulina tenuis* Bornemann 1855]
Fig. 17 × 25; Fig. 23 × 20. ZF1339. Fig. 22 × 20. ZF1350. From
Challenger Sta. 185, Torres Strait, Pacific (155fm.).
Fig. 21 × 35. From *Challenger* Sta. 166, West coast of New Zealand
(275fm.). ZF1349.

Referred by Brady (1884) to *Cristellaria obtusata* Reuss, sp. (Fig. 17)
and *C. tenuis* Bornemann, sp. (Figs 21–23), by Thalmann (1932) first
to *Lenticulina obtusata* (Reuss) (Fig. 17) and *L. tenuis* (Bornemann)
(Figs 21–23) and later (1933a) to *Marginulina tenuis* Bornemann (Figs
21–23 only), and by Barker (1960) to *Lenticulina obtusata* (Reuss)?
(Fig. 17) and *Marginulina tenuis* Bornemann? (Figs 21–23). Barker
(1960) noted that 'the "*Lenticulina obtusata* (Reuss)" of fig. 17 may
be an abnormal development' of *Marginulina tenuis* (Bornemann).

Figures 18–19 *Astacolus bradyi* (Cushman 1917)

[*Vaginulina bradyi* Cushman 1917]
× 17. From *Challenger* Sta. 185, Torres Strait, Pacific (155fm.).
ZF2585.

Referred by Brady (1884) to *Vaginulina bruckenthali* Neugeboren, by
Thalmann (1932) and Barker (1960) to *V. bradyi* Cushman, and by
Leroy (1944b) to *V.* aff. *bruckenthali* (Neugeboren). The vestigial
planispire indicates placement in *Astacolus* rather than *Vaginulina*.

Figure 20 *Amphicoryna bradii* (Silvestri 1902)

[*Nodosariopsis bradii* Silvestri 1902]
× 60. From *Challenger* Sta. 279C, off Tahiti, Pacific (620fm.).

Referred by Brady (1884) to an 'intermediate specimen with
Vaginuline commencement, and final Nodosarian chamber', by
Thalmann first (1932) to *Amphicoryne* (sic) n. sp. and later (1942) to
Nodosariopsis bradii Silvestri, and by Barker (1960) to *Amphicoryna
bradii* (Silvestri).

Figures 24–25 *Vaginulinopsis albatrossi* (Cushman 1923)

[*Cristellaria albatrossi* Cushman 1923]
× 20. From *Challenger* Sta. 23, West Indies (450fm.). ZF1340.

Referred by Brady (1884) to *Cristellaria obtusata* var. *subalata*, by
Thalmann first (1932) to *Lenticulina subalata* (Brady) and later (1937)
to *Vaginulinopsis subalata* (Brady), and by Barker (1960) to *Lenticulina
albatrossi* (Cushman). *Cristellaria obtusata* var. *subalata* Brady 1884 is
a junior homonym of *C. subalata* Reuss 1854.

Plate 67 (LXVII)

Figures 1–3 *Trifarina bradyi* Cushman 1923

Figs 1–2 × 75. From *Challenger* Sta. 33, off Bermuda (435fm.).
ZF2309.

Fig. 3 × 75. From *Challenger* Sta. 192, Ki Islands, Pacific (129fm.). 1959.5.11.193.

Referred by Brady (1884) to *Rhabdogonium tricarinata* d'Orbigny, sp., and by Thalmann (1932), Barker (1960), and Belford (1966) to *Trifarina bradyi* Cushman. This is the type-species by original designation of the genus *Trifarina* Cushman 1923.

Figures 4–6 *Trifarina reussi* (Cushman 1913)

[*Triplasia reussi* Cushman 1913]
× 75. From *Challenger* Sta. 192, Ki Islands, Pacific (129fm.). ZF2308.

Referred by Brady (1884) to *Rhabdogonium minutum* Reuss, and by Thalmann (1932), Barker (1960), and van Marle (1991) to *Trifarina reussi* (Cushman).

Figure 7 *Vaginulinopsis tasmanica* Parr 1950

× 25. From *Challenger* Sta. 33, off Bermuda (435fm.). ZF1344.

Referred by Brady (1884) to *Cristellaria schloenbachi* Reuss (an essentially Early Cretaceous species), by Thalmann first (1932) to *Lenticulina schloenbachi* (Reuss) and later (1933a) to *Astacolus schloenbachi* (Reuss), and by Barker (1960) (albeit tentatively) to *Vaginulinopsis tasmanica* Parr.

Figure 8 *Astacolus pacificus* (Cushman and Hanzawa 1936)

[*Polymorphinella pacifica* Cushman and Hanzawa 1936]
× 60. From *Challenger* Sta. 185, Torres Strait, Pacific (155fm.). ZF1348.

Referred by Brady (1884) to *Cristellaria schloenbachi* Reuss, 'anomalous specimen', by Thalmann first (1932) to *Lenticulina schloenbachi* (Reuss), later (1933a) to *Astacolus schloenbachi* (Reuss), and later still (1937) to *Polymorphinella pacifica* Cushman and Hanzawa, and by Barker (1960) to *Vaginulinopsis pacificus* (Cushman and Hanzawa). *Polymorphinella* Cushman and Hanzawa 1936 is regarded by the present author as a junior synonym of *Astacolus* de Montfort 1808.

Figure 9 *Saracenaria volpicelli* (Costa 1855)

[*Cristellaria volpicelli* Costa 1855]
× 50. From *Challenger* Sta. 335, South Atlantic (1425fm.). ZF1333.

Referred by Brady (1884) to *Cristellaria italica* var. *volpicelli* Costa, and by Thalmann (1932) and Barker (1960) to *Saracenaria italica* var. *volpicelli* (Costa).

Figures 10–12 *Vaginulina americana* Cushman 1923

× 15. From *Challenger* Sta. 24, West Indies (390fm.). ZF2589.

Referred by Brady (1884) to *Vaginulina linearis* Montagu, sp., and by Thalmann (1932) and Barker (1960) to *V. americana* Cushman.

Figures 13–14 *Vaginulina spinigera* Brady 1881

Fig. 13 × 17. From *Porcupine* Sta. 51, Faroe Channel, North Atlantic (440fm.). ZF2593.
Fig. 14 × 17. From *Porcupine* Sta. 23, off NW Ireland, Atlantic (630fm.). ZF2592.

Referred by Brady (1884), Thalmann (1932), Barker (1960), and Hofker (1976) to *Vaginulina spinigera* Brady.

Figures 15–16 *Planularia patens* (Brady 1884)

[*Vaginulina patens* Brady 1884]
× 75. From *Challenger* Sta. 209, Philippines (95–100fm.). ZF2591.

Referred by Brady (1884) and Barker (1960) to *Vaginulina*, and by Thalmann (1932) to *Planularia*. The ICZN has recently ruled in favour of the conservation of *Planularia* Defrance 1826 (Anon 1990a).

Figure 17 *Astacolus insolitus* (Schwager 1866)

[*Cristellaria insolita* Schwager 1866]
× 60. From *Challenger* Sta. 192, Ki Islands, Pacific (129fm.). ZF1316.

Referred by Brady (1884) to *Cristellaria crepidula* Fichtel and Moll, sp., by Thalmann first (1932) to *Robulus* (?*Planularia*) *crepidula* (Fichtel and Moll) and later (1937) to *Astacolus crepidulus* (Fichtel and Moll), by Barker (1960) to *A. insolitus* (Schwager), and by Hofker (1976) to *Vaginulina insolitus* (Schwager). This species has been neotypified and reillustrated by Srinivasan and Sharma (1980), who refer to it as *Astacolus insolitus* (Schwager).

Figure 18 *Planularia* sp. nov.

× 25. From *Challenger* Sta. 162, Bass Strait, Pacific (38–40fm.). ZF1334.

Referred by Brady (1884) to *Cristellaria lata* Cornuel, sp., by Thalmann (1932) to *Planularia* n. sp., and by Barker (1960) to *Astacolus* sp. nov. The ICZN has recently ruled in favour of the conservation of *Planularia* Defrance 1826 (Anon 1990a).

Figure 19 *Astacolus* sp. nov.

× 50. From *Challenger* Sta. 75, off the Azores, Atlantic (450fm.). ZF1317.

Referred by Brady (1884) to *Cristellaria crepidula* Fichtel and Moll, sp., by Thalmann first (1932) to *Robulus* (?*Planularia*) n. sp. and later (1933a) to *Astacolus* n. sp., and by Barker (1960) to *Astacolus* sp. nov.

Figure 20 *Astacolus crepidulus* (Fichtel and Moll 1798)

[*Nautilus crepidula* Fichtel and Moll 1798]
Fig. 20a × 50. From *Challenger* Sta. 23, West Indies (450fm.). ZF1318.
Fig. 20b × 50. From *Challenger* Sta. 33, off Bermuda (435fm.). ZF1319.

Referred by Brady (1884) to *Cristellaria crepidula* Fichtel and Moll, sp., by Thalmann (1932) to *Robulus* (?*Planularia*) *subarcuatula* (Williamson) (*sic*), and by Barker (1960) and van Marle (1991) to *Astacolus crepidula* (Fichtel and Moll). This is the type-species, by synonymy with the originally designated *Astacolus crepidulatus* de Montfort 1808, of the genus *Astacolus* de Montfort 1808. The long-lost holotype has been discovered and illustrated by Rögl and Hansen (1984).

Plate 68 (LXVIII)

Figures 1–2 *Astacolus crepidulus* (Fichtel and Moll 1798)

[*Nautilus crepidula* Fichtel and Moll 1798]
Fig. 1 × 50; Fig. 2 × 100.

Referred by Brady (1884) to *Cristellaria*, by Thalmann (1932) to *Planularia,* and by Thalmann (1933a), Barker (1960), and van Marle (1991) to *Astacolus*. This is the type-species, by synonymy with the originally designated *A. crepidulatus* de Montfort 1808, of the genus *Astacolus* de Montfort 1808 (see also Plate 67, Fig. 20).

Figures 3–4 *Planularia australis* Chapman 1915

Fig. 3 × 50. From *Challenger* Sta. 209, Philippines (95–100fm.). ZF1351.
Fig. 4 × 50. From *Challenger* Sta. 167, West coast of New Zealand (150fm.). ZF1352.

Referred by Brady (1884) to *Cristellaria tricarinella* Reuss, by Thalmann (1932) to *Planularia tricarinella* (Reuss), by Barker (1960) to *P. australis* Chapman var., and by van Marle (1991) to *P. australis* Chapman. The ICZN has recently ruled in favour of the conservation of *Planularia* Defrance 1826 (Anon 1990a).

Figures 5–9 *Planularia siddalliana* (Brady 1881)

[*Cristellaria siddalliana* Brady 1881]
× 35. From *Challenger* Sta. 174C, off Fiji, Pacific (210fm.). ZF1346.

Referred by Brady (1884) to *Cristellaria,* and by Thalmann (1932) and Barker (1960) to *Planularia.* The ICZN has recently ruled in favour of the conservation of *Planularia* Defrance 1826 (Anon 1990a).

Figure 10 *Planularia cassis* (Fichtel and Moll 1798)

[*Nautilus cassis* Fichtel and Moll 1798]
× 30. From *Challenger* Sta. 174C, off Fiji, Pacific (210fm.). ZF1310.

Referred by Brady (1884) to *Cristellaria cassis* Fichtel and Moll, sp., by Thalmann (1932) to *Planularia cassis* nov. var., and by Barker (1960) to *P. cassis* var. nov. Thalmann. A new varietal name appears unnecessary. *Nautilus cassis* Fichtel and Moll 1798 is the type-species by original designation of the genus *Linthuris* de Montfort 1808. The ICZN has recently ruled in favour of the suppression of *Linthuris* de Montfort 1808 and the conservation of *Planularia* Defrance 1826 (Anon 1990a). The long-lost holotype of *Nautilus cassis* Fichtel and Moll 1798 has been discovered and illustrated by Rögl and Hansen (1984).

Figures 11–16 *Neolenticulina variabilis* (Reuss 1850)

[*Cristellaria variabilis* Reuss 1850]
Figs 11–12, 16 × 50. From *Porcupine* Sta. ?, North Atlantic (?). ZF1353.
Fig. 13 × 50. From *Challenger* Sta. 24, West Indies (390fm.). ZF1354.
Fig. 14 × 50. ZF1355. Fig. 15 × 50. ZF1356. From *Challenger* Sta. VIII, off Gomera, Canaries (620fm.).

Referred by Brady (1884) to *Cristellaria variabilis* Reuss, and by Thalmann (1932), Barker (1960), Srinivasan and Sharma (1980),

Hofker (1976), van Morkhoven *et al.* (1986), and van Marle (1991) to *Lenticulina peregrina* (Schwager). *Cristellaria peregrina* Schwager 1866 is regarded by the present author as a junior synonym of *C. variabilis* Reuss 1850.

Figure 17 *Saracenaria caribbeana* Hofker 1976

× 15–20. From *Challenger* Sta. 24, West Indies (390fm.). ZF1330.

Referred by Brady (1884) to *Cristellaria italica* Defrance, sp., by Thalmann (1932), Leroy (1944b), Barker (1960), and van Marle (1991) to *Saracenaria italica* Defrance, and by Hofker (1976) to *S. caribbeana* n. sp.

Figures 18, 20–23 *Saracenaria italica* Defrance 1824

Figs 18, 21–22 × 15–20. ZF1331. Figs 20, 23 × 15–20. ZF1332. From *Challenger* Sta. 174C, off Fiji, Pacific (210fm.).

Referred by Brady (1884) to *Cristellaria italica* Defrance, sp., and by Thalmann (1932), Barker (1960), Hofker (1976), and van Marle (1991) to *Saracenaria italica* Defrance. This is the type-species by original monotypy of the genus *Saracenaria* Defrance 1824. The robuline slit is atypical.

Figure 19 *Saracenaria latifrons* (Brady 1884)

[*Cristellaria latifrons* Brady 1884]
× 20. From *Challenger* Sta. 167, West coast of New Zealand (150fm.). ZF1335.

Referred by Brady (1884) to *Cristellaria,* and by Thalmann (1932), and Barker (1960), and Hofker (1976) to *Saracenaria.*

Plate 69 (LXIX)

Figures 1–4 *Lenticulina* sp. (aberrant)

× 20. From *Challenger* Sta. 135, Tristan d'Acunha (100–150fm.). ZF1302.

Referred by Brady (1884) to 'wild-growing forms' of *Cristellaria articulata* Reuss, sp., by Thalmann (1932) to *Lenticulina articulata* (Reuss), and by Barker (1960) to *L.* (*Robulus*) sp., ('abnormal specimens of doubtful identity').

Figure 5 *Lenticulina pliocaena* (Silvestri 1898)

[*Polymorphina pliocaena* Silvestri 1898]
× 40. From *Porcupine* Sta. 42, SW of Ireland, Atlantic (862fm.). ZF1347.

Referred by Brady (1884) to *Cristellaria* sp., by Thalmann (1932) to *Robulus* n. sp., and by Thalmann (1933a) and Barker (1960) to *Robulus pliocaenicus* (sic) (Silvestri). *Robulus* de Montfort 1808 is regarded by the present author as a junior synonym of *Lenticulina* Larmarck 1804.

Figures 6–7 *Lenticulina convergens* (Bornemann 1855)

[*Cristellaria convergens* Bornemann 1855]
Fig. 6 × 40. From *Challenger* Sta. 224, North Pacific (1850fm.). ZF1311.

Fig. 7 × 50. From *Challenger* Sta. 300, North of Juan Fernandez (1375fm.). ZF1312.

Referred by Brady (1884) to *Cristellaria*, by Thalmann (1932) to *Robulus*, and by Barker (1960) and later Hermelin (1989) to *Lenticulina*.

Figures 8–9 *Lenticulina gibba* (d'Orbigny 1839)

[*Cristellaria gibba* d'Orbigny 1839]
× 25. From *Challenger* Sta. 24, West Indies (390fm.). ZF1329.

Referred by Brady (1884) to *Cristellaria*, and by Thalmann (1932), Barker (1960), and van Marle (1991) to *Lenticulina*.

Figures 10–12 *Lenticulina atlantica* (Barker 1960)

[*Robulus atlanticus* Barker 1960]
Fig. 10a × 16. ZF1303. Fig. 10b × 16. ZF1304. From *Challenger* Sta. 24, West Indies (390fm.).
Figs 11–12 × 35. From *Challenger* Sta. 135, off Tristan d'Acunha (100–150fm.). ZF1305.

Referred by Brady (1884) to *Cristellaria articulata* Reuss, sp., by Thalmann (1932, 1933a) to *Robulus lucidus* (Cushman), by Barker (1960) to *R. atlanticus* nom. nov., and by Hermelin (1989) to *Lenticulina atlantica* (Barker). *Cristellaria lucida* Cushman 1923 is a secondary (subjective) junior homonym of *C. lucida* (Silvestri 1880). *Robulus* de Montfort 1808 is regarded as a junior synonym of *Lenticulina* Lamarck 1804.

Figure 13 *Lenticulina thalmanni* (Hessland 1943)

[*Robulus thalmanni* Hessland 1943]
× 12. From *Challenger* Sta. 24, West Indies (390fm.). ZF1343.

Referred by Brady (1884) to *Cristellaria rotulata* Lamarck, sp., by Thalmann first (1932) to *Lenticulina rotulata* (Lamarck) and later (1933a) to *Robulus* n. sp., and by Barker (1960) to *R. thalmanni* Hessland. *Robulus* is regarded as a junior synonym of *Lenticulina*.

Figures 14–16 *Lenticulina vortex* (Fichtel and Moll 1798)

[*Nautilus vortex* Fichtel and Moll 1798]
× 30. From *Challenger* Sta. 174C, off Fiji, Pacific (210fm.). ZF1357.

Referred by Brady (1884) to *Cristellaria*, by Thalmann (1932) and Barker (1960) to *Robulus*, and by van Marle (1991) to *Lenticulina*. The long-lost holotype of this species has been discovered and illustrated by Rögl and Hansen (1984). *Robulus* is regarded as a junior synonym of *Lenticulina*.

Figure 17 *Lenticulina orbicularis* (d'Orbigny 1826)

[*Robulina orbicularis* d'Orbigny 1826]
× 30. From *Challenger* Sta. 174C, off Fiji, Pacific (210fm.). ZF1357.

Referred by Brady (1884) to *Cristellaria*, by Thalmann (1932), Leroy (1944a), and Barker (1960) (albeit tentatively) to *Robulus*, and by van Marle (1991) to *Lenticulina*. *Robulus* is regarded as a junior synonym of *Lenticulina*.

Plate 70 (LXX)

Figure 1 *Lenticulina inornata* (d'Orbigny 1846)

[*Robulina inornata* d'Orbigny 1846]
× 40. From *Challenger* Sta. 174C, off Fiji, Pacific (210fm.). ZF1315.

Referred by Brady (1884) to *Cristellaria crassa* d'Orbigny, by Thalmann (1932) to *Lenticulina crassa* (d'Orbigny), and by Thalmann (1933a) and Barker (1960) to *Robulus crassus* (d'Orbigny). *Robulus* de Montfort 1808 is regarded by the present author as a junior synonym of *Lenticulina* Lamarck 1804. *Cristellaria crassa* d'Orbigny 1846 is regarded by Papp and Schmid (1985) as a junior synonym of *Robulina inornata* d'Orbigny 1846. This species has been lectotypified and reillustrated by Papp and Schmid (1985), who refer it to *Lenticulina*.

Figure 2 *Lenticulina nitida* (d'Orbigny 1826)

[*Cristellaria nitida* d'Orbigny 1826]
× 30. From *Challenger* Sta. 209, Philippines, North Pacific (95–100fm.). ZF1338.

Referred by Brady (1884) to *Cristellaria*, and by Thalmann (1932) and Barker (1960) to *Robulus*. *Robulus* de Montfort 1808 is regarded by the present author as a junior synonym of *Lenticulina* Lamarck 1804.

Figure 3 *Vaginulinopsis reniformis* (d'Orbigny 1846)

[*Cristellaria reniformis* d'Orbigny 1846]
× 25. From *Challenger* Sta. 24, West Indies (390fm.). ZF1342.

Referred by Brady (1884) to *Cristellaria*, by Thalmann (1932) to *Lenticulina*, and by Thalmann (1933a) and Barker (1960) to *Astacolus*. This species has been lectotypified by Papp and Schmid (1985).

Figures 4–6 *Lenticulina iota* (Cushman 1923)

[*Cristellaria iota* Cushman 1923]
Fig. 4 × 16; Fig. 5 × 15. ZF1320. Fig. 6 × ??. ZF1321. From *Challenger* Sta. 174C, off Kandavu, Fiji (210fm.).

Referred by Brady (1884) to *Cristellaria cultrata* de Montfort, sp., by Thalmann (1932) to *Robulus iota* (Cushman), and by Thalmann (1933a), Barker (1960), Hofker (1976), and van Marle (1991) to *Lenticulina iota* (Cushman). This species has been amended by Thomas (1988).

Figures 7–8 *Lenticulina denticulifera* (Cushman 1913)

[*Cristellaria denticulifera* Cushman 1913]
× 20. From *Challenger* Sta. 174C, off Fiji, Pacific (210fm.). ZF1322.

Referred by Brady (1884) to *Cristellaria cultrata* de Montfort, sp., by Thalmann (1932) to *Robulus denticuliferus* (Cushman), and by Thalmann (1933a) and Barker (1960) to *Lenticulina denticulifera* (Cushman).

Figures 9–12 *Lenticulina calcar* (Linné 1767)

[*Nautilus calcar* Linné 1767]
Figs 9–10 × 35. From *Challenger* Sta. 209, Philippines (95–100fm.). ZF1306.

Fig. 11 × 35. From *Challenger* Sta. 23, West Indies (450fm.). 1959.5.5.272.

Fig. 12 × 35. From *Challenger* Sta. 24, West Indies (390fm.). ZF1307.

Referred by Brady (1884) to *Cristellaria*, by Thalmann (1932) and Leroy (1944b) to *Robulus*, and by Thalmann (1933a), Barker (1960), Hofker (1976), and van Marle (1991) to *Lenticulina*. This species has been amended by Thomas (1988).

Figures 13–14, ?15 *Lenticulina formosa* (Cushman 1923)

[*Cristellaria formosa* Cushman 1923]
Figs 13–14 × 25. From *Challenger* Sta. 23, West Indies (450fm.). ZF1308.
Fig. 15 × 35. From *Challenger* Sta. 174C, off Fiji, Pacific (210fm.). ZF1309.

Referred by Brady (1884) to *Cristellaria calcar* Linné, sp., by Thalmann first (1932) to *Robulus formosus* (Cushman) (Figs 13–14) and *R. calcar* (Linné) (Fig. 15), and later (1933a) to *Lenticulina formosa* (Cushman) (Figs 13–14) and *L. calcar* (Linné) (Fig. 15), and by Barker (1960) (with some doubt as to the identity of Fig. 15) to *Lenticulina formosa* (Cushman).

Figure 16 *Lenticulina papillosa* (Fichtel and Moll 1798)

[*Nautilus papillosus* Fichtel and Moll 1798]
× 20. From *Challenger* Sta. 174C, off Fiji, Pacific (210fm.). ZF1341.

Referred by Brady (1884) to *Cristellaria*, by Thalmann (1932) to *Robulus*, and by Thalmann (1933a) and Barker (1960) to *Lenticulina*. The long-lost holotype of this species has been discovered and illustrated by Rögl and Hansen (1984).

Figures 17–18 *Lenticulina submamilligera* (Cushman 1917)

[*Cristellaria submamilligera* Cushman 1913]
× 20. From *Challenger* Sta. 174C, off Fiji, Pacific (210fm.). ZF1337.

Referred by Brady (1884) to *Cristellaria mamilligera* Karrer, by Thalmann (1932) to *Robulus submamilligerus* (Cushman), and by Thalmann (1933a) and Barker (1960) to *Lenticulina submamilligera* (Cushman).

Plate 71 (LXXI)

Figures 1–3 *Lenticulina echinata* (d'Orbigny 1846)

[*Robulina echinata* d'Orbigny 1846]
Fig. 1 × 30. From *Challenger* Sta. 209, Philippines (95fm.). ZF1306.
Figs 2–3 × 20. From *Challenger* Sta. 174C, off Fiji, Pacific (210fm.). ZF1325.

Referred by Brady (1884) to *Cristellaria echinata* d'Orbigny, sp., by Thalmann (1932) to *Robulus echinatus* (d'Orbigny), by Barker (1960) to *Lenticulina papillosoechinata* (Fornasini), and by Hofker (1976) to *L. echinata* (d'Orbigny). *Cristellaria papillosoechinata* Fornasini 1906 is regarded by the present author as a junior synonym of *C. echinata* d'Orbigny 1846. *Robulus* de Montfort 1808 is regarded by the present author as a junior synonym of *Lenticulina* Lamarck 1804.

Figures 4–5 *Vaginulinopsis subaculeata* (Cushman 1923)

[*Cristellaria subaculeata* Cushman 1923]
× 25. From *Challenger* Sta. 24, West Indies (390fm.). ZF1300.

Referred by Brady (1884) to *Cristellaria aculeata* d'Orbigny, by Thalmann (1932) to *Robulus subaculeatus* (Cushman), by Barker (1960) to *Marginulinopsis subaculeatus* (Cushman), and by Hofker (1976) to *Lenticulina subaculeata* (Cushman). This is the type-species by original designation of the genus *Percultazonaria* Loeblich and Tappan 1986, regarded by the present author as a junior synonym of *Vaginulinopsis* Silvestri 1904.

Figures 6–7 *Planularia gemmata* (Brady 1881)

[*Cristellaria gemmata* Brady 1881]
Fig. 6a × 35. ZF1326. Fig. 6b × 35. ZF1327. From *Challenger* Sta. 174C, off Fiji, Pacific (210fm.).
Fig. 7 × 35. From *Challenger* Sta. 209, Philippines (95–100fm.). ZF1328.

Referred by Brady (1884) to *Cristellaria*, by Thalmann (1932), Barker (1960), and van Marle (1991) to *Planularia*, and by Leroy (1944b) to *Hemicristellaria*. The ICZN has recently ruled in favour of the conservation of *Planularia* Defrance 1826 (Anon 1990a). Van Marle (1991) regards *P. gemmata* Brady 1881 as a *nomen nudum*, but gives no justification.

Figure 8 *Lenticulina* sp. nov.

× 35. From *Challenger* Sta. 174C, off Fiji, Pacific (210fm.). ZF1314.

Referred by Brady (1884) to *Cristellaria costata* Fichtel and Moll, sp. (see below), by Thalmann (1932) to *Robulus bradyi* (Cushman), by Leroy (1944a) to *R. costatus* (Fichtel and Moll), and by Barker (1960) to *R.* sp. nov. *Cristellaria bradyi* Cushman 1917 is a junior homonym of *C. bradyi* Goes 1894. *Robulus* is a junior synonym of *Lenticulina*.

Figure 9 *Lenticulina anaglypta* (Loeblich and Tappan 1987)

[*Spincterules anaglyptus* Loeblich and Tappan 1987]
× 35. From *Challenger* Sta. 185, Torres Strait, Pacific (155fm.). ZF1313.

Referred by Brady (1884) to *Cristellaria costata* Fichtel and Moll, sp., by Thalmann (1932), Leroy (1944a) and Barker (1960) to *Robulus costatus* (Fichtel and Moll), and by van Marle (1991) (erroneously citing Pl. 70, Fig. 9 instead of Pl. 71, Fig. 9) to *Lenticulina costata* (Fichtel and Moll). *Nautilus costatus* Fichtel and Moll 1798 is a primary junior homonym of *N.* (*Orthoceras*) *costatus* Batsch 1791. It has been renamed *Spincterules anaglyptus* Loeblich and Tappan 1987. This is the type-species by original designation of the genus *Spincterules* Loeblich and Tappan 1987, regarded by the present author as a junior synonym of *Lenticulina* Lamarck 1804. Its long-lost holotype has been discovered and illustrated by Rögl and Hansen (1984).

Figure 10 *Astacolus* sp. (aberrant)

× 50. From *Challenger* Sta. 162, Bass Strait, Pacific (38–40fm.).

Referred by Brady (1884) to a 'dimorphous specimen, the earlier chambers arranged as in *Cristellaria*, the later ones as in *Polymor-*

phina', by Thalmann first (1932) to 'abnormes individuum' and later (1933*a*) to 'deformierter' *Astacolus* sp., and by Barker (1960) to *Astacolus* sp., abnormal specimen.

Figures 11–12 *Globulina gibba* (Deshayes 1830)

[*Polymorphina (Globuline) gibba* Deshayes 1830]
Fig. 11 × 50. From *Challenger* Sta. 162, Bass Strait, Pacific (38fm.). ZF2143.
Fig. 12 × 40. From Dogs Bay, Connemara, Ireland. ZF2136.

Referred by Brady (1884) to *Polymorphina lactea* Walker and Jacob, sp. (Fig. 11) and *P. gibba* d'Orbigny, sp. (Fig. 12), and by Thalmann (1932), Barker (1960) (Figs 11–12), and van Marle (1991) (Fig. 11 only) to *Globulina gibba* (d'Orbigny). *Polymorphina (Globuline) gibba* d'Orbigny 1826 was a *nomen nudum* (Tubbs (ICZN), pers. comm.). The name was validated by the description by Deshayes in a volume of the seldom-read *Encyclopédie Méthodique* (the BMNH copy of which resides in a basement vault adjoining the billiard room!) published in 1830. Loeblich and Tappan (1964, 1987) regard *Polymorphina (Globuline) gibba* d'Orbigny 1826 (*sic*) as the type-species by subsequent designation of the genus *Globulina* d'Orbigny 1839. Poag and Skinner (1968) regard *Globulina caribaea* d'Orbigny 1839 as the type.

Figure 13 *Globulina inaequalis* Reuss 1850

× 75. From *Challenger* Sta. 162, Bass Strait, Pacific (38fm.). ZF2125.

Referred by Brady (1884) to *Polymorphina amygdaloides* Reuss, and by Thalmann (1932) and Barker (1960) to *Globulina inaequalis* (Reuss). *Polymorphina amygdaloides* Reuss 1851 is a junior synonym of *Globulina inaequalis* Reuss 1850 (Cushman and Ozawa 1930).

Figure 14 *Pyrulina gutta* (d'Orbigny 1839)

[*Polymorphina (Pyruline) gutta* d'Orbigny 1839]
× 75. From *Challenger* Sta. 241, North Pacific (2300fm.). ZF2139.

Referred by Brady (1884) to *Polymorphina lactea* Walker and Jacob, sp., and by Thalmann (1932) and Barker (1960) to *Pyrulina gutta* (d'Orbigny). The latter is the type-species by original monotypy of the genus *Pyrulina* d'Orbigny 1839.

Figure 15 *Globulotuba* sp. nov. (1)

× 50. From *Porcupine* Sta. 84, Faroe Channel, North Atlantic (155fm.). ZF2158.

Referred by Brady (1884) to *Polymorphina sororia* Reuss, and by Thalmann (1932) and Barker (1960) to *Globulina minuta* (Roemer) (see below). This form is clearly entosolenian, however, and therefore a glandulinid rather than a polymorphinid.

Figure 16 *Globulina minuta* (Roemer 1838)

[*Polymorphina minuta* Roemer 1838]
× 50. From *Challenger* Sta. 205A, Hong Kong, Pacific (7fm.). ZF2159.

Referred by Brady (1884) to *Polymorphina sororia* Reuss, and by Thalmann (1932) and Barker (1960) to *Globulina minuta* (Roemer).

Polymorphina sororia Reuss 1863 is a junior synonym of *P. minuta* Roemer 1838 (Cushman and Ozawa 1930).

Figures 17–19 *Pyrulina fusiformis* (Roemer 1838)

[*Polymorphina fusiformis* Roemer 1838]
Fig. 17 × 60. From *Challenger* Sta. 146, Southern Ocean (1375fm.). ZF2161.
Fig. 18 × 60. ZF2164. Fig. 19 × 60. ZF2163. From *Porcupine* Sta. 18, West of Ireland, Atlantic (183fm.).

Referred by Brady (1884) to *Polymorphina sororia* var. *cuspidata*, and by Thalmann (1932) and Barker (1960) to *Pyrulina fusiformis* (Roemer).

Plate 72 (LXXII)

Figures 1–2 *Pyrulina angusta* (Egger 1857)

[*Polymorphina angusta* Egger 1857]
Fig. 1 × 45–60. From *Challenger* Sta. 332, South Atlantic (2200fm.). ZF2127.
Fig. 2 × 45–60. From *Challenger* Sta. 346, South Atlantic (2350fm.). ZF2128.

Referred by Brady (1884) to *Polymorphina angusta* Egger, by Thalmann (1932) to *Pyrulina fusiformis* (Roemer), and by Barker (1960) to *P. angusta* (Egger).

Figure 3 *Globulotuba* sp. nov. (2)

× 45–60. From *Challenger* Sta. 300, Juan Fernandez (1375fm.). ZF2126.

Referred by Brady (1884) to *Polymorphina angusta* Egger, by Thalmann (1932) to *Pyrulina fusiformis* (Roemer), and by Barker (1960) to *P. angusta* (Egger). This form is clearly entosolenian, however, and therefore a glandulinid rather than a polymorphinid.

Figure 4 *Pyrulina fusiformis* (Roemer 1838)

[*Polymorphina fusiformis* Roemer 1838]
× 50. From *Porcupine* Sta. 2, West of Ireland, Atlantic. ZF2162.

Referred by Brady (1884) to *Polymorphina sororia* var. *cuspidata*, and by Thalmann (1932), Barker (1960), and van Marle (1991) to *Pyrulina fusiformis* (Roemer) (see also Plate 71, Figs 17–19).

Figure 5 *Pyrulina cylindroides* (Roemer 1838)

[*Polymorphina cylindroides* Roemer 1838]
× 45. From *Challenger* Sta. 296, South Pacific (1825fm.). ZF2140.

Referred by Brady (1884) to *Polymorphina lanceolata* Reuss, and by Thalmann (1932) and Barker (1960) to *Pyrulina cylindroides* (Roemer). This species has been referred by Anderson (1936) to *Apiopterina* Zborzewski 1834, regarded by Loeblich and Tappan (1987) and by the present author as a foraminiferal genus of uncertain status.

Figure 6 *Globulotuba* sp. nov. (3)

× 45. From *Challenger* Sta. 146, Southern Ocean (1375fm.). ZF2161.

Referred by Brady (1884) to *Polymorphina lanceolata* Reuss, and by Thalmann (1932) and Barker (1960) to *Pyrulina cylindroides* (Roemer) (see above). This form is clearly entosolenian, however, and therefore a glandulinid rather than a polymorphinid.

Figures 7–8 *Guttulina ovata* (d'Orbigny 1826)

[*Polymorphina (Globuline) ovata* d'Orbigny 1826]
Fig. 7 × 25. ZF2148. Fig. 8 × 25. ZF2149. From *Challenger* Sta. 24, West Indies (390fm.).

Referred by Brady (1884) to *Polymorphina ovata* d'Orbigny, and by Thalmann (1932) and Barker (1960) to *Pseudopolymorphina ovalis* Cushman and Ozawa. Both Thalmann (1932) and Barker (1960) evidently attributed Brady's specimens to *Pseudopolymorphina ovalis* Cushman and Ozawa 1930 on the assumption that *Polymorphina ovata* d'Orbigny 1846 was distinct from *P. ovata* d'Orbigny 1826. Papp and Schmid (1985) have subsequently shown this assumption to have been erroneous. Papp and Schmid (1985) have also lectotypified and reillustrated d'Orbigny's species.

Figures 9–10, ?11 *Pseudopolymorphina* sp. nov.

Fig. 9 × 18; Fig. 10 × 25. From *Challenger* Sta. 162, Bass Strait, Pacific (38–40fm.). ZF2130.
Fig. 11 × 18. From Dogs Bay, Connemara, Ireland (shore sand). 1959.5.11.1.

Referred by Brady (1884) to *Polymorphina compressa* d'Orbigny, and by Thalmann (1932) and Barker (1960) to *Pseudopolymorphina ligua* (Roemer). Both Thalmann (1932) and Barker (1960) evidently attributed Brady's specimens to *Pseudopolymorphina ligua* (Roemer) on the assumption that *Polymorphina compressa* d'Orbigny 1846 was a junior synonym of *P. ligua* Roemer 1838. In fact, *Polymorphina compressa* d'Orbigny is clearly a bolivinitid, as recently shown by Papp and Schmid (1985). *Polymorphina ligua* Roemer, as figured by Cushman and Ozawa (1930), also appears distinct from the form figured by Brady.

Figures 12–15 *Sigmoidella elegantissima* (Parker and Jones in Brady, Parker, and Jones 1870)

[*Polymorphina elegantissima* Parker and Jones, in Brady, Parker, and Jones 1870]
Fig. 12a × 50. From *Challenger* Sta. 192, Ki Islands, Pacific (129fm.). ZF2132.
Fig. 12b × 50. From *Challenger* Sta. 162, Bass Strait, Pacific (38–40fm.). ZF2133.
Fig. 13 × 50. From *Challenger* Sta. 185, Torres Strait, Pacific (155fm.). ZF2135.
Figs 14–15 × 50. From *Challenger* Sta. 192, Ki Islands, Pacific (129fm.). ZF2134.

Referred by Brady (1884) to *Polymorphina elegantissima* Parker and Jones, by Thalmann (1932) and Barker (1960) to *Sigmoidella elegantissima* (Parker and Jones) var. nov. (Fig. 12), *S. elegantissima* (Parker and Jones) (Fig. 13) and *Guttulina (Sigmoidina) pacifica* (Cushman and Ozawa) (Figs 14–15), and by Whittaker and Hodgkinson (1979) to *Sigmoidella elegantissima* (Parker and Jones) (Figs 12–?13) and *Guttulina pacifica* (Cushman and Ozawa) (Figs 14–15). *Sigmoidella (Sigmoidina) pacifica* Cushman and Ozawa 1928 is

regarded by the present author as a junior synonym of *Polymorphina elegantissima* Parker and Jones 1865. This is the type-species, by synonymy with the originally designated *Sigmoidella (Sigmoidina) pacifica* Cushman and Ozawa 1928, of *Sigmoidella (Sigmoidina)* Cushman and Ozawa 1928, regarded by the present author as a junior synonym of *Sigmoidella* Cushman and Ozawa 1928. It has been lectotypified by Hodgkinson (1992).

Figures 16–17 *Sigmoidella seguenzana* (Brady 1884)

[*Polymorphina seguenzana* Brady 1884]
Fig. 16 × 40. ZF2156. Fig. 17 × 40. ZF2157. From *Challenger* Sta. 192, Ki Islands, Pacific (129fm.).

Referred by Brady (1884) to *Polymorphina*, by Thalmann (1932) and Barker (1960) to *Guttulina (Sigmoidina)*, and by van Marle (1991) to *Guttulina*. *Sigmoidella (Sigmoidina)* Cushman and Ozawa 1928 is regarded by the present author as a junior synonym of *Sigmoidella* Cushman and Ozawa 1928 (see above).

Figure 18 *Pseudopolymorphina dawsoni* (Cushman and Ozawa 1930)

[*Guttulina dawsoni* Cushman and Ozawa 1930]
× 18. From *Challenger* Sta. 162, Bass Strait, Pacific (38–40fm.). ZF2166.

Referred by Brady (1884) to *Polymorphina thouini* d'Orbigny, and by Thalmann (1932) and Barker (1960) to *Guttulina dawsoni* Cushman and Ozawa.

Figures 19–20 *Guttulina communis* (d'Orbigny 1826)

[*Polymorphina (Guttuline) communis* d'Orbigny 1826]
Fig. 19 × 50. ZF2129. Fig. 20 × 40. ZF2150. From *Challenger* Sta. 162, Bass Strait, Pacific (38–40fm.).

Referred by Brady (1884) to *Polymorphina communis* d'Orbigny (Fig. 19) and *P. problema* d'Orbigny (Fig. 20), and by Thalmann (1932) and Barker (1960) to *Guttulina problema* d'Orbigny (Figs 19–20). Papp and Schmid (1985) have synonymized *Guttulina* (sic) *problema* d'Orbigny with *G.* (sic) *communis* d'Orbigny, arguing that the validation date of the former was 1846 and that of the latter was 1825 [1826]. *Polymorphina (Guttuline) communis* d'Orbigny 1826 is the type-species by subsequent designation of the genus *Guttulina* d'Orbigny 1839 (Loeblich and Tappan 1964, 1987).

Plate 73 (LXXIII)

Figure 1 *Guttulina communis* (d'Orbigny 1826)

[*Polymorphina (Guttuline) communis* d'Orbigny 1826]
× 40. From *Challenger* Sta. 162, Bass Strait, Pacific (38–40fm.). ZF2151.

Referred by Brady (1884) to *Polymorphina problema* d'Orbigny, and by Thalmann (1932) and Barker (1960) to *Guttulina problema* (d'Orbigny) (see also Plate 72, Figs 19–20).

Figures 2–3 *Guttulina yabei* Cushman and Ozawa 1930

Fig. 2 × 40. From *Challenger* Sta. 162, Bass Strait, Pacific (38–40fm.). ZF2146.

Fig. 3 ×40. From *Challenger* Sta. 163C, Port Jackson, Australia (6fm.). ZF2145.

Referred by Brady (1884) to *Polymorphina oblonga* d'Orbigny, and by Thalmann (1932) and Barker (1960) to *Guttulina yabei* Cushman and Ozawa.

Figure 4 *Guttulina austriaca* d'Orbigny 1846

×40. From *Challenger* Sta. 192, Ki Islands, Pacific (129fm.). ZF2147.

Referred by Brady (1884) to *Polymorphina oblonga* d'Orbigny, and by Thalmann (1932) and Barker (1960) to *Guttulina austriaca* d'Orbigny. This species has recently been lectotypified by Papp and Schmid (1985). Brady's specimen differs from the lectotype essentially only in being elongate rather than elongate-ovate in outline, and in having its greatest width slightly below mid-height.

Figures 5–6 *Pyrulina* sp. nov.

Fig. 5 ×30. From *Challenger* Sta. 296, South Pacific (1825fm.). ZF2140.
Fig. 6 ×30. From *Challenger* Sta. 224, North Pacific (1850fm.). ZF2154.

Referred by Brady (1884) to *Polymorphina rotundata* Bornemann, sp., and by Thalmann (1932) and Barker (1960) to *Globulina* sp. nov.

Figures 7–8 *Globulina rotundata* (Bornemann 1855)

[*Guttulina rotundata* Bornemann 1855]
×30. From *Challenger* Sta. 145, Prince Edward Island, South Pacific (50–150fm.). ZF2155.

Referred by Brady (1884) to *Polymorphina*, and by Thalmann (1932) and Barker (1960) to *Globulina*.

Figures 9–10 *Globulina myristiformis* (Williamson 1858)

[*Polymorphina myristiformis* Williamson 1858]
×60–75. From Mounts Bay, Cornwall. ZF2144.

Referred by Brady (1884) to *Polymorphina myristiformis* Williamson, and by Thalmann (1932) and Barker (1960) to *Globulina gibba* var. *myristiformis* (Williamson). Interestingly, Murray (1971*b*) figures under this designation a specimen with a cribrate aperture.

Figures 11–13 *Globulina regina* (Brady, Parker, and Jones 1870)

[*Polymorphina regina* Brady, Parker, and Jones 1870]
Figs 11–12 ×35–50. From *Challenger* Sta. 163C, Port Jackson, Australia (6fm.). ZF2152.
Fig. 13 ×35–50. From *Challenger* Sta. 185, Torres Strait, Pacific (155fm.). ZF2153.

Referred by Brady (1884) to *Polymorphina*, and by Thalmann (1932, 1933*a*) and Barker (1960) to *Guttulina*. In 1932, Thalmann erroneously attributed this species to Brady rather than to Brady, Parker, and Jones. The following year he rectified his error. *Polymorphina regina* Brady, Parker, and Jones has been lectotypified by Hodgkinson (1992).

Figures 14–17 Fistulose polymorphinids

Fig. 14 ×50. From *Challenger* Sta. 24, West Indies (390fm.). ZF2138.
Fig. 15 ×30. From *Challenger* Sta. 246, North Pacific (2050fm.). ZF2160.
Fig. 16 ×35. From *Challenger* Sta. 162, Bass Strait, Pacific (38–40fm.). ZF2137.
Fig. 17 ×50. From *Challenger* Sta. 33, Bermuda, Atlantic (435fm.). ZF2131.

Referred by Brady (1884) to *Polymorphina lactea* (Fig. 14), *P. sororia* (Fig. 15), *P. gibba* (Fig. 16) and *P. compressa* (Fig. 17), and by Thalmann (1932) and Barker (1960) to 'Polymorphinidae, formae fistulosae'.

Figures 18–19 *Francuscia extensa* (Cushman 1923)

[*Polymorphina extensa* Cushman 1923]
Fig. 18 ×50–60. From *Challenger* Sta. 283, South Pacific (2075fm.). ZF2141.
Fig. 19 ×50–60. From *Challenger* Sta. 338, South Atlantic (1990fm.). ZF2142.

Referred by Brady (1884) to *Polymorphina longicollis* Brady, by Thalmann (1932) to *Pyrulina extensa* Cushman and Ozawa, and by Thalmann (1933*a*) and Barker (1960) to *Pyrulina extensa* (Cushman). The appropriate generic assigmnet of this species would appear to be to *Francuscia* McCulloch 1981 (a new name for *Frankia* McCulloch 1977, not Brunchorst 1886). *Mitrapolymorphina* Loeblich and Tappan 1986 is regarded by the present author as a junior synonym of *Francuscia* McCulloch 1981. *Polymorphina longicollis* Brady 1881 is a junior homonym of *P. longicollis* Karrer 1870.

Plate 74 (LXXIV)

Figures 1–3 *Uvigerina canariensis* d'Orbigny 1839

Fig. 1 ×50. From *Challenger* Sta. 323, South Atlantic (1900fm.). ZF2571.
Fig. 2 ×60. From *Challenger* Sta. 309, West coast of Patagonia (40–140fm.). ZF2571.
Fig. 3 ×75. From *Challenger* Sta. 33, off Bermuda (435fm.). ZF2573.

Referred by Brady (1884), Barker (1960), and van Marle (1991) to *Uvigerina canariensis* d'Orbigny, and by Thalmann (1932) to *U. farinosa* Hantken.

Figures 4–7 *Uvigerina hollicki* Thalmann 1950

Figs 4–5 ×50. From *Challenger* Sta. 185, Torres Strait, Pacific (155fm.). ZF2581.
Figs 6–7 ×45. From *Challenger* Sta. 151, Heard Island, South Pacific (75fm.). ZF2582.

Referred by Brady (1884) and Thalmann (1932) to *Uvigerina tenuistriata* Reuss, by Thalmann (1950) to *U. hollicki* (a new name for *U. peregrina* var. *bradyana* Cushman 1923, not *U. bradyana* Fornasini 1900), and by Barker (1960) to *U. cushmani* Todd.

Figures 8–10 *Uvigerina schwageri* Brady 1884

Figs 8–9 ×30. From *Challenger* Sta. 174C, off Fiji, Pacific (210fm.). ZF2579.

Fig. 10 × 30. From *Challenger* Sta. 209, Philippines (95–100fm.). ZF2580.

Referred by Brady (1884), Thalmann (1932), Leroy (1944b), Barker (1960), and Borsetti *et al.* (in van der Zwaan *et al.* 1986) to *Uvigerina*, and by Belford (1966) to *Euuvigerina*. Hofker (1951) suggested that this species might represent the megalospheric generation of *Euuvigerina aculeata* (d'Orbigny) (see Plate 75, Figs 1–3), but Belford (1966) maintained it as separate.

Figures 11–12 *Uvigerina mediterranea* Hofker 1932

× 50. From *Challenger* Sta. 24, West Indies. ZF2577.

Referred by Brady (1884) to *Uvigerina pygmaea* d'Orbigny, by Thalmann (1932) to *U. mediterranea* Hofker 1932 (regarded by Høglund (1947) as a junior synonym of *U. peregrina* Cushman 1923, but by the present author as distinct), by Barker (1960) to *Euuvigerina peregrina* (Cushman), and by van Marle (1991) to *Uvigerina peregrina* Cushman.

Figures 13–14 *Uvigerina bifurcata* d'Orbigny 1839

× 60. From *Challenger* Sta. 232, South of Japan, North Pacific (345fm.). ZF2578.

Referred by Brady (1884) to *Uvigerina pygmaea* d'Orbigny, and by Thalmann (1932) and Barker (1960) to *U. bifurcata* d'Orbigny.

Figures 15–16 *Trifarina angulosa* (Williamson 1858)

[*Uvigerina angulosa* Williamson 1858]
× 60. From *Challenger* Sta. 145, Prince Edward Island. ZF2563.

Referred by Brady (1884) to *Uvigerina*, and by Thalmann (1932) and Barker (1960) to *Angulogerina*. This is the type-species by original designation of the genus *Angulogerina* Cushman 1927, regarded by the present author as a junior synonym of *Trifarina* Cushman 1923.

Figures 17–18 *Trifarina carinata* (Cushman 1927)

[*Angulogerina carinata* Cushman 1927]
× 60. From *Challenger* Sta. 300, North of Juan Fernandez, East Pacific (1375fm.). ZF2564.

Referred by Brady (1884) to *Uvigerina angulosa* Williamson (see above), and by Thalmann (1932) and Barker (1960) to *Angulogerina carinata* var. *bradyana* Cushman (Fig. 17) and *A. carinata* Cushman (Fig. 18). The variety *bradyana* appears superfluous. Indeed, there must be some doubt as to whether *carinata* is more than subspecifically distinct from *angulosa* (see above).

Figures 19–20 *Trifarina spinipes* (Brady 1881)

[*Uvigerina spinipes* Brady 1881]
× 60. From *Challenger* Sta. 135, off Tristan d'Acunha, South Atlantic (100–150fm.). ZF2565.

Referred by Brady (1884) to *Uvigerina angulosa* var. *spinipes* Brady, and by Thalmann (1932) and Barker (1960) to *Angulogerina carinata* var. *spinipes* (Brady). There is some doubt as to whether *spinipes* is more than subspecifically distinct from *angulosa* (see above).

Figures 21–23 *Siphouvigerina porrecta* (Brady 1879)

[*Uvigerina porrecta* Brady 1879]
× 70. From *Challenger* Sta. 185, Torres Strait, Pacific (155fm.). ZF2576.

Referred by Brady (1884), Thalmann (1932), and van Marle (1991) to *Uvigerina*, and by Barker (1960) to *Neouvigerina*. *Neouvigerina* Hofker 1951 is regarded by the present author as a junior synonym of *Siphouvigerina* Parr 1950.

Figures 24–26 *Uvigerina bradyana* Fornasini 1900

Figs 24–25 × 40–55. From *Challenger* Sta. 191A, Ki Islands, Pacific (580fm.). ZF2584.
Fig. 26 × 40–55. From *Challenger* Sta. 300, North of Juan Fernandez, East Pacific (1375fm.). ZF2583.

Referred by Brady (1884) to *Uvigerina* sp., 'intermediate specimens connecting *U. pygmaea* d'Orbigny with *U. aculeata* d'Orbigny', by Thalmann (1932) tentatively to *U. mediterranea* Hofker (Fig. 24) and *U. peregrina* var. *parvula* Cushman (Figs 25–26), by Barker (1960) to *U. bradyana* Fornasini, and by Hermelin (1989) to *U. peregrina* Cushman.

Plate 75 (LXXV)

Figures 1–3 *Uvigerina aculeata* d'Orbigny 1846

Figs 1, 3 ZF2562. Fig. 2 ZF2561. × 35. From *Challenger* Sta. 191A, Ki Islands, Pacific (580fm.).

Referred by Brady (1884) and Thalmann (1932) to *Uvigerina*, and by Barker (1960), Vella (1961), and Belford (1966) to *Euuvigerina*. This is the type-species by original designation of the genus *Euuvigerina* Thalmann 1952, regarded by Revets (pers. comm.) as a junior synonym of *Uvigerina* d'Orbigny 1826. Papp and Schmid (1985) placed *Uvigerina aculeata* d'Orbigny 1846 in *U. grilli* Schmid 1971, but are not followed here.

Figures 4–5 *Uvigerina brunnensis* Karrer 1877

× 45. From *Challenger* Sta. 149I, Kerguelen Island, South Pacific (120fm.). ZF2570.

Figures 6–9 *Uvigerina auberiana* d'Orbigny 1839

Fig. 6 × 50. ZF2566. Fig. 7 × 50. ZF2567. Fig. 9 × 40. ZF2562. From *Challenger* Sta. 191A, Ki Islands, Pacific (580fm.).
Fig. 8 × 50. From *Challenger* Sta. 323, South Atlantic (1900fm.). ZF2568.

Referred by Brady (1884) to *Uvigerina asperula* Czjzek (Figs 6–8) and *U. asperula* var. *auberiana* d'Orbigny (Fig. 9), and by Thalmann (1932) and Barker (1960) to *U. asperula* Czjzek (Figs 6–9).

Figures 10–11 *Siphouvigerina ampullacea* (Brady 1884)

[*Uvigerina asperula* var. *ampullacea* Brady 1884]
× 60. From *Challenger* Sta. 344, Ascension Island, South Atlantic (420fm.). ZF2569.

Referred by Brady (1884) to *Uvigerina asperula* var. *ampullacea*, by

Thalmann first (1932) to *U. ampullacea* (Brady) and later (1937) to *U. asperula* var. *proboscidea* (Schwager), and by Barker (1960) and Vella (1961) to *Neouvigerina ampullacea* (Brady). This is the type-species by original designation of the genus *Neouvigerina* Hofker 1951, regarded herein as a junior synonym of *Siphouvigerina* Parr 1950.

Figures 12–14 *Siphouvigerina interrupta* (Brady 1879)

[*Uvigerina interrupta* Brady 1879]
Figs 12–13 × 70. From *Challenger* Sta. 217A, off Papua, Pacific (37fm.). ZF2574.
Fig. 14 × 70. From *Challenger* Sta. 174C, off Fiji, Pacific (210fm.). ZF2575.

Referred by Brady (1884) and Thalmann (1932) to *Uvigerina*, by Barker (1960) to *Neouvigerina*, and by Belford (1966) to *Siphouvigerina*. *Neouvigerina* is herein regarded as a junior synonym of *Siphouvigerina*.

Figures 15–17 *Siphogenerina columellaris* (Brady 1881)

[*Sagrina columellaris* Brady 1881]
Figs 15–16 × 45. From *Challenger* Sta. VIII, off the Canaries, Atlantic (620fm.). ZF2347.
Fig. 17 × 45. From *Challenger* Sta. 162, Bass Strait, Pacific (38–40fm.). ZF2348.

Referred by Brady (1884) to *Sagrina columellaris* Brady, by Thalmann first (1932) to *Siphogenerina columellaris* (Brady) and later (1937) to *S. zitteli* (Karrer), and by Barker (1960), Belford (1966), and van Marle (1991) to *Rectobolivina columellaris* (Brady). van Marle (1991) regards *Sagrina columellaris* Brady 1881 as a *nomen nudum*, but gives no justification.

Figures 18–20 *Rectobolivina bifrons* (Brady 1881)

[*Sagrina bifrons* Brady 1881]
Figs 18–19 ZF2345. Fig. 20 ZF2346. × 50. From *Challenger* Sta. 232, off Japan (345fm.).

Referred by Brady (1884) to *Sagrina*, by Thalmann first (1932) to *Siphogenerina* and later (1942) to *Rectobolivina*, and also by Barker (1960), Belford (1966), and van Marle (1991) to *Rectobolivina*. Van Marle (1991) regards *Sagrina bifrons* Brady 1881 as a *nomen nudum*, but gives no justification. This is the type-species by original designation of the genus *Rectobolivina* Cushman 1927. *Rectobolivina* Cushman 1927 is regarded by the present author as differing from *Siphogenerina* Schlumberger 1882 principally in its compressed rather than inflated test.

Figures 21–22 *Siphogenerina raphanus* (Parker and Jones 1865)

[*Uvigerina* (*Sagrina*) *raphanus* Parker and Jones 1865]
Fig. 21 × 60. From *Challenger* Sta. 162, Bass Strait, Pacific (38–40fm.). ZF2353.
Fig. 22 × 60. From *Challenger* Sta. 218A, Admiralty Islands, Pacific (16–25fm.). ZF2354.

Referred by Brady (1884) to *Sagrina*, by Thalmann (1932) and Barker (1960) to *Siphogenerina*, and by Whittaker and Hodgkinson (1979) and Hermelin (1989) to *Rectobolivina*. This species has been lectotypified by Hodgkinson (1992).

Figures 23–?24 *Siphogenerina indica* Leroy 1941

Fig. 23a × 60.
Fig. 23b × 60. From *Challenger* Sta. 209, Philippines (95–100fm.). ZF2355.
Fig. 24 × 45. From *Challenger* Sta. 279A, off Tahiti, Pacific (420fm.). ZF2356.

Referred by Brady (1884) to *Sagrina raphanus* Parker and Jones, sp., by Thalmann (1932) and Barker (1960) to *Siphogenerina raphanus* (Parker and Jones) (see above), by Whittaker and Hodgkinson (1979) (?Fig. 23 only) and Hermelin (1989) to *Rectobolivina raphana* (Parker and Jones), and by van Marle (1991) to *R. indica* (Leroy) (Fig. 23 only). Figure 24 is close to *Siphogenerina raphanus* var. *tropica* Cushman. This is the type-species by original designation of the genus *Rectuvigerinella* Saidova 1975, regarded by Loeblich and Tappan (1987) as a junior synonym of *Siphogenerina* Schlumberger 1882.

Figures 25–26 *Siphogenerina striata* var. *curta* Cushman 1926

Fig. 25 × 45. From *Challenger* Sta. 192, Ki Islands, Pacific (129fm.). ZF2357.
Fig. 26 × 45. From *Challenger* Sta. 209, Philippines (95–100fm.). ZF2358.

Referred by Brady (1884) to *Sagrina striata* Schwager, sp., by Thalmann (1932) to *S. striata* (Schwager), by Barker (1960) to *Siphogenerina striata* var. *curta* Cushman, by Whittaker and Hodgkinson (1979) to *Rectobolivina striata* var. *curta* (Cushman), and by van Marle (1991) to *Rectuvigerina striata* (Schwager). Srinivasan and Sharma (1980) specifically excluded the present forms from *R. striata* (Schwager).

Plate 76 (LXXVI)

Figures 1–3 *Siphogenerina dimorpha* (Parker and Jones 1865)

[*Uvigerina* (*Sagrina*) *dimorpha* Parker and Jones 1865]
Fig. 1 × 60. From *Challenger* Sta. 191A, Ki Islands, Pacific (580fm.). ZF2349.
Figs 2–3 × 60. From *Challenger* Sta. 279A, off Tahiti, Pacific (420fm.). ZF2350.

Referred by Brady (1884) to *Sagrina dimorpha* Parker and Jones, sp., by Thalmann (1932) to *Siphogenerina dimorpha* var. *pacifica* Cushman, by Barker (1960) to *Rectobolivina dimorpha* var. *pacifica* (Cushman), and by Belford (1966) and van Marle (1991) to *R. dimorpha* (Parker and Jones). This species has been lectotypified by Hodgkinson (1992). Cushman's varietal name appears superfluous.

Figures 4–7 *Siphogenerina* sp. nov.

Fig. 4 × 70. From *Challenger* Sta. 195A, off Amboyna, Pacific (15–20fm.). ZF2360.
Fig. 5 ZF2361; Figs 6–7 ZF2362. × 70. From *Challenger* Sta. 217A, off Papua, Pacific (37fm.).

Referred by Brady (1884) to *Sagrina virgula* Brady, by Thalmann (1932) to *Siphogenerina virgula* (Brady), and by Barker (1960) to *Rectobolivina* (?) *virgula* (Brady). The present forms appear distinct from *Nodogenerina virgula* (Brady) (see below) in terms of their initially biserial chamber arrangement, which indicates possible placement in *Siphogenerina*.

Figure 8 *Nodogenerina virgula* (Brady 1879)

[*Sagrina virgula* Brady 1879]
× 50–70. From *Challenger* Sta. 120, off Pernambuco, Atlantic (675fm.). ZF2363. Lectotype designated by Loeblich and Tappan (1964) for *Nodogenerina bradyi* Cushman 1927.

Referred by Brady (1884) to *Sagrina virgula* Brady, by Thalmann (1932) to *Nodogenerina bradyi* Cushman, by Barker (1960) and van Marle (1991) to *Stilostomella bradyi* (Cushman) and by Haig (1993) to *Amphimorphina virgula* (Brady). This is the type-species, by synonymy with the originally designated *N. bradyi* Cushman 1927, of the genus *Nodogenerina* Cushman 1927. Brady's figured specimen is herein refigured on Supplementary Plate 2. It clearly shows that the aperture lacks the tooth characteristic of *Stilostomella* Guppy 1894.

Figures 9–10 *Nodogenerina antillea* (Cushman 1923)

[*Nodosaria antillea* Cushman 1923]
× 50–70. From *Challenger* Sta. 192, Ki Islands, Pacific (129fm.). 1959.5.11.369–370.

Referred by Brady (1884) to *Sagrina virgula* Brady, by Thalmann (1932) to *Nodogenerina antillea* (Cushman), by Barker (1960) and van Marle (1991) to *Stilostomella antillea* (Cushman), and by Hermelin (1989) (Figs 9?, 10) to *Siphonodosaria lepidula* (Schwager). The specimen figured by Brady as Figure 9 is herein refigured on Supplementary Plate 2.

Figures 11–16 *Nubeculina divaricata* (Brady 1879)

[*Sagrina divaricata* Brady 1879]
Figs 11–13 ZF2004. Lectotype designated by Loeblich and Tappan (1964) ex this slide, now in ZF3615. Figs 14–15 ZF2005. × 70. From *Challenger* Sta. 217A, off Papua, Pacific (37fm.).
Fig. 16 × 70. From *Challenger* Sta. 172, Friendly Islands, Pacific (18fm.). ZF2006.

Referred by Brady (1884) to *Nubecularia*, and by Thalmann (1932) and Barker (1960) to *Nubeculina*. This is the type-species by original designation of the genus *Nubeculina* Cushman 1924.

Figures 17–19 *Millettia tessellata* (Brady 1884)

[*Sagrina* (?) *tessellata* Brady 1884]
Fig. 17 × 75 (lectotype designated by Loeblich and Tappan 1955);
Fig. 18 × 75; Fig. 19 × 150. From *Challenger* Sta. 219A, Admiralty Islands, Pacific (17fm.). ZF2359.

Referred by Brady (1884) to *Sagrina* (?), by Thalmann (1932) and Barker (1960) to *Schubertia*, and by Revets (1992) to *Millettia*. This is the type-species by original designation of the genus *Millettia* Schubert 1911 (substitute name for *Milletia* Wright 1889, not Duncan 1889). This genus has been reviewed by Revets (1992). *Schubertia* Silvestri 1912 was an unnecessary replacement name for *Millettia* Schubert 1911, and also incidentally a junior homonym of *Schubertia* Gistl 1848.

Figures 20–21 *Glandulonodosaria annulata* (Brady 1884)

[*Sagrina* (?) *annulata* Brady 1884]
Fig. 20 × 60; Fig. 21 × 100. From *Challenger* Sta. 260A, off Honolulu reefs, Pacific (40fm.). ZF2344.

Referred by Brady (1884) to *Sagrina* (?), and by Thalmann (1932) and Barker (1960) tentatively to *Siphogenerina*. The specimen figured by Brady as Figure 20 is herein refigured on Supplementary Plate 1.

Figures 22–28 *Ramulina globulifera* Brady 1879

Figs 22–23, 27 × 35. ZF2253. Fig. 24 × 35. ZF2255. Fig. 28 × 35. 1959.5.11.173–176. From *Challenger* Sta. 174C, off Fiji, Pacific (210fm.).
Fig. 26 × 35. From *Challenger* Sta. 23, West Indies (450fm.). ZF2254.

Plate 77 (LXXVII)

Globigerina bulloides d'Orbigny 1826

× 200.

This is the type-species by subsequent designation of the genus *Globigerina* d'Orbigny 1826.

Plate 78 (LXXVIII)

Orbulina universa d'Orbigny 1839

× 40.

This is the type-species by original monotypy of the genus *Orbulina* d'Orbigny 1839.

Plate 79 (LXXIX)

Figures 1–2 *Globigerina* sp.

Referred by Brady (1884) to *Globigerina bulloides* var. *triloba* Reuss, by Thalmann (1932) to *G. triloba* Reuss, and by Barker (1960) to *G.* sp. nov.?

Figures 3–7 *Globigerina bulloides* d'Orbigny 1826

Figs 3–5 × 50. Fig. 6 × 50. From Shetland (40–60fm.). ZF1467. Fig. 7 × 50. From *Challenger* Sta. 232, South of Japan (345fm.). ZF1468.

This is the type-species by subsequent designation of the genus *Globigerina* d'Orbigny 1826.

Figures 8–10 *Globorotalia* (*Globoconella*) *inflata* (d'Orbigny 1839)

[*Globigerina inflata* d'Orbigny 1839]
Figs 8–9 × 50. From *Challenger* Sta. 232, South of Japan (345fm.). ZF1487.
Fig. 10 × 50. From *Challenger* Sta. 64, North Atlantic (2750fm.). ZF1492.

Referred by Brady (1884), Thalmann (1932), and Barker (1960) to *Globigerina*. This species was referred by Parker (1960) and later Saito *et al.* (1981) to *Globorotalia*, by Banner and Blow (1967) to *Globorotalia* (*Turborotalia*), and by Kennett and Srinivasan (1983) to *Globorotalia* (*Globoconella*).

Figures 11–12, 16 *Globigerinoides ruber* (d'Orbigny 1839)

[*Globigerina rubra* d'Orbigny 1839]
Fig. 11 ZF1499. × 50. From *Challenger* Sta. 338, South Atlantic (1990fm.).
Fig. 12 × 50. From *Challenger* Sta. 344, Ascension Island, South Atlantic (420fm.).
Fig. 16 × 50. From *Challenger* Sta. 346, South Atlantic (2350fm.). ZF1502.

Referred by Brady (1884) and Thalmann (1932) to *Globigerina rubra* d'Orbigny, and by Barker (1960) to *Globigerinoides rubra* (d'Orbigny). This is the type-species by original designation of the genus *Globigerinoides* Cushman 1927. All genera ending in the suffix *-oides* should be regarded as masculine and hence the specific ending should be *-er* rather than *-ra* (see discussion under *Eggerelloides scaber* (Williamson), Plate 47, Figs 15–17).

Figures 13–15 *Globigerinoides pyramidalis* (van den Broeck 1876)

[*Globigerina bulloides* var. *rubra* subvar. *pyramidalis* van den Broeck 1876]
× 50. From *Challenger* Sta. 338, South Atlantic (1990fm.). ZF1501.

Referred by Brady (1884) and Thalmann (1932) to *Globigerina rubra* d'Orbigny, by Barker (1960) to *Globigerinoides rubra* (d'Orbigny), and by Saito *et al.* (1981) to *G. pyramidalis* (van den Broeck).

Figure 17 *Neogloboquadrina eggeri* (Rhumbler 1901)

[*Globigerina eggeri* Rhumbler 1901]
× 50. From *Challenger* Sta. 300, North of Juan Fernandez, East Pacific (1375fm.). ZF1482.

Referred by Brady (1884) to *Globigerina dubia* Egger, by Thalmann (1932), Barker (1960), and Banner and Blow (1960*a*) to *G. eggeri* (Rhumbler), and by Saito *et al.* (1981) to *Neogloboquadrina eggeri* (Rhumbler).

Plate 80 (LXXX)

Figures 1–5 *Globigerinoides conglobatus* (Brady 1879)

[*Globigerina conglobata* Brady 1879]
Figs 1, 3 × 50. From *Challenger* Sta. 338, South Atlantic (1990fm.). ZF1470. Lectotype (1959.4.13.7) designated by Banner and Blow (1960*a*) ex this slide.
Fig. 2 × 50. From *Challenger* Sta. 64, North Atlantic (2750fm.). ZF1471.
Fig. 4 × 50. From *Challenger* Sta. 181, NE of Australia (surface). 1957.2.6.2.
Fig. 5 × 50. From *Challenger* Sta. 224 (1850fm.). ZF1473.

Referred by Brady (1884) to *Globigerina,* and by Thalmann (1932), Barker (1960), Banner and Blow (1960*a*), Stainforth *et al.* (1975), and Saito *et al.* (1981) to *Globigerinoides.*

Figures 6–10 *Beella digitata* (Brady 1879) emend. Banner and Blow 1959

[*Globigerina digitata* Brady 1879]
Figs 6–7, 9 × 50. From *Challenger* Sta. 276, South Pacific (2350fm.). ZF1477.

Fig. 8 × 50; Fig. 10 × 50. Lectotype designated by Banner and Blow (1959). 1958.8.18.3. (ZF1479). From *Challenger* Sta. 338, South Atlantic (1990fm.).

Referred by Brady (1884) and Thalmann (1932) to *Globigerina,* by Barker (1960, text) to *Hastigerinella* and by Barker (1960, addenda) (following Banner and Blow (1959, 1960*b*)) to *Globorotalia* (*Beella*). This is the type-species by original designation of *Globorotalia* (*Beella*) Banner and Blow 1960.

Figures 11–17 *Globigerinoides sacculifer* (Brady 1877)

[*Globigerina sacculifera* Brady 1877]
Fig. 11 1959.4.13.8 (ex ZF1503). Figs 13–14 ZF1503. Fig. 17 ZF1505. × 50. From *Challenger* Sta. 224, North Pacific (1850fm.).
Fig. 12 × 50. From *Challenger* Sta. 24, West Indies (390fm.). ZF1504. Figs 15–16 × 50.

Referred by Brady (1884) to *Globigerina,* and by Thalmann (1932), Leroy (1944*a*), Barker (1960), and Saito *et al.* (1981) to *Globigerinoides.* This species has been lectotypified by Banner and Blow (1960*a*).

Figures 18–21 *Globigerinella aequilateralis* (Brady 1879)

[*Globigerina aequilateralis* Brady 1879]
Fig. 18 × 50. From *Challenger* Sta. 33, off Bermuda (435fm.). ZF1463.
Fig. 19 × 50. ZF1564. Lectotype designated by Banner and Blow (1960*b*) (1959.6.11.1) ex this slide. Fig. 21 × 50. From *Challenger* Sta. 224, North Pacific (1850fm.). ZF1465. From *Challenger* Sta. 224, North Pacific (1850fm.).
Fig. 20 × 50. Pacific (surface). ZF1466.

Referred by Brady (1884) to *Globigerina aequilateralis* Brady, by Thalmann (1932), Leroy (1944*a,b*), Saito *et al.* (1981), by Kennett and Srinivasan (1983) to *Globigerinella aequilateralis* (Brady), by Barker (1960, text) to *Hastigerina aequilateralis* (Brady), by Banner and Blow (1960*b*) and later Barker (1960, addenda) to *H.* (*H.*) *siphonifera* d'Orbigny 1839 (with *Globigerina aequilateralis* Brady 1879 in synonymy), and by Parker (1962) to *Globigerinella siphonifera* (d'Orbigny). *Globigerina aequilateralis* Brady 1879 is the type-species by original designation of the genus *Globigerinella* Cushman 1927. Loeblich and Tappan (1987) have argued that it is not a junior synonym of *Globigerina siphonifera* d'Orbigny 1839, as some authors had earlier conjectured.

Plate 81 (LXXXI)

Figure 1 *Globoquadrina conglomerata* (Schwager 1866)

[*Globigerina conglomerata* Schwager 1866]
× 50. From *Challenger* Sta. 155, Antarctic (1300fm.). ZF1483.

Referred by Brady (1884) to *Globigerina dutertrei* d'Orbigny, and by Thalmann (1932) and Barker (1960) to *G. conglomerata* Schwager. This species was referred to the genus *Globoquadrina* by Parker (1962) and later Srinivasan and Sharma (1980) and Saito *et al.* (1981). It has been neotypified by Banner and Blow (1960*b*).

Figures 2–3 *Globigerinoides sacculifer* (Brady 1877)

[*Globigerina sacculifera* Brady 1877]
× 50. From *Challenger* Sta. 344, Ascension Island, South Atlantic (420fm.). ZF1469.

Referred by Brady (1884) to *Globigerina bulloides* var. *triloba* Reuss, by Thalmann first (1932) to *G. triloba* Reuss and later (1937) to *Globigerina tricamerata* Tolmachoff, also by Barker (1960) to *G. tricamerata* Tolmachoff, and by Saito *et al.* (1981) to *Globigerinoides sacculifer* (Brady) (Fig. 2 only) (with *Globigerina tricamerata* Tolmachoff in synonymy). This species has been lectotypified by Banner and Blow (1960*b*).

Figures 4–5 *Globigerinoides ruber* (d'Orbigny 1839)

[*Globigerina rubra* d'Orbigny 1839]
Fig. 4 × 50. From *Challenger* Sta. 338, South Atlantic (1990fm.).
Fig. 5 × 50. From *Challenger* Sta. 85, Canaries, Atlantic (1125fm.).

Referred by Brady (1884), Thalmann (1932), and Barker (1960) to *Globigerina helicina* d'Orbigny, and by Saito *et al.* (1981) to *Globigerinoides ruber* (d'Orbigny) (see also Plate 79, Figs 11, 12, 16).

Figures 6–7 *Globigerinoides* sp. (aberrant)

Fig. 6 × 50. (?) From *Challenger* Sta. 338, South Atlantic (1990fm.). ZF1507.
Fig. 7 × 50. From *Challenger* Sta. 185, Torres Strait, Pacific (155fm.). ZF1508.

Referred by Brady (1884), Thalmann (1932), and Barker (1960) to *Globigerina* sp. ('wild-growing, monstrous forms').

Figures 8–27 *Orbulina universa* d'Orbigny 1839

Fig. 8 × 50. From *Challenger* Sta. 246, North Pacific (2050fm.). ZF2047.
Fig. 9 × 50. (?) From *Challenger* Sta. 24, West Indies (390fm.). ZF2048.
Fig. 10 × 50. From off Seaham, Durham, England (40–48fm.). ZF2050.
Fig. 11 × 50. (?) From off the Shetlands, Scotland (50–60fm.). ZF2051.
Figs 12–14, 18–19, 24–25, 27 × 50. Fig. 23 × 100.
Fig. 15 × 50. ZF2053. Fig. 22 × 100. From surface tows. ZF2059.
Fig. 16 × 50. ZF2054.
Fig. 17 × 50. ZF2055.
Fig. 20 × 50. From *Porcupine* Sta. 36, South of Ireland (725fm.). ZF2057.
Fig. 21 × 50. From *Challenger* Sta. 78, East of the Azores, Atlantic (1000fm.). ZF2058.
Fig. 26 × 50. From *Challenger* Sta. 302, West coast of Patagonia (1450fm.). ZF2060.

Referred by Brady (1884), Thalmann (1932), and Barker (1960) to *Orbulina universa* d'Orbigny (Figs 8–26) and *O. porosa* (Terquem) (Fig. 27). *O. universa* d'Orbigny 1839 is the type-species by original monotypy of the genus *Orbulina* d'Orbigny 1839. Leroy (1944*a*) erroneously cited Plate 78 (*sic*), Figs 8–26 as *O. universa* d'Orbigny.

Plate 82 (LXXXII)

Figures 1–3 *Orbulina universa* d'Orbigny 1839

Fig. 1 × 50.
Fig. 2 × 50. From *Challenger* Sta. 120, off Pernambuco, Atlantic (675fm.).
Fig. 3 × 50. From *Challenger* Sta. 166, West coast of New Zealand (275fm.).

Referred by Brady (1884), Thalmann (1932), and Barker (1960) to *Orbulina universa* d'Orbigny. This is the type-species by original monotypy of the genus *Orbulina* d'Orbigny 1839. Figure 2 is close to *Biorbulina bilobata* Blow 1956, regarded by most modern authors as a junior synonym of *Orbulina universa* d'Orbigny. Since *B. bilobata* is the type-species by original designation of the genus *Biorbulina* Blow 1956, *Biorbulina* is regarded by most modern authors as a junior synonym of *Orbulina*.

Figure 4 *Globigerinoides sacculifer* (Brady 1877)

[*Globigerina sacculifera* Brady 1877]
× 50. From *Challenger* Sta. 36, North Atlantic (surface). ZF1506.

Referred by Brady (1884) and Thalmann (1932) to *Globigerina*, and by Thalmann (1937), Barker (1960), and Whittaker and Hodgkinson (1979) to *Globigerinoides*. This species has been lectotypified by Banner and Blow (1960*b*) (see also Plates 80 and 81).

Figure 5 *Globigerinoides conglobatus* (Brady 1879)

[*Globigerina conglobata* Brady 1879]
× 50. No locality (surface). ZF1472.

Referred by Brady (1884) to *Globigerina*, and by Thalmann (1932), Barker (1960), Stainforth *et al.* (1975), and Saito *et al.* (1981) to *Globigerinoides*. This species has been lectotypified by Banner and Blow (1960*b*) (see also Plate 80, Figs 1–5).

Figures 6–7 *Bolliella adamsi* (Banner and Blow 1959)

[*Hastigerinella* (*Bolliella*) *adamsi* Banner and Blow 1959]
Fig. 6 Holotype (Banner and Blow 1959) 1958.8.18.1 (ZF1480); Fig. 7 Paratype (Banner and Blow 1959) 1958.8.18.2 (ZF1481). From *Challenger* Sta. 191A, Ki Islands, Pacific. × 50.

Referred by Brady (1884) and Thalmann (1932) to *Globigerina digitata* Brady, by Thalmann (1933*a*) and Barker (1960, text) to *Hastigerinella digitata* (Brady), and by Banner and Blow (1959, 1960*b*) and later Barker (1960, addenda) to *Hastigerinella* (*Bolliella*) *adamsi*. This is the type-species by original designation of *Hastigerina* (*Bolliella*) Banner and Blow 1959. Parker (1962) and Saito *et al.* (1981) synonymized *Hastigerina* (*Bolliella*) Banner and Blow 1959 with *Globigerinella* Cushman 1927, but Loeblich and Tappan (1964, 1987) have reinstated it. Kennett and Srinivasan (1983) have rejected the hypothesis that the closely related *Protentella prolixa* Lipps 1964, the type-species by original designation of the genus *Prontentella* Lipps 1964, might be synonymous.

Figures 8–9 *Globigerina bradyi* Wiesner 1931

Fig. 8 × 50. Lectotype designated by Banner and Blow (1960*a*) 1959.5.11.738 (ex ZF1509). Fig. 9 × 50. ZF1509. From *Challenger* Sta. 144, Southern Ocean (1570fm.).

Referred by Brady (1884) to *Globigerina* sp., by Thalmann first (1932) to *Globigerina* sp., later (1933*a*) to *G. elevata* d'Orbigny, and later still (1937) to *G. bradyi* Wiesner, also by Barker (1960), Banner and Blow (1960*a*), and Jenkins *et al.* (1986) to *G. bradyi* Wiesner, and by Saito *et al.* (1981) to *Globigerinita uvula* (Ehrenberg) (with *Globigerina bradyi* Wiesner in synonymy). Kennett and Srinivasan (1983) also regarded *Globigerina bradyi* Wiesner as a junior synonym of *Globigerinita uvula* (Ehrenberg), but the original figures of that species are ambiguous (Jenkins *et al.* 1986).

Figure 10 *Neogloboquadrina blowi* (Rögl and Bolli 1973)

[*Neogloboquadrina dutertrei* subsp. *blowi* Rögl and Bolli 1973] × 50. From *Challenger* Sta. 191A, Ki Islands, Pacific (580fm.). Lectotype designated by Banner 1965.2.1.1.

Referred by Brady (1884) to *Globigerina cretacea* d'Orbigny (?), by Thalmann first (1932) to *Globigerinella subcretacea* (Chapman) and later (1933*a*) to *G. subcretacea* (Lomnicki), by Barker (1960) to *Hastigerina subcretacea* (Lomnicki), by Rögl and Bolli (1973) to *Neogloboquadrina dutertrei* subsp. *blowi* (a new name for *Globigerina subcretacea* Chapman 1902, not Lomnicki 1901), and by Saito *et al.* (1981) to *N. blowi* Rögl and Bolli.

Figure 11 *Hedbergella planispira* (Tappan 1940)

[*Globigerina planispira* Tappan 1940] × 50. From the White Chalk, Iowa, USA (for comparison). ZF1474.

Referred by Brady (1884) and Barker (1960) to *Globigerina cretacea* d'Orbigny, and by Thalmann (1932) to *Globigerinella cretacea* (d'Orbigny). This fossil specimen was figured for comparison with the Recent one above.

Figure 12 *Marginotruncana marginata* (Reuss 1845)

[*Rosalina marginata* Reuss 1845] × 50. Fossil.

Referred by Brady (1884) to *Globigerina linneiana* d'Orbigny, sp., and by Thalmann (1932) and Barker (1960) to *Globotruncana linneiana* (d'Orbigny).

Figures 13–20 *Candeina nitida* d'Orbigny 1839

Fig. 13 × 50. From *Challenger* Sta. 33, the Philippines (surface). ZF1238.
Fig. 14 × 50. From *Challenger* Sta. 338, South Atlantic (1990fm.). ZF1239.
Figs 15–17 × 50. From *Challenger* Sta. 280, South Pacific (1940fm.). ZF1240.
Fig. 18 × 50. ZF1241. Fig. 19 × 50. 1959.5.5.205. Fig. 20 × 100. ZF1242. From *Challenger* Sta. 120, off Pernambuco, Atlantic (675fm.).

This is the type-species by original monotypy of the genus *Candeina* d'Orbigny 1839.

Plate 83 (LXXXIII)

Figures 1–6 *Hastigerina pelagica* (d'Orbigny 1839), emend. Banner and Blow 1960

[*Nonionina pelagica* d'Orbigny 1839]
Fig. 1 × 43. Fig. 2 × 47. Fig. 3 × 50. Fig. 4 × 50 ZF1562. Neotype designated by Banner and Blow (1960*b*). Figs 5–6 × 50. ZF1563. From *Challenger* Sta. 338, South Atlantic (1990fm.).

Referred by Brady (1884), Thalmann (1932), and Saito *et al.* (1981) to *Hastigerina pelagica* (d'Orbigny 1839), Barker (1960, text) to *H. murrayi* Thomson 1876, and by Banner and Blow (1960*b*) and later Barker (1960, addenda) to *H.* (*H.*) *pelagica* (d'Orbigny 1839) (with *H. murrayi* Thomson 1876 in synonymy) (Figs 1–4, 6 only). This is the type-species, by synonymy with the originally designated *Hastigerina murrayi* Thomson 1876, of the genus *Hastigerina* Murray 1876.

Figures 7?, 8 *Hastigerina parapelagica* Saito and Thompson 1976

× 50. From *Challenger* Sta. 338, South Atlantic (1990fm.). ZF1564.

Referred by Brady (1884) and Thalmann (1932) to *Hastigerina pelagica* (d'Orbigny) (see above), by Barker (1960, text) to *H. murrayi* Thomson, and by Saito *et al.* (1981) tentatively to *Hastigerina pelagica* (d'Orbigny) (Fig. 7) and to *H. parapelagica* Saito and Thompson (Fig. 8).

Plate 84 (LXXXIV)

Figures 1–5, ?6–7 *Sphaeroidina bulloides* Deshayes 1832

Fig. 1 × 50. From *Challenger* Sta. 323, South Atlantic (1900fm.). ZF2365.
Fig. 2 × 50. ZF2366. Fig. 5 × 50. ZF2368. Figs 6–7 × 50. 1959.5.11.403–406. From *Challenger* Sta. 120, off Pernambuco, Atlantic (675fm.).
Figs 3–4 × 50. From *Challenger* Sta. 279A, off Tahiti, Pacific (420fm.). ZF2367.

Referred by Brady (1884), Thalmann (1932, 1933*a*), Leroy (1944*a*, *b*), Barker (1960), Hermelin (1989) (Figs 1–2?, 5 only), and van Marle (1991) (Figs 1–2 only) to *Sphaeroidina bulloides* d'Orbigny. *Sphaeroidina bulloides* d'Orbigny 1826 was a *nomen nudum* (Tubbs (ICZN), pers. comm.). The name was validated by the description by Deshayes in a volume of the seldom-read *Encyclopédie Méthodique* (the BMNH copy of which resides in a basement vault adjoining the billiard room!) published in 1832. This is the type-species by original monotypy of the genus *Sphaeroidina* d'Orbigny 1826.

Figures 8–11 *Sphaeroidinella dehiscens* (Parker and Jones 1865)

[*Sphaeroidina bulloides* var. *dehiscens* Parker and Jones 1865]
Figs 8–9 ZF2369. Fig. 11 ZF2371. × 50. From *Challenger* Sta. 224, North Pacific (1850fm.).
Fig. 10 × 50. From *Challenger* Sta. 345, off Ascension Island. ZF2370.

Referred by Brady (1884) to *Sphaeroidina*, and by Thalmann (1932) and Barker (1960) to *Sphaeroidinella*. This is the type-species by original designation of the genus *Sphaeroidinella* Cushman 1927. It has been lectotypified by Bolli *et al.* (1957).

Figures 12–13 *Pullenia bulloides* (d'Orbigny 1846)

[*Nonionina bulloides* d'Orbigny 1846]
Fig. 12 × 50. From *Challenger* Sta. 224, North Pacific (1850fm.). ZF2201.
Fig. 13 × 50. From *Challenger* Sta. 332, South Atlantic (2200fm.). ZF2202.

Referred by Brady (1884), Thalmann (1932, 1933*a*), and Leroy (1944*b*) to *Pullenia sphaeroides* d'Orbigny, sp., and by Barker (1960), Hermelin (1989), and van Marle (1991) to *P. bulloides* (d'Orbigny) (see also Text-figure 18). *Nonionina bulloides* d'Orbigny 1846 is the type-species of the genus *Pullenia* Parker and Jones 1862 (*N. bulloides* d'Orbigny 1826 and *N. sphaeroides* d'Orbigny 1826 being *nomina nuda*). It has been lectotypified by Papp and Schmid (1985).

Figures 14–15 *Pullenia quinqueloba* (Reuss 1851)

[*Nonionina quinqueloba* Reuss 1851]
Fig. 14 × 50. From *Challenger* Sta. 149I, Kerguelen Island, South Pacific (120fm.). ZF2199.
Fig. 15 × 50. From *Challenger* Sta. 145, Prince Edward Island, South Pacific (50–150fm.).

Referred by Brady (1884), Thalmann (1932), Leroy (1944*b*), and van Marle (1991) to *Pullenia quinqueloba* (Reuss), and by Thalmann (1933*a*) and Barker (1960) (following Heron-Allen and Earland 1932) to *P. subcarinata* (d'Orbigny).

Figures 16–20 *Pulleniatina obliquiloculata* (Parker and Jones 1865)

[*Pullenia obliquiloculata* Parker and Jones 1865]
Fig. 16 ZF2196. Fig. 20 ZF2198. × 50. From *Challenger* Sta. 224, North Pacific (2740fm.).
Fig. 17 × 50. From *Challenger* Sta. 5, SW of the Canaries, Atlantic (2740fm.). ZF2197.
Fig. 18 × 50. From *Challenger* Sta. 215, W Pacific (surface). 1957.2.6.2.
Fig. 19 × 50. From *Challenger* Sta. 344, Ascension Island, Mid Atlantic (420fm.).

Referred by Brady (1884) to *Pullenia*, and by Thalmann (1932) and Barker (1960) to *Pulleniatina*. This is the type-species by original designation of the genus *Pulleniatina* Cushman 1927. It has been lectotypified (more correctly, neotypified (Hodgkinson 1992)) by Bolli *et al.* (1957).

Plate 85 (LXXXV)

Figures 1–4 *Spirillina vivipara* Ehrenberg 1843

Fig. 1 × 100. From *Challenger* Sta. 145, Prince Edward Island, South Pacific (50–150fm.). ZF2383.
Fig. 2a × 100. From *Challenger* Sta. 260A, Honolulu reefs, Pacific (40fm.). ZF2384.
Fig. 2b × 100. From *Challenger* Sta. 149I, Kerguelen Island, South Pacific (120fm.). ZF2382.

Fig. 3 × 100. From *Challenger* Sta. 219A, Admiralty Islands, Pacific (17fm.). ZF2385.
Fig. 4 × 100. From *Challenger* Sta. 279C, off Tahiti, Pacific (620fm.). ZF2386.

Referred by Brady (1884), Thalmann (1932), Barker (1960), and van Marle (1991) to *Spirillina vivipara* Ehrenberg. This is the type-species by original monotypy of the genus *Spirillina* Ehrenberg 1843.

Figure 5 *Mychostomina revertens* (Rhumbler 1906)

[*Spirillina vivipara* var. *revertens* Rhumbler 1906]
× 100. From *Challenger* Sta. 219A, Admiralty Islands, Pacific (17fm.). ZF2385.

Referred by Brady (1884) and van Marle (1991) to *Spirillina vivipara* Ehrenberg (see above), and by Thalmann (1932) and Barker (1960) to *S. vivipara* var. *revertens* Rhumbler. This is the type-species by subsequent designation of the genus *Mychostomina* Berthelin 1881, regarded by Loeblich and Tappan (1964) as a junior synonym of *Spirillina* Ehrenberg 1843, but by Smith and Isham (1974) and Loeblich and Tappan (1987) as a valid genus.

Figures 6–7 *Spirillina obconica* Brady 1879

Fig. 6 × 120. From *Challenger* Sta. 145, Prince Edward Island, South Pacific (50–150fm.). ZF2379.
Fig. 7 × 120. From *Challenger* Sta. 149I, Kerguelen Island, South Pacific (120fm.). ZF2380.

Figures 8–11 *Spirillina inaequalis* Brady 1879

Figs 8, 10 × 120. From *Challenger* Sta. 260A, Honolulu reefs, Pacific (40fm.).
Figs 9, 11 × 120. From *Challenger* Sta. 219A, Admiralty Islands, Pacific (17fm.). ZF2375.

Figures 12–16 *Spirillina tuberculata* Brady 1878, emend. Brady 1879

Figs 12–13, 15 × 50; Figs 14, 16 × 45. From *Challenger* Sta. 149D, Kerguelen Island, South Pacific (20–60fm.).

Figure 17 *Spirillina denticulata* (Brady 1884)

[*Spirillina limbata* var. *denticulata* Brady 1884]
× 50. From *Challenger* Sta. 162, Bass Strait, Pacific (20–60fm.). ZF2378.

Referred by Brady (1884) and Thalmann (1932) to *Spirillina limbata* var. *denticulata*, and by Barker (1960) to *S. denticulata* (Brady).

Figures 18–21 *Spirillina limbata* Brady 1879

Fig. 18 × 60. From *Challenger* Sta. 145, Prince Edward Island, South Pacific (50–150fm.). ZF2376.
Figs 19–20 × 50. From *Challenger* Sta. 120, off Pernambuco, Atlantic (675fm.). ZF2377. Fig. 21 × 60.

Referred by Brady (1884), Thalmann (1932), Barker (1960), and van Marle (1991) to *Spirillina limbata* Brady.

Figures 22–25 *Spirillina decorata* Brady 1884

Figs 22, 24–25 × 50. From *Challenger* Sta. 120, off Pernambuco, Atlantic (675fm.). ZF2373.

Fig. 23 × 50. From *Challenger* Sta. 85, off the Canaries, Atlantic (1125fm.). ZF2372.

Plate 86 (LXXXVI)

Figures 1–7 *Patellina corrugata* Williamson 1858

Fig. 1 × 120. From *Challenger* Sta. 162, Bass Strait, Pacific (38fm.). ZF2066.

Fig. 2 × 120. From *Challenger* Sta. 174A, off Fiji, Pacific (255fm.). ZF2067.

Figs 3–5 × 120. From *Challenger* Sta. 145, Prince Edward Island, South Pacific (50–150fm.). 1959.5.5.876–899.

Fig. 6 × 150. From *Challenger* Sta. 149D, Kerguelen Island, South Pacific (20–60fm.). ZF2068.

Fig. 7 × 150. From *Challenger* Sta. 219A, Admiralty Islands, Pacific (17fm.). ZF2069.

Referred by Brady (1884), Thalmann (1932), Barker (1960), and van Marle (1991) to *Patellina corrugata* Williamson. This is the type-species by original monotypy of the genus *Patellina* Williamson, 1858.

Figure 8 *Rosalina bradyi* (Cushman 1915)

[*Discorbina globularis* var. *bradyi* Cushman 1915]
× 75. From *Challenger* Sta. 205A, off Hong Kong, Pacific (7fm.). ZF1392.

Referred by Brady (1884) to *Discorbina globularis* d'Orbigny, sp., by Thalmann (1932) to *D. globularis* var. *bradyi* Cushman, and by Barker (1960) and Whittaker and Hodgkinson (1979) to *Rosalina bradyi* (Cushman).

Figure 9 *Rosalina vilardeboana* d'Orbigny 1839

× 75. From *Challenger* Sta. 151, Heard Island, South Pacific (75fm.). ZF1427.

Referred by Brady (1884) and Thalmann (1932) to *Discorbina*, and by Barker (1960) and van Marle (1991) to *Rosalina*.

Figures 10–11 *Discorbinella araucana* (d'Orbigny 1839)

[*Rosalina araucana* d'Orbigny 1839]
Fig. 10 × 75–90. From *Challenger* Sta. 149I, Kerguelen Island, South Pacific (120fm.). ZF1382.
Fig. 11 × 75–90. From *Challenger* Sta. 162, Bass Strait, Pacific (38–40fm.). ZF1383.

Referred by Brady (1884) to *Discorbina*, by Thalmann (1932) to *Discorbis*, and by Barker (1960) to *Discopulvinulina* (?).

Figure 12 *Rosalina* sp. nov. (1)

× 75. From *Challenger* Sta. 168, NE coast of New Zealand (1100fm.). ZF1428.

Referred by Brady (1884) to *Discorbina vilardeboana* d'Orbigny, sp., by Thalmann (1932) to *Discorbis vilardeboana* (d'Orbigny), and by Barker (1960) to *Rosalina* sp. nov.?. The generic identity of this species is somewhat questionable.

Figure 13 *Rosalina globularis* d'Orbigny 1826

× 50. From *Challenger* Sta. 33, off Bermuda, Atlantic (435fm.). ZF1393.

Referred by Brady (1884) to *Discorbina*, by Thalmann (1932) to *Discorbis*, and by Barker (1960) to *Rosalina*. This is the type-species by subsequent designation of the genus *Rosalina* d'Orbigny 1826. This genus has been reviewed by Sellier de Civrieux (1976) and Hansen and Revets (1992). The latter authors refigure the holotype of *R. globularis* d'Orbigny 1826 from the Golfe de Gascoigne (Bay of Biscay). Brady's specimen appears to differ from the holotype essentially only in its plano-convex rather than biconvex outline, although its apertural characteristics may also be different (it is difficult to be certain).

Plate 87 (LXXXVII)

Figure 1 *Strebloides advenus* (Cushman 1922)

[*Discorbis advena* Cushman 1922]
× 100. From *Challenger* Sta. 218A, Admiralty Islands, Pacific (16–27fm.). ZF1415.

Referred by Brady (1884) to *Discorbina rosacea* d'Orbigny, sp., by Thalmann (1932) to *Discorbis advena* Cushman, by Barker (1960) to *Conorboides advena* (Cushman), and by Hansen and Revets (1992) to *Strebloides advenus* (Cushman). This is the type-species by original designation of the genus *Strebloides* Bermúdez and Seiglie 1963. It has been lectotypified by Loeblich and Tappan (1987).

Figure 2 *Lamellodiscorbis* sp. nov.

× 35. From shore sand, Melbourne, Australia or *Challenger* Sta. 187A, off Booby Island, Pacific (8fm.). ZF1425.

Referred by Brady (1884) to the Eocene species *Discorbina vesicularis* Lamarck, sp., by Thalmann first (1932) to *Discorbis vesicularis* (Lamarck) and later (1937) to *D. vesicularis* var. *dimidiata* (Jones and Parker), and by Barker (1960) to *Discorbis* sp. nov. (text) and *D.* sp. nov. (1) (index). *Discorbina dimidiata* Jones and Parker 1862 is the type-species by original designation of the genus *Lamellodiscorbis* Bermúdez 1952. Brady's form is distinct (Hornibrook and Vella 1954)

Figure 3 *Valvulineria rugosa* (d'Orbigny 1839)

[*Rosalina rugosa* d'Orbigny 1839]
× 60. From *Challenger* Sta. 185, Torres Strait, Pacific (155fm.). ZF1432.

Referred by Brady (1884) to *Discorbina*, by Thalmann (1932) to *Discorbis*, and by Barker (1960) to *Valvulineria*.

Figure 4 *Discorbis* sp. nov.

× 50. From *Challenger* Sta. 162, Bass Strait, Pacific (38fm.). ZF1417.

Referred by Brady (1884) to *Discorbina rosacea* d'Orbigny, sp., by Thalmann (1932) and van Marle (1991) to *Discorbis rosacea* (d'Orbigny), and by Barker (1960) to *Discorbis* sp. nov. (text) and *D.* sp. nov. (2) (index).

Figures 5–7 *Rosalina australis* (Parr 1932)

[*Discorbis australis* Parr 1932]
Fig. 5 × 75. From *Challenger* Sta. 162, Bass Strait, Pacific (38fm.). ZF1422.
Fig. 6 × 100.
Fig. 7 × 100. From *Challenger* Sta. 172, off Tongatabu, Fiji, Pacific (18fm.). ZF1423.

Referred by Brady (1884) to *Discorbina valvulata* d'Orbigny, sp., by Thalmann (1932), and Barker (1960, text), and van Marle (1991) to *Discorbis australis* Parr, and by Barker (1960, addenda) to *Discopulvinulina australis* (Parr).

Figure 8 *Neoeponides auberii* (d'Orbigny 1839)

[*Rosalina auberii* d'Orbigny 1839]
× 60. From *Challenger* Sta. 352A, off Cape Verde Islands, Atlantic (11fm.). ZF1421.

Referred by Brady (1884) to the Eocene species *Discorbina turbo* d'Orbigny, by Thalmann (1932) to *Discorbis mira* Cushman, by Barker (1960) to *Discorbina* (?) *mira* (Cushman), and by Hofker (1964, 1976) to *Rotorbinella mira* (Cushman). *Discorbis mira* Cushman 1922 is a junior synonymy of *Rosalina auberii* d'Orbigny 1839, the type-species by original designation of the genus *Rotorbis* Sellier de Civrieux 1977 (Loeblich and Tappan 1987). *Rotorbis* Sellier de Civrieux 1977 is regarded by Loeblich and Tappan (1987) as a junior synonym of *Neoeponides* Reiss 1960, but by Hansen and Revets (1992) as a valid genus. *Neoeponides* Reiss 1960 has been commented upon by Hottinger *et al.* (1990).

Plate 88 (LXXXVIII)

Figure 1 *Gavelinopsis lobatula* (Parr 1950)

[*Discorbis lobatulus* Parr 1950]
× 100. From *Challenger* Sta. 279C, off Tahiti, Pacific (620fm.). ZF1394.

Referred by Brady (1884) to *Discorbina isabelleana* d'Orbigny, sp., by Thalmann (1932) to *Discorbis isabelleana* (d'Orbigny), and by Barker (1960), Hermelin (1989), and van Marle (1991) to *Gavelinopsis lobatulus* (Parr). Douglas and Sliter (1965) have synonymized *Gavelinopsis* Hofker 1951 with *Rotorbinella* Bandy 1965. Loeblich and Tappan (1964, 1987), however, regard the genera as distinct.

Figure 2 *Rosalina* sp. nov. (2)

× 100. From *Challenger* Sta. 233B, Inland Sea, Japan (15fm.). ZF1429.

Referred by Brady (1884) to *Discorbina vilardeboana* d'Orbigny, sp., by Thalmann (1932) to *Discorbis vilardeboana* (d'Orbigny), and by Barker (1960) to *Rosalina* sp. nov.

Figure 3 *Glabratella patelliformis* (Brady 1884)

[*Discorbina patelliformis* Brady 1884]
× 100. From *Challenger* Sta. 219A, Admiralty Islands, Pacific (17fm.). ZF1408.

Referred by Brady (1884) to *Discorbina*, by Thalmann first (1932) to *Discorbis* and later (1937) to *Conorbina*, by Collins (1958) to *Conorbella*, and by Barker (1960) to *Pileolina* (?). This species was referred by Saidova (1975) to *Discorbinoides*.

Figure 4 *Neoconorbina marginata* Hofker 1951

× 90. From *Challenger* Sta. 172, Friendly Islands, Pacific (18fm.). ZF1398.

Referred by Brady (1884) to *Discorbina orbicularis* Terquem, sp., by Thalmann (1932) to *Discorbis orbicularis* (Terquem), and by Barker (1960) to *Neoconorbina terquemi* (Rzehak) (see below). Barker (1960) noted the similarity to both *N. marginata* Hofker and *N. pacifica* Hofker.

Figures 5–8 *Neoconorbina terquemi* (Rzehak 1888)

[*Discorbina terquemi* Rzehak 1888]
Fig. 5 × 100. From *Challenger* Sta. 33, off Bermuda, Atlantic (435fm.). ZF1401.
Fig. 6 × 50; Fig. 7 × 75. From *Challenger* Sta. 185, Torres Strait, Pacific (155fm.). ZF1402.
Fig. 8 × 60. From *Challenger* Sta. 163B, off Port Jackson, Australia (2–10fm.). ZF1403.

Referred by Brady (1884) to *Discorbina orbicularis* Terquem, sp., by Thalmann (1932) to *Discorbis orbicularis* (Terquem), and by Barker (1960) and van Marle (1991) to *Neoconorbina terquemi* (Rzehak). Barker (1960) noted that Brady's Figures 6–8 were 'variants' of *N. terquemi* (Rzehak), and that his Figures 6–7 could also be compared to *N. rostrata* Poag 1966, and his Figure 8 to *N. pacifica* Hofker 1951. *Discorbina terquemi* Rzehak 1888, a new name for *Rosalina orbicularis* Terquem 1876, not d'Orbigny 1850, is the type-species by original designation of the genus *Neoconorbina* Hofker 1951.

Figure 9 *Rosalina eximia* (Hantken 1875)

[*Discorbina eximia* Hantken 1875]
× 50. From *Challenger* Sta. 185, Torres Strait, Pacific (155fm.). ZF1391.

Referred by Brady (1884) to *Discorbina*, by Thalmann (1932) to *Discorbis*, and by Barker (1960) to *Rosalina* (?).

Figure 10 *Glabratella pulvinata* (Brady 1884)

[*Discorbina pulvinata* Brady 1884]
× 100. From *Challenger* Sta. 219A, Admiralty Islands, Pacific (17fm.). ZF1413.

Referred by Brady (1884) to *Discorbina*, by Thalmann (1932) to *Discorbis*, by Hofker (1951) and Collins (1958) to *Conorbella*, and by Barker (1960) and Sen Gupta and Schafer (1973) to *Glabratella*. This is the type-species by original designation of the genus *Conorbella* Hofker 1951. Loeblich and Tappan first (1964) synonymized *Conorbella* Hofker 1951 with *Glabratella* Dorreen 1948, but later (1987) reinstated it.

Plate 89 (LXXXIX)

Figure 1 *Glabratella patelliformis* (Brady 1884)

[*Discorbina patelliformis* Brady 1884]
× 100. From *Challenger* Sta. 187, South of Papua, Pacific (6fm.). ZF1407.

See Plate 88, Fig. 3.

Figures 2–4 *Glabratella australensis* (Heron-Allen and Earland 1932)

[*Discorbis australensis* Heron-Allen and Earland 1932]
Fig. 2 × 100. From shore sand, Port Elizabeth, Algoa Bay, South Africa. ZF1411.
Fig. 3 × 100. ZF1412. Fig. 4 × 100. ZF1410. From *Challenger* Sta. 163B, Port Jackson, Australia (2–10fm.).

Referred by Brady (1884) to *Discorbina pileolus* d'Orbigny, sp., by Thalmann (1932) to *Discorbis pileolus* (d'Orbigny), by Barker (1960) to *Pileolina australensis* (Heron-Allen and Earland), and by van Marle (1991) to *Glabratella australensis* (Heron-Allen and Earland).

Figures 5–7 *Glabratella tabernacularis* (Brady 1881)

[*Discorbina tabernacularis* Brady 1881]
Figs 5–6 × 100. From *Challenger* Sta. 187, South of Papua, Pacific (6fm.). ZF1419.
Fig. 7 × 100. From *Challenger* Sta. 219A, Admiralty Islands, Pacific (17fm.). ZF1420.

Referred by Brady (1884) to *Discorbina*, by Thalmann (1932) to *Discorbis*, by Collins (1958) to *Conorbella* (Fig. 7 only), and by Barker (1960) to *Pileolina*. This species was referred by Saidova (1975) to *Discorbinoides*.

Figures 8–9 *Planoglabratella opercularis* (d'Orbigny 1839)

[*Rosalina opercularis* d'Orbigny 1839]
Fig. 8a × 100. From *Challenger* Sta. 185, Torres Strait, Pacific (155fm.). ZF1396.
Figs 8b, 9 × 100. From Curtis Strait, Queensland, Australia. ZF1397.

Referred by Brady (1884) to *Discorbina*, by Thalmann (1932) to *Discorbis*, by Collins (1958) to *Conorbella*, and by Barker (1960) to *Pileolina* (?). *Rosalina opercularis* d'Orbigny 1826 was a *nomen nudum* (Tubbs (ICZN), pers. comm.). This is the type-species by original designation of the genus *Truyolsia* González-Donoso 1968, regarded by Loeblich and Tappan (1987) as a junior synonym of *Planoglabratella* Seiglie and Bermúdez 1965. It was referred by Saidova (1975) to *Discorbinoides*.

Figures 10–12 *Discorbinella bertheloti* (d'Orbigny 1839)

[*Rosalina bertheloti* d'Orbigny 1839]
Fig. 10 × 90; Fig. 11 × 60. From *Challenger* Sta. 209, Philippines (95–100fm.). ZF1384.
Fig. 12 × 100. From *Porcupine* Sta. 6, West of Ireland, Atlantic (90fm.). ZF1385.

Referred by Brady (1884) to *Discorbina bertheloti* d'Orbigny, sp., by Thalmann (1932) to *Discorbis suppertheloti* Cushman (Fig. 10) and *D.*

bertheloti (d'Orbigny) (Figs 11–12), by Barker (1960) to *Discopulvinulina suppertheloti* (Cushman) (Fig. 10) and *D. bertheloti* (d'Orbigny) (Figs 11–12), by Belford (1966) to *Discorbinella suppertheloti* (Cushman) (Fig. 10) and *D. bertheloti* (d'Orbigny) (Figs 11–12), and also by Hermelin (1989) and van Marle (1991) to *Discorbinella bertheloti* (d'Orbigny) (Figs 11–12). This is the type-species by original designation of the genus *Discopulvinulina* Hofker 1951, regarded by Loeblich and Tappan (1964, 1987) as a junior synonym of *Discorbinella* Cushman and Martin 1935. The relationship of this genus to *Hanzawaia* Asano 1944 requires investigation.

Plate 90 (XC)

Figure 1 *Planodiscorbis* sp. nov.

Fig. 1a × 75. From *Porcupine* Sta. 17, West of Ireland, Atlantic (1230fm.). ZF1386.
Fig. 1b × 75. From *Challenger* Sta. VIII, off the Canaries, Atlantic (620fm.). ZF1387.

Referred by Brady (1884) to *Discorbina bertheloti* var. *baconica* Hantken, by Thalmann (1932) to *Discorbis baconica* (Hantken), and by Barker (1960) to *Discopulvinulina baconica* (Hantken)?. Hantken's is an Eocene species.

Figures 2–3 *Planodiscorbis rarescens* (Brady 1884)

[*Discorbina rarescens* Brady 1884]
× 75. From *Challenger* Sta. 185, Torres Strait, Pacific (155fm.). ZF1414. Lectotype designated by Loeblich and Tappan (1964) ex this slide, now in ZF3648.

Referred by Brady (1884) to *Discorbina*, by Thalmann (1932) to *Discorbis*, and by Barker (1960) and Hansen and Revets (1992) to *Planodiscorbis*. This is the type-species by original designation of the genus *Planodiscorbis* Bermúdez 1952. Hansen and Revets (1992) have argued that this genus is of uncertain status.

Figure 4 *Planodiscorbis circularis* (Sidebottom 1912)

[*Discorbina circularis* Sidebottom 1912]
× 100. From *Challenger* Sta. 185, Torres Strait, Pacific (155fm.).

Referred by Brady (1884) to *Discorbina*, 'carinate form possibly related to the last named species', by Thalmann (1932) to *Discorbis rarescens* (Brady), and by Barker (1960) to *Planodiscorbis circularis* (Sidebottom).

Figures 5–6, 9–12 *Neoglobratella wiesneri* (Parr 1932)

[*Discorbis wiesneri* Parr 1932]
Fig. 5 × 40. 1959.12.23.14. Fig. 6 × 40. ZF1405. Fig. 10 × 100. ZF1399. Fig. 11 × 100. ZF1400. Fig. 12 × 90. ZF1406. From *Challenger* Stas. 149D (20–60fm.) and 149E (28fm.), Kerguelen Island, South Pacific (20–60fm.). Fig. 9 × 90.

Referred by Brady (1884) to *Discorbina parisiensis* d'Orbigny, sp., by Thalmann (1932) to *Discorbis parisiensis* (d'Orbigny), and by Barker (1960) to *Pileolina* (?) *wiesneri* (Parr). This is the type-species by original designation of the genus *Neoglabratella* Seiglie and Bermúdez 1965. It is also the type-species by original designation of the genus

Orbignyella Saidova 1975. Thus, *Orbignyella* is an isotypic junior synonym of *Neoglabratella*.

Figures 7–?8 *Tretomphaloides concinnus* (Brady 1884)

[*Discorbina concinna* Brady 1884]
× 100. From *Challenger* Sta. 279C, off Papeete, Tahiti, Pacific (620fm.). ZF1390. Lectotype designated by Banner *et al.* (1987) in Slide ZF4386.

Referred by Brady (1884) to *Discorbina*, by Thalmann first (1932) to *Discorbis* and later (1937) to *Tretomphalus*, by Barker (1960) to *Rosalina*, by Banner *et al.* (1985) to *Neoconorbina* (*Tretomphaloides*), and by Hansen and Revets (1992) to *Neoconorbina*. This is the type-species by original designation of *Neoconorbina* (*Tretomphaloides*) Banner, Pereira and Desai 1985. This erstwhile subgenus was raised to the rank of genus by Loeblich and Tappan (1987).

Plate 91 (XCI)

Figure 1 *Epistomaroides polystomelloides* (Parker and Jones 1865)

[*Discorbina turbo* var. *polystomelloides* Parker and Jones 1865]
× 30. From *Challenger* Sta. 186, Wednesday Island, Pacific (7–8fm.). ZF1409.

Referred by Brady (1884) to *Discorbina*, by Thalmann (1932) to *Rotalia,* and by Barker (1960) and Whittaker and Hodgkinson (1979) to *Epistomaroides*. This is the type-species by original designation of the genus *Epistomaroides* Uchio 1952. It has been lectotypified (more correctly, neotypified (Hodgkinson 1992)) by Loeblich and Tappan (1964).

Figure 2 *Planulinoides biconcavus* (Parker and Jones in Carpenter 1862)

[*Discorbina biconcava* Parker and Jones in Carpenter 1862]
× 60. From *Challenger* Sta. 162, Bass Strait, Pacific (38fm.). ZF1388.

Referred by Brady (1884) to *Discorbina*, by Thalmann (1932, 1933*a*) to *Discorbis* (erroneously in 1932 to *D. biconcara* as opposed to *D. biconcava*), by Barker (1960) to *Discorbinella*, and by van Marle (1991) to *Planulinoides*. This is the type-species by original designation of the genus *Planulinoides* Parr 1941. It has been lectotypified by Loeblich and Tappan (1964).

Figure 3 *Heronallenia lingulata* (Burrows and Holland 1895)

[*Discorbina lingulata* Burrows and Holland 1895]
× 60. From *Challenger* Sta. 162, Bass Strait, Pacific (38fm.). ZF1389.

Referred by Brady (1884) to *Discorbina biconcava* Jones and Parker, and by Thalmann (1932), Barker (1960), Hermelin (1989), and van Marle (1991) to *Heronallenia lingulata* (Burrows and Holland).

Figure 4 *Valvulineria minuta* (Schubert 1904)

[*Discorbina rugosa* var. *minuta* Schubert 1904]
× 90. From *Challenger* Sta. 191A, Ki Islands, Pacific (580fm.). ZF1418.

Referred by Brady (1884) to *Discorbina rugosa* d'Orbigny, sp., by Thalmann first (1932) to *Discorbis rugosa* (d'Orbigny) and later

(1942) to *D. rugosa* var. *minuta* Schubert, by Barker (1960) to *Valvulineria rugosa* var. *minuta* (Schubert), and by Belford (1966) to *Rotamorphina minuta* (Schubert). *Rotamorphina* Finlay 1939 is regarded by the present author as a junior synonym of *Valvulineria* Cushman 1926.

Figures 5, 8 *Baggina bradyi* (Brotzen 1936)

[*Valvulineria bradyi* Brotzen 1936]
Fig. 5 × 100. From *Challenger* Sta. 209, Philippines (95–100fm.). ZF1380.
Fig. 8 × 50. From *Challenger* Sta. 185, Torres Strait, Pacific (155fm.). ZF1381.

Referred by Brady (1884) to *Discorbina allomorphinoides* Reuss, sp., by Thalmann (1932) to *Valvulineria allomorphinoides* (Reuss), and by Thalmann (1937) and Barker (1960) to *V. bradyi* Brotzen. The presence of a poreless area on the final chamber indicates placement in *Baggina*.

Figure 6 *Ceratocancris* sp. nov.

× 50. From *Challenger* Sta. 279C, off Tahiti, Pacific (620fm.). ZF1433.

Referred by Brady (1884) to *Discorbina saulcyi* d'Orbigny, sp., by Thalmann (1932) to *Discorbis* sp.?, and by Barker (1960) to *Lamarckina* sp. nov. The generic identity of this species is somewhat questionable.

Figure 7 *Lamarckina ventricosa* (Brady 1884)

[*Discorbina ventricosa* Brady 1884]
× 50. From *Challenger* Sta. 33, off Bermuda, Atlantic (435fm.). ZF1424.

Referred by Brady (1884) to *Discorbina*, and by Thalmann (1932), Barker (1960), Belford (1966), and van Marle (1991) to *Lamarckina*.

Figure 9 *Baggina* sp. nov.

× 50. From *Challenger* Sta. 344, Ascension Island, Atlantic (420fm.). ZF1395.

Referred by Brady (1884) tentatively to *Discorbina obtusa* d'Orbigny, sp., by Thalmann (1932) to *Discorbis obtusa* (d'Orbigny)?, and by Barker (1960) to *Valvulineria* (?) sp. nov. The generic identity of this species is somewhat questionable.

Plate 92 (XCII)

Figure 1 *Planorbulina mediterranensis* d'Orbigny 1826

× 35–45. From *Challenger* Sta. 162, Bass Strait, Pacific (38fm.). ZF2117.

Referred by Brady (1884), Thalmann (1932), Barker (1960), and van Marle (1991) to *Planorbulina mediterranensis* d'Orbigny. This is the type-species by subsequent designation of the genus *Planorbulina* d'Orbigny 1826.

Figures 2–3 *Planorbulina distoma* Terquem 1876

Fig. 2 ×35–45. From *Challenger* Sta. 33, off Bermuda, Atlantic (435fm.). ZF2118.
Fig. 3 ×35–45. From *Porcupine* Sta. 23, NW of Ireland, Atlantic (450fm.). ZF2119.

Referred by Brady (1884), Thalmann (1932), Barker (1960), and van Marle (1991) to *Planorbulina mediterranensis* d'Orbigny, and by Haynes (1937a) to *P. distoma* Terquem.

Figure 4 *Planorbulina acervalis* Brady 1884

×35. From *Challenger* Sta. 187A, Booby Island, Pacific (8fm.). ZF2114.

Referred by Brady (1884), Thalmann (1932), Barker (1960), and Hofker (1976) to *Planorbulina acervalis*.

Figures 5–6 *Planorbulinella larvata* (Parker and Jones 1865)

[*Planorbulina larvata* Parker and Jones 1865]
Fig. 5 ×35. ZF2115. Fig. 6 ×35. ZF2116. From *Challenger* Sta. 172, Friendly Islands, Pacific (18fm.).

Referred by Brady (1884) to *Planorbulina larvata* Parker and Jones, and by Thalmann (1932), Leroy (1944b), and Barker (1960) to *Planorbulinella larvata* (Parker and Jones). This is the type-species by original designation of the genus *Planorbulinella* Cushman 1927.

Figures 7–9 *Cibicides refulgens* de Montfort 1808

Figs 7, 9 ×30. From *Porcupine* Sta. 36, SW of Ireland, Atlantic (725fm.). 1959.5.11.595–596.
Fig. 8 ×30. From *Challenger* Sta. 305, West coast of Patagonia, Pacific (120fm.). ZF2543.

Referred by Brady (1884) to *Truncatulina*, and by Thalmann (1932), Barker (1960), and van Marle (1991) to *Cibicides*. This is the type-species by original designation of the genus *Cibicides* de Montfort 1808.

Figure 10 *Cibicides lobatulus* (Walker and Jacob 1798)

[*Nautilus lobatulus* Walker and Jacob 1798]
×60. From *Challenger* Sta. 172, Friendly Islands, Pacific (18fm.). ZF2532.

Referred by Brady (1884) to *Truncatulina*, and by Thalmann (1932) and Barker (1960) to *Cibicides*. This is the type-species, by synonymy with the originally designated *L. vulgaris* Fleming 1828, of the genus *Lobatula* Fleming 1828, regarded by the present author as a junior synonym of *Cibicides* de Montfort 1808. Its variability has been reviewed by Cooper (1965) and Nyholm (1961).

Plate 93 (XCIII)

Figures 1, 4–5 *Cibicides lobatulus* (Walker and Jacob 1798)

[*Nautilus lobatulus* Walker and Jacob 1798]
Fig. 1 ×60. From *Challenger* Sta. 354A, Cape Verde Islands, Atlantic (11fm.). ZF2531.

Fig. 4 ×40. From *Challenger* Sta. 246, North Pacific (2050fm.). ZF2533.
Fig. 5 ×40. From *Challenger* Sta. 300, North of Juan Fernandez, Pacific (1375fm.). ZF2543.

Referred by Brady (1884) to *Truncatulina*, and by Thalmann (1932), Barker (1960), Haynes (1973a) (Fig. 1 only), and van Marle (1991) to *Cibicides*. This is the type-species of the genus *Lobatula* Fleming 1828, regarded herein as a junior synonym of *Cibicides* de Montfort 1808 (see Plate 92, Fig. 10).

Figure 2 *Parvicarinina altocamerata* (Heron-Allen and Earland 1922)

[*Truncatulina tenuimargo* var. *altocamerata* Heron-Allen and Earland 1922]
×60–100. From *Challenger* Sta. 166, West coast of New Zealand (275fm.). ZF2551.

Referred by Brady (1884) to *Truncatulina tenuimargo* (see below), by Thalmann first (1932) to *Cibicides tenuimargo* (Brady), later (1933a) to *Cibicides tenuimargo* var. *altocamerata* (Heron-Allen and Earland) and later still (1937) to *Laticarinina tenuimargo* (Brady), by Barker (1960) and Belford (1966) to *Parvicarinina altocamerata* (Heron-Allen and Earland), and by van Marle (1991) to *Laticarinina altocamerata* (Heron-Allen and Earland). This is the type-species by original designation of the genus *Parvicarinina* Finlay 1940, regarded by Loeblich and Tappan (1964, 1987) as a junior synonym of *Laticarinina* Galloway and Wissler 1927.

Figure 3 *Parvicarinina tenuimargo* (Brady 1884)

[*Truncatulina tenuimargo* Brady 1884]
×60–100. From *Challenger* Sta. 174A, off Fiji, Pacific (255fm.). ZF2552.

Referred by Brady (1884) to *Truncatulina*, by Thalmann first (1932) to *Cibicides* and later (1937) to *Laticarinina*, and by Barker (1960), Belford (1966), Perelis and Reiss (1975 (1976)), and van Marle (1991) to *Cibicides*.

Figure 6 *Dyocibicides biserialis* Cushman and Valentine 1930

×50. From *Challenger* Sta. 166, West coast of New Zealand (275fm.). ZF2557.

Referred by Brady (1884) to *Truncatulina variabilis* d'Orbigny, and by Thalmann (1932) and Barker (1960) to *Dyocibicides biserialis* Cushman and Valentine. This is the type-species by original designation of the genus *Dyocibicides* Cushman and Valentine 1930.

Figure 7 *Karreria uniserialis* (Thalmann 1933)

[*Dyocibicides uniserialis* Thalmann 1933]
×50. From *Challenger* Sta. 305, W. coast of Patagonia (120fm.). ZF2556.

Referred by Brady (1884) to *Truncatulina variabilis* d'Orbigny, by Thalmann first (1932) to *Dyocibicides* n. sp. and later (1933a) to *D. uniserialis* nov. spec., and also by Barker (1960) to *D. uniserialis* Thalmann. The late uniserial rather than biserial chamber arrangement indicates placement within *Karreria* rather than *Dyocibicides*.

Figures 8–9 *Cibicidoides wuellerstorfi* (Schwager 1866)

[*Anomalina wuellerstorfi* Schwager 1866]
Fig. 8 × 50–60. From *Porcupine* Sta. 21, West of Ireland, Atlantic. ZF2559.
Fig. 9 × 50–60. From *Challenger* Sta. 166, West coast of New Zealand (275fm.). ZF2558.

Referred by Brady (1884) to *Truncatulina wuellerstorfi* Schwager, sp., by Thalmann first (1932) to *Planulina wuellerstorfi* (Schwager) and later (1937) to *P. bradii* Tolmachoff (Fig. 8 only), by Leroy (1944*b*) and Srinivasan and Sharma (1980) to *Cibicides wuellerstorfi* (Schwager) (Fig. 9 only), by Barker (1960) to *Planulina bradii* Tolmachoff (Fig. 8) and *P. wuellerstorfi* (Schwager) (Fig. 9), also by Belford (1966) and van Marle (1991) to *P. wuellerstorfi* (Schwager) (Fig. 9 only), and by Hermelin (1989) to *Cibicidoides wuellerstorfi* (Schwager) (Figs 8–9). *Anomalina wuellerstorfi* Schwager 1866 is the type-species by original designation of the genus *Fontbotia* González-Donoso and Linares 1970, regarded by the present author as a junior synonym of *Cibicidoides* Thalmann 1939. It has been neotypified by Srinivasan and Sharma (1980). *Planulina bradii* Tolmachoff 1934 appears synonymous.

Figures 10–11 *Planulina ariminensis* d'Orbigny 1826

× 55–60. From *Challenger* Sta. 122, SE of Pernambuco, Atlantic. ZF1081.

Referred by Brady (1884) to *Anomalina*, and by Thalmann (1932), Barker (1960), and van Marle (1991) to *Planulina*. This is the type-species by subsequent designation of the genus *Planulina* d'Orbigny 1826.

Plate 94 (XCIV)

Figure 1 *Planulina foveolata* (Brady 1884)

[*Anomalina foveolata* Brady 1884]
× 60. From *Challenger* Sta. 33, off Bermuda, Atlantic (435fm.). ZF1082.

Referred by Brady (1884) to *Anomalina*, and by Thalmann (1932), Barker (1960), and van Morkhoven *et al.* (1986) to *Planulina*.

Figures 2–3 *Anomalinoides colligerus* (Chapman and Parr 1937)

[*Anomalina colligera* Chapman and Parr 1937]
Fig. 2 × 55. From *Challenger* Sta. 174C, off Fiji, Pacific (210fm.). ZF1079.
Fig. 3 × 90. From *Challenger* Sta. 217A, off Papua, Pacific (37fm.). ZF1080.

Referred by Brady (1884) and Leroy (1944*b*) to *Anomalina ammonoides* Reuss, sp., by Barker (1960) to *A. colligera* Chapman and Parr, and by Belford (1966) and van Marle (1991) to *Anomalinoides colligerus* (Chapman and Parr).

Figures 4–5 *Cibicidoides globulosus* (Chapman and Parr 1937)

[*Anomalina globulosa* Chapman and Parr 1937]
Fig. 4 × 40–45. From *Challenger* Sta. 344, Ascension Island, Atlantic (420fm.). ZF1083.

Fig. 5 × 40–45. From *Challenger* Sta. 246, North Pacific (2050fm.). ZF1084.

Referred by Brady (1884) to *Anomalina grosserugosa* Gumbel, sp., by Thalmann (1932, 1933*a*) to *A.* n. sp., by Barker (1960) to *A. globulosa* Chapman and Parr, and by van Morkhoven *et al.* (1986), Hermelin (1989), and van Marle (1991) to *Anomalinoides globulosus* (Chapman and Parr).

Figure 6 *Anomalinella rostrata* (Brady 1881)

[*Truncatulina rostrata* Brady 1881]
× 50. From *Challenger* Sta. 217A, off Papua, Pacific (2740fm.). ZF2549.

Referred by Brady (1884) to *Truncatulina*, and by Thalmann (1932), Leroy (1944*b*), Barker (1960), Belford (1966), and van Marle (1991) to *Anomalinella*. Van Marle (1991) regards *Truncatulina rostrata* Brady 1881 as a *nomen nudum*, but gives no justification. This is the type-species by original designation of the genus *Anomalinella* Cushman 1927.

Figure 7 *Turborotalita humilis* (Brady 1884)

[*Truncatulina humilis* Brady 1884]
× 100. From *Challenger* Sta. 276, South Pacific (2350fm.). ZF2529. Lectotype (Banner and Blow 1960*a*) from *Challenger* Sta. 5, SW of the Canaries, Atlantic (2740fm.). 1959.10.2.1 (ex. ZF2530).

Referred by Brady (1884) to *Truncatulina*, by Thalmann (1932) to *Anomalina*, by Barker (1960) to *Valvulineria* (?), by Parker (1962) to *Globigerinita*, and by Blow and Banner (1962) and Saito *et al.* (1981) to *Turborotalita*. This is the type-species by original designation of the genus *Turborotalita* Blow and Banner 1962, emend. Holmes 1984.

Figure 8 *Cibicidoides cicatricosus* (Schwager 1866)

[*Anomalina cicatricosa* Schwager 1866]
× 45. From *Challenger* Sta. 120, off Pernambuco, Atlantic (675fm.). ZF2521.

Referred by Brady (1884) to *Truncatulina akneriana* d'Orbigny, sp., by Thalmann (1932), Barker (1960), and Srinivasan and Sharma (1980) to *Cibicides cicatricosus* (Schwager), and by van Morkhoven *et al.* (1986) to *Cibicidoides cicatricosus* (Schwager). This species has been neotypified and reillustrated by Srinivasan and Sharma (1980).

Figure 9 *Cibicidoides pachyderma* (Rzehak 1886)

[*Truncatulina pachyderma* Rzehak 1886]
× 35. From *Porcupine* Sta. 42, SW of Ireland, Atlantic (862fm.). ZF2554.

Referred by Brady (1884) to *Truncatulina ungeriana* d'Orbigny, sp., by Thalmann (1932) and Barker (1960) to *Cibicides pseudoungerianus* (Cushman), by Belford (1966) to *Heterolepa mediocris* (Finlay), by van Morkhoven *et al.* (1986) to *Cibicidoides pachyderma* (Rzehak) (with *Cibicides mediocris* Finlay and *C. pseudoungerianus* of various authors in synonymy), and by van Marle (1991) to *Cibicidoides mediocris* (Finlay) (on p. 134) and *Planulina ungeriana* (d'Orbigny) (on p. 206). *Truncatulina pachyderma* Rzehak has been lectotypified and reillustrated by van Morkhoven *et al.* (1986).

Plate 95 (XCV)

Figures 1–3 *Neoeponides praecinctus* (Karrer 1868)

[*Rotalina praecincta* Karrer 1868]
Fig. 1 × 30. From *Challenger* Sta. 209, Philippines, Pacific (95–100fm.). ZF2538.
Fig. 2 × 30. From *Challenger* Sta. 217A, off Papua, Pacific (37fm.). ZF2537.
Fig. 3 × 35. From *Challenger* Sta. 219A, Admiralty Islands, Pacific (17fm.). ZF2540.

Referred by Brady (1884) to *Truncatulina praecincta* Karrer, sp., by Thalmann (1932) and Leroy (1944a,b) to *Eponides praecinctus* (Karrer), by Barker (1960) to *Cibicides praecinctus* (Karrer), by Belford (1966) to '*Eponides*' *praecinctus* (Karrer), and by van Marle (1991) to *Cibicidoides dutemplei* (d'Orbigny) *s.l.* Tentatively transferred by the present author to *Neoeponides* Reiss 1960. This genus has been commented upon by Hottinger *et al.* (1990).

Figure 4 *Cibicidoides robertsonianus* (Brady 1881)

[*Planorbulina (Truncatulina) robertsoniana* Brady 1881]
× 50. From *Challenger* Sta. 24, West Indies (390fm.). ZF2547.

Referred by Brady (1884) to *Truncatulina*, by Thalmann (1932) and Barker (1960) to *Cibicides,* and by van Morkhoven *et al.* (1986), Hermelin (1989), and van Marle (1991) to *Cibicidoides.* Van Marle (1991) regards *Truncatulina robertsoniana* Brady 1881 as a *nomen nudum,* but gives no justification.

Figure 5 *Gyroidina bradyi* (Trauth 1918)

[*Truncatulina bradyi* Trauth 1918]
× 60. From *Challenger* Sta. 323, South Atlantic (1900fm.). ZF2523.

Referred by Brady (1884) to *Truncatulina dutemplei* d'Orbigny, sp., by Thalmann first (1932) to *Cibicides lobatulus* (Walker and Jacob) and later (1942) to *C. bradyi* (Trauth), also by Barker (1960) to *C. bradyi* (Trauth), by Belford (1966) to *Parrelloides bradyi* (Trauth), and by van Morkhoven *et al.* (1986), Hermelin (1989), and van Marle (1991) to *Cibicidoides bradyi* (Trauth). *Parrelloides* Hofker 1956 (and, incidentally, *Heterolepa* Franzenau 1884) is regarded by the present author as a junior synonym of *Gyroidina* d'Orbigny 1826, emend. Hansen 1967 (which differs from *Cibicidoides* Thalmann 1939 in possessing an aperture more or less restricted to the umbilical side rather than in the equatorial position). *Cibicides hyalina* Hofker 1951, the type-species by original designation of the genus *Parrelloides* Hofker 1956, is regarded as a likely junior synonym of the present species.

Figure 6 *Cibicidoides mundulus* (Brady, Parker, and Jones 1888)

[*Truncatulina mundula* Brady, Parker, and Jones 1888]
× 40. From *Challenger* Sta. 296, South Pacific (1825fm.). ZF2560.

Referred by Brady (1884) to *Truncatulina* sp., 'near *T. haidingerii*', by Thalmann first (1932) to *Cibicides mundulus* (Brady, Parker, and Jones) and later (1937) to *Cibicidoides mundulus* (Brady, Parker, and Jones), and also by Barker (1960) and Hermelin (1989) to *Cibicidoides mundulus* (Brady, Parker, and Jones). This is the type-species by original designation of the genus *Cibicidoides* Thalmann 1939, which is herein interpreted as differing from *Gyroidina* d'Orbigny 1826,

emend. Hansen 1967 in having the aperture in the equatorial position. It has been lectotypified by Loeblich and Tappan (1955).

Figure 7 *Cibicidoides subhaidingerii* (Parr 1950)

[*Cibicides subhaidingerii* Parr 1950]
× 30. From *Challenger* Sta. 185, Torres Strait, Pacific (155fm.). ZF2528.

Referred by Brady (1884) to *Truncatulina haidingerii* d'Orbigny, sp., by Thalmann (1932) to *Eponides haidingerii* (d'Orbigny), by Barker (1960) to *Cibicides subhaidingerii* Parr, by Belford (1966) to '*Eponides*' *subhaidingerii* (Parr), and by van Morkhoven *et al.* (1986) to *Cibicidoides subhaidingerii* (Parr).

Figure 8 *Ioanella tumidula* (Brady 1884)

[*Truncatulina tumidula* Brady 1884]
× 100. From *Challenger* Sta. 5, North Atlantic (2740fm.). ZF2553.

Referred by Brady (1884) to *Truncatulina,* and by Thalmann (1932) and Barker (1960) to *Eponides.* This is the type-species by original designation of the genus *Ioanella* Saidova 1975.

Figures 9–10 *Osangulariella umbonifera* (Cushman 1933)

[*Pulvinulinella umbonifera* Cushman 1933]
Fig. 9 × 60. From *Challenger* Sta. 64, North Atlantic (2750fm.). ZF2541. This specimen designated as lectotype of *Eponides bradyi* Earland 1934 by Loeblich and Tappan (1987).
Fig. 10 × 60. From *Challenger* Sta. 5, North Atlantic (2740fm.). ZF2542.

Referred by Brady (1884) to *Truncatulina pygmaea* Hantken, by Thalmann (1932) to *Eponides pygmaeus* (Hantken) (Fig. 9) and *Pulvinulinella bradyana* Cushman (Fig. 10), by Thalmann (1937), Barker (1960), and van Marle (1991) to *Eponides bradyi* Earland, by Phleger *et al.* (1953) to *Epistominella* (?) *umbonifera* (Cushman) (with *Eponides bradyi* Earland tentatively in synonymy), and by Hermelin (1989) to *Nuttallides umbonifera* (Cushman) (also with *E. bradyi* Earland in synonymy). *Pulvinulinella umbonifera* Cushman 1933 is the type-species, by synonymy with the originally designated *Eponides bradyi* Earland 1934, of the genus *Osangulariella* Saidova 1975.

Figure 11 *Oridorsalis umbonata* (Reuss 1851)

[*Rotalina umbonata* Reuss 1851]
× 75. From *Challenger* Sta. 305, West coast of Patagonia, Pacific (120fm.). ZF2550.

Referred by Brady (1884) to *Truncatulina tenera,* by Thalmann (1932) to *Eponides umbonatus* (Reuss), by Barker (1960) to *E.* (?) *tenera* (Brady), by Leroy and Hodgkinson (1975) and Pflum and Frerichs (1976) to *Oridorsalis tener* (Brady), and by Hermelin (1989) to *O. umbonatus* (Reuss) (with *Truncatulina tenera* Brady in synonymy). Brady's figured specimens lack the supplementary spiral sutural apertures characteristic of *Oridorsalis* Andersen 1961, but these features are developed (albeit variably) in unfigured specimens from *Challenger* and comparable deep-water material examined by the present author. I do not agree with the suggestion by McCulloch (1977) that a new genus is required for Brady's forms.

Plate 96 (XCVI)

Figure 1 *Trochulina rosea* (Guérin-Méneville 1832, ex d'Orbigny 1826)

[*Rotalia (Rotalie) rosea* Guérin-Méneville 1832, ex d'Orbigny 1826] × 70. From shore sand, Cuba. ZF2548.

Referred by Brady (1884) to *Truncatulina*, by Thalmann (1932) to *Rotalia*, and by Barker (1960) to *Discorbina*. *Discorbina* Parker and Jones 1862 is an isotypic junior synonym of *Trochulina* d'Orbigny 1839. *Rotalia rosea* d'Orbigny 1826 was a *nomen nudum* (Tubbs (ICZN), pers. comm.). The name was validated by the publication of a figure identified as *Rotalia rosea* d'Orbigny by Guérin-Méneville in the seldom-read *Iconographie du Règne Animal de G. Cuvier*, published in parts (livraisons) between 1829 and 1843 (Cowan 1971). The part containing the relevant (mollusc) plates, believed to be part 17, was issued in January of 1832 (Cowan 1971).

Figure 2 *Neoeponides margaritifer* (Brady 1881)

[*Truncatulina margaritifera* Brady 1881] × 28. From *Challenger* Sta. 209, Philippines, Pacific (95–100fm.). ZF2537.

Referred by Brady (1884) to *Truncatulina*, by Thalmann (1932) to *Rotalia*, by Barker (1960) to *Cibicides*, and by Belford (1966) to '*Eonides*'. The generic identity of this species is somewhat questionable in that it appears to possess supplementary apertures (Belford 1966). It is tentatively referred by the present author to *Neoeponides* Reiss 1960. This genus has been commented upon by Hottinger *et al.* (1990).

Figure 3 *Osangularia bengalensis* (Schwager 1866)

[*Anomalina bengalensis* Schwager 1866] × 90. From *Challenger* Sta. 191A, Ki Islands, Pacific (580fm.). ZF2552.

Referred by Brady (1884) to *Truncatulina culter* Parker and Jones, sp., by Thalmann (1932) to *Pulvinulinella culter* (Parker and Jones), by Barker (1960) and Srinivasan and Sharma (1980) to *Osangularia bengalensis* (Schwager), and by Hermelin (1989) and van Marle (1991) to *O. culter* (Parker and Jones). *Anomalina bengalensis* Schwager 1866 is the type-species by original designation of the genus *Parrella* Finlay 1939, which is a junior homonym of *Parrella* Ginsburg 1938, and a junior synonym of *Osangularia* Brotzen 1940 (Loeblich and Tappan 1987). It has been neotypified by Srinivasan and Sharma (1980).

Figure 4 *Siphoninella soluta* (Brady 1884)

[*Truncatulina soluta* Brady 1884] × 100. From *Challenger* Sta. 24, West Indies (390fm.).

Referred by Brady (1884) to *Truncatulina*, and by Thalmann (1932) and Barker (1960) to *Siphoninella*. This is the type-species by original designation of the genus *Siphoninella* Cushman 1927.

Figures 5–7 *Siphonina tubulosa* Cushman 1924

Fig. 5 × 100. ZF2544. Figs 6–7 Magnification not given. ZF2545. From *Challenger* Sta. 174A, off Kandavu, Fiji (185fm.).

Referred by Brady (1884) to *Truncatulina reticulata* Czjzek, sp., and by Thalmann (1932), Barker (1960), and van Marle (1991) to *Siphonina tubulosa* Cushman.

Figure 8 *Siphonina bradyana* Cushman 1927

× 70. From *Challenger* Sta. 24, West Indies (390fm.). ZF2546.

Referred by Brady (1884) to *Truncatulina reticulata* Czjzek, sp., and by Thalmann (1932), Barker (1960), and van Marle (1991) to *S. bradyana* Cushman.

Figures 9–14 *Siphoninoides echinatus* (Brady 1879)

[*Planorbulina echinata* Brady 1879]
Fig. 9 × 100. From *Challenger* Sta. 185, Torres Strait, Pacific (155fm.). ZF2524.
Figs 10, 13 × 100. ZF2525. Fig. 14 × 100. ZF2527. From *Challenger* Sta. 260A, off Honolulu, Pacific (40fm.).
Figs 11–12 × 100. From *Challenger* Sta. 219A, Admiralty Islands, Pacific (17fm.). ZF2526.

Referred by Brady (1884) to *Truncatulina*, and by Thalmann (1932), Barker (1960), Whittaker and Hodgkinson (1979), and van Marle (1991) to *Siphoninoides*. This is the type-species by original designation of the genus *Siphoninoides* Cushman 1927.

Plate 97 (XCVII)

Figures 1–2 *Discoanomalina coronata* (Parker and Jones 1857)

[*Anomalina coronata* Parker and Jones 1857]
Fig. 1 × 40. From *Challenger* Sta. 145, Prince Edward Island, South Pacific (50–150fm.). 1959.5.5.87.
Fig. 2 × 50. From *Challenger* Sta. VIII, off the Canaries, Atlantic (620fm.).

Referred by Brady (1884) and Thalmann (1932) to *Anomalina*, and by Barker (1960) to *Paromalina*. This species has recently been lectotypified by Hodgkinson (1992). *Paromalina* Loeblich and Tappan 1957 was regarded by Medioli and Scott (1978) and later Loeblich and Tappan (1987) as a junior synonym of *Discoanomalina* Asano 1951.

Figures 3–6 *Discoanomalina semipunctata* (Bailey 1851)

[*Rotalina semipuncta* Bailey 1851]
Fig. 3 × 35–45. From *Challenger* Sta. 174C, off Fiji, Pacific (210fm.). ZF1086.
Figs 4, 6 × 35–45. From *Challenger* Sta. 164A, off Sydney, Australia (410fm.). ZF1089.
Fig. 5 × 35–45. From *Challenger* Sta. 24, West Indies (390fm.). ZF1085.

Referred by Brady (1884) to *Anomalina polymorpha* Costa, by Thalmann (1932) and Barker (1960, text) to *A. semipunctata* (Bailey), and by Barker (1960, addenda) to *Discoanomalina semipunctata* (Bailey). This is the type-species, by synonymy with the originally designated *D. japonica* Asano 1951, of the genus *Discoanomalina* Asano 1951. It is also the type-species, by synonymy with the originally designated *P. bilateralis* Loeblich and Tappan 1957, of the genus *Paromalina* Loeblich and Tappan 1957. *Paromalina* is an isotypic junior synonym of *Discoanomalina*.

Figure 7 *Discoanomalina vermiculata* (d'Orbigny 1839)

[*Truncatulina vermiculata* d'Orbigny 1839]
×45. From *Challenger* Sta. 33, off Bermuda, Atlantic (435fm.). ZF1088.

Referred by Brady (1884) to *Anomalina polymorpha* Costa?, by Thalmann first (1932) to *A. semipunctata* (Bailey) and later (1933a) to *A. vermiculata* (d'Orbigny), and by Barker (1960) also to *A. vermiculata* (d'Orbigny).

Figures 8–14 *Carpenteria proteiformis* (Goes 1882)

[*Carpenteria balaniformis* var. *proteiformis* Goes 1882]
Figs 8, 10 ×15–17. ZF1248. Fig. 11 ×18. ZF1250. Figs 12, 14 ×50. ZF1251. From *Challenger* Sta. 24, West Indies (390fm.).
Fig. 9 ×15–17. From *Challenger* Sta. 33, off Bermuda (435fm.). ZF1249.
Fig. 13 ×50. From *Challenger* Sta. 219A, Admiralty Islands, Pacific (17fm.). ZF1252.

Plate 98 (XCVIII)

Figures 1–12 *Rupertina stabilis* (Wallich 1877)

[*Rupertia stabilis* Wallich 1877]
Figs 1, 6 ×40. ZF2334. Figs 2–3, 5 ×40. Figs 4, 8, 9 ×40. ZF2335. Fig. 7 ×40. ZF2336. Figs 10–11 ×40. ZF2337. Fig. 12 ×40. ZF2338. From *Porcupine* Sta. 57, Faroe Channel ('cold area'), North Atlantic (632fm.).

Referred by Brady (1884), Thalmann (1932), and Barker (1960) to *Rupertia*. This is the type-species by original designation of the genus *Rupertina* Loeblich and Tappan 1961, a new name for *Rupertia* Wallich 1877, not Gray 1865.

Figures 13, 15–16 *Carpenteria monticularis* Carter 1877

×35. From *Challenger* Sta. 344, Ascension Island, Atlantic (420fm.). ZF1246.

Figures 14, 17 *Carpenteria balaniformis* Gray 1858

×30–40. From *Challenger* Sta. 344, Ascension Island, Atlantic (420fm.). ZF1245.

This is the type-species by original monotypy of the genus *Carpenteria* Gray 1858.

Plate 99 (XCIX)

Figures 1–5 *Carpenteria monticularis* Carter 1877

Figs 1, 4 ×12. Fig. 2 ×12. ZF1247. Figs 3, 5 ×12. ZF1244. From *Challenger* Sta. 201, Philippines, Pacific (102fm.).

Referred by Brady (1884) and Barker (1960) to *Carpenteria monticularis* Carter (see Plate 98, Figs 13, 15–16), and by Thalmann (1932) to *C. utricularis* Carter (see below and also Plate 100, Figs 1–4).

Figures 6–7 *Carpenteria utricularis* (Carter 1876)

[*Polytrema utriculare* Carter 1876]
×15. From *Challenger* Sta. 216A(?), Admiralty Islands, Pacific (16–25fm.). ZF1253.

Plate 100 (C)

Figures 1–4 *Carpenteria utricularis* (Carter 1876)

[*Polytrema utriculare* Carter 1876]
Fig. 1 ×15. ZF1253. Fig. 2 ×15. 1959.5.5.239. Figs 3–4 ×15. ZF1254. From *Challenger* Sta. 218A, Admiralty Islands, Pacific (16–25fm.).

See also Plate 99, Figs 6–7.

Figures 5–9 *Miniacina miniacea* (Pallas 1766)

[*Millepora miniacea* Pallas 1766]
Fig. 5 ×20. From *Challenger* Sta. 219A, Admiralty Islands, Pacific (17fm.). ZF2187.
Figs 6–7 ×20. From *Challenger* Sta. 172, Friendly Islands, Pacific (18fm.). ZF2185.
Fig. 8 ×20. From *Challenger* Sta. 218A, Admiralty Islands, Pacific (16–25fm.). ZF2184.
Fig. 9 ×50. From *Challenger* Sta. 33, off Bermuda, Atlantic (435fm.). ZF2186.

Referred by Brady (1884) and Thalmann (1932) to *Polytrema*, and by Thalmann (1933a) and Barker (1960) to *Miniacina*. This is the type-species by original designation of the genus *Miniacina* Galloway 1933. *Polytrema* Risso 1826 is a bryozoan genus.

Plate 101 (CI)

Figure 1 *Miniacina miniacea* (Pallas 1766)

[*Millepora miniacea* Pallas 1766]
×40. ZF2188.

See Plate 100, Figs 5–9.

Figures 2–3 *Miniacina alba* (Carter 1877)

[*Polytrema miniacea* var. *album* Carter 1877]
×30

Referred by Brady (1884) to *Polytrema miniacea* var. *alba* Carter, by Thalmann first (1932) to *P. alba* (Carter) and later (1933a) to *Miniacina alba* (Carter), and also by Barker (1960) to *M. alba* (Carter).

Figures 4–7 *Baculogypsina sphaerulata* (Parker and Jones 1860)

[*Orbitolina concava* var. *sphaerulata* Parker and Jones 1860]
Figs 4–6 ×30. From *Challenger* Sta. 173A, off Fiji, Pacific (12fm.). ZF2488.
Fig. 7 ×40. From Wednesday Island, Pacific. ZF2489.

Referred by Brady (1884) to *Tinoporus baculatus* de Montfort, and by Thalmann (1932) and Barker (1960) to *Baculogypsina sphaerulata* (Parker and Jones). This is the type-species by original monotypy of the genus *Baculogypsina* Sacco 1893. It has been lectotypified by Loeblich and Tappan (1964).

Figure 8 *Sphaerogypsina globulus* (Reuss 1848)

[*Ceriopora globulus* Reuss 1848]
× 30. From *Challenger* Sta. 260A, off Honolulu reefs, Pacific (40fm.). ZF1510.

Referred by Brady (1884), Thalmann (1932), and Whittaker and Hodgkinson (1979) to *Gypsina,* and by Thalmann (1933*a*), Barker (1960), Hofker (1976) to *Sphaerogypsina.* This is the type-species by original designation of the genus *Sphaerogypsina* Galloway 1933.

Figures 9–12 *Discogypsina vesicularis* (Parker and Jones 1860)

[*Orbitolina concava* var. *vesicularis* Parker and Jones 1860]
Figs 9–11 × 30. From *Challenger* Sta. 172, Friendly Islands, Pacific (18fm.). ZF1517.
Fig. 12 × 40. From Wednesday Island, Pacific. ZF1514.

Referred by Brady (1884), Thalmann (1932, 1933*a*), Leroy (1944*b*), and Barker (1960) to *Gypsina vesicularis* (Parker and Jones). This species has recently been lectotypified by Hodgkinson (1992).

Plate 102 (CII)

Figures 1–6 *Acervulina inhaerens* Schultze 1854

Figs 1, 5 × 30–40. ZF1511. Fig. 6 × 40. ZF1513. From Dogs Bay, Connemara, Ireland.
Fig. 2 × 30–40. From *Challenger* Sta. 172, Friendly Islands, Pacific (18fm.). ZF1512.
Fig. 3 × 30–40. From *Challenger* Sta. 162, Bass Strait, Pacific (38fm.). ZF1516.
Fig. 4 × 30–40. From *Challenger* Sta. 187, off Papua, Pacific (6fm.). 1959.5.5.405.

Referred by Brady (1884) to *Gypsina,* and by Thalmann (1932), Barker (1960), and van Marle (1991) to *Acervulina.* This is the type-species by subsequent designation of the genus *Acervulina* Schultze 1854.

Figures 7, 8, 12 *Cymbaloporetta plana* (Cushman 1924)

[*Tretomphalus bulloides* var. *planus* Cushman 1924]
Fig. 7 × 60. From the Pacific (surface) ZF1368.
Fig. 8 × 60. From *Challenger* Sta. 260A, Honolulu reefs, Pacific (40fm.). ZF1367.
Fig. 12 × 60. From *Challenger* Sta. 185, Torres Strait, Pacific (155fm.). ZF1371.

Referred by Brady (1884) to *Cymbalopora* (*Tretomphalus*) *bulloides* d'Orbigny, sp., by Thalmann (1932) to *Tretomphalus bulloides* (d'Orbigny), and by Thalmann (1937) and Barker (1960) to *T. planus* (Cushman) (see also Text-figure 20). This is the type-species by original designation of the genus *Pseudotretomphalus* Hofker 1979, regarded by Loeblich and Tappan (1987) as a junior synonym of *Cymbaloporetta* Cushman 1928.

Figure 9 *Millettiana millettii* (Heron-Allen and Earland 1915)

[*Cymbalopora millettii* Heron-Allen and Earland 1915]
× 60. From *Challenger* Sta. 260A, Honolulu reefs, Pacific (40fm.). ZF1367.

Referred by Brady (1884) to *Cymbalopora* (*Tretomphalus*) *bulloides* d'Orbigny, sp., by Thalmann first (1932) to *Tretomphalus bulloides* (d'Orbigny) and later (1937) to *T. millettii* (Heron-Allen and Earland), and also by Barker (1960) to *T. millettii* (Heron-Allen and Earland). This is the type-species by original designation of *Cymbaloporetta* (*Millettiana*) Banner, Pereira, and Desai 1985. This erstwhile subgenus has been raised to the rank of genus by Loeblich and Tappan (1987).

Figures 10–11 *Cymbaloporetta atlantica* (Cushman 1934)

[*Tretomphalus atlanticus* Cushman 1934]
× 60. From *Challenger* Sta. 33, off Bermuda, Atlantic (435fm.). ZF1369.

Referred by Brady (1884) to *Cymbalopora* (*Tretomphalus*) *bulloides* d'Orbigny, sp., by Thalmann first (1932) to *Tretomphalus bulloides* (d'Orbigny) and later (1937) to *T. atlanticus* Cushman, and also by Barker (1960) to *T. atlanticus* Cushman. This species was referred by Banner *et al.* (1985) to *Cymbaloporetta.*

Figure 13 *Cymbaloporetta squammosa* (d'Orbigny 1839)

[*Rosalina squammosa* d'Orbigny 1839]
× 50. From *Challenger* Sta. 219A, Admiralty Islands, Pacific (17fm.). ZF1372.

Referred by Brady (1884) to *Cymbalopora poeyi* d'Orbigny, sp., by Thalmann first (1932) to *Cymbaloporetta squammosa* (d'Orbigny) and later (1933*a*) to *Cymbalopora squammosa* (d'Orbigny), and by Barker (1960) and Whittaker and Hodgkinson (1979) to *Cymbaloporetta squammosa* (d'Orbigny). This is the type-species by original designation of the genus *Cymbaloporetta* Cushman 1928.

Figure 14 *Cymbaloporetta bradyi* (Cushman 1915)

[*Cymbalopora poeyi* var. *bradyi* Cushman 1915]
Figs 14a–c × 50. From *Challenger* Sta. 185, Torres Strait, Pacific (155fm.). ZF1373.
Fig. 14d × 50. From *Challenger* Sta. 217A, Papua (37fm.). ZF1374.

Referred by Brady (1884) to *Cymbalopora poeyi* var., by Thalmann first (1932) to *Cymbaloporetta bradyi* (Cushman) and later (1933*a*) to *Cymbalopora bradyi* (Cushman), and by Barker (1960), Whittaker and Hodgkinson (1979), and van Marle (1991) to *Cymbaloporetta bradyi* (Cushman).

Figures 15–18 *Cymbaloporella tabellaeformis* (Brady 1884)

[*Cymbalopora tabellaeformis* Brady 1884]
Fig. 15 × 40. ZF1375. Fig. 18 × 40. ZF1379. From *Challenger* Sta. 174C, off Fiji, Pacific (210fm.).
Fig. 16 × 40. From *Challenger* Sta. 192, Ki Islands, Pacific (129fm.). ZF1377.
Fig. 17 × 40. From *Challenger* Sta. 209, Philippines, Pacific (95–100fm.). ZF1378.

Referred by Brady (1884) to *Cymbalopora*, and by Thalmann (1932) and Barker (1960) to *Cymbaloporella*. This is the type-species by original designation of the genus *Cymbaloporella* Cushman 1927.

Plate 103 (CIII)

Figures 1–2 *Globorotalia (Menardella) menardii* (Parker, Jones, and Brady 1865, ex d'Orbigny 1826)

[*Pulvinulina repanda* var. *menardii* Parker, Jones, and Brady 1865 (ex *Rotalia (Rotalie) menardii* d'Orbigny 1826)]
Fig. 1 × 35. From *Challenger* Sta. 224, North Pacific (1850fm.). ZF2226.
Fig. 2 × 35.

Referred by Brady (1884) to *Pulvinulina menardii* Parker, Jones, and Brady, by Thalmann (1932) and Leroy (1944a) to *Globorotalia menardii* (d'Orbigny) (*sic*), by Barker (1960), Stainforth *et al.* (1975), and Saito *et al.* (1981) to *Globorotalia menardii* (Parker, Jones, and Brady), and by Parker (1962) to *G. cultrata* (d'Orbigny). *Pulvinulina repanda* var. *menardii* is the type-species by original designation of the subgenus *Globorotalia (Menardella)* Bandy 1972, regarded by Kennett and Srinivasan (1983) as valid, but by Loeblich and Tappan (1987) as a junior synonym of *Globorotalia* Cushman 1927. It has been lectotypified by Stainforth *et al.* (1978) (see also discussion in Loeblich and Tappan 1987).

Figure 3 *Globorotalia (Menardella) menardii* var. *fimbriata* (Brady 1884)

[*Pulvinulina menardii* var. *fimbriata* Brady 1884]
× 35. From *Challenger* Sta. 24, West Indies (390fm.). Lectotype designated by Banner and Blow (1960a) 1959.7.1.2 (ex ZF2227).

Referred by Brady (1884) to *Pulvinulina menardii* var. *fimbriata*, by Thalmann (1932) and Barker (1960) to *Globorotalia menardii* var. *fimbriata* (Brady), by Banner and Blow (1960a) to *G. (G.) menardii* subsp. *fimbriata* (Brady), and by Saito *et al.* (1981) to *G. fimbriata* (Brady).

Figures 4–6 *Globorotalia (Globorotalia) tumida* (Brady 1877)

[*Pulvinulina menardii* var. *tumida* Brady 1877]
Fig. 4 × 35. From *Challenger* Sta. 224, North Pacific (1850fm.). ZF2248.
Fig. 5 × 35. From *Challenger* Sta. 276, North of Tahiti, Pacific (2350fm.). ZF2249.
Fig. 6 × 35.

Referred by Brady (1884) to *Pulvinulina tumida* Brady, by Thalmann first (1932) to *Globorotalia menardii* var. *tumida* (Brady) and later (1933a) to *G. tumida* (Brady), also by Barker (1960), Parker (1962), Stainforth *et al.* (1975), and Saito *et al.* (1981) to *G. tumida* (Brady), and by Kennett and Srinivasan (1983) to *G. (G.) tumida* (Brady). This is the type-species by original designation of the genus *Globorotalia* Cushman 1927.

Figure 7 *Globorotalia (Obandyella) scitula* (Brady 1882)

[*Pulvinulina scitula* Brady 1882]
× 35. From *Challenger* Sta. 302, West coast of Patagonia, East Pacific (1450fm.).

Referred by Brady (1884) to *Pulvinulina patagonica* d'Orbigny, sp., by Thalmann (1932), Barker (1960), Parker (1962), Stainforth *et al.* (1975), and Saito *et al.* (1981) to *Globorotalia scitula* (Brady), by Banner and Blow (1960a) to *Turborotalia scitula* (Brady), and by Kennett and Srinivasan (1983) to *G. (Hirsutella) scitula* (Brady). *Pulvinulina scitula* Brady 1882 has been lectotypified by Banner and Blow (1960a). *Globorotalia (Hirsutella)* Bandy 1972 is a junior homonym of *Hirsutella* Cooper and Muir-Wood 1951. Forms previously referred to *Hirsutella sensu* Bandy 1972 should be transferred to *Obandyella* Haman, Huddlestone, and Donaghue 1981 (see below).

Figures 8–10 *Globorotalia (Obandyella) hirsuta* (d'Orbigny 1839)

[*Rotalina hirsuta* d'Orbigny 1839]
Fig. 8 × 35. From *Challenger* Sta. 300, North of Juan Fernandez, East Pacific (1375fm.). ZF2206.
Fig. 9 × 35. From *Challenger* Sta. 33, off Bermuda, Atlantic (435fm.). ZF2207.
Fig. 10 × 35.

Referred by Brady (1884) to *Pulvinulina canariensis* d'Orbigny, sp., by Thalmann first (1932) to *Globorotalia canariensis* (d'Orbigny) and later (1933a) to *G. hirsuta* (d'Orbigny), and also by Barker (1960), Stainforth *et al.* (1975), and Saito *et al.* (1981) to *G. hirsuta* (d'Orbigny). This is the type-species by original designation of *Obandyella* Haman, Huddlestone, and Donaghue 1981, substitute name for *Globorotalia (Hirsutella)* Bandy 1972, not *Hirsutella* Cooper and Muir-Wood 1951. Loeblich and Tappan (1987) synonymized *Obandyella* Haman, Huddlestone, and Haman 1981 with *Globorotalia* Cushman 1927. I prefer to retain it as a subgenus.

Figures 11–12 *Globorotalia (Truncorotalia) crassaformis* (Galloway and Wissler 1927)

[*Globigerina crassaformis* Galloway and Wissler 1927]
× 40. From *Challenger* Sta. 5, North Atlantic (2740fm.). ZF2209.

Referred by Brady (1884) to *Pulvinulina crassa* d'Orbigny, sp., by Thalmann (1932) to *Globorotalia crassula* Cushman and Stewart, by Barker (1960) to *G. punctulata* (d'Orbigny), and by Saito *et al.* (1981) to *G. crassaformis* (Galloway and Wissler) (Figs 11–12), and also erroneously to *G. oceanica* Cushman and Bermúdez (Fig. 12 only). This species was referred by Kennett and Srinivasan (1983) to the subgenus *Globorotalia (Truncorotalia)* Cushman and Bermúdez 1949, raised to the rank of genus by Loeblich and Tappen (1987).

Figures 13–14 *Alabaminoides exiguus* (Brady 1884)

[*Pulvinulina exigua* Brady 1884]
Fig. 13 × 40. From *Challenger* Sta. 332, South Atlantic (2200fm.). ZF2215.
Fig. 14 × 40. From *Challenger* Sta. 160, Southern Ocean (2600fm.). ZF2214.

Referred by Brady (1884) to *Pulvinulina*, by Thalmann (1932) to *Eponides*, and by Barker (1960), Hermelin (1989), and van Marle (1991) to *Epistominella*.

Plate 104 (CIV)

Figures 1–2 *Globorotalia (Truncorotalia) truncatulinoides* (d'Orbigny 1839)

[*Rotalina truncatulinoides* d'Orbigny 1839]
Fig. 1 ×40. From *Challenger* Sta. 335, South Atlantic (1425fm.). ZF2228.
Fig. 2 ×40.

Referred by Brady (1884) to *Pulvinulina micheliniana* d'Orbigny, sp., and by Thalmann (1932), Barker (1960), and Saito *et al.* (1981) to *Globorotalia truncatulinoides* (d'Orbigny). This is the type-species by original designation of the subgenus *Globorotalia (Truncorotalia)* Cushman and Bermúdez 1949, raised to the rank of genus by Loeblich and Tappan (1987).

Figures 3–11 *Laticarinina pauperata* (Parker and Jones 1865)

[*Pulvinulina repanda* var. *menardii* subvar. *pauperata* Parker and Jones 1865]
Fig. 3 ×35. ZF2239. Figs 4, 11 ×35. ZF2240. Fig. 6 ×35. ZF2242. From *Challenger* Sta. 120, off Pernambuco, Atlantic (675fm.).
Figs 5, 10 ×35. From *Challenger* Sta. 335, South Atlantic (1425fm.). ZF2241.
Figs 7–9 ×35. From *Challenger* Sta. 146, Southern Ocean (1375fm.). ZF2239.

Referred by Brady (1884) to *Pulvinulina pauperata* Parker and Jones, by Thalmann (1932), Belford (1966), van Morkhoven *et al.* (1986), and van Marle (1991) to *Laticarinina pauperata* (Parker and Jones), and by Thalmann (1942) (Figs 4–5 only), Barker (1960), and Hermelin (1989) to *L. halophora* (Stache). Hornibrook (1971) has shown *Robulina halophora* Stache 1864 to belong in the nodosariid genus *Planularia*. *Pulvinulina repanda* var. *menardii* subvar. *pauperata* Parker and Jones 1865 is the type-species by original designation of the genus *Laticarinina* Galloway and Wissler 1927. It has been lectotypified by Loeblich and Tappan (1964).

Figures 12–16 *Favocassidulina favus* (Brady 1877)

[*Pulvinulina favus* Brady 1877]
Figs 12, 14 ×40; Fig. 15 ×80. From *Challenger* Sta. 300, North of Juan Fernandez, Pacific (1375fm.). ZF2216.
Fig. 13 ×40; Fig. 16 ×80. From *Challenger* Sta. 283, South Pacific (2075fm.). ZF2217.

Referred by Brady (1884) to *Pulvinulina*, by Thalmann (1932) to *Cassidulina*, and by Barker (1960), Belford (1966), Hermelin (1989), and van Marle (1991) to *Favocassidulina*. Van Marle (1991) regards *Cassidulina favus* Brady 1877 as a *nomen nudum*, but gives no justification. This is the type-species by original designation of the genus *Favocassidulina* Loeblich and Tappan 1957. It has been well illustrated by Nomura (1984).

Figure 17 *Rosalina* sp. nov. (3)

×16. From *Challenger* Sta. 24, West Indies (390fm.). ZF2244.

Referred by Brady (1884) to *Pulvinulina punctulata* d'Orbigny, sp., and by Thalmann (1932) and Barker (1960) to *Eponides punctulatus* (d'Orbigny).

Figure 18 *Cribroeponides cribrorepandus* (Asano and Uchio 1951)

[*Poroeponides cribrorepandus* Asano and Uchio 1951]
×30. From *Challenger* Sta. 354A, Vigo Bay, Spain, Atlantic (11fm.). ZF2245.

Referred by Brady (1884) to *Pulvinulina repanda* Fichtel and Moll, sp., by Thalmann first (1932) to *Eponides repandus* (Parker and Jones) (*sic*) and later (1933*a*) to *E. repandus* (Fichtel and Moll), and by Barker (1960) to '*Eponides repandus* (Fichtel and Moll)'. *Poroeponides cribrorepandus* Asano and Uchio 1951 is the type-species by original designation of the genus *Cribroeponides* Shchedrina 1964, regarded by Loeblich and Tappan (1987) as a junior synonym of *Eponides* de Montfort 1808. Whittaker and Hodgkinson (1979) retained it in *Poroeponides* Cushman 1944.

Figure 19 *Eponides repandus* (Fichtell and Moll 1798)

[*Nautilus repandus* Fichtel and Moll 1798]
×30. From *Porcupine* Sta. AA, off Loch Scavaig, Scotland, Atlantic (45–60fm.). ZF2246.

Referred by Brady (1884) to *Pulvinulina repanda* var. *concamerata* Montagu, sp., by Thalmann first (1932) to *Eponides repandus* var. *concameratus* (Williamson) (*sic*) and later (1933*a*) to *Discorbis concamerata* (Montagu), by Barker (1960) to '*Eponides repandus* (F. and M.)' var. *concamerata* (Williamson), by Haynes (1973*a*) to *Eponides repandus* var. *concamerata* (Montagu), and by van Marle (1991) to *E. repandus* (Fichtel and Moll). *Nautilus repandus* Fichtel and Moll 1798 is the type-species by original designation of the genus *Eponides* de Montfort 1808. Brady's specimen is close to the long-lost holotype of this species, discovered and illustrated by Rögl and Hansen (1984). The ICZN have ruled in favour of the rediscovered holotype replacing the neotype designated by Loeblich and Tappan (1962) (Anon 1990*b*).

Plate 105 (CV)

Figure 1 *Mississippina concentrica* (Parker and Jones in Brady 1864)

[*Pulvinulina concentrica* Parker and Jones in Brady 1864]
×35. From *Challenger* Sta. 174C, off Kandavu, Fiji (210fm.). ZF2208.

Referred by Brady (1884) to *Pulvinulina*, by Thalmann (1932) to *Eponides*, and by Barker (1960) to *Mississippina*. This species has been lectotypified by Hodgkinson (1992).

Figure 2 *Oridorsalis umbonata* (Reuss 1851)

[*Rotalina umbonata* Reuss 1851]
×50. From *Challenger* Sta. 308, West coast of Patagonia, East Pacific (175fm.). ZF2250.

Referred by Brady (1884) to *Pulvinulina*, by Thalmann (1932) and Barker (1960) to *Eponides*, and by Belford (1966), Hermelin (1989), and van Marle (1991) to *Oridorsalis*.

Figures 3–6 *Hoeglundina elegans* (d'Orbigny 1826)

[*Rotalia (Turbinuline) elegans* d'Orbigny 1826]
Fig. 3 ×40. From a *Porcupine* station in the North Atlantic (?). ZF2234.

Fig. 4 × 35. From *Challenger* Sta. 135, Tristan d'Acunha, Atlantic (100–150fm.). ZF2211.

Fig. 5 × 35. From *Challenger* Sta. 24, West Indies (390fm.). ZF2212.

Fig. 6 × 35. From *Challenger* Sta. 174C, Fiji, Pacific (210fm.). ZF2213.

Referred by Brady (1884) to *Pulvinulina partschiana* d'Orbigny, sp. (Fig. 3) and *P. elegans* d'Orbigny, sp. (Figs 4–6), by Thalmann (1932) to *Epistomina elegans* (d'Orbigny), by Barker (1960) to *Hoglundina* (sic) *elegans* (d'Orbigny), and by Belford (1966), van Morkhoven *et al.* (1986), and van Marle (1991) to *Hoeglundina elegans* (d'Orbigny) (see also Text-figure 21). This is the type-species by original designation of the genus *Hoeglundina* Brotzen 1948.

Figure 7 *Neoeponides procerus* (Brady 1881)

[*Pulvinulina procera* Brady 1881]
× 35. From *Challenger* Sta. 174C, off Fiji, Pacific (210fm.). ZF2243.

Referred by Brady (1884) to *Pulvinulina*, by Thalmann (1932) to *Discorbis*, by Leroy (1944b) and van Marle (1991) to *Eponides*, and by Barker (1960) to *Eponides* (?). Van Marle (1991) regards *Pulvinulina procera* Brady 1881 as a *nomen nudum*, but gives no justification. The present author tentatively assigns it to the genus *Neoeponides* Reiss 1960. This genus has been commented upon by Hottinger *et al.* (1990).

Figures 8–9 *Buccella frigida* (Cushman 1921)

[*Pulvinulina frigida* Cushman 1921]
Fig. 8 × 50. From *Challenger* Sta. 313, Magellan Strait, Antarctic (55fm.). ZF2220.

Fig. 9 × 50. From Cape Frazer, Latitude 79 deg., 45 min. (North Polar Expedition, 1875–76). ZF2221.

Referred by Brady (1884) to *Pulvinulina karsteni* Reuss, sp., by Thalmann (1932) to *Eponides frigidus* (Cushman), and by Barker (1960) to *Buccella frigida* (Cushman).

Plate 106 (CVI)

Figure 1 *Neoeponides berthelotianus* (d'Orbigny 1839)

[*Rotalina berthelotiana* d'Orbigny 1839]
× 35. From *Challenger* Sta. 189, South of New Guinea, Pacific (25–29fm.). ZF2205.

Referred by Brady (1884) to *Pulvinulina*, by Thalmann (1932) and Barker (1960) to *Eponides*, and by Belford (1966) and van Marle (1991) to *Neoeponides*. *Neoeponides* has been commented upon by Hottinger *et al.* (1960).

Figures 2–3 *Poroeponides lateralis* (Terquem 1878)

[*Rosalina lateralis* Terquem 1878]
Fig. 2 × 35–40. From *Challenger* Sta. 172, Friendly Islands, Pacific (18fm.). ZF2222.

Figs 3a, c × 35–40. ZF2223. Fig. 3b × 35–40. ZF2224. From the Gulf of Suez (7–8fm.).

Referred by Brady (1884) to *Pulvinulina*, by Thalmann (1932) to *Eponides*, and by Barker (1960), and Whittaker and Hodgkinson

(1979) to *Poroeponides*. This is the type-species by original designation of the genus *Poroeponides* Cushman 1944. Hottinger *et al.* (1991a) have demonstrated conclusively that *Poroeponides* is distinct from *Eponides*.

Figure 4 *Cancris auriculus* (Fichtel and Moll 1798)

[*Nautilus auricula* Fichtel and Moll 1798]
× 40. From *Challenger* Sta. 142, off South Africa, Atlantic (150fm.). ZF2230.

Referred by Brady (1884) to *Pulvinulina oblonga* Williamson, by Thalmann (1932) to *Cancris auriculus* (Fichtel and Moll), and by Barker (1960) to *C. oblongus* (Williamson)? Van Marle (1991) noted that 'the specimens shown in 4a–c [belong] to *C. auriculus*'.

Figure 5 *Cancris oblongus* (Williamson 1858)

[*Rotalina oblonga* Williamson 1858]
× 40. From *Challenger* Sta. 24, West Indies (390fm.). ZF2203.

Referred by Brady (1884) to *Pulvinulina auricula* Fichtel and Moll, sp., by Thalmann (1932) and Rögl and Hansen (1984) to *Cancris auriculus* (Fichtel and Moll), and by Barker (1960) and van Marle (1991) to *C. oblongus* (Williamson)? This is the type-species, by synonymy with the originally designated *C. auriculatus* de Montfort 1808, of the genus *Cancris* de Montfort 1808. It has been lectotypified by Rögl and Hansen (1984). Brady's figure is very similar to the lectotype.

Figure 6 *Baggina indica* (Cushman 1921)

[*Pulvinulina indica* Cushman 1921]
× 40–50. From *Challenger* Sta. 217A, off Papua, Pacific (37fm.). ZF2218.

Referred by Brady (1884) to *Pulvinulina hauerii* d'Orbigny, sp., by Thalmann (1932) to *Baggina* (?) *indica* (Cushman), by Barker (1960) to *Cancris indicus* (Cushman), and by Belford (1966) and van Marle (1991) to *Baggina indica* (Cushman).

Figure 7 *Baggina philippinensis* (Cushman 1921)

[*Pulvinulina philippinensis* Cushman 1921]
× 40–50. From *Challenger* Sta. 192, Ki Islands, Pacific (129fm.). ZF2219.

Referred by Brady (1884) to *Pulvinulina hauerii* d'Orbigny, sp., by Thalmann first (1932) to *Valvulineria philippinensis* (Cushman) and later (1942) to *Cancris philippinensis* (Cushman), and by Barker (1960) to *Baggina philippinensis* (Cushman).

Figure 8 *Ceratocancris scaber* (Brady 1884)

[*Pulvinulina oblonga* var. *scabra* Brady 1884]
× 50. From *Challenger* Sta. 185, Torres Strait, Pacific (155fm.). ZF2231.

Referred by Brady (1884) to *Pulvinulina oblonga* var. *scabra*, and by Thalmann (1932), Barker (1960), and Belford (1966) to *Lamarckina scabra* (Brady). The generic identity of this species is questionable.

Figure 9 *Rotalinoides gaimardii* (Fornasini 1906)

[*Turbinulina gaimardii* Fornasini 1906]
× 30. From *Challenger* Sta. 188, South of New Guinea, Pacific (28fm.). ZF2324.

Referred by Brady (1884) to *Rotalia papillosa*, by Thalmann (1932, 1933*a*, 1937) to *R. gaimardi* (d'Orbigny), by Collins (1958) (erroneously citing Plate CLV, Fig. 9) to *Streblus papillosus* (Brady), by Barker (1960) to *S. gaimardii* (d'Orbigny), by Huang (1964) and Belford (1966) to *Pseudorotalia gaimardii* (d'Orbigny), and by Billman *et al.* (1980) to *Asterorotalia gaimardii* (d'Orbigny). *Rotalia papillosa* Brady 1884 is homonymous with *R. papillosa* d'Orbigny 1850 and synonymous with *Turbinulina gaimardii* Fornasini 1906 [= *Rotalia (Turbinulina) gaimardii* d'Orbigny 1826 (*nomen nudum*)]. It is the type-species by original designation of the genus *Rotalinoides* Saidova 1975.

Plate 107 (CVII)

Figures 1, 3 *Rotalinoides compressiusculus* (Brady 1884)

[*Rotalia papillosa* var. *compressiuscula* Brady 1884]
Fig. 1 × 30. From *Challenger* Sta. 233B, Inland Sea, Japan (14fm.). ZF2325. Fig. 3 × 50.

Figure 1 was referred by Brady (1884) to *Rotalia papillosa* var. *compressiuscula*, by Thalmann first (1932, 1933*a*) to *R. papillosa* var. *compressiuscula* Brady and later (1937) to *R. gaimardii* var. *compressiuscula* (Brady), by Barker (1960) to *Streblus gaimardii* var. *compressiuscula* (Brady), and by Billman *et al.* (1980) and van Marle (1991) to *Asterorotalia gaimardii* (d'Orbigny). Figure 3 was referred by Brady (1884) and Thalmann (1932, 1933*a*) to *Rotalia becarrii* (Linne), and by Barker (1960) to *Streblus* cf. *catesbyanus* (d'Orbigny).

Figure 2 *Challengerella bradyi* Billman, Hottinger, and Oesterle 1980

× 50. From *Challenger* Sta. 187A, Booby Island, Torres Strait, Pacific (8fm.). ZF2317.

Referred by Brady (1884) and Thalmann (1932, 1933*a*) to *Rotalia becarrii* (Linne), by Barker (1960) to the Miocene species *Streblus beccarrii* var. *koeboensis* (Leroy), and by Billman *et al.* (1980) to *Challengerella bradyi*. This is the type-species by original designation of the genus *Challengerella* Billman, Hottinger, and Oesterle 1980.

Figure 4 *Gyroidina broeckhiana* (Karrer 1878)

[*Rotalia broeckhiana* Karrer 1878]
× 50. From *Challenger* Sta. 191A, Ki Islands, New Guinea, Pacific (580fm.). ZF2318.

Referred by Brady (1884) and Thalmann (1932) to *Rotalia*, by Barker (1960) and Belford (1966) to *Gyroidina*, and by Hermelin (1989) to *Gyroidinoides*. Saidova (1975) referred this species to the new genus *Gyroidinus*, regarded by Loeblich and Tappan (1987) as a junior synonym of *Gyroidina* d'Orbigny 1826.

Figure 5 *Ammonia falsobeccarii* (Rouvillois 1974)

[*Pseudoeponides falsobeccarii* Rouvillois 1974]
× 50. From *Porcupine* Sta. 18, West of Ireland, Atlantic (183fm.). ZF2322.

Referred by Brady (1884) and later Thalmann (1932, 1933*a*) to 'a passage form' between *Rotalia beccarii* and *R. orbicularis*, by Barker (1960) to *Streblus batavus* Hofker, and by Hermelin (1989) to *Gyroidinoides orbicularis* (d'Orbigny). The form figured by Brady appears identical to *Pseudoeponides falsobeccarii* Rouvillois 1974 (Haynes, pers. comm.; Whittaker, pers. comm.), herein transferred to the genus *Ammonia* Brunnich 1772. This species was originally described from the Bay of Biscay (Rouvillois 1974), and appears to be widely distributed throughout the Lusitanian Province. It has also been described (as *Ammonia* sp. A) from Late Quaternary interglacial deposits of the Somerset Levels (Kidson *et al.* 1978; Haynes, pers. comm.).

Figures 6–7 *Gyroidinoides soldanii* (d'Orbigny 1826)

[*Gyroidina soldanii* d'Orbigny 1826]
Fig. 6 × 40. From *Challenger* Sta. 302, South Pacific (1450fm.). ZF2330.
Fig. 7 × 40. From *Challenger* Sta. 246, North Pacific (2050fm.). ZF2331.

Referred by Brady (1884) and Thalmann (1932) to *Rotalia soldanii* (d'Orbigny), by Thalmann (1937), Barker (1960, text), and van Marle (1991) to *Gyroidina neosoldanii* Brotzen, by Barker (1960, addenda) to *Gyroidinoides neosoldanii* (Brotzen), and by the present author to *G. soldanii* (d'Orbigny). This is the type-species by original designation of the genus *Hansenisca* Loeblich and Tappan 1987, regarded by the present author as a junior synonym of *Gyroidinoides* Brotzen 1942.

Figure 8 *Polystomellina clathrata* (Brady 1884)

[*Rotalia clathrata* Brady 1884]
× 50. From *Challenger* Sta. 162, Bass Strait, Pacific (38–40fm.). ZF2320.

Referred by Brady (1884) and Thalmann (1932, 1933*a*) to *Rotalia*, and by Barker (1960) to *Notorotalia*. *Notorotalia* Finlay 1939 is regarded by the present author as a junior synonym of *Polystomellina* Yabe and Hanzawa 1923.

Figure 9 *Polystomellina patagonica* (Parr 1950)

[*Notorotalia patagonica* Parr 1950]
× 50. From *Challenger* Sta. 305, West coast of Patagonia, East Pacific (120fm.). ZF2321. Paratype re-registered as 1974.5.14.1.

Referred by Brady (1884) and Thalmann (1932, 1933*a*) to *Rotalia clathrata* (see above), and by Barker (1960) to *N. patagonica* Parr. *Notorotalia* is regarded as a junior synonym of *Polystomellina*.

Plate 108 (CVIII)

Figure 1 *Pararotalia* sp.

× 50. From *Challenger* Sta. 186, Wednesday Island, Pacific (7–8fm.). ZF2326.

Referred by Brady (1884) to *Rotalia papillosa* var. *compressiuscula,* by Thalmann first (1932, 1933*a*) to *R. papillosa* var. *compressiuscula* Brady and later (1937) to *R. gaimardii* var. *compressiuscula* (Brady), and by Barker (1960) to *Streblus gaimardii* var. *compressiuscula* (Brady) (see Plate 107, Figs 1, 3).

Figure 2 *Pararotalia venusta* (Brady 1884)

[*Rotalia venusta* Brady 1884]
× 50. From *Challenger* Sta. 187A, Booby Island, Torres Strait, Pacific (8fm.). ZF2332.

Referred by Brady (1884) and Thalmann (1932, 1933*a*) to *Rotalia,* and by Barker (1960) to *Calcarina.*

Figures 3–?4 *Pararotalia stellata* (de Férussac 1827)

[*Calcarina stellata* de Férussac 1827]
Fig. 3 × 50. From *Challenger* Sta. 217A, off Papua, Pacific (37fm.). Fig. 4 × 50. From *Challenger* Sta. 187, off New Guinea, Pacific (6fm.).

Referred by Brady (1884) to *Rotalia calcar* d'Orbigny, sp., by Thalmann (1932), Barker (1960), and van Marle (1991) to *Calcarina calcar* d'Orbigny, by Whittaker and Hodgkinson (1979) to *Pararotalia calcar* (d'Orbigny), and by Billman *et al.* (1980) to *Asterorotalia dentata* (Parker and Jones) (Fig. 4 only). Loeblich and Tappan (1964) regard *Calcarina calcar* d'Orbigny 1826 as invalid. In contrast, Hansen (1981) has put forward a petition to the ICZN requesting its retention. Pending acceptance of this request, Brady's specimens are tentatively referred herein to *Pararotalia stellata* (de Férussac 1827). *Calcarina nicobarensis* Schwager 1866 appears synonymous. *C. calcar* d'Orbigny 1839 has been referred by Hottinger *et al.* (1991*b*) to the genus *Neorotalia* Bermúdez 1952.

Figures 5–7 *Calcarina spengleri* (Gmelin 1791)

[*Nautilus spengleri* Gmelin 1791]
Fig. 5 × 30–35. ZF1237. Fig. 6 × 30. Fig. 7 × 30–35. 1959.12.23.1. From *Challenger* Sta. 218A, Admiralty Islands, Pacific (16–25fm.).

Referred by Brady (1884) and Thalmann (1932) to *Calcarina spengleri* (Gmelin) (Figs 5, 7) and *C. defrancii* d'Orbigny (Fig. 6), by Barker (1960) to *Tinoporus spengleri* (Gmelin), and by van Marle (1991) to *Calcarina calcar* (Gmelin) (Figs 5, 7). The ICZN have ruled in favour of the conservation of *Calcarina* d'Orbigny 1826 and the suppression of *Tinoporus* de Montfort 1808 (Anon 1990*c*). *Nautilus spengleri* Gmelin 1791 is the type-species by subsequent designation of the genus *Calcarina* d'Orbigny 1826. It has been neotypified by Hansen (1981).

Figures 8–9 *Calcarina hispida* Brady 1876

× 30. From the Loo Choo Islands, North Pacific (washed from algae). ZF1236.

Referred by Brady (1884), Thalmann (1932), and Whittaker and Hodgkinson (1979) to *Calcarina,* and by Barker (1960) to *Tinoporus.* The ICZN have ruled in favour of the conservation of *Calcarina* and the suppression of *Tinoporus* (Anon 1990*c*).

Plate 109 (CIX)

FIgures 1–2 *Fijinonion fijiense* (Cushman and Edwards 1937)

[*Astrononion fijiense* Cushman and Edwards 1937]
Fig. 1 × 50. 1959.5.5.663. Fig. 2 × 50. ZF1990. From *Challenger* Sta. 174C, off Fiji, Pacific (210fm.).

Referred by Brady (1884) to *Nonionina asterizans* Fichtel and Moll, sp., by Thalmann first (1932) to *Nonion asterizans* (Fichtel and Moll) and later (1937) to *Astrononion italicum* Cushman and Edwards, and by Barker (1960) to *A. fijiense* Cushman and Edwards. This is the type-species by original designation of *Astrononion* (*Fijinonion*) Hornibrook 1964, which erstwhile subgenus was raised to the rank of genus by Loeblich and Tappan (1987).

Figures 3–4 *Astrononion stelligerum* (d'Orbigny 1839)

[*Nonionina stelligera* d'Orbigny 1839]
× 75. From *Porcupine* Sta. 67–68, East of the Shetlands, North Atlantic (64–75fm.). ZF1998.

Referred by Brady (1884) to *Nonionina,* by Thalmann (1932) to *Nonion,* and by Barker (1960) to *Astrononion.* This is the type-species by original designation of the genus *Astrononion* Cushman and Edwards 1937.

Figure 5 *Laminononion tumidum* (Cushman and Edwards 1937)

[*Astrononion tumidum* Cushman and Edwards 1937]
× 60. From *Challenger* Sta. 344, Ascension Island, Atlantic (420fm.). ZF1999.

Referred by Brady (1884) to *Nonionina stelligera* d'Orbigny, by Thalmann (1932) to *Nonion stelligerum* nov. var., and by Barker (1960) to *Astrononion tumidum* Cushman and Edwards. This is the type-species by original designation of *Astrononion* (*Laminononion*) Hornibrook 1964, which erstwhile subgenus was raised to the rank of genus by Loeblich and Tappan (1987).

Figures 6–?7 *Haynesina germanica* (Ehrenberg 1840)

[*Nonionina germanica* Ehrenberg 1840]
× 75. From *Challenger* Sta. 163C, off Sydney, Australia (6fm.). ZF1993.

Referred by Brady (1884) to *Nonionina depressula* Walker and Jacob, sp., by Thalmann first (1932) to *Nonion umbilicatulum* (Walker and Jacob) and later (1933*a*) to *N. depressulum* (Walker and Jacob), also by Barker (1960) and van Marle (1991) to *N. depressulum* (Walker and Jacob), and by Banner and Culver (1978) to *Haynesina germanica* (Ehrenberg) (Fig. 6 only). This is the type-species by original designation of the genus *Haynesina* Banner and Culver 1978.

Figures 8–9 *Melonis affinis* (Reuss 1851)

[*Nonionina affinis* Reuss 1851]
Fig. 8 × 70. From *Challenger* Sta. 24, West Indies (290fm.). ZF2003. Fig. 9 × 70. From *Challenger* Sta. 276, off Tahiti, Pacific (2350fm.). ZF2002.

Referred by Brady (1884) to *Nonionina umbilicatula* Walker and Jacob, sp., by Thalmann (1932, 1933*a*) to *N. umbilicatulum* (Walker and

Jacob), by Barker (1960) to *Gavelinonion barleeanum* (Williamson), by Hermelin (1989) to *Melonis barleeanum* (Williamson), and by van Marle (1991) and the present author to *M. affinis* (Reuss).

Figures 10–11 *Melonis pompilioides* (Fichtel and Moll 1798)

[*Nautilus pompilioides* Fichtel and Moll 1798]
× 75. From *Porcupine* Sta. 20, West of Ireland, East Atlantic (1443fm.). ZF1995.

Referred by Brady (1884) to *Nonionina*, by Thalmann (1932) and Barker (1960) (albeit tentatively) to *Nonion*, and by Belford (1966) and van Marle (1991) to *Melonis*. This is the type-species, by synonymy with the originally designated *M. etruscus* de Montfort 1808, of the genus *Melonis* de Montfort 1808. Its long-lost holotype has been discovered and illustrated by Rögl and Hansen (1984). Brady's Figure 10, which is the holotype of *Melonis sphaeroides* Voloshinova 1958, differs little from the holotype of *M. pompilioides* (Fichtel and Moll 1798). *M. sphaeroides* Voloshinova is therefore regarded by the present author as a junior synonym (ecophenotypic variant) of *M. pompilioides* (Fichtel and Moll).

Figures 12–13 *Nonion fabum* (Fichtel and Moll 1798)

[*Nautilus faba* Fichtel and Moll 1798]
× 70. From *Challenger* Sta. 354A, Vigo Bay, East Atlantic (11fm.). ZF1991.

Referred by Brady (1884) to *Nonionina boueana* d'Orbigny, by Thalmann (1932) to *Nonion boueanum* (d'Orbigny), by Barker (1960) to *Nonion* cf. *asterizans* (Fichtel and Moll), and by the present author to *N. fabum* (Fichtel and Moll). Brady's figures are close to the illustrated holotype of *Nautilus faba* Fichtel and Moll 1798 (Rögl and Hansen 1984). The ICZN have ruled in favour of this being designated the type-species of the genus *Nonion* de Montfort 1808 (Anon 1990d). This is because the former type-species, *Nautilus incrassatus* Fichtel and Moll 1798, has been shown to belong in the genus *Anomalinoides* Brotzen 1942 (Rögl and Hansen 1984).

Figures 14–15 *Nonion commune* (d'Orbigny 1846)

[*Nonionina communis* d'Orbigny 1846]
× 75. From *Challenger* Sta. 306, West coast of Patagonia, East Pacific (345fm.). ZF1996.

Referred by Brady (1884) to *Nonionina scapha* Fichtel and Moll, sp., by Thalmann (1932) to *Nonion scaph-a* (Fichtel and Moll), and by Barker (1960) and van Marle (1991) to *N. scaph-um* (Fichtel and Moll). *Nautilus scapha* Fichtel and Moll 1798 is a primary homonym of *N. scapha* Wulfen 1791. Rögl and Hansen (1984) regard most records of *scapha* in the sense of Fichtel and Moll (not Wulfen) as referable to *Nonion commune* (d'Orbigny). This species has been lectotypified by Papp and Schmid (1985).

Figure 16 *Nonionella bradii* (Chapman 1917)

[*Nonion scapha* var. *bradii* Chapman 1917]
× 75. From *Challenger* Sta. 217A, off Papua, Pacific (37fm.). ZF1997.

Referred by Brady (1884) to *Nonionina scapha* Fichtel and Moll, sp., by Thalmann (1932) to *Nonion scapha* (Fichtel and Moll), and by Barker (1960) and van Marle (1991) to *Nonionella bradii* (Chapman).

Figures 17–19 *Nonionella turgida* (Williamson 1858)

[*Rotalina turgida* Williamson 1858]
Figs 17–18 × 75. From *Porcupine* Sta. 67–68, East of the Shetlands, Atlantic (64–75fm.). ZF2000.
Fig. 19 × 75. From *Challenger* Sta. 167, West coast of New Zealand (150fm.). ZF2001.

Referred by Brady (1884) to *Nonionina*, and by Thalmann (1932), Barker (1960), Haynes (1973a), and van Marle (1991) to *Nonionella*.

Figures 20–21 *Haynesina orbicularis* (Brady 1881)

[*Nonionina orbicularis* Brady 1881]
× 50. From *Porcupine* Sta. 57, Faroe Channel ('cold area'), North Atlantic (632fm.). ZF1994. Fig. 20 designated lectotype by Banner and Culver (1978)

Referred by Brady (1884) to *Nonionina*, by Thalmann (1932) to *Nonion*, by Barker (1960) to *Elphidium*, by Banner and Culver (1978) and Buzas *et al.* (1985) to *Haynesina*, and by Cole (1981) to *Protelphidium*.

Figure 22 *Cribrononion kerguelenense* (Parr 1950)

[*Elphidium kerguelenense* Parr 1950]
Fig. 22a × 75. ?.
Fig. 22b × 75. From *Challenger* Sta. 315A, Stanley Harbour, Falkland Islands, South Atlantic (6fm.). ZF2179.

Referred by Brady (1884) to *Polystomella striatopunctata* Fichtel and Moll, sp., by Thalmann first (1932) to *Elphidium granulosum* (Galloway and Wissler) and later (1942) to *E. subgranulosum* Asano, and by Barker (1960) to *Elphidium poeyanum* (d'Orbigny) (Fig. 22a) and *E.* cf. *kerguelenense* Parr? (Fig. 22b).

Figure 23 *Cribrononion incertum* (Williamson 1858)

[*Polystomella umbilicatula* var. *incerta* Williamson 1858]
× 75. From *Challenger* Sta. 46, North Atlantic (1350fm.). ZF2180.

Referred by Brady (1884) to *Polystomella striatopunctata* Fichtel and Moll, sp., by Thalmann (1932) to *E. angulatum* (Egger), and by Barker (1960) (albeit tentatively) and van Marle (1991) to *E. incertum* (Williamson).

Plate 110 (CX)

Figure 1 *Elphidium advenum* (Cushman 1922)

[*Polystomella advena* Cushman 1922]
Fig. 1a × 75. From *Challenger* Sta. 187A, Booby Island, Torres Strait, Pacific (8fm.). ZF2181.
Fig. 1b × 75. From *Challenger* Sta. 189, South of Papua, Pacific (25–29fm.). ZF2182.

Referred by Brady (1884) to *Polystomella subnodosa* Munster, sp., by Thalmann (1932) to *Elphidium adven-um* (Cushman), and by Barker (1960) and van Marle (1991) to *E. adven-a* (Cushman).

Figures 2–5 *Elphidiella arctica* (Parker and Jones in Brady 1864)

[*Polystomella arctica* Parker and Jones in Brady 1864]
Fig. 2 × 30. From Smith Sound (210fm.). North Polar Expedition, 1875–76. ZF2167.
Fig. 3 × 30. From *Porcupine* Sta. 57, Faroe Channel ('cold area'), North Atlantic (632fm.). ZF2168.
Fig. 4 × 30; Fig. 5 × 100. From Cape Frazer (80fm.). North Polar Expedition, 1875–76. ZF2169.

Referred by Brady (1884) to *Polystomella arctica* Parker and Jones, by Thalmann (1932) to *Elphidium arcticum* (Parker and Jones), and by Barker (1960) to *Elphidiella arctica* (Parker and Jones). This is the type-species by original designation of the genus *Elphidiella* Cushman 1936. It has been lectotypified by Hodgkinson (1992).

Figures 6–7 *Elphidium crispum* (Linné 1758)

[*Nautilus crispus* Linné 1758]
× 40. From *Challenger* Sta. 186, Wednesday Island, Pacific (7–8fm.). ZF2172.

Referred by Brady (1884) to *Polystomella*, and by Thalmann (1932), Barker (1960), and van Marle (1991) to *Elphidium*. This is the type-species by subsequent designation of the genera *Themeon* de Montfort 1808 and *Polystomella* Larmarck 1822, regarded by the present author as junior synonyms of *Elphidium* de Montfort 1808. Its long-lost holotype has been discovered and illustrated by Rögl and Hansen (1984).

Figures 8, 11 *Elphidium macellum* (Fichtel and Moll 1798)

[*Nautilus macellus* Fichtel and Moll 1798]
Fig. 8 × 45–60. From *Challenger* Sta. 163C, off Sydney, Australia (6fm.). ZF2175.
Fig. 11 × 45–60. From *Challenger* Sta. 142A, off South Africa (15–20fm.). ZF2177.

Referred by Brady (1884) to *Polystomella macella* Fichtel and Moll, sp., by Thalmann (1932) to *Elphidium crispum* (Linne), and by Barker (1960) and van Marle (1991) to *E. macellum* (Fichtel and Moll). This is the type-species by original designation of the genus *Elphidium* de Montfort 1808. Its long-lost holotype has been discovered and illustrated by Rögl and Hansen (1984). It is also the type-species by original designation of the genus *Geophonus* de Montfort 1808. *Geophonus* is an isotypic junior synonym of *Elphidium*.

Figure 9 *Elphidium lessonii* (d'Orbigny 1839)

[*Polystomella lessonii* d'Orbigny 1839]
× 45–60. From *Challenger* Sta. 315A, Falkland Islands, South Atlantic (6fm.). ZF2176.

Referred by Brady (1884) to *Polystomella macella* Fichtel and Moll, sp. and by Thalmann (1932) and Barker (1960) to *Elphidium lessonii* (d'Orbigny).

Figure 10 *Elphidium aculeatum* (d'Orbigny 1846)

[*Polystomella aculeata* d'Orbigny 1846]
× 60. From *Challenger* Sta. 315A, Falkland Islands, South Atlantic (6fm.). ZF2176.

Referred by Brady (1884) to *Polystomella macella* ('young specimen'), by Thalmann (1932) to *Elphidium aculeatum* (d'Orbigny), and by Barker (1960) to *E. macellum* var. *aculeatum* (Silvestri). *Polystomella macella* var. *aculeata* Silvestri 1901 appears to be both homonymous and synonymous with *P. aculeata* d'Orbigny 1846. *Polystomella aculeata* d'Orbigny 1846 has been lectotypified by Papp and Schmid (1985).

Figure 12 *Parrellina verriculata* (Brady 1881)

[*Polystomella verriculata* Brady 1881]
× 65. From *Challenger* Sta. 162, Bass Strait, Pacific (38–40fm.). ZF2183.

Referred by Brady (1884) to *Polystomella*, by Thalmann (1932) to *Elphidium*, and by Barker (1960) to *Parrellina*.

Figures 13–15 *Parrellina imperatrix* (Brady 1881)

[*Polystomella imperatrix* Brady 1881]
Figs 13–14 × 30. From Storm Bay, Tasmania. ZF2174.
Fig. 15 × 30. From *Challenger* Sta. 163B, off Sydney, Australia (2–10fm.). ZF2173.

Referred by Brady (1884) to *Polystomella*, by Thalmann (1932) to *Elphidium*, and by Barker (1960) to *Parrellina*. This is the type-species by original designation of the genus *Parrellina* Thalmann 1951 (new name for *Elphidioides* Parr 1950, not *Elphidioides* Cushman 1945).

Figures 16–17 *Cellanthus craticulatus* (Fichtel and Moll 1798)

[*Nautilus craticulatus* Fichtel and Moll 1798]
× 30–40. From *Challenger* Sta. 218A, Admiralty Islands, Pacific (16–25fm.). ZF2170.

Referred by Brady (1884) to *Polystomella*, by Thalmann (1932) and Barker (1960) to *Elphidium*, and by Whittaker and Hodgkinson (1979) to *Cellanthus*. This is the type-species by original designation of the genus *Cellanthus* de Montfort 1808. Its long-lost holotype has been discovered and illustrated by Rögl and Hansen (1984).

Plate 111 (CXI)

Figure 1 *Amphistegina bicirculata* Larsen 1976

× 30. From *Challenger* Sta. 218A, Admiralty Islands, Pacific (16–25fm.). ZF1068.

Referred by Brady (1884) and van Marle (1991) to *Amphistegina lessonii* d'Orbigny, by Thalmann (1932) to *A. radiata* (Fichtel and Moll), by Barker (1960) to *A. quoyii* d'Orbigny?, and by Larsen (1976) to *A. bicirculata*.

Figures 2, 4–7 *Amphistegina lessonii sensu* Parker, Jones and Brady 1865

Fig. 2 × 30. From *Challenger* Sta. 33, off Bermuda, Atlantic (435fm.). ZF1072.
Fig. 4 × 30. ZF1076. Fig. 7 × 30. ZF1077. From *Challenger* Sta. 352, Cape Verde Islands, Atlantic (11fm.).
Fig. 5 × 30. From *Challenger* Sta. 218A, Admiralty Islands, Pacific (16–25fm.). ZF1070.

Fig. 6 × 30. From *Challenger* Sta. 172, Friendly Islands, Pacific (18fm.). ZF1071.

Referred by Brady (1884) to *Amphistegina lessonii* d'Orbigny, by Thalmann (1932) to *A. radiata* (Fichtel and Moll), by Barker (1960) to *A. gibbosa* d'Orbigny (Figs 2, 4, 7) and *A. lessonii* d'Orbigny (Figs 5–6), by O'Herne (1974) to *A. quoii* d'Orbigny (Fig. 4) and *A. lessonii* d'Orbigny (Fig. 5), by Larsen (1976) to *A. lessonii* d'Orbigny (with *A. gibbosa* d'Orbigny in synonymy) (Figs 5, 7), and also by van Marle (1991) to *A. lessonii* d'Orbigny (Figs 4–6). *Amphistegina lessonii* d'Orbigny 1826 was a *nomen nudum* (Tubbs (ICZN), pers. comm.) (d'Orbigny's 1826 figures identified as such being *A. quoii*). The name was validated either by the publication of a figure identified as *Amphistegina lessonii* d'Orbigny (and stated to be 'd'aprés d'Orbigny') by Guérin-Méneville in the seldom-read *Iconographie du Règne Animal de G. Cuvier*, published in parts (livraisons) between 1829 and 1843 (Cowan 1971), or by the description provided by Deshayes in the equally seldom-read *Encyclopédie Méthodique* of 1830. In view of this confusion, *Amphistegina lessonii* is herein interpreted in the sense of Parker *et al.* (1865). These authors based their concept of the species (their drawing) on the three-dimensional model fashioned by d'Orbigny and supplied along with his 1826 *Tableau Méthodique* . . . . The model was unfortunately in itself not valid for the purpose of taxonomic description (Tubbs (ICZN), pers. comm.). *Amphistegina lessonii* d'Orbigny 1826 (*sic*) has been neotypified by Larsen (1977).

Figure 3 *Amphistegina radiata* (Fichtel and Moll 1798)

[*Nautilus radiatus* Fichtel and Moll 1798]
× 30. From *Challenger* Sta. 173A, off Fiji, Pacific (12fm.). ZF1075.

Referred by Brady (1884) and van Marle (1991) to *Amphistegina lessonii* d'Orbigny, by Thalmann (1932), O'Herne (1974), Larsen (1976), and Rögl and Hansen (1984) to *A. radiata* (Fichtel and Moll), and by Barker (1960) to *A. quoyii* d'Orbigny? *A. radiata* (Fichtel and Moll) is the type-species, by synonymy with the originally designated *Amphistegina quoii* d'Orbigny 1826, of the genus *Amphistegina* d'Orbigny 1826. Its long-lost holotype has been discovered and illustrated by Rögl and Hansen (1984).

Figure 8 *Cycloclypeus carpenteri* Brady 1881

× 35. From *Challenger* Sta. 174C, off Fiji, Pacific (210fm.). ZF1366.

Referred by Brady (1884) to *Cycloclypeus guembelianus* Brady, and by Thalmann (1932), Barker (1960), and Adams and Frame (1979) to *C. (C.) carpenteri* Brady. Loeblich and Tappan (1987) regard *C. guembelianus* Brady 1881 as a junior synonym of *C. carpenteri* Brady 1881. These authors have put forward a petition to the ICZN proposing that *C. carpenteri* Brady be designated type-species of the genus *Cycloclypeus* Carpenter 1856.

Plate 112 (CXII)

Figures 1–2 *Hyalinea balthica* (Schroeter 1783)

[*Nautilus balthicus* Schroeter 1783]
× 50. From *Porcupine* Sta. 11, West of Ireland, Atlantic (1630fm.). ZF2014.

Referred by Brady (1884) to *Operculina ammonoides* Gronovius, sp., by Thalmann first (1932) to *Anomalina balthica* (Gronovius) (*sic*) and later (1933*a*) to *A. balthica* (Schroeter), and by Barker (1960), Belford (1966), van Morkhoven *et al.* (1986), and van Marle (1991) to *Hyalinea balthica* (Schroeter). This is the type-species by original designation of the genus *Hyalinea* Hofker 1951.

Figures 3–9 *Operculina complanata* (Defrance 1822)

[*Lenticulites complanatus* Defrance 1822]
Figs 3, 8 × 12. ZF2015. Fig. 6 × 12. From *Challenger* Sta. 195, off Amboyna, Pacific (15–20fm.).
Fig. 4 × 12. From *Challenger* Sta. 218A, Admiralty Islands, Pacific (16–25fm.). ZF2016.
Fig. 5 × 12. From *Challenger* Sta. 172, Friendly Islands, Pacific (18fm.). ZF2017.
Fig. 9 × 12. From *Challenger* Sta. 186, Flinders Passage, Pacific (7–8fm.). ZF2018.

Referred by Brady (1884) to *Operculina complanata* Defrance, sp. (Figs 3–5, 8) and *O. complanata* var. *granulosa* Leymerie (Figs 6–7, 9), by Thalmann (1932) to *O. complanata* (Defrance) (Figs 3–9), and by Barker (1960) (albeit tentatively) and van Marle (1991) to *O. ammonoides* (Gronovius) (*sic*). *Lenticulites complanatus* Defrance 1822 is the type-species by subsequent designation of the genus *Operculina* d'Orbigny 1826. Brady's specimens are close to those figured as *Operculina complanata* (Defrance) by Loeblich and Tappan (1964, 1987) from the mid-Tertiary of France and Japan. *Nautilus ammonoides* Schroter 1783 is the type-species by original designation of the genus *Neooperculinoides* Golev 1961, regarded by Loeblich and Tappan first (1964) as a junior synonym of *Nummulites* Lamarck 1801 and later (1987) as a junior synonym of *Assilina* d'Orbigny 1839, and by the present author as a junior synonym of *Operculina* d'Orbigny 1826. *Nautilus ammonoides* Gronovius 1781 is invalid.

Figure 10 *Operculina gaimardi* Deshayes 1832

× 12. From *Challenger* Sta. 218A, Admiralty Islands, Pacific (16–25fm.). ZF2019.

Referred by Brady (1884) to *Operculina complanata* var. *granulosa* Leymerie, by Thalmann (1932) to *O. complanata* (Defrance) (see above), and by Barker (1960) to *O. gaimardi* d'Orbigny?. Smout and Eames (1960) attribute this species to Deshayes 1832.

Figures 11–13 *Operculinella cumingii* (Carpenter 1860)

[*Amphistegina cumingii* Carpenter 1860]
Fig. 11 × 18. From the China Sea (dredged).
Figs 12–13 × 18. From *Challenger* Sta. 218A, Admiralty Islands, Pacific (16–25fm.). ZF2013.

Referred by Brady (1884) to *Nummulites cumingii* Carpenter, sp., by Thalmann first (1932) to *Operculinella cumingii* (Carpenter) and later (1933*a*) to *Verbeekia cumingii* (Carpenter), and by Barker (1960) and van Marle (1991) to *Operculina ammonoides* (Gronovius) (*sic*) (see also Text-figure 22). *Amphistegina cumingii* Carpenter 1860 is the type-species by original designation of the genera *Verbeekia* Silvestri 1908, *Operculinella* Yabe 1918 and *Pseudonummulites* Silvestri 1937. *Verbeekia* Silvestri 1908 is a primary junior homonym of *Verbeekia* Fritsch 1877. Note also that *Operculinella* Yabe 1908 is a subjective junior synonym of *Palaeonummulites* Schubert 1908.

Figures 14–16, ?17–18 *Heterostegina depressa* d'Orbigny 1826

Fig. 14 × 12. From *Challenger* Sta. 219A, Admiralty Islands, Pacific (17fm.). ZF1576.

Fig. 15 × 12. From *Challenger* Sta. 260A, off Honolulu, Pacific (40fm.).

Fig. 16 × 15. From *Challenger* Sta. 172, Friendly Islands, Pacific (18fm.). ZF1575.

Figs 17–18 × 25. From *Challenger* Sta. 218A, Admiralty Islands, Pacific (16–25fm.). ZF1577.

Referred by Brady (1884) to *Heterostegina depressa* d'Orbigny, by Thalmann (1932) to *H. antillarum* d'Orbigny (Figs 14–16) and *H. costata* d'Orbigny (Figs 17–18), by Barker (1960) to *H. depressa* d'Orbigny (Figs 14–16) and *H.* sp., juv. (Figs 17–18), and also by van Marle (1991) to *H. depressa* d'Orbigny (Figs 14–16 only). This is the type-species by subsequent designation of the genus *Heterostegina* d'Orbigny 1826.

Figures 19–20 *Heterostegina curva* Moebius 1880

× 25. From *Challenger* Sta. 260A, off Honolulu, Pacific (40fm.). ZF1578.

Referred by Brady (1884), Thalmann (1932), and Barker (1960) to *Heterostegina* cf. *curva* Moebius.

Plate 113 (CXIII)

Figure 1 *Sahulia conica* (d'Orbigny 1839)

[*Textularia conica* d'Orbigny 1839]
× 50. From *Challenger* Sta. 205A, off Hong Kong, Pacific (7fm.). ZF2444.

Referred by Brady (1884), Thalmann (1932), and Barker (1960) to *Textularia conica* d'Orbigny, 'short variety' (see also Plate 43, Figs 13–14). Langer (1992) states that this form differs from his *Sahulia lutzei* sp. nov. in its 'rather fine agglutination, the height of its chambers and in possessing an distinct rim bordering the entire apertural opening'.

Figure 2 *Tortoplectella crispata* (Brady 1884)

[*Textularia crispata* Brady 1884]
× 60. From *Challenger* Sta. 185, Torres Strait, Pacific (155fm.). ZF2447.

Referred by Brady (1884) and Thalmann (1932, 1933a) to *Textularia,* and by Barker (1960) to *Siphotextularia.* This is the type-species by original designation of the genus *Tortoplectella* Loeblich and Tappan 1985, a probable textulariid (Revets, pers. comm.).

Figures 3–5 *Siphoniferoides transversarius* (Brady 1884)

[*Textularia transversaria* Brady 1884]
Fig. 3 ZF2467. Fig. 4 ZF2468. Fig. 5 ZF2469. × 75. From *Challenger* Sta. 185, Torres Strait, Pacific (155fm.).

Referred by Brady (1884) and Thalmann (1932, 1933a) to *Textularia,* and by Barker (1960) to *Gaudryina* (*Siphogaudryina*).

Figure 6 *Geminospira bradyi* Bermúdez 1952

× 100. From *Challenger* Sta. 185, Torres Strait, Pacific (155fm.). ZF1208.

Referred by Brady (1884) to *Bulimina convoluta* Williamson, also by Thalmann first (1932) to *B. convoluta* Williamson, later (1933a) to *Cancris convolutus* (Williamson), and later still (1937) to *Pseudobulimina convoluta* (Williamson), by Collins (1958) and Belford (1966) to *Geminospira bradyi* Bermúdez, and by Barker (1960) to *Pseudobulimina* sp. nov. (?).

Figure 7 *Pseudobrizalina strigosa* (Brady 1884)

[*Bolivina lobata* var. *strigosa* Brady 1884]
× 80. From *Challenger* Sta. 185, Torres Strait, Pacific (155fm.). ZF1186.

Referred by Brady (1884) to *Bolivina lobata* var. *strigosa,* by Thalmann first (1932) to *Bifarina strigosa* (Brady) and later (1942) to *Loxostomum strigosum* (Brady), and also by Barker (1960, text) to *Loxostomum strigosum* (Brady) and later (1960, addenda) to *Loxostomoides* (?) *strigosum* (Brady). This species was referred by Saidova (1975) to *Sagrinella.*

Figure 8 *Globocassidulina pacifica* (Cushman 1925)

[*Cassidulina pacifica* Cushman 1925]
× 75. From *Challenger* Sta. 185, Torres Strait, Pacific (155fm.). ZF1258.

Referred by Brady (1884) and Thalmann (1932, 1933a) to *Cassidulina calabra* (Seguenza), and by Thalmann (1942) and Barker (1960) to *C. pacifica* Cushman. This is the type-species by original designation of the genus *Cushmanulla* Saidova 1975, regarded by Loeblich and Tappan (1987) as a junior synonym of *Burseolina* Seguenza 1880 and by the present author as a probable junior synonym of the genus *Globocassidulina* Voloshinova 1960.

Figure 9 *Orthoplecta clavata* (Brady 1884)

[*Cassidulina* (*Orthoplecta*) *clavata* Brady 1884]
× 100. From *Challenger* Sta. 219A, Admiralty Islands, Pacific (17fm.). ZF2064 (holotype).

Referred by Brady (1884) to *Cassidulina* (*Orthoplecta*), and by Thalmann (1932) and Barker (1960) to *Orthoplecta.* This is the type-species by original monotypy of *Cassidulina* (*Orthoplecta*) Brady 1884. It has been reviewed by Revets and Whittaker (1991).

Figure 10 *Ehrenbergina pupa* (d'Orbigny 1839)

[*Cassidulina pupa* d'Orbigny 1839]
× 80. From *Challenger* Sta. 305, West coast of Patagonia, East Pacific (120fm.). ZF1435.

Referred by Brady (1884), Thalmann (1932), and Barker (1960) to *Ehrenbergina pupa* (d'Orbigny) (see also Plate 55, Fig. 1).

Figure 11 *Saracenaria latifrons* (Brady 1884)

[*Cristellaria latifrons* Brady 1884]
× 35. From *Challenger* Sta. 166, West coast of New Zealand, Pacific (275fm.). ZF1336.

Referred by Brady (1884) to *Cristellaria*, and by Thalmann (1932) and Barker (1960) to *Saracenaria*.

Figure 12 *Planularia magnifica* Thalmann 1933

× 35. From *Challenger* Sta. 174C, off Fiji, Pacific (210fm.). ZF1323.

Referred by Brady (1884) to *Cristellaria dentata* Karrer, by Thalmann first (1932) to *Planularia* n. sp. and later (1933a) to *P. magnifica* nov. spec., and by Barker (1960) to *P. magnifica* Thalmann.

Figure 13 *Amphicoryna* sp. nov. (2)

× 50. From *Challenger* Sta. 174C, off Fiji, Pacific (210fm.). ZF1067.

Referred by Brady (1884) to *Amphicoryne* (*sic*) sp.?, by Thalmann (1932) to *A.* nov. spec., and by Barker (1960) to *Amphicoryna* sp. nov. Thalmann.

Figure 14 *Millettia limbata* (Brady 1884)

[*Sagrina limbata* Brady 1884]
× 100. From *Challenger* Sta. 185, Torres Strait, Pacific (155fm.). ZF2351.

Referred by Brady (1884) to *Sagrina*, by Thalmann first (1932) to *Siphogenerina* and later (1942) to *Schubertia*, also by Barker (1960) to *Schubertia*, and by Revets (1992) to *Millettia*. *Schubertia* Silvestri 1912 is a junior synonym of *Millettia* Schubert 1911.

Figure 15 *Articularia scrobiculata* (Brady 1884)

[*Miliolina scrobiculata* Brady 1884]
× 50. From shore sand, Tamatavé, Madagascar. ZF1895.

Referred by Brady (1884) to *Miliolina,* and by Thalmann (1932) and Barker (1960) to *Articulina.* The apparently quinqueloculine rather than milioline initial chamber arrangement indicates placement in *Articularia* rather than *Articulina.*

Figure 16 *Massilina* sp. nov.

×45. From *Challenger* Sta. 218A, Admiralty Islands, Pacific (16–25fm.). ZF1881.

Referred by Brady (1884) to *Miliolina pygmaea* Reuss, sp., and by Thalmann (1932) and Barker (1960) to *Quinqueloculina pygmaea* Reuss. Barker (1960) noted that 'Brady's figures are very different from those of Reuss, and moreover, seem closer to *Massilina* than to *Quinqueloculina*'.

Figure 17 *Adelosina* sp. nov.

×30. From *Challenger* Sta. 185, Torres Strait, Pacific (155fm.). ZF1864.

Referred by Brady (1884) to *Miliolina ferussacii* d'Orbigny, sp., by Thalmann (1932) to *Quinqueloculina ferussacii* d'Orbigny, by Barker (1960) to *Q.* sp. nov., and by the present author to *Adelosina* sp. nov. *Adelosina* differs from *Quinqueloculina* externally, principally in the terminal placement of the aperture (see also discussions on Plates 6 and 9).

Figures 18–19 *Cornuspiroides striolatus* (Brady 1882)

[*Cornuspira striolata* Brady 1882]
Fig. 18 × 5; Fig. 19 × 20. From *Knight Errant* Sta. 8, Faroe Channel ('cold area'), North Atlantic (540fm.). 1959.5.5.271.

Referred by Brady (1884) to *Cornuspira,* and by Thalmann (1932) and Barker (1960) to *Cornuspiroides.* This is the type-species by original designation of the genus *Cornuspiroides* Cushman 1928.

Figure 20 *Cornuspira crassisepta* Brady 1882

× 60. From *Knight Errant* Sta. 7, Faroe Channel ('warm area'), North Atlantic (530fm.). ZF1289.

Referred by Brady (1884), Thalmann (1932), and Barker (1960) to *Cornuspira,* and by Zheng (1988) to *Cyclogyra.* The ICZN have ruled for the retention of *Cornuspira* Schultze 1854 and the suppression of *Cyclogyra* Wood 1842 (Melville 1978).

Figure 21 *Cornuspira lacunosa* Brady 1884

× 60. From *Challenger* Sta. 185, Torres Strait, Pacific (155fm.). ZF1298.

Referred by Brady (1884), Thalmann (1932, 1933a), and Barker (1960) to *Cornuspira,* and by Zheng (1988) to *Cornuspiroides.*

Plate 114 (CXIV)

Figure 1 *Triloculina terquemiana* (Brady 1884)

[*Miliolina terquemiana* Brady 1884]
× 60. From shore sand, Tamatavé, Madagascar. ZF1904.

Referred by Brady (1884) to *Miliolina,* and by Thalmann (1932) and Barker (1960) to *Triloculina.*

Figure 2 *Triloculina bertheliniana* (Brady 1884)

[*Miliolina bertheliniana* Brady 1884]
× 100. From shore sand, Tamatavé, Madagascar. ZF1850.

Referred by Brady (1884) to *Miliolina,* and by Thalmann (1932) and Barker (1960) to *Triloculina.*

Figure 3 *Planispirinoides bucculentus* (Brady 1884)

[*Miliolina bucculenta* Brady 1884]
× 20. From *Porcupine* Sta. 57, Faroe Channel ('cold area'), North Atlantic (632fm.). ZF1856.

Referred by Brady (1884) to *Miliolina,* by Thalmann first (1932) to *Triloculina,* later (1933a) to *Planispira,* and later still (1937) to *Miliolinella,* and by Barker (1960), Zheng (1988), and van Marle (1991) to *Planispirinoides.* This is the type-species by original designation of the genus *Planispirinoides* Parr 1950.

Figures 4–7 *Fischerina communis* (Seguenza 1880)

[*Planispirina communis* Seguenza 1880]
Figs 4–5, 7 × 20. From a *Lightning* station off the Faroes, North Atlantic. 1959.5.5.970–972.

Fig. 6 × 20. From a *Porcupine* station off the Faroes, North Atlantic. ZF2104.

Referred by Brady (1884), Thalmann (1932), and Barker (1960) to *Planispirina*. This is the type-species by subsequent designation of the genus *Planispirina* Seguenza 1880, regarded by the present author as a junior synonym of *Fischerina* Terquem 1878.

Figure 8 *Fissurina laevigata* Reuss 1850

× 70. ZF1695.

Referred by Brady (1884) and Thalmann (1932) to *Lagena*, and by Leroy (1944a), Barker (1960), and van Marle (1991) to *Fissurina*. This is the type-species by original monotypy of the genus *Fissurina* Reuss 1850.

Figure 9 *Cushmanina spiralis* (Brady 1884)

[*Lagena spiralis* Brady 1884]
× 120. From *Challenger* Sta. 185, Torres Strait, Pacific (155fm.). ZF1760.

Referred by Brady (1884), Thalmann (1932), and Barker (1960) to *Lagena*. Barker (1960) questioned the generic assignment. Albani and Yassini (1989) have assigned the closely related species *Lagena tasmaniae* Quilty 1974 to the genus *Cushmanina* Jones 1984.

Figure 10 *Fissurina brevis* (Brady 1884)

[*Lagena formosa* var. *brevis* Brady 1884]
× 100. From *Challenger* Sta. 185, Torres Strait, Pacific (155fm.). ZF1662.

Referred by Brady (1884) to *Lagena formosa* var. *brevis*, also by Thalmann (1932) to *L. formosa* var. *brevis* Brady, and by Barker (1960) to *Fissurina formosa* var. *brevis* (Brady). This form appears specifically distinct from *Fissurina formosa* (Schwager) (see Plate 60, Figs 18–19).

Figure 11 *Laculatina quadrangularis* (Brady 1884)

[*Lagena quadrangularis* Brady 1884]
× 100. From *Challenger* Sta. 185, Torres Strait, Pacific (155fm.). ZF1736.

Referred by Brady (1884) and Thalmann (1932) to *Lagena*, and by Barker (1960) to *Fissurina* (?). This species was referred by Patterson and Richardson (1988) to *Laculatina*.

Figure 12 *Frondicularia compta* var. *villosa* Heron-Allen and Earland 1924

× 60. From *Challenger* Sta. 185, Torres Strait, Pacific (155fm.). ZF1442.

Referred by Brady (1884) and Leroy (1944b) to *Frondicularia archiaciana* d'Orbigny, by Thalmann (1932) to *F. compta* Brady, and by Barker (1960) to *F. compta* var. *villosa* Heron-Allen and Earland.

FIgure 13 *Vaginulina arquata* (Brady 1884)

[*Vaginulina legumen* var. *arquata* Brady 1884]
× 15. From *Challenger* Sta. 24, West Indies (390fm.). ZF2588.

Referred by Brady (1884) to *Vaginulina legumen* var. *arquata*, also by Thalmann (1932) and Barker (1960) to *V. legumen* var. *arquata* Brady, and by Hofker (1976) to *V. spinigera* Brady.

Figure 14 *Vaginulinopsis* sp. nov.

× 18, From *Challenger* Sta. 185, Torres Strait, Pacific (155fm.). ZF1358.

Referred by Brady (1884) to *Cristellaria wetherelli* Jones, sp., by Thalmann first (1932) to *Robulus semilituus* (Montagu) and later (1933a) to *Hemicristellaria semilituus* (Montagu), and by Barker (1960) to *Vaginulinopsis* sp. nov.

Figures 15–16 *Planularia magnifica* var. *falciformis* Thalmann 1937

× 10. From *Knight Errant* Sta. 7, Faroe Channel, North Atlantic (530fm.). 1959.5.5.289–290.

Referred by Brady (1884) to *Cristellaria compressa* d'Orbigny, by Thalmann first (1932) to *Planularia* n. sp., later (1933a) to *P. magnifica* var. *elongata* nov. var., and later still (1937) to *P. magnifica* var. *falciformis* var. nov., and by Barker (1960) to *P. magnifica* var. *falciformis* Thalmann. *Planularia magnifica* var. *elongata* Thalmann 1933 was a primary homonym of *P. elongata* d'Orbigny 1826.

Figure 17 *Saracenaria altifrons* (Parr 1950)

[*Lenticulina altifrons* Parr 1950]
× 70. From *Challenger* Sta. 185, Torres Strait, Pacific (155fm.). ZF1301.

Referred by Brady (1884) to *Cristellaria acutauricularis* Fichtel and Moll, sp., by Thalmann (1932) to *Robulus acutauricularis* (Fichtel and Moll), and by Barker (1960) to *Lenticulina altifrons* Parr.

Figure 18 *Rectuvigerina nicoli* Mathews 1945

× 70. From *Challenger* Sta. 142, Off South Africa. ZF2352.

Referred by Brady (1884) to *Sagrina nodosa* Parker and Jones, by Thalmann (1932) to *Siphogenerina nodosa* (Parker and Jones), and by Barker (1960) to *Rectuvigerina nicoli* Matthes.

Figures 19–20 *Neogloboquadrina pachyderma* (Ehrenberg 1861)

[*Aristerospira pachyderma* Ehrenberg 1861]
Figs 19a, c ZF1494. Fig. 19b ZF1493. Fig. 20 ZF1497. × 50. From *Knight Errant* Sta. 18, Faroe Channel, North Atlantic (375fm.).

Referred by Brady (1884) and Barker (1960) to *Globigerina*, by Thalmann (1932) to *Globigerinoides*, and by Saito *et al.* (1981) to *Neogloboquadrina*. *Globigerina bulloides* var. *borealis* Brady 1881 appears synonymous, though Banner and Blow (1960a) regarded the two forms as distinct.

Figure 21 *Globotruncana linneiana* (d'Orbigny 1839)

[*Rosalina linneiana* d'Orbigny 1839]
No magnification given (copied from d'Orbigny's figures).

Plate 115 (CXV)

Figure 1 *Neoeponides schreibersii* (d'Orbigny 1846)

[*Rotalina schreibersii* d'Orbigny 1846]
× 30. From *Challenger* Sta. 185, Torres Strait, Pacific (155fm.). ZF2247.

Referred by Brady (1884) to *Pulvinulina schreibersi* d'Orbigny, sp., by Thalmann (1932) to *Eponides schreibersi* (d'Orbigny), and by Barker (1960) to *E. schreibersiana* (sic) (d'Orbigny). *Rotalina schreibersii* d'Orbigny 1846 is the type-species by original designation of the genus *Neoeponides* Reiss 1960. It has been lectotypified by Papp and Schmid (1985). *Neoeponides* has been commented upon by Hottinger *et al.* (1990).

Figure 2 *Cyclocibicides vermiculatus* (d'Orbigny 1826)

[*Planorbulina vermiculata* d'Orbigny 1826]
× 30. From Cagliari, Sardinia, Mediterranean ('shallow water'). ZF2252.

Referred by Brady (1884) to *Pulvinulina*, by Thalmann (1932) tentatively to *Annulocibicides*, and by Barker (1960) to *Cyclocibicides*. This is the type-species by original designation of the genus *Cyclocibicides* Cushman 1927.

Figure 3 *Planopulvinulina dispansa* (Brady 1884)

[*Pulvinulina dispansa* Brady 1884]
× 14. From off Madeira, Atlantic (dredged). ZF2210. Lectotype designated by Loeblich and Tappan (1964) (ZF3641) ex this slide.

Referred by Brady (1884) to *Pulvinulina*, and by Thalmann (1932) and Barker (1960) to *Planopulvinulina*. This is the type-species by subsequent designation of the genus *Planopulvinulina* Schubert 1921.

Figures 4–5 *Cibicides lobatulus* (Walker and Jacob 1798)

[*Nautilus lobatulus* Walker and Jacob 1798]
Fig. 4 × 30. ZF2535. Fig. 5 × 30. ZF2536. From *Challenger* Sta. 78, SE of the Azores, Atlantic (1000fm.).

Referred by Brady (1884) to *Truncatulina*, and by Thalmann (1932) and Barker (1960) to *Cibicides*. This is the type-species, by synonymy with the originally designated *L. vulgaris* Fleming 1928, of the genus *Lobatula* Fleming 1928, regarded by the present author as a junior synonym of *Cibicides* de Montfort 1808 (see also Plates 92 and 93).

Figure 6 *Gyroidina orbicularis* (*sensu* Parker, Jones, and Brady 1865)

[*Rosalina orbicularis* Parker, Jones, and Brady 1865]
× 50. From *Challenger* Sta. 142, off South Africa (150fm.). ZF2323.

Referred by Brady (1884) to *Rotalina orbicularis* d'Orbigny, sp., by Thalmann (1932), Barker (1960, text), Belford (1966), and van Marle (1991) to *Gyroidina orbicularis* (d'Orbigny), and by Barker (1960, addenda) to *Gyroidinoides altiformis* (Stewart and Stewart). *Gyroidina orbicularis* d'Orbigny 1826 is the type-species by subsequent designation of the genus *Gyroidina* d'Orbigny 1826, emend. Hansen 1967. Banner and Clarke (1962) correctly (according to Tubbs (ICZN) (pers. comm.)) interpreted it as a *nomen nudum*, and *Rosalina orbicularis* Parker, Jones, and Brady 1865 to be the first validation of the name. They isolated a specimen (British Museum (Natural History) register number 64.2.13.352) from one of Parker, Jones, and Brady's syntypic localities

(the Recent of south-west Ireland), intending to lectotypify it in a subsequent publication, but unfortunately never did so. Subsequently, Hansen (1967), evidently unaware of the unavailability of *Gyroidina orbicularis* d'Orbigny 1826 or of Banner and Clarke's work, did designate a specimen (Mineralogisk Museum (Copenhagen) (MMH) No. 10.319) from another of Brady, Parker, and Jones's syntypic localities (the Recent of Rimini in the Adriatic) as lectotype for '*Gyroidina orbicularis* d'Orbigny, 1826'. Hansen's lectotype is valid, and must therefore serve as the model for the species *orbicularis* and the genus of which it is type, *Gyroidina*, emend., which is herein interpreted as differing from *Cibicidoides* Thalmann 1939 in having the aperture more or less restricted to the umbilical side rather than in the equatorial position. Brady's specimens are specifically distinct from Hansen's, but conform closely to *Gyroidina orbicularis* in the sense of, among others, Parker *et al.* (1865) and Banner and Clarke (1962). Parker *et al.* (1865) based their concept of the species (their drawing) on the three-dimensional model, fashioned by d'Orbigny and supplied along with his 1826 '*Tableau Méthodique*', which model was unfortunately in itself not valid for the purpose of taxonomic description (Tubbs (ICZN), pers. comm.).

Figure 7 *Pseudorotalia schroeteriana* (Parker and Jones in Carpenter 1862)

[*Rotalia schroeteriana* Parker and Jones in Carpenter 1862]
× 25. From the China Sea (dredged). ZF2329.

Referred by Brady (1884) and Thalmann (1932, 1933a) to *Rotalia*, by Barker (1960) to *Streblus*, and by Huang (1964), Belford (1966), Whittaker and Hodgkinson (1979), Billman *et al.* (1980), and van Marle (1991) to *Pseudorotalia*. This is the type-species by original designation of the genus *Pseudorotalia* Reiss and Merling 1958. It has been lectotypified by Hodgkinson (1992).

Figure 8 *Asterorotalia pulchella* (d'Orbigny 1839)

[*Rotalina* (*Calcarina*) *pulchella* d'Orbigny 1839]
× 50. From the Straits of Banca, East Indies, Pacific. ZF2327.

Referred by Brady (1884) to *Rotalia pulchella* d'Orbigny, sp., also by Thalmann (1932) to *R. pulchella* (d'Orbigny) and later (1933a,b) to *R. trispinosa* n. sp., by Barker (1960), Huang (1964), and Billman *et al.* (1980) to *Asterorotalia trispinosa* (Thalmann), and by Whittaker and Hodgkinson (1979) to *A. pulchella* (d'Orbigny). As pointed out by Loeblich and Tappan (1987), *Rotalia trispinosa* Thalmann 1933 is a superfluous name, *Rotalina* (*Calcarina*) *pulchella* d'Orbigny 1839 being neither a primary nor a secondary homonym of *R.* (*R.*) *pulchella* d'Orbigny 1826 (which was a *nomen nudum* not validated until 1865). *R.* (*C.*) *pulchella* d'Orbigny 1839 is the type-species by original monotypy of the genus *Asterorotalia* Hofker 1950.

Figure 9 *Nonion armatum* (Brady 1884)

[*Nonionina boueana* var. *armata* Brady 1884]
× 60. From shore sand, Tamatavé, Madagascar, Indian Ocean. ZF1992.

Referred by Brady (1884) to *Nonionina boueana* var. *armata*, and by Thalmann (1932) and Barker (1960) to *Nonion boueanum* var. *armatum* (Brady). Papp and Schmid (1985) have shown that the species *Nonionina boueana* d'Orbigny 1846 belongs in the genus *Hanzawaia* Asano 1944.

Appendix 1
Suggested suprageneric classification of the *Challenger* foraminifera

Figures in parentheses are numbers of recognized constituent species.

The classification essentially follows that of Haynes (1981). Note, however, that the families Bolivinitidae and Cassidulinidae have been interpreted in a broader sense (the latter so as to include the family Islandiellidae), and also that the superfamily Komokiacea (families Baculellidae and Komokiidae) and the families Keramosphaeridae, Millettiidae, Glabratellidae, and Epistomariidae have been added.

It is interesting to note how similar Haynes's scheme is to Brady's, formulated nearly a century earlier (1884). In essence, Haynes's Order Allogromiida corresponds to Brady's Family Gromiidae, his Astrorhizida to Brady's Astrorhizidae, his Lituolida to Brady's Lituolidae and Textularidae (part), his Miliolida to Brady's Miliolidae, his Nodosariida to Brady's Lagenidae, his Buliminida to Brady's Textularidae (part), his Robertinida and Rotaliida to Brady's Chilostomellidae, Rotalidae, and Nummulinidae, and his Globigerinida to Brady's Globigerinidae.

The only main philosophical (as opposed to hierarchical and nomenclatorial) difference between the two schemes lies in the different interpretation of the importance of the agglutinated wall structure in classification. Brady noted that while 'it is a fact that . . . there are certain groups which are exclusively arenaceous, and some which are calcareous and perforate, there are yet others in which no such uniform rule obtains'. He therefore placed both agglutinated and calcareous perforate taxa in his Family Textularidae (the former in the Subfamily Textularinae, the latter in the subfamily Bulimininae), and stated that 'the minute structure of the test has been abandoned as an exclusive basis for the primary division of the Order'. In this thinking he was clearly influenced by Carpenter (1862), and his division of the foraminifera into the suborders 'Perforata' and 'Imperforata'. Haynes, in contrast, maintained a high-level (ordinal) distinction between the agglutinated and calcareous perforate taxa. The appropriate weight to attach to the perforation of the agglutinated foraminifera in classification has remained the subject of debate to this day (see, for instance, Loeblich and Tappan 1989 and Banner *et al.* 1991).

Order Astrorhizida (65)
Superfamily Ammodiscacea (62)
Family Saccamminidae (24)
Hemisphaerammina (1), *Lagenammina* (4), *Pelosina* (3), *Pilulina* (1), *Psammosphaera* (1), *Saccammina* (2), *Sagenina* (1), *Sorosphaera* (1), *Storthosphaera* (1), *Technitella* (4), *Tholosina* (2), *Thurammina* (3).

Family Astrorhizidae (29)
Archimerismus (1), *Astrorhiza* (4), *Bathysiphon* (2), *Botellina* (1), *Dendrophrya* (3), *Halyphysema* (2), *Hippocrepina* (1), *Hyperammina* (4), *Jaculella* (2), *Marsipella* (2), *Rhabdammina* (6), *Saccorhiza* (1).

Family Ammodiscidae (9)
Ammodiscus (2), *Ammolagena* (1), *Glomospira* (1), *Tolypammina* (2), *Turritellella* (2), *Usbekistania* (1).

Superfamily Komokiacea (3)
Family Baculellidae (1)
Catena (1).

Family Komokiidae (2)
Rhizammina (1), *Testulosiphon* (1).

Order Lituolida (127)
Superfamily Lituolacea (83)

Family Lituotubidae (3)
Lituotuba (1), *Trochamminoides* (2).

Family Lituolidae (57)
Ammobaculites (5), *Ammoscalaria* (2), *Ammotium* (1), *Bdelloidina* (1), *Buzasina* (2), *Conglophragmium* (1), *Cribrostomoides* (1), *Cyclammina* (4), *Cystammina* (2), *Discammina* (1), *Eratidus* (1), *Evolutinella* (1), *Ginesina* (1), *Glaphyrammina* (1), *Haplophragmoides* (2), *Hormosina* (4), *Hormosinella* (4), *Loeblichopsis* (2), *Placopsilina* (2), *Protoschista* (1), *Recurvoides* (1), *Reophax* (10), *Subreophax* (2), *Triplasia* (1), *Veleroninoides* (4).

Family Textulariidae (23)
Bigenerina (2), *Planctostoma* (1), *Sahulia* (2), *Septotextularia* (1), *Siphoniferoides* (2), *Siphotextularia* (3), *Spiroplectammina* (2), *Spiroplectella* (1), *Spiroplectinella* (1), *Spirorutilus* (1), *Spirotextularia* (1), *Textularia* (4), *Tortoplectella* (1), *Vulvulina* (1).

Superfamily Ataxophragmiacea (44)
Family Trochamminidae (12)
Adercotryma (1), *Carterina* (1), *Paratrochammina* (1), *Polystomammina* (1), *Portatrochammina* (1), *Tritaxis* (3), *Trochammina* (2), *Trochamminella* (2).

Family Verneuilinidae (12)
Dorothia (5), *Globotextularia* (1), *Pseudogaudryina* (3), *Textulariella* (1), *Verneuilinulla* (2).

Family Valvulinidae (6)
Clavulina (4), *Cylindroclavulina* (1), *Martinottiella* (1).

Family Eggerellidae (13)
Cribrogoesella (2), *Eggerella* (1), *Karreriella* (3), *Karrerulina* (1), *Latentoverneuilina* (1), *Liebusella* (1), *Multifidella* (1), *Pseudoclavulina* (2), *Tritaxilina* (1).

Family Ataxophragmiidae (1)
Eggerelloides (1).

Order Miliolida (130)
Superfamily Nubeculariacea (17)
Family Cornuspiridae (8)
Cornuspira (5), *Cornuspirella* (1), *Cornuspiroides* (2).

Family Nubeculariidae (9)
Fischerina (1), *Nodobaculariella* (1), *Nodophthalmidium* (1), *Nubecularia* (1), *Nubeculina* (1), *Planispirinella* (1), *Vertebralina* (3).

Superfamily Ophthalmidiacea (5)
Family Ophthalmidiidae (5)
Cornuloculina (2), *Discospirina* (1), *Spirophthalmidium* (2).

Superfamily Soritacea (14)
Family Soritidae (14)
Amphisorus (1), *Archaias* (1), *Coscinospira* (1), *Cyclorbiculina* (1), *Marginopora* (1), *Monalysidium* (1), *Parasorites* (2), *Peneroplis* (5), *Spirolina* (1).

Superfamily Miliolacea (94)
Family Miliolidae (91)
Adelosina (7), *Ammomassilina* (1), *Articularia* (3), *Articulina* (2), *Biloculinella* (1), *Cribromiliolinella* (1), *Edentostomina* (3), *Flintina* (1), *Flintinoides* (1), *Hauerina* (1), *Massilina* (3), *Miliolinella* (1), *Nummoloculina* (1), *Parrina* (1), *Planispirinoides* (1), *Polysegmentina* (1), *Proemassilina* (2), *Pseudoflintina* (1), *Pseudohauerina* (2), *Pseudomassilina* (2), *Pseudotriloculina* (1), *Ptychomiliola* (1), *Pyrgo* (9), *Pyrogoella* (2), *Quinqueloculina* (10), *Schlumbergerina* (1), *Sigmamiliolinella* (1), *Sigmoihauerina* (1), *Sigmoilina* (1), *Sigmoilopsis* (1), *Sigmopyrgo* (1), *Siphonaperta* (2), *Spiroglutina* (1), *Spiroloculina* (11), *Spirosigmoilina* (3), *Triloculina* (5), *Triloculinella* (2), *Tubinella* (2).

Family Alveolinidae (2)
Alveolinella (1), *Borelis* (1).

Family Keramosphaeridae (1)
Keramosphara (1).

Order Nodosariida (224)
Superfamily Nodosariacea (120)
Family Nodosariidae (116)
Amphicoryna (11), *Astacolus* (6), *Bifarilamellina* (1), *Dentalina* (15), *Frondicularia* (6), *Glandulonodosaria* (4), *Grigelis* (2), *Lagena* (19), *Lenticulina* (19), *Marginulina* (2), *Neolenticulina* (1), *Nodosaria* (2),

Planularia (7), *Pseudoglandulina* (3), *Saracenaria* (5), *Vaginulina* (5), *Vaginulinopsis* (8).

Family Lingulinidae (2)
Lingulina (2).

Family Plectofrondiculariidae (2)
Plectofrondicularia (2).

Superfamily Polymorphinacea (104)
Family Polymorphinidae (21)
Francuscia (1), *Globulina* (6), *Guttulina* (4), *Pseudopolymorphina* (2), *Pyrulina* (5), *Ramulina* (1), *Sigmoidella* (2).

Family Glandulinidae (83)
Anturina (1), *Cushmanina* (8), *Entosolenia* (1), *Fissurina* (45), *Galwayella* (4), *Glandulina* (1), *Globulotuba* (3), *Laculatina* (1), *Oolina* (14), *Pseudofissurina* (1), *Pseudoolina* (2), *Pseudosolenina* (1), *Sipholagena* (1).

Order Buliminida (121)
Superfamily Buliminacea (50)
Family Buliminidae (17)
Bulimina (8), *Fursenkoina* (5), *Globobulimina* (1), *Praeglobobulimina* (3).

Family Turrilinidae (5)
Buliminella (1), *Buliminoides* (1), *Elongobula* (2), *Eubuliminella* (1).

Family Uvigerinidae (25)
Rectuvigerina (1), *Siphogenerina* (6), *Siphouvigerina* (3), *Trifarina* (6), *Uvigerina* (9).

Family Pavoninidae (3)
Chrysalidinella (1), *Pavonina* (1), *Reussella* (1).

Superfamily Bolivinitacea (44)
Family Bolivinitidae s.l. (31)
Bolivina (2), *Bolivinella* (3), *Bolivinita* (1), *Brizalina* (11), *Euloxostoma* (1), *Loxostomina* (2), *Lugdunum* (2), *Parabrizalina* (1), *Pseudobrizalina* (2), *Rectobolivina* (1), *Sagrinella* (1), *Saidovina* (3), *Sigmavirgulina* (1).

Family Eouvigerinidae (13)
Nodogenerina (2), *Orthomorphina* (2), *Procerolagena* (5), *Stilostomella* (4).

Superfamily Cassidulinacea (25)
Family Pleurostomellidae (8)
Parafissurina (3), *Pleurostomella* (5).

Family Cassidulinidae s.l. (17)
Cassidulina (4), *Cassidulinoides* (3), *Ehrenbergina* (4), *Evolvocassidulina* (1), *Favocassidulina* (1), *Globocassidulina* (2), *Orthoplecta* (1), *Sphaeroidina* (1).

Superfamily Incertae Sedis (2)
Family Millettiidae (2)
Millettia (2).

---

*Arguably a bolivinitid.

Order Robertinida (11)
Superfamily Ceratobuliminacea (6)
Family Epistominidae (2)
Hoeglundina (1), *Mississippina* (1).

Family Ceratobuliminidae (4)
Ceratobulimina (1), *Ceratocancris* (2), *Lamarckina* (1).

Superfamily Robertinacea (5)
Family Robertinidae (5)
Geminospira (1), *Robertina* (2), *Robertinoides* (2).

Order Rotaliida (smaller) (149)
Superfamily Spirillinacea (10)
Family Spirillinidae (9)
Alanwoodia (1), *Mychostomina* (1), *Spirillina* (7).

Family Patellinidae (1)
Patellina (1).

Superfamily Discorbacea (49)
Family Discorbidae (27)
Buccella (1), *Discorbinella* (2), *Discorbis* (1), *Gavelinopsis* (1), *Lamellodiscorbis* (1), *Laticarinina* (1), *Neoconorbina* (2), *Parvicarinina* (2), *Patellinella* (1), *Planodiscorbis* (3), *Planulinoides* (1), *Rosalina* (8), *Strebloides* (1), *Tretomphaloides* (1), *Trochulina* (1).

Family Cancrisidae (5)
Cancris (2), *Gyroidinoides* (1), *Valvulineria* (2).

Family Bagginidae (4)
Baggina (4).

Family Glabratellidae (7)
Glabratella (4), *Heronallenia* (1), *Neoglabratella* (1), *Planoglabratella* (1).

Family Cymbaloporidae (6)
Cymbaloporella (1), *Cymbaloporetta* (4), *Millettiana* (1).

Superfamily Asterigerinacea (22)
Family Eponididae (11)
Cribroeponides (1), *Eponides* (1), *Ioanella* (1), *Neoeponides* (6), *Oridorsalis* (1), *Poroeponides* (1).

Family Epistomariidae (1)
Epistomaroides (1).

Family Alabaminidae (3)
Alabaminoides (1), *Osangularia* (1), *Osangulariella* (1).

Family Siphoninidae (4)
Siphonina (2), *Siphoninella* (1), *Siphoninoides* (1).

Family Asterigerinidae (3)
Amphistegina (3),

Superfamily Orbitoidacea (38)
Family Anomalinidae (24)

Anomalinella (1), *Anomalinoides* (1), *Cibicides* (2), *Cibicidoides* (7), *Cyclocibicides* (1), *Discoanomalina* (3), *Dyocibicides* (1), *Gyroidina* (3), *Hyalinea* (1), *Karreria* (1), *Planopulvinulina* (1), *Planulina* (2).

Family Planorbulinidae (4)
Planorbulina (3), *Planorbulinella* (1).

Family Acervulinidae (3)
Acervulina (1), *Discogypsina* (1), *Sphaerogypsina* (1).

Family Homotrematidae (7)
Carpenteria (4), *Miniacina* (2), *Rupertina* (1).

Superfamily Nonionacea (30)
Family Chilostomellidae (5)
Allomorphina (1), *Chilostomella* (2), *Pullenia* (2).

Family Nonionidae (7)
Melonis (2), *Nonion* (3), *Nonionella* (2).

Family Elphidiidae (18)
Astrononion (1), *Cellanthus* (1), *Cribrononion* (2), *Elphidiella* (1), *Elphidium* (5), *Fijinonion* (1), *Haynesina* (2), *Laminononion* (1), *Parrellina* (2), *Polystomellina* (2).

Order Rotaliida (larger) (18)
Superfamily Rotaliacea (12)
Family Rotaliidae (9)
Ammonia (1), *Asterorotalia* (1), *Challengerella* (1), *Pararotalia* (3), *Pseudorotalia* (1), *Rotalinoides* (2).

Family Calcarinidae (3)
Baculogypsina (1), *Calcarina* (2).

Superfamily Nummulitacea (6)
Family Nummulitidae (3)
Operculina (2), *Operculinella* (1).

Family Cycloclypeidae (3)
Cycloclypeus (1), *Heterostegina* (2).

Order Globigerinida (30)
Superfamily Globigerinacea (26)
Family Globigerinidae (15)
Beella (1), *Candeina* (1), *Globigerina* (3), *Globigerinoides* (6), *Orbulina* (1), *Pulleniatina* (1), *Sphaeroidinella* (1), *Turborotalita* (1).

Family Globorotaliidae (11)
Globoquadrina (1), *Globorotalia* (8), *Neogloboquadrina* (2).

Superfamily Hantkeninacea (4)
Family Hastigerinidae (4)
Bolliella (1), *Globigerinella* (1), *Hastigerina* (2).

Appendix 2
Summary ecological data pertaining to selected *Challenger* foraminifera (species arranged in taxonomic order)

The following data pertain to those benthonic species which Brady is regarded as having interpreted correctly (and therefore whose distributions he is regarded as having documented accurately).

Note on the biogeographic data

Unless stated otherwise, the *Challenger* foraminifera, especially the deeper water species, can be considered cosmopolitan in terms of their geographic distribution (at least in low to moderate latitudes). Note, however, that seventy-five species appear endemic to the Atlantic Ocean and its borderlands, and one hundred and forty-five (in other words, about twice as many) to the Indo-Pacific. Of the species apparently endemic to the Atlantic and its borderlands, thirty-six (48 per cent) are agglutinating foraminifera (Astrorhizida and Lituolida) and thirty-nine (52 per cent) calcareous. Nine (12 per cent) are porcelaneous forms (Miliolida), sixteen (21 per cent) Nodosariida, five (6 per cent) Buliminida, and nine (12 per cent) rotaliiforms (Rotaliida). In contrast, of the species apparently endemic to the Indo-Pacific, only eleven (8 per cent) are agglutinating and one hundred and thirty-four (92 per cent) calcareous. Twenty-eight (19 per cent) are porcelaneous forms (Miliolida) (chiefly associated with reefal and peri-reefal environments in tropical latitudes), thirty-four (23 per cent) Nodosariida, thirty-four (23 per cent) Buliminida, one (1 per cent) an aragonitic form (Robertinida), and thirty-five (24 per cent) rotaliiforms (Rotaliida).

Note on the bathymetric data

Many of the *Challenger* foraminifera appear to have broad bathymetric preferences. As Brady did not differentiate between live and dead individuals, the likelihood is that some of the depth limits he quotes have been artificially extended by post-mortem transportation of tests. Depths are quoted in fathoms (fm.) throughout (1fm. equals 1.83m.).

Family Allogromiidae
Nodellum membranaceum 1900–3950fm.

Family Saccamminidae
Hemisphaerammina bradyi Great Britain; Shallow water (25–33fm.).
Lagenammina ampullacea Southern Ocean; 120fm.
Pelosina cylindrica 620–2900fm.
Pelosina rotundata 350–1675fm.
Pelosina variabilis 50–1250fm.
Pilulina jeffreysii N. Atlantic; 630–1476fm.
Psammosphaera fusca 45–2750fm.
Saccammina socialis 1263–2050fm.
Sagenina frondescens S. Pacific; 16–35fm.
Sorosphaera confusa 542–2900fm.
Storthosphaera albida Atlantic; 180–1900fm.
Technitella melo 420–1215fm.
Technitella raphanus Fiji; 210fm.
Tholosina bulla 1366–2160fm.
Tholosina vesicularis 630–1443fm.

Thurammina albicans S. Atlantic; 1900fm.
Thurammina compressa N. Atlantic; 630fm.
Thurammina papillata 350–2425fm.

Family Astrorhizidae
Archimerismus subnodosus 20–2600fm.
Astrorhiza arenaria Atlantic; 150–650fm.
Astrorhiza crassatina Faroes; 640fm.
Astrorhiza granulosa N. Atlantic; 630–1000fm.
Astrorhiza limnicola N. Atlantic borderlands; 10–30fm.
Botellina labyrinthica Faroe Channel; 440fm.
Dendrophrya arborescens Arctic, Atlantic; 20–350fm.
Dendrophrya erecta Coasts of Great Britain.
Dendrophrya radiata Coasts of Great Britain.
Halyphysema ramulosa Coasts of Europe; Shallow water (to 15fm.).
Halyphysema tumanowiczii Coasts of Europe; Shallow water (to 25fm.).
Hippocrepina indivisa Arctic, N. Atlantic; 10–20fm.
Hyperammina friabilis 350–1425fm.

Jaculella acuta 60–2900fm.
Jaculella obtusa N. Atlantic; 350–542fm.
Marsipella cylindrica 210–1900fm.
Marsipella elongata Predominantly N. Atlantic; 54–900fm.
Rhabdammina cornuta 350–1215fm.
Rhabdammina discreta 20–2475fm (shallower in higher latitudes).
Saccorhiza ramosa 60–3000fm.
Testulosiphon indivisus 210–540fm.

Family Ammodiscidae
Ammodiscus tenuis 400–1350fm.
Ammolagena clavata 100–2000fm.
Glomospira gordialis 50–2000fm.
Turritellella shoneana Estuaries to abyssal plains (3550fm.).
Turritellella spectabilis N. Atlantic; 358–1900fm.
Usbekistania charoides Estuaries to abyssal plains (2452fm.).

Family Komokiidae
Rhizammina algaeformis 630–2900fm.

Family Lituotubidae
Lituotuba lituiformis Caribbean, S. Atlantic; 390–900fm.

Family Lituolidae
Ammoscalaria pseudospiralis N. Atlantic; 30–370fm.
Ammoscalaria tenuimargo 530–3950fm.
Ammotium cassis N. Atlantic borderlands; 5–20fm.
Bdelloidina aggregata S. Pacific; Shallow water.
Buzasina galeata 1825–2750fm.
Buzasina ringens 1600–2750fm.
Conglophragmium coronatum 390–3950fm.
Cyclammina cancellata 250–1000fm.
Cyclammina pusilla S. Atlantic, Southern Ocean; 1675–1900fm.
Cyclammina rotundidorsata 1100–1900fm.
Cystammina pauciloculata 173–3950fm.
Discammina compressa Caribbean; 390–450fm.
Eratidus foliaceus 1070–2750fm.
Evolutinella rotulata 2740–3150fm.
Glaphyrammina americana 40–1900fm.
Hormosina bacillaris 420–1750fm.
Hormosina normani 1380–2900fm.
Hormosina pilulifera 400–2435fm.
Hormosinella carpenteri Predominantly N. Atlantic; 350–1940fm.
Hormosinella distans 355–2775fm.
Hormosinella guttifera Atlantic; 540–1900fm.
Hormosinella ovicula 1000–3950fm.
Loeblichopsis cylindrica Atlantic, Southern Ocean; 1570–1750fm.
Loeblichopsis sabulosa Cold area, Faroe Channel; 530–540fm.
Protoschista findens Gaspé Bay, 15–20fm.
Recurvoides turbinatus 1425–2350fm.
Reophax dentaliniformis 1000–3000fm.
Reophax fusiformis 40–1443fm.
Reophax spiculifer S. Pacific; 255–620fm.
Subreophax aduncus 540–2900fm.
Subreophax monile S. Atlantic; 350fm.
Triplasia variabilis Fiji; 210fm.
Veleroninoides scitulus 400–2900fm.

Family Textulariidae
Bigenerina cylindrica Atlantic; 40–1230fm.
Bigenerina nodosaria 27–1630fm.
Planctostoma luculenta Atlantic; 350–675fm.
Sahulia conica Tropical; Shallow water.
Septotextularia rugosa Tropical; 7–30fm.
Siphoniferoides siphoniferus Tropical; 15–40fm.
Siphoniferoides transversarius S. Pacific; 155–255fm.
Spirorutilus carinatus Philippines; 95fm.
Spiroplectammina biformis 55–1900fm. (deeper in lower latitudes).
Spiroplectammina sp. nov. Raine Island; 155fm.
Spiroplectella earlandi Raine Island, Ki Islands; 140–155fm.
Textularia pseudogramen Shallow water.
Vulvulina pennatula Subtropical; 350–675fm.

Family Trochamminidae
Adercotryma glomeratum 14–2740fm. (deeper in lower latitudes).
Carterina spiculotesta Tropical to subtropical; Shallow water.
Polystomammina nitida Arctic, Atlantic, Southern Ocean; 50–200fm.
Tritaxis challengeri 13–1100fm.
Tritaxis fusca 7–600fm.
Trochammina inflata Coasts of NW Europe; Shallow water.
Trochamminella conica Shallow water to 1375fm.

Family Verneuilinidae
Dorothia bradyana Equatorial Atlantic and Pacific; 155–390fm.
Dorothia pseudofiliformis 390–620fm.
Dorothia scabra Caribbean; 390–450fm.
Globotextularia anceps Arctic, Atlantic; 390–2200fm.
Pseudogaudryina sp. nov. (1) 11–675fm.
Textulariella barrettii Caribbean, S. Atlantic; 100–435fm.
Tortoplectella crispata S. Pacific; 155fm.

Family Valvulinidae
Clavulina pacifica 8–350fm.

Family Eggerellidae
Cribrogoesella robusta Caribbean, S. Atlantic; 350–390fm.
Eggerella bradyi 129–3125fm.
Karreriella bradyi 129–2050fm.
Karreriella novangliae 210–2300fm.
Karrerulina conversa 1000–3950fm.
Latentoverneuilina indiscreta Fiji; 210fm.
Pseudoclavulina tricarinata Raine Island; 155fm.

Family Ataxophragmiidae
Eggerelloides scaber <10–50fm.

Family Cornuspiridae
Cornuspira carinata 70–1630fm.
Cornuspira crassisepta Warm area, Faroe Channel; 530fm.
Cornuspira involvens 7–1900fm.
Cornuspira lacunosa Bass Strait; 155fm.
Cornuspiroides striolatus Cold area, Faroe Channel; 540fm.

Family Nubeculariidae
Fischerina communis Faroes; 170fm.
Nodobaculariella convexiuscula Tropical Pacific; 16–155fm.

Nodophthalmidium simplex Equatorial Atlantic and Pacific; 15–390fm.

Nubeculina divaricata Tropical Pacific; 18–155fm.

Nubecularia lucifuga Tropical Pacific; 18fm.

Planispirinella exigua Tropical; <25–620fm.

Vertebralina insignis Equatorial Atlantic and Pacific; 18–390fm.

Vertebralina striata Shallow water (6–40fm.).

Family Ophthalmidiidae

Cornuloculina inconstans 100–2300fm.

Cornuloculina tumidula Caribbean; 390fm.

Discospirina italica N. Atlantic, Mediterranean; 100–1443fm. (<2.8 deg. C).

Family Soritidae

Amphisorus hemprichii Tropical; Shallow water.

Marginopora vertebralis Tropical-subtropical; Shallow water (7–450fm.).

Family Miliolidae

Adelosina bicornis Shoreline to 50fm.

Adelosina granulocostata Tropical; Shoreline to shallow water.

Adelosina pulchella <100fm.

Adelosina sp. nov. Canaries, Caribbean, Tropical Pacific; to 155fm.

Articularia lineata Caribbean, Tropical Pacific; 210–435fm.

Articularia scrobiculata Indian Ocean, Tropical Pacific; to 17fm.

Cribromiliolinella subvalvularis S. Pacific; 1100fm.

Edentostomina cultrata Tropical Indian and Pacific Oceans; 2–27fm.

Edentostomina rupertiana Tropical; Shoreline to shallow water (6–20fm.).

Flintina bradyana Shoreline to 14fm.

Flintinoides labiosa Pacific; Shallow water, to 3950fm.

Hauerina fragilissima Tropical Pacific; 3–620fm.

Massilina amygdaloides Pacific; 147–565fm.

Massilina sp. nov. Vigo Bay, Tropical Pacific; Shallow water to 580fm.

Miliolinella subrotunda Southern Ocean, Pacific; 28–150fm.

Nummoloculina contraria 40–2160fm.

Parrina bradyi Southern Ocean, Pacific; 17–50fm.

Planispirinoides bucculentus 20–1785fm.

Polysegmentina circanata Tropical Pacific; 3–11fm.

Proemassilina arenaria Tropical Pacific; 95–210fm.

Pseudoflintina triquetra S. Pacific; 37–155fm.

Pseudomassilina australis Shallow water.

Pseudomassilina macilenta Tropical Pacific; 17–40fm.

Ptychomiliola separans Tropical Pacific; 8–155fm.

Pyrgo comata 15–2900fm.

Pyrgo denticulata Equatorial Atlantic and Pacific; 11–610fm.

Pyrgo denticulata var. *striolata* Tropical Pacific; 6–8fm.

Pyrgo laevis Shallow water to 1215fm.

Pyrgo lucernula 300–1000fm.

Pyrgo murrhina 1180–1900fm.

Pyrgo serrata 580–1750fm.

Pyrgoella irregularis 350–1415fm.

Pyrgoella sphaera Shallow water to 1000fm.

Quinqueloculina auberiana Atlantic; 245–2435fm.

Quinqueloculina boueana Shoreline to shallow water.

Quinqueloculina lamarckiana Tropical Pacific; 6–95fm.

Quinqueloculina parkeri Tropical Pacific; Shallow water to 420fm.

Quinqueloculina seminulum Shallow water to 3000fm.

Quinqueloculina tropicalis Caribbean, Tropical Pacific; Shallow water (shoreline to 37fm.).

Schlumbergerina alveoliniformis Tropical; Shoreline to 420fm.

Sigmamiliolinella australis Shoreline to 150fm.

Sigmoihauerina bradyi Tropical Pacific; 6–155fm.

Sigmoilina sigmoidea Atlantic; 300–900fm.

Sigmoilopsis schlumbergeri 28–1630fm.

Siphonaperta crassatina S. Pacific; 38fm.

Siphonaperta sp. nov. Shallow water to 440fm.

Spiroloculina angulata Tropical; Shallow water to 500fm.

Spiroloculina antillarum S. Atlantic; 350fm.

Spiroloculina communis Shallow water to 100fm.

Spiroloculina henbesti Littoral to 2000fm.

Spiroloculina robusta Caribbean; 390fm.

Spiroloculina tenuiseptata Mediterranean, Tropical Pacific; 580–1200fm.

Spirosigmoilina tenuis Shallow water to 2750fm.

Triloculina bertheliniana Indian Ocean; Shoreline to shallow water (7fm.).

Triloculina terquemiana Indian Ocean (Ceylon, Madagascar); Shoreline.

Triloculina transversestriata Indo-Pacific; Shallow water to 155fm.

Triloculina tricarinata 6–2350fm.

Triloculina trigonula Shoreline to 100fm.

Triloculinella obliquinodus Shallow water.

Triloculinella sublineata Tropical Pacific; 15–25fm.

Tubinella inornata Southern Ocean, S. Pacific; 50–150fm.

Tubinella funalis Southern Ocean, S. Pacific; 20–150fm.

Family Alveolinidae

Alveolinella quoyi Tropical-subtropical; Shallow water (to 155fm.).

Borelis melo Tropical; Shallow water (to 40fm.).

Family Keramosphaeridae

Keramosphara murrayi S. Pacific; 1950fm.

Family Nodosariidae

Amphicoryna intercellularis Bermuda; 435fm.

Amphicoryna proxima 40–450fm.

Amphicoryna scalaris 2–1630fm.

Amphicoryna sublineata Atlantic; 350–435fm.

Amphicoryna substriatula Tahiti; 420fm.

Astacolus bradyi Torres Strait; 155fm.

Dentalina advena <400fm.

Dentalina albatrossi 120–450fm.

Dentalina ariena 345–2350fm.

Dentalina catenulata Pacific; 95–155fm.

Dentalina filiformis 50–450fm.

Dentalina flintii To 2000fm.

Dentalina inflexa 95–1400fm.

Dentalina plebeia Bermuda; 435fm.

Frondicularia compta Bass Strait, Torres Strait; 38–155fm.

Frondicularia kiensis Pacific; 40–155fm.

Frondicularia millettii Pacific; 155fm.

Frondicularia robusta Pacific; 40–129fm.

Frondicularia sagittula Atlantic; 84–435fm.

Glandulonodosaria ambigua Ki Islands; 129fm.
Glandulonodosaria annulata Azores, Honolulu; 40–450fm.
Glandulonodosaria calomorpha 6–2200fm.
Glandulonodosaria sp. nov. Ki Islands; 129fm.
Grigelis semirugosa Caribbean, Tropical Pacific; 95–450fm.
Lagena crenata 15–2425fm.
Lagena quinquelaterata 50–2350fm.
Lenticulina atlantica Atlantic; 100–390fm.
Lenticulina calcar 95–450fm.
Lenticulina convergens 16–2740fm.
Lenticulina echinata Pacific; 95–210fm.
Lenticulina gibba < 500fm.
Lenticulina inornata Fiji; 210fm.
Lenticulina nitida Philippines; 95fm.
Lenticulina orbicularis 38–450fm.
Lenticulina papillosa 210–390fm.
Lenticulina submamilligera Pacific; 95–210fm.
Lenticulina thalmanni 345–2200fm.
Lenticulina vortex 90–435fm.
Marginulina obesa 15–2740fm.
Neolenticulina variabilis 50–1630fm.
Nodosaria lamnulifera To 1400fm.
Planularia australis Pacific; 95–155fm.
Planularia cassis 90–210fm.
Planularia gemmata Pacific; 95–210fm.
Planularia magnifica Fiji; 210fm.
Planularia magnifica var. *falciformis* Atlantic; 300–1000fm.
Planularia patens Pacific; 95–155fm.
Planularia siddalliana Fiji; 210fm.
Planularia sp. nov. Bass Strait; 38fm.
Pseudoglandulina comatula 390–450fm.
Pseudoglandulina glanduliniformis 37–1360fm.
Saracenaria latifrons 275–390fm.
Saracenaria volpicelli S. Atlantic; 1425fm.
Sipholagena hertwigiana 150–2600fm.
Vaginulina americana Atlantic; 15–435fm.
Vaginulina spinigera 100–1200fm.
Vaginulinopsis albatrossi Atlantic; To 630fm.
Vaginulinopsis reniformis 150–2050fm.
Vaginulinopsis tenuis 150–2350fm.
Vaginulinopsis subaculeata Atlantic; 390–450fm.
Vaginulinopsis sp. nov. 155–350fm.

Family Lingulinidae
Lingulina seminuda 350–1200fm.

Family Plectofrondiculariidae
Plectofrondicularia advena 129–1240fm.
Plectofrondicularia helenae Ki Islands; 129fm.

Family Polymorphinidae
Francuscia extensa 1100–2435fm.
Globulina myristiformis Atlantic; 30–630fm.
Globulina regina Pacific; 6–155fm.
Guttulina austriaca To 2050fm.
Guttulina ovata Caribbean; 390fm.
Guttulina problema To 155fm.
Pseudopolymorphina dawsoni 38–90fm.

Pseudopolymorphina sp. nov. 5–600fm.
Pyrulina fusiformis Atlantic, Southern Ocean; 808–1443fm.
Ramulina globulifera 95–435fm.
Sigmoidella elegantissima 7–580fm.
Sigmoidella seguenzana Pacific; 2–129fm.

Family Glandulinidae
Cushmanina desmophora 390–2350fm.
Cushmanina feildeniana 80–2300fm.
Cushmanina quadralata 2200–2600fm.
Cushmanina spiralis Torres Strait; 155fm.
Cushmanina striatopunctata 55–2750fm.
Cushmanina torquata Pacific; 1375fm.
Entosolenia lineata 20–150fm.
Fissurina alveolata var. *semisculpta* Southern Ocean; 1375fm.
Fissurina annectens 20–410fm.
Fissurina brevis Torres Strait; 155fm.
Fissurina clathrata Pacific; 580–800fm.
Fissurina formosa var. *comata* N. Pacific; 1850fm.
Fissurina formosa var. *favosa* N. Pacific; 1850fm.
Fissurina laevigata 2–3125fm.
Fissurina radiatomarginata Pacific; 37–155fm.
Fissurina schulzeana Torres Strait; 155fm.
Fissurina semimarginata 50–2350fm.
Fissurina seminiformis 1000–2350fm.
Fissurina siliqua 1086–2350fm.
Fissurina squamosoalata N. Atlantic; 173–1445fm.
Fissurina squamosomarginata 422–1100fm.
Fissurina staphyllearia 2200–2750fm.
Fissurina unguiculata S. Atlantic; 2200fm.
Fissurina wrightiana Admiralty Islands; Shallow water.
Galwayella oblonga To 620fm.
Galwayella trigonomarginata 90–2300fm.
Galwayella trigonoornata To 1070fm.
Glandulina ovula 28–1360fm.
Laculatina quadrangularis Torres Strait; 155fm.
Oolina exsculpta 800–2350fm.
Oolina ovum N. Pacific; 2300fm.
Oolina seminuda 1300–2350fm.
Oolina truncata 1825–2740fm.
Oolina variata Bass Strait; 38fm.
Pseudoolina multicosta S. Atlantic; 2350fm.

Family Buliminidae
Bulimina aculeata 1000–2740fm.
Bulimina elongata Atlantic; 630–1425fm.
Bulimina mexicana 95–2435fm.
Bulimina marginata Temperate; To 1630fm.
Bulimina rostrata 580–1570fm.
Bulimina subornata S. Pacific; 345–800fm.
Bulimina sp. nov. 90–2375fm.
Eubuliminella exilis 350–1500fm.
Fursenkoina complanata 10–3000fm.
Fursenkoina pauciloculata Pacific; 3–129fm.
Fursenkoina texturata 129–2350fm.
Praeglobobulimina ovata 15–2200fm.
Praeglobobulimina spinescens Ki Islands; 580fm.

Family Turrilinidae
Buliminella elegantissima 2–410fm.
Buliminoides williamsonianus Pacific; 2–155fm.

Family Uvigerinidae
Rectuvigerina nicoli Southern Ocean; 150fm.
Siphogenerina columellaris 6–1125fm.
Siphogenerina dimorpha 50–620fm.
Siphogenerina raphanus 2–260fm.
Siphogenerina striata var. *curta* 3–350fm.
Siphouvigerina ampullacea 350–675fm.
Siphouvigerina interrupta Pacific; 37–1375fm.
Siphouvigerina porrecta 12–1850fm.
Trifarina bradyi 12–1360fm.
Trifarina lepida N. Atlantic; 1240fm.
Trifarina reussi Ki Islands; 129fm.
Trifarina spinipes Atlantic; 110–150fm.
Uvigerina aculeata 580–1900fm.
Uvigerina auberiana 37–2335fm.
Uvigerina brunnensis Pacific, Southern Ocean; 120–245fm.
Uvigerina canariensis 150–1900fm.
Uvigerina hollicki Pacific, Southern Ocean; 40–155fm.
Uvigerina schwageri Pacific; 95–210fm.

Family Pavoninidae
Chrysalidinella dimorpha Tropical; Shallow water (7–155fm.).
Pavonina flabelliformis Subtropical; 2–390fm.
Reussella spinulosa 7–100fm.

Family Bolivinitidae *s.l.*
Bolivina decussata S. Pacific; 1375–1450fm.
Bolivina robusta 7–1900fm.
Bolivinita quadrilatera 410–1070fm.
Brizalina alata 50–800fm.
Brizalina earlandi 2–2750fm.
Brizalina nitida S. Pacific; 38–155fm.
Brizalina pygmaea 50–620fm.
Brizalina semicostata S. Pacific; 15–155fm.
Brizalina spathulata N. Atlantic; 96–1180fm.
Brizalina subaenariensis var. *mexicana* 13–1630fm.
Brizalina subreticulata S. Pacific, Southern Ocean; 130–1570fm.
Brizalina subtenuis Fiji; 255fm.
Euloxostoma bradyi 95–1125fm.
Loxostomina limbata 7–200fm.
Loxostomina mayori 6–420fm.
Lugdunum hantkenianum Pacific; 130–420fm.
Lugdunum schwagerianum S. Pacific; 7–155fm.
Parabrizalina porrecta 37–420fm.
Pseudobrizalina lobata S. Pacific; 16–155fm.
Pseudobrizalina strigosa S. Pacific; 16–155fm.
Rectobolivina bifrons Japan; 345fm.
Sagrinella jugosa Indo-Pacific; 15–155fm.
Saidovina amygdalaeformis S. Pacific; 95–1070fm.
Saidovina karreriana 345–675fm.
Saidovina subangularis S. Pacific; 95–155fm.
Sigmavirgulina tortuosa 2–420fm.

Family Eouvigerinidae
Orthomorphina challengeriana Ki Islands; 129fm.

Orthomorphina jedlitschkai Ki Islands; 129fm.
Procerolagena distomamargaritifera 38–275fm.
Stilostomella abyssorum S. Pacific; 1825fm.
Stilostomella consorbrina 129–1375fm.
Stilostomella fistuca Ki Islands; 129fm.
Stilostomella retrorsa Ki Islands; 580fm.

Family Millettiidae
Millettia limbata Torres Strait; 155fm.
Millettia tessellata Pacific; 17–155fm.

Family Pleurostomellidae
Parafissurina botelliformis 1450–2350fm.
Pleurostomella brevis Ki Islands; 129fm.
Pleurostomella recens Ki Islands; 129fm.
Pleurostomella sp. nov. (2) 1375–2350fm.

Family Cassidulinidae *s.l.*
Cassidulinoides parkerianus S. Pacific; 45–175fm.
Ehrenbergina hystrix S. Pacific; 1940–2925fm.
Ehrenbergina pupa Atlantic; 13–1035fm.
Favocassidulina favus Pacific; 1375–2600fm.
Globocassidulina pacifica 155–610fm.
Globocassidulina subglobosa 12–2950fm.
Orthoplecta clavata Admiralty Islands; 17fm.
Sphaeroidina bulloides 85–2600fm.

Family Epistominidae
Mississippina concentrica 15–1000fm.

Family Ceratobuliminidae
Ceratobulimina jonesiana Ki Islands; 580fm.
Ceratocancris scaber 17–1000fm.
Ceratocancris sp. nov. Pacific; 100–620fm.
Lamarckina ventricosa 155–620fm.

Family Robertinidae
Geminospira bradyi 40–155fm.
Robertina subcylindrica 155–1070fm.
Robertinoides oceanicus Ki Islands; 580fm.

Family Spirillinidae
Alanwoodia campanaeformis Torres Strait; 155fm.
Spirillina decorata 6–1125fm.
Spirillina denticulata Pacific; 17–155fm.
Spirillina inaequalis Pacific; 12–155fm.
Spirillina limbata 6–1425fm.
Spirillina obconica Pacific, Southern Ocean; 17–150fm.
Spirillina tuberculata 20–400fm.

Family Patellinidae
Patellina corrugata 7–620fm.

Family Discorbidae
Buccella frigida 13–220fm.
Discorbinella araucana Shallow water.
Discorbinella bertheloti Generally < 500fm.
Gavelinopsis lobatula Tahiti; 620fm.

Lamellodiscorbis sp. nov. Pacific; To 210fm.
Laticarinina pauperata 129–2550fm.
Patellinella inconspicua S. Pacific; 17–345fm.
Planodiscorbis rarescens Pacific; 95–155fm.
Rosalina australis Shallow water.
Rosalina eximia Torres Strait; 155fm.
Tretomphaloides concinnus Tropical Pacific; 15–610fm.
Trochulina rosea Atlantic, Caribbean; Shoreline.

Family Cancrisidae
Cancris oblongus Littoral to 500fm.
Cancris auriculus Littoral to 500fm.
Gyroidinoides soldanii 300–2000fm.

Family Bagginidae
Baggina bradyi Pacific; 2–155fm.
Baggina sp. nov. Atlantic; 28–420fm.

Family Glabratellidae
Glabratella australensis 2–20fm.
Glabratella patelliformis Pacific; 6–150fm.
Glabratella pulvinata Pacific; 6–17fm.
Glabratella tabernacularis 2–255fm.
Planoglabratella opercularis 2–155fm.

Family Cymbaloporidae
Cymbaloporella tabellaeformis Tropical Indian and Pacific Oceans; 12–610fm.

Family Eponididae
Ioanella tumidula N. Atlantic; 2740fm.
Neoeponides auberii 2–420fm.
Neoeponides berthelotianus 16–25fm.
Neoeponides margaritifer Pacific; 10–125fm.
Neoeponides praecinctus Predominantly Pacific; 15–255fm.
Neoeponides procerus Pacific; 3–210fm.
Neoeponides schreibersii 3–345fm.
Poroeponides lateralis Shoreline to 28fm.
Oridorsalis umbonata 166–3125fm.

Family Epistomariidae
Epistomaroides polystomelloides Pacific; 6–8fm.

Family Alabaminidae
Alabaminoides exiguus 15–2600fm.
Osangularia bengalensis 680–1525fm.
Osangulariella umbonifera 1450–3125fm.

Family Siphoninidae
Siphoninella soluta Caribbean; 390fm.
Siphoninoides echinatus 2–155fm.

Family Anomalinidae
Anomalinella rostrata Pacific; 16–25fm.
Anomalinoides colligerus 37–1350fm.
Cibicides lobatulus Shallow water to 3000fm.
Cibicides refulgens 50–2400fm.
Cibicidoides globulosus 345–2050fm.

Cibicidoides pachyderma 37–2600fm.
Cibicidoides robertsonianus Atlantic; 55–1785fm.
Cibicidoides subhaidingerii 90–1776fm.
Cibicidoides wuellerstorfi 345–2435fm.
Cyclocibicides vermiculatus Mediterranean; Shoreline.
Discoanomalina coronata 30–1035fm.
Gyroidina bradyi 1070–1900fm.
Gyroidina broeckhiana Ki Islands; 580fm.
Planopulvinulina dispansa Madeira, Atlantic; 390fm.
Planulina ariminensis 47–2200fm.
Planulina foveolata Bermuda; 345fm.

Family Planorbulinidae
Planorbulina acervalis Red Sea, Indo-Pacific; <20–30fm.
Planorbulinella larvata Tropical Pacific; 15–210fm.

Family Acervulinidae
Acervulina inhaerens Shallow water.
Sphaerogypsina globulus Littoral to 400fm.

Family Homotrematidae
Carpenteria monticularis 16–620fm.
Carpenteria proteiformis 17–435fm.
Carpenteria utricularis 16–350fm.
Miniacina alba Shallow water.
Miniacina miniacea Shallow water (to 1000fm).
Rupertina stabilis N. Atlantic; 5–1375fm.

Family Chilostomellidae
Allomorphina pacifica Pacific; 345–620fm.
Pullenia bulloides 300–2750fm.
Pullenia quinqueloba 20–2750fm.

Family Nonionidae
Melonis affinis 30–3125fm.
Melonis pompilioides 1000–2421fm.
Nonion armatum Shoreline.
Nonion commune 7–1375fm.
Nonion fabum 7–2000fm.
Nonionella turgida 11–1630fm.

Family Elphidiidae
Cellanthus craticulatus 3–40fm.
Elphidiella arctica Arctic; 7–210fm.
Elphidium advenum Pacific; 6–28fm.
Elphidium crispum Shallow water (to 355fm.).
Elphidium macellum Low latitudes; Shallow water.
Fijinonion fijiense Pacific; 155–255fm.
Haynesina germanica <50fm.
Haynesina orbicularis 7–112fm.
Parrellina imperatrix Pacific; 2–10fm.
Parrellina verriculata Bass Strait; 38fm.

Family Rotaliidae
Asterorotalia pulchella Pacific; 7–255fm.
Pseudorotalia schroeteriana Indo-Pacific; <60fm.
Pararotalia stellata Shallow water.
Pararotalia venusta Indo-Pacific; 2–345fm.

Rotalinoides gaimardii Pacific; 2–37fm.

Family Calcarinidae
Baculogypsina sphaerulata Pacific; 6–155fm.
Calcarina hispida Pacific; 3–155fm.

Family Nummulitidae
Operculinella cumingii Red Sea, Pacific; 10–25fm.

Family Cycloclypeidae
Cycloclypeus carpenteri Fiji; 210fm.

Appendix 3
Summary ecological data pertaining to selected *Challenger* foraminifera (species arranged in alphabetic order)

See preamble to Appendix 2.

Acervulina inhaerens Shallow water.
Adelosina bicornis Shoreline to 50fm.
Adelosina granulocostata Tropical; Shoreline to shallow water.
Adelosina pulchella <100fm.
Adelosina sp. nov. Canaries, Caribbean, Tropical Pacific; To 155fm.
Adercotryma glomeratum 14–2740fm. (deeper in lower latitudes).
Alabaminoides exiguus 15–2600fm.
Alanwoodia campanaeformis Torres Strait; 155fm.
Allomorphina pacifica Pacific; 345–620fm.
Alveolinella quoyi Tropical-subtropical; Shallow water (to 155fm.).
Ammodiscus tenuis 400–1350fm.
Ammolagena clavata 100–2000fm.
Ammoscalaria pseudospiralis N. Atlantic; 30–370fm.
Ammoscalaria tenuimargo 530–3950fm.
Ammotium cassis N. Atlantic borderlands; 5–20fm.
Amphicoryna intercellularis Bermuda; 435fm.
Amphicoryna proxima 40–450fm.
Amphicoryna scalaris 2–1630fm.
Amphicoryna sublineata Atlantic; 350–435fm.
Amphicoryna substriatula Tahiti; 420fm.
Amphisorus hemprichii Tropical; Shallow water.
Anomalinella rostrata Pacific; 16–25fm.
Anomalinoides colligerus 37–1350fm.
Archimerismus subnodosus 20–2600fm.
Articularia lineata Caribbean, Tropical Pacific; 210–435fm.
Articularia scrobiculata Indian Ocean, Tropical Pacific; To 17fm.
Astacolus bradyi Torres Strait; 155fm.
Asterorotalia pulchella Pacific; 7–255fm.
Astrorhiza arenaria Atlantic; 150–650fm.
Astrorhiza crassatina Faroes; 640fm.
Astrorhiza granulosa N. Atlantic; 630–1000fm.
Astrorhiza limnicola N. Atlantic borderlands; 10–30fm.

Baculogypsina sphaerulata Pacific; 6–155fm.
Baggina bradyi Pacific; 2–155fm.
Baggina sp. nov. Atlantic; 28–420fm.
Bdelloidina aggregata S. Pacific; Shallow water.
Bigenerina cylindrica Atlantic; 40–1230fm.
Bigenerina nodosaria 27–1630fm.
Bolivina decussata S. Pacific; 1375–1450fm.
Bolivina robusta 7–1900fm.
Bolivinita quadrilatera 410–1070fm.
Borelis melo Tropical; Shallow water (to 40fm.).
Botellina labyrinthica Faroe Channel; 440fm.
Brizalina alata 50–800fm.

Brizalina earlandi 2–2750fm.
Brizalina nitida S. Pacific; 38–155fm.
Brizalina pygmaea 50–620fm.
Brizalina semicostata S. Pacific; 15–155fm.
Brizalina spathulata N. Atlantic; 96–1180fm.
Brizalina subaenariensis var. *mexicana* 13–1630fm.
Brizalina subreticulata S. Pacific, Southern Ocean; 130–1570fm.
Brizalina subtenuis Fiji; 255fm.
Buccella frigida 13–220fm.
Bulimina aculeata 1000–2740fm.
Bulimina elongata Atlantic; 630–1425fm.
Bulimina marginata Temperate; To 1630fm.
Bulimina mexicana 95–2435fm.
Bulimina rostrata 580–1570fm.
Bulimina subornata S. Pacific; 345–800fm.
Bulimina sp. nov. 90–2375fm.
Buliminella elegantissima 2–410fm.
Buliminoides williamsonianus Pacific; 2–155fm.
Buzasina galeata 1825–2750fm.
Buzasina ringens 1600–2750fm.

Calcarina hispida Pacific; 3–155fm.
Cancris auriculus Littoral to 500fm.
Cancris oblongus Littoral to 500fm.
Carpenteria moniticularis 16–620fm.
Carpenteria proteiformis 17–435fm.
Carpenteria utricularis 16–350fm.
Carterina spiculotesta Tropical to subtropical; Shallow water.
Cassidulinoides parkerianus S. Pacific; 45–175fm.
Cellanthus craticulatus 3–40fm.
Ceratobulimina jonesiana Ki Islands; 580fm.
Ceratocancris scaber 17–1000fm.
Ceratocancris sp. nov. Pacific; 100–620fm.
Chrysalidinella dimorpha Tropical; Shallow water (7–155fm.).
Cibicides lobatulus Shallow water to 3000fm.
Cibicides refulgens 50–2400fm.
Cibicidoides globulosus 345–2050fm.
Cibicidoides pachyderma 37–2600fm.
Cibicidoides robertsonianus Atlantic; 55–1785fm.
Cibicidoides subhaidingerii 90–1776fm.
Cibicidoides wuellerstorfi 345–2435fm.
Clavulina pacifica 8–350fm.
Conglophragmium coronatum 390–3950fm.
Cornuloculina inconstans 100–2300fm.
Cornuloculina tumidula Caribbean; 390fm.
Cornuspira carinata 70–1630fm.
Cornuspira crassisepta Warm area, Faroe Channel; 530fm.

Cornuspira involvens 7–1900fm.
Cornuspira lacunosa Bass Strait; 155fm.
Cornuspiroides striolatus Cold area, Faroe Channel; 540fm.
Cribrogoesella robusta Caribbean, S. Atlantic; 350–390fm.
Cribromiliolinella subvalvularis S. Pacific; 1100fm.
Cushmanina desmophora 390–2350fm.
Cushmanina feildeniana 80–2300fm.
Cushmanina quadralata 2200–2600fm.
Cushmanina spiralis Torres Strait; 155fm.
Cushmanina striatopunctata 55–2750fm.
Cushmanina torquata Pacific; 1375fm.
Cyclammina cancellata 250–1000fm.
Cyclammina pusilla S. Atlantic, Southern Ocean; 1675–1900fm.
Cyclammina rotundidorsata 1100–1900fm.
Cyclocibicides vermiculatus Mediterranean; Shoreline.
Cycloclypeus carpenteri Fiji; 210fm.
Cymbaloporella tabellaeformis Tropical Indian and Pacific Oceans; 12–610fm.
Cystammina pauciloculata 173–3950fm.

Dendrophrya arborescens Arctic, Atlantic; 20–350fm.
Dendrophrya erecta Coasts of Great Britain.
Dendrophrya radiata Coasts of Great Britain.
Dentalina advena <400fm.
Dentalina albatrossi 120–450fm.
Dentalina ariena 345–2350fm.
Dentalina catenulata Pacific; 95–155fm.
Dentalina filiformis 50–450fm.
Dentalina flintii To 2000fm.
Dentalina inflexa 95–1400fm.
Dentalina plebeia Bermuda; 435fm.
Discammina compressa Caribbean; 390–450fm.
Discoanomalina coronata 30–1035fm.
Discorbinella araucana Shallow water.
Discorbinella bertheloti Generally <500fm.
Discospirina italica N. Atlantic, Mediterranean; 100–1443fm. (<2.8 deg. C).
Dorothia bradyana Equatorial Atlantic and Pacific; 155–390fm.
Dorothia pseudofiliformis 390–620fm.
Dorothia scabra Caribbean; 390–450fm.

Endentostomina cultrata Tropical Indian and Pacific Oceans; 2–27fm.
Edentostomina rupertiana Tropical; Shoreline to shallow water (6–20fm.).
Eggerella bradyi 129–3125fm.
Eggerelloides scaber <10–50fm.
Ehrenbergina hystrix S. Pacific; 1940–2925fm.
Ehrenbergina pupa Atlantic; 13–1035fm.
Elphidiella arctica Atlantic; 7–210fm.
Elphidium advenum Pacific; 6–28fm.
Elphidium crispum Shallow water (to 355fm.).
Elphidium macellum Low latitudes; Shallow water.
Entosolenia lineata 20–150fm.
Epistomaroides polystomelloides Pacific; 6–8fm.
Eratidus foliaceus 1070–2750fm.
Eubuliminella exilis 350–1500fm.
Euloxostoma bradyi 95–1125fm.
Evolutinella rotulata 2740–3150fm.

Favocassidulina favus Pacific; 1375–2600fm.
Fijinonion fijiense Pacific; 155–255fm.
Fischerina communis Faroes; 170fm.
Fissurina alveolata var. *semisculpta* Southern Ocean; 1375fm.
Fissurina annectens 20–410fm.
Fissurina brevis Torres Strait; 155fm.
Fissurina clathrata Pacific; 580–800fm.
Fissurina formosa var. *comata* N. Pacific; 1850fm.
Fissurina formosa var. *favosa* N. Pacific; 1850fm.
Fissurina laevigata 2–3125fm.
Fissurina radiatomarginata Pacific; 37–155fm.
Fissurina schulzeana Torres Strait; 155fm.
Fissurina semimarginata 50–2350fm.
Fissurina seminiformis 1000–2350fm.
Fissurina siliqua 1086–2350fm.
Fissurina squamosoalata N. Atlantic; 173–1445fm.
Fissurina squamosomarginata 422–1100fm.
Fissurina staphyllearia 2200–2750fm.
Fissurina unguiculata S. Atlantic; 2200fm.
Fissurina wrightiana Admiralty Islands; Shallow water.
Flintina bradyana Shoreline to 14fm.
Flintinoides labiosa Pacific; Shallow water to 3950fm.
Francuscia extensa 1100–2435fm.
Frondicularia compta Bass Strait, Torres Strait; 38–155fm.
Frondicularia kiensis Pacific; 40–155fm.
Frondicularia millettii Pacific; 155fm.
Frondicularia robusta Pacific; 40–129fm.
Frondicularia sagittula Atlantic; 84–435fm.
Fursenkoina complanata 10–3000fm.
Fursenkoina pauciloculata Pacific; 3–129fm.
Fursenkoina texturata 129–2350fm.

Galwayella oblonga To 620fm.
Galwayella trigonomarginata 90–2300fm.
Galwayella trigonoornata To 1070fm.
Gavelinopsis lobatula Tahiti; 620fm.
Geminospira bradyi 40–155fm.
Glabratella australensis 2–20fm.
Glabratella patelliformis Pacific; 6–150fm.
Glabratella pulvinata Pacific; 6–17fm.
Glabratella tabernacularis 2–255fm.
Glandulina ovula 28–1360fm.
Glandulonodosaria ambigua Ki Islands; 129fm.
Glandulonodosaria annulata Azores, Honolulu; 40–450fm.
Glandulonodosaria calomorpha 6–2200fm.
Glandulonodosaria sp. nov. Ki Islands; 129fm.
Glaphyrammina americana 40–1900fm.
Globocassidulina pacifica 155–610fm.
Globocassidulina subglobosa 12–2950fm.
Globotextularia anceps Arctic, Atlantic; 390–2200fm.
Globulina myristiformis Atlantic; 30–630fm.
Globulina regina Pacific; 6–155fm.
Glomospira gordialis 50–2000fm.
Grigelis semirugosa Caribbean, Tropical Pacific; 95–450fm.
Guttulina austriaca To 2050fm.
Guttulina ovata Caribbean; 390fm.
Guttulina problema To 155fm.
Gyroidina bradyi 1070–1900fm.

Gyroidina broeckhiana Ki Islands; 580fm.
Gyroidinoides soldanii 300–2000fm.

Halyphysema ramulosa Coasts of Europe; Shallow water (to 15fm.).
Halyphysema tumanowiczii Coasts of Europe; Shallow water (to 25fm.).
Hauerina fragilissima Tropical Pacific; 3–620fm.
Haynesina germanica < 50fm.
Haynesina orbicularis 7–112fm.
Hemisphaerammina bradyi Great Britain; Shallow water (25–33fm.).
Hippocrepina indivisa Arctic, N. Atlantic; 10–20fm.
Hormosina bacillaris 420–1750fm.
Hormosina normani 1380–2900fm.
Hormosina pilulifera 400–2435fm.
Hormosinella carpenteri Predominantly N. Atlantic; 350–1940fm.
Hormosinella distans 355–2775fm.
Hormosinella guttifera Atlantic; 540–1900fm.
Hormosinella ovicula 1000–3950fm.
Hyperammina friabilis 350–1425fm.

Ioanella tumidula N. Atlantic; 2740fm.

Jaculella acuta 60–2900fm.
Jaculella obtusa N. Atlantic; 350–542fm.

Karreriella bradyi 129–2050fm.
Karreriella novangliae 210–2300fm.
Karrerulina conversa 1000–3950fm.
Keramosphara murrayi S. Pacific; 1950fm.

Laculatina quadrangularis Torres Strait; 155fm.
Lagena crenata 15–2425fm.
Lagena quinquelaterata 50–2350fm.
Lagenammina ampullacea Southern Ocean; 120fm.
Lamarckina ventricosa 155–620fm.
Lamellodiscorbis sp. nov. Pacific; To 210fm.
Latentoverneuilina indiscreta Fiji; 210fm.
Laticarinina pauperata 129–2550fm.
Lenticulina atlantica Atlantic; 100–390fm.
Lenticulina calcar 95–450fm.
Lenticulina convergens 16–2740fm.
Lenticulina echinata Pacific; 95–210fm.
Lenticulina gibba < 500fm.
Lenticulina inornata Fiji; 210fm.
Lenticulina nitida Philippines; 95fm.
Lenticulina orbicularis 38–450fm.
Lenticulina papillosa 210–390fm.
Lenticulina submamilligera Pacific; 95–210fm.
Lenticulina thalmanni 345–2200fm.
Lenticulina vortex 90–435fm.
Lingulina seminuda 350–1200fm.
Lituotuba lituiformis Caribbean, S. Atlantic; 390–900fm.
Loeblichopsis cylindrica Atlantic, Southern Ocean; 1570–1750fm.
Loeblichopsis sabulosa Cold area, Faroe Channel; 530–540fm.
Loxostomina limbata 7–200fm.
Loxostomina mayori 6–420fm.
Lugdunum hantkenianum Pacific; 130–420fm.
Lugdunum schwagerianum S. Pacific; 7–155fm.

Marginopora vertebralis Tropical-subtropical; Shallow water (7–450fm.).
Marginulina obesa 15–2740fm.
Marsipella cylindrica 210–1900fm.
Marsipella elongata Predominantly N. Atlantic; 54–900fm.
Massilina amygdaloides Pacific; 147–565fm.
Massilina sp. nov. Vigo Bay, Tropical Pacific; Shallow water to 580fm.
Melonis affinis 30–3125fm.
Melonis pompilioides 1000–2421fm.
Miliolinella subrotunda Southern Ocean, Pacific; 28–150fm.
Millettia limbata Torres Strait; 155fm.
Millettia tessellata Pacific; 17–155fm.
Miniacina alba Shallow water.
Miniacina miniacea Shallow water (to 1000fm.).
Mississippina concentrica 15–1000fm.

Neoeponides auberii 2–420fm.
Neoeponides berthelotianus 16–25fm.
Neoeponides margaritifer Pacific; 10–125fm.
Neoeponides praecinctus Predominantly Pacific; 15–255fm.
Neoeponides procerus Pacific; 3–210fm.
Neoeponides schreibersii 3–345fm.
Neolenticulina variabilis 50–1630fm.
Nodellum membranaceum 1900–3950fm.
Nodobaculariella convexiuscula Tropical Pacific; 16–155fm.
Nodophthalmidium simplex Equatorial Atlantic and Pacific; 15–390fm.
Nodosaria lamnulifera; To 1400fm.
Nonion armatum Shoreline.
Nonion commune 7–1375fm.
Nonion fabum 7–200fm.
Nonionella turgida 11–1630fm.
Nubecularia lucifuga Tropical Pacific; 18fm.
Nubeculina divaricata Tropical Pacific; 18–155fm.
Nummoloculina contraria 40–2160fm.

Oolina exsculpta 800–2350fm.
Oolina ovum N. Pacific; 2300fm.
Oolina seminuda 1300–2350fm.
Oolina truncata 1825–2740fm.
Oolina variata Bass Strait; 38fm.
Operculinella cumingii Red Sea, Pacific; 10–25fm.
Oridorsalis umbonata 166–3125fm.
Orthomorphina challengeriana Ki Islands; 129fm.
Orthomorphina jedlitschkai Ki Islands; 129fm.
Orthoplecta clavata Admiralty Islands; 17fm.
Osangularia bengalensis 680–1525fm.
Osangulariella umbonifera 1450–3125fm.

Parabrizalina porrecta 37–420fm.
Parafissurina botelliformis 1450–2350fm.
Pararotalia stellata Shallow water.
Pararotalia venusta Indo-Pacific; 2–345fm.
Parrellina imperatrix Pacific; 2–10fm.
Parrellina verriculata Bass Strait; 38fm.
Parrina bradyi Southern Ocean, Pacific; 17–50fm.
Patellina corrugata 7–620fm.

Patellinella inconspicua S. Pacific; 17–345fm.
Pavonina flabelliformis Subtropical; 2–390fm.
Pelosina cylindrica 620–2900fm.
Pelosina rotundata 350–1675fm.
Pelosina variabilis 50–1250fm.
Pilulina jeffreysii N. Atlantic; 630–1476fm.
Planctostoma luculenta Atlantic; 350–675fm.
Planispirinella exigua Tropical; <25–620fm.
Planispirinoides bucculentus 20–1785fm.
Planodiscorbis rarescens Pacific; 95–155fm.
Planoglabratella opercularis 2–155fm.
Planopulvinulina dispansa Madeira, Atlantic; 390fm.
Planorbulina acervalis Red Sea, Indo-Pacific; <20–30fm.
Planorbulinella larvata Tropical Pacific; 15–210fm.
Planularia australis Pacific; 95–155fm.
Planularia cassis 90–210fm.
Planularia gemmata Pacific; 95–210fm.
Planularia magnifica Fiji; 210fm.
Planularia magnifica var. *falciformis* Atlantic; 300–1000fm.
Planularia patens Pacific; 95–155fm.
Planularia siddalliana Fiji; 210fm.
Planularia sp. nov. Bass Strait; 38fm.
Planulina ariminensis 47–2200fm.
Planulina foveolata Bermuda; 345fm.
Plectofrondicularia advena 129–1240fm.
Plectofrondicularia helenae Ki Islands; 129fm.
Pleurostomella brevis Ki Islands; 129fm.
Pleurostomella recens Ki Islands; 129fm.
Pleurostomella sp. nov. (2) 1375–2350fm.
Polysegmentina circanata Tropical Pacific; 3–11fm.
Polystomammina nitida Arctic, Atlantic, Southern Ocean; 50–200fm.
Poroeponides lateralis Shoreline to 28fm.
Praeglobobulimina ovata 15–2200fm.
Praeglobobulimina spinescens Ki Islands; 580fm.
Procerolagena distomamargaritifera 38–275fm.
Proemassilina arenaria Tropical Pacific; 95–210fm.
Protoschista findens Gaspé Bay, 15–20fm.
Psammosphaera fusca 45–2750fm.
Pseudobrizalina lobata S. Pacific; 16–155fm.
Pseudobrizalina strigosa S. Pacific; 16–155fm.
Pseudoclavulina tricarinata Raine Island; 155fm.
Pseudoflintina triquetra S. Pacific; 37–155fm.
Pseudogaudryina sp. nov. (1) 11–675fm.
Pseudoglandulina comatula 390–450fm.
Pseudoglandulina glanduliniformis 37–1360fm.
Pseudomassilina australis Shallow water.
Pseudomassilina macilenta Tropical Pacific; 17–40fm.
Pseudoolina multicosta S. Atlantic; 2350fm.
Pseudopolymorphina dawsoni 38–90fm.
Pseudopolymorphina sp. nov. 5–600fm.
Pseudorotalia schroeteriana Indo-Pacific; <60fm.
Ptychomiliola separans Tropical Pacific; 8–155fm.
Pullenia bulloides 300–2750fm.
Pullenia quinqueloba 20–2750fm.
Pyrgo comata 15–2900fm.
Pyrgo denticulata Equatorial Atlantic and Pacific; 11–610fm.
Pyrgo denticulata var. *striolata* Tropical Pacific; 6–8fm.
Pyrgo laevis Shallow water to 1215fm.

Pyrgo lucernula 300–1000fm.
Pyrgo murrhina 1180–1900fm.
Pyrgo serrata 580–1750fm.
Pyrgoella irregularis 350–1415fm.
Pyrgoella sphaera Shallow water to 1000fm.
Pyrulina fusiformis Atlantic, Southern Ocean; 808–1443fm.

Quinqueloculina auberiana Atlantic; 245–2435fm.
Quinqueloculina boueana Shoreline to shallow water.
Quinqueloculina lamarckiana Tropical Pacific; 6–95fm.
Quinqueloculina parkeri Tropical Pacific; Shallow water to 420fm.
Quinqueloculina seminulum Shallow water to 3000fm.
Quinqueloculina tropicalis Caribbean, Tropical Pacific; Shallow water (shoreline to 37fm.).

Ramulina globulifera 95–435fm.
Rectobolivina bifrons Japan; 345fm.
Rectuvigerina nicoli Southern Ocean; 150fm.
Recurvoides turbinatus 1425–2350fm.
Reophax dentaliniformis 1000–3000fm.
Reophax fusiformis 40–1443fm.
Reophax spiculifer S. Pacific; 255–620fm.
Reussella spinulosa 7–100fm.
Rhabdammina cornuta 350–1215fm.
Rhabdammina discreta 20–2475 (shallower in higher latitudes).
Rhizammina algaeformis 630–2900fm.
Robertina subcylindrica 155–1070fm.
Robertinoides oceanicus Ki Islands; 580fm.
Rosalina australis Shallow water.
Rosalina eximia Torres Strait; 155fm.
Rotalinoides gaimardii Pacific; 2–37fm.
Rupertina stabilis N. Atlantic; 5–1375fm.

Saccammina socialis 1263–2050fm.
Saccorhiza ramosa 60–3000fm.
Sagenina frondescens S. Pacific; 16–35fm.
Sagrinella jugosa Indo-Pacific; 15–155fm.
Sahulia conica Tropical; Shallow water.
Saidovina amygdalaeformis S. Pacific; 95–1070fm.
Saidovina karreriana 345–675fm.
Saidovina subangularis S. Pacific; 95–155fm.
Saracenaria latifrons 275–390fm.
Saracenaria volpicelli S. Atlantic; 1425fm.
Schlumbergerina alveoliniformis Tropical; Shoreline to 420fm.
Septotextularia rugosa Tropical; 7–30fm.
Sigmamiliolinella australis Shoreline to 150fm.
Sigmavirgulina tortuosa 2–420fm.
Sigmoidella elegantissima 7–580fm.
Sigmoidella seguenzana Pacific; 2–129fm.
Sigmoihauerina bradyi Tropical Pacific; 6–155fm.
Sigmoilina sigmoidea Atlantic; 300–900fm.
Sigmoilopsis schlumbergeri 28–1630fm.
Siphogenerina columellaris 6–1125fm.
Siphogenerina dimorpha 50–620fm.
Siphogenerina raphanus 2–260fm.
Siphogenerina striata var. *curta* 3–350fm.
Sipholagena hertwigiana 150–2600fm.
Siphonaperta crassatina S. Pacific; 38fm.

Siphonaperta sp. nov. Shallow water to 440fm.

Siphoniferoides siphoniferus Tropical; 15–40fm.

Siphoniferoides transversarius S. Pacific; 155–255fm.

Siphoninella soluta Caribbean; 390fm.

Siphoninoides echinatus 2–155fm.

Siphouvigerina ampullacea 350–675fm.

Siphouvigerina interrupta Pacific; 37–1375fm.

Siphouvigerina porrecta 12–1850fm.

Sorosphaera confusa 542–2900fm.

Sphaerogypsina globulus Littoral to 400fm.

Sphaeroidina bulloides 85–2600fm.

Spirillina decorata 6–1125fm.

Spirillina denticulata Pacific; 17–155fm.

Spirillina inaequalis Pacific; 12–155fm.

Spirillina limbata 6–1425fm.

Spirillina obconica Pacific, Southern Ocean; 17–150fm.

Spirillina tuberculata 20–400fm.

Spiroloculina angulata Tropical; Shallow water to 500fm.

Spiroloculina antillarum S. Atlantic; 350fm.

Spiroloculina communis Shallow water to 100fm.

Spiroloculina henbesti Littoral to 2000fm.

Spiroloculina robusta Caribbean; 390fm.

Spiroloculina tenuiseptata Mediterranean, Tropical Pacific; 580–1200fm.

Spiroplectammina biformis 55–1900fm. (deeper in lower latitudes).

Spiroplectammina sp. nov. Raine Island; 155fm.

Spiroplectella earlandi Raine Island, Ki Islands; 140–155fm.

Spirorutilus carinatus Philippines; 95fm.

Spirosigmoilina tenuis Shallow water to 2750fm.

Stilostomella abyssorum S. Pacific; 1825fm.

Stilostomella consobrina 129–1375fm.

Stilostomella fistuca Ki Islands; 129fm.

Stilostomella retrorsa Ki Islands; 580fm.

Storthosphaera albida Atlantic; 180–1900fm.

Subreophax aduncus 540–2900fm.

Subreophax monile S. Atlantic; 350fm.

Technitella melo 420–1215fm.

Technitella raphanus Fiji; 210fm.

Testulosiphon indivisus 210–540fm.

Textularia pseudogramen Shallow water.

Textulariella barrettii Caribbean, S. Atlantic; 100–435fm.

Tholosina bulla 1366–2160fm.

Tholosina vesicularis 630–1443fm.

Thurammina albicans S. Atlantic; 1900fm.

Thurammina compressa N. Atlantic; 630fm.

Thurammina papillata 350–2425fm.

Tortoplectella crispata S. Pacific; 155fm.

Tretomphaloides concinnus Tropical Pacific; 15–610fm.

Trifarina bradyi 12–1360fm.

Trifarina lepida N. Atlantic; 1240fm.

Trifarina reussi Ki Islands; 129fm.

Trifarina spinipes Atlantic; 110–150fm.

Triloculina bertheliniana Indian Ocean; Shoreline to shallow water (7fm.).

Triloculina terquemiana Indian Ocean (Ceylon, Madagascar); Shoreline.

Triloculina transversestriata Indo-Pacific; Shallow water to 155fm.

Triloculina tricarinata 6–2350fm.

Triloculina trigonula Shoreline to 100fm.

Triloculinella obliquinodus Shallow water.

Triloculinella sublineata Tropical Pacific; 15–25fm.

Triplasia variabilis Fiji; 210fm.

Tritaxis challengeri 13–1100fm.

Tritaxis fusca 7–600fm.

Trochammina inflata Coasts of NW Europe; Shallow water.

Trochamminella conica Shallow water to 1375fm.

Trochulina rosea Atlantic, Caribbean; Shoreline.

Tubinella funalis Southern Ocean, S. Pacific; 20–150fm.

Tubinella inornata Southern Ocean, S. Pacific; 50–150fm.

Turritellella shoneana Estuaries to abyssal plains (3350fm.).

Turritellella spectabilis N. Atlantic; 358–1900fm.

Usbekistania charoides Estuaries to abyssal plains (2452fm.).

Uvigerina aculeata 580–1900fm.

Uvigerina auberiana 37–2335fm.

Uvigerina brunnensis Pacific, Southern Ocean; 120–245fm.

Uvigerina canariensis 150–1900fm.

Uvigerina hollicki Pacific, Southern Ocean; 40–155fm.

Uvigerina schwageri Pacific; 95–210fm.

Vaginulina americana Atlantic; 15–435fm.

Vaginulina spinigera 100–1200fm.

Vaginulinopsis albatrossi Atlantic; To 630fm.

Vaginulinopsis reniformis 150–2050fm.

Vaginulinopsis subaculeata Atlantic; 390–450fm.

Vaginulinopsis tenuis 150–2350fm.

Vaginulinopsis sp. nov. 155–350fm.

Veleroninoides scitulus 400–2900fm.

Vertebralina insignis Equatorial Atlantic and Pacific; 18–390fm.

Vertebralina striata Shallow water (6–40fm.).

Vulvulina pennatula Subtropical; 350–675fm.

Appendix 4
Stratigraphic ranges of selected *Challenger* foraminifera (species arranged in alphabetic order)

Compiled from various sources.

Acerulina inhaerens Pleistocene–Recent.
Adercotryma glomeratum Middle Eocene–Recent.
Alabaminoides exiguus Miocene–Recent.
Allomorphina pacifica Miocene–Recent.
Alveolinella quoyi Middle Miocene–Recent.
Ammodiscus tenuis Eocene?–Recent.
Ammolagena clavata Late Palaeocene–Recent.
Amphicoryna scalaris Miocene–Recent.
Amphicoryna sublineata Pleistocene–Recent.
Amphistegina lessonii Miocene–Recent.
Anomalinella rostrata Miocene–Recent.
Anomalinoides colligerus Miocene–Recent.
Astacolus bradyi Miocene?–Recent.
Astacolus crepidulus Miocene–Recent.
Astacolus insolitus Miocene–Recent.
Asterorotalia pulchella Pliocene–Recent.

Baculogypsina sphaerulata Pleistocene–Recent.
Baggina indica Miocene–Recent.
Beella digitata Late Pliocene–Recent.
Bifarilamellina advena Miocene–Recent.
Bolivina robusta Miocene–Recent.
Bolivinella elegans Pleistocene–Recent.
Bolivinita quadrilatera Miocene–Recent.
Bolliella adamsi Pleistocene–Recent.
Borelis melo Middle Miocene–Recent.
Brizalina alata Miocene–Recent.
Brizalina subreticulata Miocene–Recent.
Bulimina aculeata Miocene–Recent.
Bulimina elongata Miocene–Recent.
Bulimina marginata Late Miocene–Recent.
Bulimina mexicana Miocene–Recent.
Bulimina rostrata Oligocene–Recent.
Bulimina subornata Pliocene–Recent.
Buliminoides williamsonianus Late Miocene–Recent.
Buzasina galeata Maastrichtian–Recent.

Calcarina hispida Pliocene–Recent.
Calcarina spengleri Late Pliocene–Recent.
Cancris auriculus Miocene–Recent.
Candeina nitida Late Miocene–Recent.
Cassidulina crassa Miocene–Recent.
Cassidulina laevigata var. *carinata* Middle Miocene–Recent.
Cassidulina teretis Pliocene–Recent.
Cassidulinoides bradyi Miocene–Recent.

Cellanthus craticulatus Miocene–Recent.
Ceratobulimina jonesiana Miocene–Recent.
Ceratocancris scaber Miocene–Recent.
Challengerella bradyi Pliocene–Recent.
Chilostomella oolina Miocene–Recent.
Chrysalidinella dimorpha Pliocene–Recent.
Cibicides lobatulus Middle Miocene–Recent.
Cibicides refulgens Late Miocene–Recent.
Cibicidoides cicatricosus Miocene–Recent.
Cibicidoides globulosus Oligocene–Recent.
Cibicidoides mundulus Oligocene–Recent.
Cibicidoides pachyderma Oligocene–Recent.
Cibicidoides robertsonianus Middle? Miocene–Recent.
Cibicidoides subhaidingerii Late Eocene?–Recent.
Cibicidoides wuellerstorfi Miocene–Recent.
Clavulina pacifica Pliocene–Recent.
Conglophragmium coronatum Santonian–Recent.
Cribroeponides cribrorepandus Pliocene–Recent.
Cribrostomoides subglobosus Santonian–Recent.
Cushmanina desmorphora Middle Miocene–Recent.
Cushmanina feildeniana Pliocene–Recent.
Cushmanina plumigera Miocene–Recent.
Cushmanina stelligera Pliocene–Recent.
Cyclammina cancellata Miocene–Recent.
Cyclammina rotundidorsata Eocene?–Recent.
Cycloclypeus carpenteri Pleistocene–Recent.
Cymbaloporetta bradyi Miocene?–Recent.
Cymbaloporetta plana Pliocene–Recent.
Cymbaloporetta squammosa Pliocene–Recent.
Cystammina pauciloculata Campanian–Recent.

Dentalina albatrossi Miocene–Recent.
Dentalina aphelis Miocene–Recent.
Dentalina ariena Miocene–Recent.
Dentalina elegans Miocene–Recent.
Discoanomalina semipunctata Pleistocene–Recent.
Discogypsina vesicularis Miocene–Recent.
Discorbinella bertheloti Middle? Miocene–Recent.
Dorothia scabra Middle Eocene–Recent.

Eggerella bradyi Oligocene–Recent.
Ehrenbergina hystrix Pleistocene–Recent.
Ehrenbergina pacifica Miocene–Recent.
Ehrenbergina trigona Miocene–Recent.
Elphidium crispum Miocene–Recent.
Elphidium macellum Miocene–Recent.
Eratidus foliaceus Middle Eocene–Recent.

Epistomaroides polystomelloides Pliocene–Recent.
Eubuliminella exilis Pliocene–Recent.
Evolutinella rotulata Eocene–Recent.

Favocassidulina favus Middle Miocene–Recent.
Fissurina annectens Late? Miocene–Recent.
Fissurina auriculata Miocene–Recent.
Fissurina auriculata var. *duplicata* Pliocene–Recent.
Fissurina clathrata Pliocene–Recent.
Fissurina fimbriata Miocene–Recent.
Fissurina formosa Pliocene–Recent.
Fissurina incomposita Miocene–Recent.
Fissurina laevigata Miocene–Recent.
Fissurina orbignyana Miocene–Recent.
Fissurina radiata Middle Miocene–Recent.
Fissurina radiatomarginata Pliocene–Recent.
Fissurina seminiformis Miocene–Recent.
Fissurina staphyllearia Pliocene–Recent.
Fissurina submarginata Miocene–Recent.
Flintina bradyana Miocene–Recent.
Frondicularia compta var. *villosa* Miocene–Recent.
Fursenkoina bradyi Oligocene–Recent.

Gavelinopsis lobatula Miocene–Recent.
Geminospira bradyi Pliocene–Recent.
Glabratella australensis Late Miocene–Recent.
Glandulina ovula Miocene–Recent.
Globigerina bradyi Late Oligocene–Recent.
Globigerina bulloides Middle Miocene–Recent.
Globigerinella aequilateralis Middle Miocene–Recent.
Globigerinoides conglobatus Late Miocene–Recent.
Globigerinoides pyramidalis Pleistocene–Recent.
Globigerinoides ruber Miocene–Recent.
Globigerinoides sacculifer Miocene–Recent.
Globobulimina pacifica Miocene–Recent.
Globocassidulina pacifica Miocene–Recent.
Globocassidulina subglobosa Oligocene–Recent.
Globoquadrina conglomerata Pliocene–Recent.
Globorotalia crassaformis Late Miocene–Recent.
Globorotalia fimbriata Pliocene–Recent.
Globorotalia hirsuta Late Pliocene–Recent.
Globorotalia inflata Late Pliocene–Recent.
Globorotalia menardii Middle Miocene–Recent.
Globorotalia scitula Middle Miocene–Recent.
Globorotalia truncatulinoides Pleistocene–Recent.
Globorotalia tumida Late Miocene–Recent.
Globulina gibba Miocene–Recent.
Globulina myristiformis Miocene–Recent.
Glomospira gordialis Valanginian–Recent.
Grigelis semirugosus Miocene–Recent.
Gyroidina bradyi Eocene–Recent.
Gyroidina broeckhiana Miocene–Recent.
Gyroidinoides soldanii Oligocene–Recent.

Hastigerina parapelagica Pleistocene?–Recent.
Hastigerina pelagica Late Miocene–Recent.
Hauerina fragilissima Pliocene–Recent.
Heronallenia lingulata Miocene–Recent.

Heterostegina depressa Pliocene–Recent.
Hoeglundina elegans Late Eocene–Recent.
Hormosina bacillaris Late Palaeocene–Recent.
Hormosina globulifera Late Palaeocene–Recent.
Hormosina pilulifera Eocene–Recent.
Hormosinella carpenteri Late Palaeocene–Recent.
Hormosinella distans Eocene–Recent.
Hormosinella guttifera Eocene–Recent.
Hormosinella ovicula Oligocene–Recent.
Hyalinea balthica Middle? Miocene–Recent.

Karreriella bradyi Late Palaeocene–Recent.
Karreriella chilostoma Oligocene–Recent.
Karreriella novangliae Pleistocene–Recent.
Karrerulina conversa Santonian?, Campanian–Recent.

Lagena aspera Eocene–Recent.
Lagena flexa Pliocene–Recent.
Lagena hispida Eocene–Recent.
Lagena hispidula Miocene–Recent.
Lagena laevis Pliocene–Recent.
Lagena meridionalis Miocene–Recent.
Lagena semilineata var. *spinigera* Miocene–Recent.
Lagena semistriata Miocene–Recent.
Lagena striata Miocene–Recent.
Lagena sulcata Miocene–Recent.
Lagenammina ampullacea Eocene–Recent.
Lagenammina difflugiformis Eocene–Recent.
Lamarckina ventricosa Miocene–Recent.
Laticarinina pauperata Oligocene–Recent.
Lenticulina anaglypta Miocene–Recent.
Lenticulina atlantica Miocene–Recent.
Lenticulina calcar Miocene–Recent.
Lenticulina convergens Miocene–Recent.
Lenticulina inornata Miocene?–Recent.
Lenticulina orbicularis Miocene–Recent.
Lingulina seminuda Pleistocene–Recent.
Lituotuba lituiformis Late Palaeocene–Recent.
Loxostomina limbata Middle Miocene–Recent.

Marginulina obesa Late Miocene–Recent.
Martinottiella communis Oligocene–Recent.
Melonis affinis Oligocene–Recent.
Melonis pompilioides Oligocene–Recent.
Millettia limbata Miocene–Recent.

Neoconorbina terquemi Pleistocene–Recent.
Neoeponides berthelotianus Miocene–Recent.
Neoeponides margaritifer Middle Miocene–Recent.
Neoeponides praecinctus Miocene–Recent.
Neoeponides procerus Miocene–Recent.
Neogloboquadrina blowi Miocene–Recent.
Neogloboquadrina eggeri Pleistocene–Recent.
Neogloboquadrina pachyderma Late Miocene–Recent.
Neolenticulina variabilis Oligocene–Recent.
Nodogenerina antillea Oligocene–Recent.

Oolina exsculpta Oligocene–Recent.

Oolina globosa Miocene?–Recent.
Oolina globosa var. *setosa* Miocene–Recent.
Oolina hexagona Pliocene–Recent.
Oolina squamosa Late Miocene–Recent.
Operculina complanata Oligocene–Recent.
Orbulina universa Middle Miocene–Recent.
Oridorsalis umbonata Oligocene–Recent.
Orthomorphina challengeriana Middle Miocene–Recent.
Orthomorphina jedlitschkai Pliocene–Recent.
Osangularia bengalensis Miocene–Recent.
Osangulariella umbonifera Miocene–Recent.

Parafissurina felsinea Middle Miocene–Recent.
Parafissurina lateralis Miocene?–Recent.
Pararotalia stellata Miocene–Recent.
Parasorites marginalis Pliocene–Recent.
Paratrochammina challengeri Maastrichtian–Recent.
Parvicarinina altocamerata Middle Miocene–Recent.
Parvicarinina tenuimargo Middle Miocene–Recent.
Patellina corrugata Late Miocene–Recent.
Pavonina flabelliformis Pliocene–Recent.
Planorbulina mediterranensis Late Miocene–Recent.
Planorbulinella larvata Miocene–Recent.
Planularia australis Late Miocene–Recent.
Planularia gemmata Miocene–Recent.
Planulina ariminensis Middle Miocene–Recent.
Planulina foveolata Pliocene–Recent.
Planulinoides biconcavus Pliocene–Recent.
Pleurostomella acuminata Oligocene–Recent.
Pleurostomella brevis Miocene–Recent.
Pleurostomella recens Miocene–Recent.
Pleurostomella sp. nov. (1) Miocene–Recent.
Pleurostomella sp. nov. (2) Pliocene–Recent.
Polysegmentina circanata Pliocene–Recent.
Poroeponides lateralis Pliocene–Recent.
Praeglobobulimina ovata Miocene–Recent.
Praeglobobulimina pupoides Miocene–Recent.
Praeglobobulimina spinescens Late Miocene–Recent.
Procerolagena clavata Pliocene–Recent.
Procerolagena clavata var. *setigera* Pliocene–Recent.
Procerolagena gracilis Miocene–Recent.
Procerolagena gracillima Late Miocene–Recent.
Proemassilina arenaria Late Miocene–Recent.
Psammosphaera fusca Turonian?–Recent.
Pseudobrizalina lobata Miocene?–Recent.
Pseudoclavulina tricarinata Miocene–Recent.
Pseudoglandulina glanduliniformis Miocene–Recent.
Pseudoolina multicostata Pliocene–Recent.
Pseudogaudryina atlantica Middle Miocene–Recent.
Pseudoglandulina comatula Pleistocene–Recent.
Pseudomassilina australis Pliocene–Recent.
Pseudorotalia schroeteriana Middle Miocene–Recent.
Pseudosolenina wiesneri Pliocene–Recent.
Pullenia bulloides Oligocene–Recent.
Pullenia quinqueloba Oligocene–Recent.
Pulleniatina obliquiloculata Pliocene–Recent.
Pyrgo comata Pleistocene–Recent.
Pyrgo denticulata Pliocene–Recent.

Pyrgo lucernula Middle Miocene–Recent.
Pyrgo murrhina Middle? Miocene–Recent.
Pyrgo serrata Miocene–Recent.
Pyrgoella irregularis Pleistocene–Recent.
Pyrulina gutta Pliocene–Recent.

Quinqueloculina lamarckiana Miocene–Recent.
Quinqueloculina parkeri Pliocene–Recent.
Quinqueloculina pseudoreticulata Miocene–Recent.
Quinqueloculina seminulum Miocene–Recent.
Quinqueloculina tropicalis Pliocene–Recent.
Quinqueloculina venusta Miocene–Recent.

Rectobolivina bifrons Middle Miocene–Recent.
Recurvoides turbinatus Santonian–Recent.
Reophax dentaliniformis Oligocene–Recent.
Reophax nodulosus Campanian–Recent.
Reussella spinulosa Miocene–Recent.
Rhabdammina abyssorum Turonian?–Recent.
Robertinoides oceanicus Late Miocene?–Recent.
Rosalina australis Middle Miocene–Recent.
Rosalina bradyi Late Miocene?–Recent.
Rosalina vilardeboana Middle Miocene–Recent.
Rotalinoides gaimardii Middle Miocene–Recent.

Saccammina sphaerica Eocene–Recent.
Sagrinella jugosa Late Miocene–Recent.
Sahulia conica Pliocene–Recent.
Saidovina karreriana Middle Miocene–Recent.
Saidovina subangularis Late Miocene–Recent.
Saracenaria italica Miocene–Recent.
Saracenaria latifrons Miocene–Recent.
Sigmavirgulina tortuosa Miocene–Recent.
Sigmoidella elegantissima Miocene–Recent.
Sigmoidella seguenzana Late Miocene–Recent.
Sigmoilopsis schlumbergeri Miocene–Recent.
Sigmopyrgo vespertilio Pleistocene–Recent.
Siphogenerina columellaris Middle Miocene–Recent.
Siphogenerina dimorpha Miocene–Recent.
Siphogenerina indica Late Pliocene–Recent.
Siphogenerina raphanus Miocene–Recent.
Siphogenerina striata var. *curta* Miocene?–Recent.
Siphonina bradyana Miocene–Recent.
Siphonina tubulosa Pleistocene–Recent.
Siphoninoides echinatus Miocene?–Recent.
Siphotextularia concava Late Miocene–Recent.
Siphotextularia rolshauseni Oligocene–Recent.
Siphouvigerina interrupta Middle Miocene–Recent.
Siphouvigerina porrecta Miocene–Recent.
Sorosphaera confusa Eocene–Recent.
Sphaerogypsina globulus Miocene–Recent.
Sphaeroidina bulloides Oligocene–Recent.
Sphaeroidinella dehiscens Pliocene–Recent.
Spirorutilus carinatus Eocene–Recent.
Spirosigmoilina pusilla Miocene–Recent.
Stilostomella abyssorum Oligocene–Recent.
Stilostomella consobrina Oligocene–Recent.
Subreophax aduncus Palaeocene–Recent.

Technitella legumen Eocene–Recent.
Textularia porrecta Miocene–Recent.
Thurammina papillata Late Palaeocene–Recent.
Tolypammina vagans Late Palaeocene–Recent.
Trifarina angulosa Middle Miocene–Recent.
Trifarina bradyi Miocene–Recent.
Trifarina reussi Middle Miocene–Recent.
Triloculina tricarinata Miocene–Recent.
Triloculina trigonula Eocene–Recent.
Turborotalita humilis Late Miocene–Recent.

Usbekistania charoides Valanginian–Recent.
Uvigerina aculeata Pliocene–Recent.

Uvigerina auberiana Pliocene–Recent.
Uvigerina canariensis Middle Miocene–Recent.
Uvigerina bradyana Miocene–Recent.
Uvigerina schwageri Miocene–Recent.

Vaginulinopsis subaculeata Pleistocene–Recent.
Vaginulinopsis sublegumen Late Miocene–Recent.
Valvulineria minuta Late Miocene?–Recent.
Veleroninoides scitulus Santonian?–Recent.
Verneuilinulla propinqua Late Palaeocene–Recent.
Vertebralina striata Pliocene–Recent.
Vulvulina pennatula Eocene–Recent.

References

Adams, C.G. (1960). A note on two important collections of foraminifera in the British Museum (Natural History). *Micropaleontology*, **6**, 417–18.

Adams, C.G. (1978). Great names in micropalaeontology. 3. Henry Bowman Brady, 1835–1891. In *Foraminifera*. Volume 3 (ed. R.H. Hedley and C.G. Adams), pp. 275–80. Academic Press, London.

Adams, C.G. and Frame, P. (1979). Observations on *Cycloclypeus* (*Cycloclypeus*) Carpenter and *Cycloclypeus* (*Katacycloclypeus*) Tan (Foraminiferida). *Bulletin of the British Museum (Natural History), Geology*, **32**, 3–17.

Adams, C.G., Harrison, C.A., and Hodgkinson, R.L. (1980). Some primary type specimens of foraminifera in the British Museum (Natural History). *Micropaleontology*, **26**, 1–16.

Al-Abdul Razzaq, S.K. and Bhalla, S.N. (1987). On the genotype of *Cribrospirolina* Haman, 1972. *Journal of Micropalaeontology*, **6**, 63–4.

Albani, A.D. and Yassini, I (1989). Taxonomy and distribution of shallow-water lagenid Foraminiferida from the south-eastern coast of Australia. *Australian Journal of Marine and Freshwater Research*, **40**, 369–401.

Anderson, G.J. (1963). Distribution patterns of Recent foraminifera of the Bering Sea. *Micropaleontology*, **9**, 305–17.

Anon (1990*a*). *Planularia* Defrance, 1826 (Foraminiferida): conserved. *Bulletin of Zoological Nomenclature*, **47**, 60–1.

Anon (1990*b*). *Nautilus repandus* Fichtel and Moll, 1798 (currently *Eponides repandus*; Foraminiferida): neotype replaced by rediscovered holotype. *Bulletin of Zoological Nomenclature*, **47**, 62.

Anon (1990*c*). *Calcarina* d'Orbigny, 1826 (Foraminiferida): conserved. *Bulletin of Zoological Nomenclature*, **47**, 57–8.

Anon (1990*d*). *Nonion* de Montfort, 1808 (Foraminiferida): *Nautilus faba* Fichtel and Moll, 1798 designated as the type species. *Bulletin of Zoological Nomenclature*, **47**, 53–4.

Atkinson, K. (1968). A taxonomic note on *Massilina carinata* (Fornasini, 1903). *Contributions from the Cushman Foundation for Foraminiferal Research*, **19**, 165–7.

Banner, F.T. (1966). Morphology, classification and stratigraphic significance of the Spirocyclinidae. *Voprosy̆ Mikropaleontologii*, **10**, 201–24 (in Russian).

Banner, F.T. and Blow, W.H. (1959). The classification and stratigraphical distribution of the Globigerinaceae. *Palaeontology*, **2**, 1–27.

Banner, F.T. and Blow, W.H. (1960*a*). Some primary types of species belonging to the superfamily Globigerinacea. *Contributions from the Cushman Foundation for Foraminiferal Research*, **11**, 1–41.

Banner, F.T. and Blow, W.H. (1960*b*). The taxonomy, morphology and affinities of the genera included in the subfamily Hastigerininae. *Micropaleontology*, **6**, 19–31.

Banner, F.T. and Blow, W.H. (1967). The origin, evolution and taxonomy of the foraminiferal genus *Pulleniatina* Cushman, 1927. *Micropaleontology*, **13**, 133–62.

Banner, F.T. and Clarke, W.J. (1962). Type species of foraminiferal genera. *Nature*, **196**, 1334–5.

Banner, F.T. and Culver, S.J. (1978). Quaternary *Haynesina* n. gen. and Paleogene *Protelphidium* Haynes; their morphology, affinities and distribution. *Journal of Foraminiferal Research*, **8**, 177–207.

Banner, F.T. and Pereira, C.P.G. (1981). Some biserial and triserial agglutinated smaller foraminifera: their wall structure and its significance. *Journal of Foraminiferal Research*, **11**, 85–117.

Banner, F.T., Pereira, C.P.G., and Desai, D. (1985). 'Tretomphaloid' float chambers in the Discorbidae and Cymbaloporidae. *Journal of Foraminiferal Research*, **15**, 159–74.

Banner, F.T., Simmons, M.D., and Whittaker, J.E. (1991). The Mesozoic Chrysalidinidae (Foraminifera, Textulariacea) of the Middle East; the Redmond (ARAMCO) Taxa and their relatives. *Bulletin of the British Museum (Natural History), Geology*, **47**, 101–52.

Barker, R.W. (1960). Taxonomic notes on the species figured by H.B. Brady in his report on the Foraminifera dredged by H.M.S. Challenger during the years 1873–1876. Accompanied by a reproduction of Brady's plates. *Special Publications, Society of Economic Paleontologists and Mineralogists*, **9**.

Belford, D.J. (1966). Miocene and Pliocene smaller foraminifera from Papua and New Guinea. *Bureau of Mineral Resources, Geology and Geophysics, Bulletin*, **79**.

Bender, H. and Hemleben, Ch. (1988). Constructional aspects in test formation in some agglutinated foraminifera. *Abhandlungen der Geologischen Bundesanstalt*, **41**, 13–22.

Berggren, W.A. and Kaminski, M.A. (1990). Abyssal agglutinates: back to basics. In *Paleoecology, biostratigraphy, paleoceanography and taxonomy of agglutinated foraminifera* (ed. C. Hemleben *et al.*), pp. 53–76. Kluwer, Dordrecht.

Billman, H., Hottinger, L., and Oesterle, H. (1980). Neogene to Recent rotaliid foraminifera from the Indo-Pacific Ocean; their canal system, their classification and their stratigraphic use. *Schweizerische Paläontologische Abhandlungen*, **101**, 71–113.

Blow, W.H. (1969). Late Middle Eocene to Recent planktonic foraminiferal biostratigraphy. In *Proceedings of the First International Conference on Planktonic Microfossils* (ed. P. Brönnimann and H.H. Renz), pp. 199–421. E.J. Brill, Leiden.

Blow, W.H. (1979). *The Cainozoic Globigerinida*. E.J. Brill, Leiden.

Blow. W.H. and Banner, F.T., in Eames, F.E., Banner, F.T., Blow, W.H., and Clarke, W.J. (1962). *Fundamentals of mid-tertiary stratigraphical correlation*. Cambridge University Press.

Boersma, A. (1984). *Handbook of common Tertiary Uvigerina*. Microclimates Press, Stony Point, New York.

Bogdanovich, A.K. (1969). To the revision of the Miliolidae with quinqueloculine and triloculine structure of tests. *Rocznik Polskiego Towarzstwa Geologicznego*, **39**, 351–8.

Bolli, H.M., Loeblich, A.R., Jr., and Tappan, H. (1957). Planktonic foraminiferal families Hantkeninidae, Orbulinidae, Globorotaliidae, and Globotruncaniddae. *Bulletin of the United States National Museum*, **215**, 3–50.

Boog Watson, W.N. (1967). Sir John Murray—a chronic student. *University of Edinburgh Journal*, **23**, 123–38.

Brady, H.B. (1863). Report on the Foraminifera. In Report on the dredging expedition to the Dogger Bank and the coasts of Northumberland (ed. H.T. Mennell). *Transactions of the Tyneside Naturalists' Field Club*, **5**, 291–4.

Brady, H.B. (1864*a*). Notes on foraminifera new to the British fauna. *Report of the British Association for the Advancement of Science*, **1863**, 100–1.

Brady, H.B. (1864*b*). Report on the Foraminifera. In Report of dredging operations on the coasts of Northumberland and Durham in July and August 1863 (ed. G.S. Brady). *Transactions of the Tyneside Naturalists' Field Club*, **6**, 193–4.

Brady, H.B. (1865*a*). Report on the Foraminifera. In Reports of deep sea dredging on the coasts of Northumberland and Durham, 1862–4 (ed. G.S. Brady). *Transactions of the Natural History Society of Northumberland, Durham and Newcastle-upon-Tyne, Old Series*, **1**, 51–82.

Brady, H.B. (1865*b*). A catalogue of the Recent foraminifera of Northumberland and Durham. *Transactions of the Natural History Society of Northumberland, Durham and Newcastle-upon-Tyne, Old Series*, **1**, 83–107.

Brady, H.B. (1878). On the reticularian and radiolarian Rhizopoda (Foraminifera and Polycystina) of the north polar expedition of 1875–76. *Annals and Magazine of Natural History*, Series 5, **1**, 425–40.

Brady, H.B. (1879*a*). Notes on some of the reticularian Rhizopoda of the "*Challenger*" expedition. 1. On new or little known arenaceous types. *Quarterly Journal of Microscopical Science, New Series*, **19**, 20–63.

Brady, H.B. (1879*b*). Notes on some of the reticularian Rhizopoda of the "*Challenger*" expedition. 2. Additions to the knowledge of porcellaneous and hyaline types. *Quarterly Journal of Microscopical Science, New Series*, **19**, 261–99.

Brady, H.B. (1881*a*). Notes on some of the reticularian Rhizopoda of the "*Challenger*" expedition. 3. 1—Classification, 2—Further notes on new species, 3—Note on *Biloculina* mud. *Quarterly Journal of Microscopical Science, New Series*, **21**, 31–71.

Brady, H.B. (1881*b*). On some Arctic foraminifera from soundings obtained on the Austro-Hungarian north polar expedition of 1872–1874. *Annals and Magazine of Natural History*, Series 5, **8**, 93–418.

Brady, H.B. (1881*c*). Über einige arktische Tiefsee-Foraminifern gesammelt während der österreichisch-ungarischen Nordpol-Expedition in der Jahren 1872–74. *Denkschriften der Kaiserlichen Akademie der Wissenschaften, Wien, Mathematisch-Naturwissenschaften Klasse*, **43**, 9–110.

Brady, H.B., in Tizard, Staff-Commander T.H. and Murray, J. (1882). Exploration of the Faroe Channel during the summer of 1880 in Her Majesty's Hired Ship "*Knight Errant*". *Proceedings of the Royal Society of Edinburgh*, **11**, 638–720.

Brady, H.B. (1884). Report on the foraminifera dredged by H.M.S. *Challenger* during the years 1873–1876. *Report of the scientific results of the voyage of H.M.S. Challenger, 1873–1876, Zoology*, **9**.

Brönnimann, P. and Whittaker, J.E. (1980*a*). A revision of *Reophax* and its type-species, with remarks on several other Recent hormosinid species (Protozoa: Foraminiferida) in the collections of the British Museum (Natural History). *Bulletin of the British Museum (Natural History), Zoology*, **39**, 259–72.

Brönnimann, P. and Whittaker, J.E. (1980*b*). A redescription of *Trochammina nana* (Brady) (Protozoa: Foraminiferida), with observations on several other Recent Trochamminidae in the collections of the British Museum (Natural History). *Bulletin of the British Museum (Natural History), Zoology*, **38**, 175–85.

Brönnimann, P. and Whittaker, J.E. (1983). *Zaninettia* n. gen., a spicular-walled remaneicid (Foraminiferida: Trochamminacea) from the Indian and South Atlantic Oceans, with remarks on the origin of the spicules. *Revue de Paléobiologie*, **2**, 13–33.

Brönnimann, P. and Whittaker, J.E. (1984*a*). A neotype for *Trochammina inflata* (Montagu) (Protozoa: Foraminiferida) with notes on the wall structure. *Bulletin of the British Museum (Natural History), Zoology*, **46**, 311–5.

Brönnimann, P. and Whittaker, J.E. (1984*b*). On the foraminiferal genera *Tritaxis* Schubert and *Trochamminella* Cushman (Protozoa: Foraminiferida). *Bulletin of the British Museum (Natural History), Zoology*, **46**, 291–302.

Brönnimann, P. and Whittaker, J.E. (1987). A revision of the foraminiferal genus *Adercotryma* Loeblich & Tappan, with a description of *A. wrighti* sp. nov. from British waters. *Bulletin of the British Museum (Natural History), Zoology*, **52**, 19–28.

Brönnimann, P. and Whittaker, J.E. (1988). *The Trochamminacea of the Discovery Reports*. British Museum (Natural History), London.

Brönnimann, P., Zaninetti, L., and Whittaker, J.E. (1983). On the classification of the Trochamminacea (Foraminiferida). *Journal of Foraminiferal Research*, **13**, 202–18.

Buckley, H.A., Elliott, C.J., Graham, N.M., Johnson, L.R., Kempe, D.R.C., Morgan, D.L., and Williams, D.B. (1979). *Catalogue of the ocean bottom deposits collection in the British Museum (Natural History). Part 1*. British Museum (Natural History), London.

Buckley, H.A., Elliott, C.J., Graham, N.M., Johnson, L.R., Kempe, D.R.C., Morgan, D.L., and Williams, D.B. (1984). *Catalogue of the ocean bottom deposits collection in the British Museum (Natural History). Part 2*. British Museum (Natural History), London.

Burstyn, H.L. (1968). Science and government in the nineteenth century: the *Challenger* expedition and its report. *Bulletin of the Institute of Oceanography, Monaco, Special Issue*, **21**, 603–13.

Burstyn, H.L. (1972). Pioneering in large-scale scientific organisation: the *Challenger* expedition and its report. 1. Launching the expedition. *Proceedings of the Second International Congress on the History of Oceanography, Edinburgh, 1972 (Challenger Expedition Centenary Volume)*, 47–62.

Buzas, M.A., Culver, S.J., and Isham, L.B. (1985). A comparison of fourteen elphidiid (Foraminiferida) taxa. *Journal of Paleontology*, **59**, 1075–90.

Carpenter, W.B. (1862). *Introduction to the study of the foraminifera*. Ray Society, London.

Carpenter, W.B. (1883). Report on the specimens of the genus *Orbitolites* collected by H.M.S. *Challenger* during the years 1873–1876. *Report of the Scientific Results of the Voyage of H.M.S. Challenger, 1873–1876, Zoology*, **7**.

Carpenter, W.B. and Brady, H.B. (1869). Description of *Parkeria* and *Loftusia*, two gigantic types of arenaceous foraminifera. *Philosophical Transactions of the Royal Society of London*, **159**, 721–54.

Cartwright, N.G., Gooday, A.J., and Jones, A.R. (1989). The morphology, internal organization, and taxonomic position of *Rhizammina algaeformis* Brady, a large, agglutinated, deep-sea foraminifer. *Journal of Foraminiferal Research*, **19**, 115–25.

Charnock, H. (1973). H.M.S. *Challenger* and the development of marine science. *Journal of Navigation*, **26**, 1–12.

Charnock, M.A. and Jones, R.W. (1990). Agglutinated foraminifera from the Palaeogene of the North Sea. In *Paleoecology, biostratigraphy, paleoceanography and taxonomy of agglutinated foraminifera* (ed. C. Hemleben *et al.*), pp. 139–244. Kluwer, Dordrecht.

Cherif, O.H. and Flick, H. (1974). On the taxonomic value of the wall structure of *Quinqueloculina*. *Micropaleontology*, **20**, 236–43.

Cifelli, R. and Richardson, S.L. (1990). A history of the classification of foraminifera (1826–1933). *Special publications of the Cushman Foundation for Foraminiferal Research*, **27**.

Cifelli, R. and Scott, G. (1986). Stratigraphic record of the Neogene globorotaliid radiation (planktonic foraminifera). *Smithsonian Contributions to Paleobiology*, **58**, 1–101.

Cole, F.E. (1981). Taxonomic notes on the bathyal zone benthonic foraminiferal species off Northeast Newfoundland. *Bedford Institute of Oceanography, Report*, **BI–R–81–7**.

Collins, A.C. (1958). Foraminifera. *British Museum (Natural History) Great Barrier Reef Expedition Scientific Reports*, **6**, 335–437.

Cooke, W.J. (1978). *Tubinella funalis* (Brady) as a sessile form, with notes on its distribution and wall structure. *Journal of Foraminiferal Research*, **8**, 42–5.

Cooper, S.C. (1965). A new morphologic variation of the foraminifer *Cubicides lobatulus*. *Contributions from the Cushman Foundation for Foraminiferal Research*, **16**, 137–40.

Cowans, C.F. (1971). On Guérin's *Iconographie*: particularly the insects. *Journal of the Society for the Bibliography of Natural History*, **6**, 18–29.

Crane, W. (1897). *Portraits of the contributors, reproduced from the photographs, presented by them to John Murray, naturalist on the "Challenger" expedition and editor of the "Challenger" reports* . . . Dulau, London.

Cushman, J.A. (1944). Foraminifera from the shallow water of the New England coast. *Special Publications of the Cushman Laboratory for Foraminiferal Research*, **12**.

Cushman, J.A. and Ozawa, Y. (1930). A monograph of the foraminiferal family Polymorphinidae, Recent and fossil. *Proceedings of the United States National Museum*, **77**, 1–195.

Davis, P.S. and Horne, D.J. (1985). George Stewardson Brady (1832–1921) and his collections at the Hancock Museum, Newcastle-upon-Tyne. *Journal of Micropalaeontology*, **4**, 141–52.

Deacon, M. (1971). The voyage of H.M.S. *Challenger*. In *Scientists and the Sea, 1650–1900*, pp. 333–65. Academic Press, London.

Deacon, M. (1972). The *Challenger* expedition and geology. *Proceedings of the Second International Congress on the History of Oceanography, Edinburgh, 1972 (Challenger Expedition Centenary Volume)*: 145–54.

Desai, D. and Banner, F.T. (1987). The evolution of Early Cretaceous Dorothiinae (Foraminiferida). *Journal of Micropalaeontology*, **6**, 13–28.

Deshayes, G.P. (1830). *Encyclopédie Méthodique: Histoire Naturelle des Vers*, **2**. Mme Agasse, Paris.

Deshayes, G.P. (1832). *Encyclopédie Méthodique: Histoire Naturelle des Vers*, **3**. Mme Agasse, Paris.

Douglas, R. and Sliter, W.V. (1965). Taxonomic revision of certain Discorbacea and Orbitoidacea (Foraminiferida). *Tulane Studies in Geology*, **3**, 149–64.

Echols, R.J. (1971). Distribution of foraminifera in sediments of the Scotia Sea area, Antarctic waters. *Antarctic oceanology, I.* (ed. J.L. Reid), pp. 93–168. American Geophysical Union, Washington, D.C.

Ellis, B.F. and Messina, A. (1940 *et seq*). *Catalogue of foraminifera*. American Museum of Natural History, New York.

Fornasini, C. (1906). Illustrazione di specie Orbignyane di 'Rotalidi' istitute nel 1826. *Memorie della Reale Academia della Scienza dell'Istituto di Bolognia, Sezzione dell Scienze Naturali*, Ser. 6, **3**, 61–70.

Gooday, A.J. (1983). Primitive Foraminifera and Xenophyophorea in IOS epibenthic sledge samples from the Northeast Atlantic. *Report, Institute of Oceanographic Sciences*, **156**.

Gooday, A.J. (1986). The genus *Rhabdammina* in the Northeast Atlantic: a new species, a redescription of *R. major* De Folin, 1887, and some speculations on species relationships. *Journal of Foraminiferal Research*, **16**, 150–60.

Gooday, A.J. (1988). The genus *Bathysiphon* (Protista, Foraminiferida) in the North-East Atlantic: a neotype for *B. filiformis* G.O. & M. Sars, 1872 and the description of a new species. *Journal of Natural History*, **22**, 95–105.

Gooday, A.J. and Nott, J.A. (1982). Intracellular barite crystals in two xenophyophores, *Aschemonella ramuliformis* and *Galatheammina* sp. (Protozoa: Rhizopoda) with comments on the taxonomy of *A. ramuliformis*. *Journal of the Marine Biological Association of the United Kingdom*, **62**, 595–605.

Guérin-Méneville, F.E. (1832). *Iconographie du Regne Animal de G. Cuvier* . . . (Livraison 17). Paris.

Haig, D.W. (1988). Miliolid foraminifera from inner neritic sand and mud facies of the Papuan Lagoon, New Guinea. *Journal of Foraminiferal Research*, **18**, 203–36.

Haig, D.W. (1993). Buliminid foraminifera from inner neritic sand and mud facies of the Papuan Lagoon, New Guinea. *Journal of Foraminiferal Research*, **23**, 162–79.

Haman, D. (1967). A taxonomic reinterpretation and emendation of the genus *Technitella* Norman 1878. *Contributions from the Cushman Foundation for Foraminiferal Research,* **18,** 27–30.

Haman, D. (1971). Morphologic variability of the genus *Technitella* Norman, 1878. *Micropaleontology,* **17,** 471–4.

Hansen, H.J. (1981). On Lorentz Spengler and a neotype for the foraminifer *Calcarina spengleri. Bulletin of the Geological Society of Denmark,* **29,** 191–201.

Hansen, H.J. and Revets, S. (1992). A revision and reclassification of the Discorbidae, Rosalinidae and Rotaliidae. *Journal of Foraminiferal Research,* **22,** 166–80.

Haynes, J.R. (1973a). Cardigan Bay Recent Foraminifera (cruises of the R.V. *Antur,* 1962–1964). *Bulletin of the British Museum (Natural History), Zoology, Supplement,* No. 4.

Haynes, J.R. (1973b). Further remarks on Cardigan Bay Foraminifera. *University College of Wales Department of Geology Publications,* **4.**

Haynes, J.R. (1981). *Foraminifera.* Macmillan, London.

Haynes, J.R. (1990). The classification of the foraminifera—a review of historical and philosophical perspectives. *Palaeontology,* **33,** 503–28.

Haynes, J.R. and Whittaker, J.E. (1990). The status of *Rotalia* Lamarck (Foraminifera) and of the Rotaliidae Ehrenberg. *Journal of Micropalaeontology,* **9,** 95–106.

Hayward, B.W. and Brazier, R.C. (1980). Taxonomy and distribution of present day *Bolivinella. Journal of Foraminiferal Research,* **10,** 102–16.

Hedley, R.H., Hurdle, C.M., and Burdett, I.D.J. (1964). *Trochammina squamata* Jones & Parker (Foraminifera), with observations on some closely related species. *New Zealand Journal of Science,* **7,** 417–26.

Herdman, Sir W. (1923). *Founders of oceanography and their work: an introduction to the science of the sea.* Edward Arnold, London.

Hermelin, J.O.R. (1989). Pliocene benthic foraminifera from the Ontong-Java Plateau (Western Equatorial Pacific Ocean): faunal response to changing paleoenvironment. *Special Publications of the Cushman Foundation for Foraminiferal Research,* **26.**

Hodgkinson, R.L. (1992). W.K. Parker's collection of foraminifera in the British Museum (Natural History). *Bulletin of the British Museum (Natural History), Geology,* **48,** 45–78.

Hofker, J. (1930). Foraminifera of the Siboga expedition. Part 2. Families Astrorhizidae, Rhizamminidae, Reophacidae, Anomalinidae, Peneroplidae. *Siboga Expeditie, Monographie IVa,* **2,** 79–170.

Hofker, J. (1951). Foraminifera of the Siboga expedition. Part 3. *Siboga Expeditie, Monographie IVa,* **3.**

Hofker, J. (1969). Recent foraminifera from Barbados. *Studies on the Fauna of Curaçao and other Caribbean Islands,* **31.**

Hofker, J. (1972). *Primitive agglutinated foraminifera.* E.J. Brill, Leiden.

Hofker, J. (1976). Further studies on Caribbean foraminifera. *Studies on the Fauna of Curaçao and other Caribbean Islands,* **49.**

Hooyberghs, H. and van de Sande, G. (1988). Remarques sur une collection des Foraminifères de Quelques Enchantillons Originaux de l'Expédition du "*Challenger*" (1873–1876), Conservée a Leuven (Belgique). *Revue de Micropaléontologie,* **30,** 261–66.

Hornibrook, N. de B. (1971). A revision of the Oligocene and Miocene foraminifera from New Zealand described by Karrer and Stache in the reports of the "Novara" expedition (1864). *Paleontological Bulletin, New Zealand Geological Survey,* **43.**

Hornibrook, N. de B. and Vella, P. (1954). Notes on the generic names of some rotaliform foraminifera. *The Micropaleontologist,* **8,** 24–8.

Hottinger, L. (1977). Foraminifères operculiniformes. *Mémoires, Musée National d'Histoire naturelle, Serie C, Sciences de la Terre,* **40,** 1–159.

Hottinger, L., Halicz, E., and Reiss, Z. (1990). Wall texture of *Spirorutilus. Journal of Foraminiferal Research,* **20,** 65–70.

Hottinger, L., Halicz, R., and Reiss, Z. (1991a). Architecture of *Eponides* and *Poroeponides* (Foraminifera) reexamined. *Micropaleontology,* **37,** 60–75.

Hottinger, L., Halicz, R., and Reiss, Z. (1991b). The foraminiferal genera *Pararotalia, Neorotalia* and *Calcarina*: taxonomic revision. *Journal of Paleontology,* **65,** 18–33.

Hottinger, L., Reiss, Z. and Halicz, E. (1990). Comments on *Neoeponides* (Foraminifera). *Revue de Paleobiologie,* **9,** 335–40.

Huang, T. (1964). *Rotalia* group from the Upper Cenozoic of Taiwan. *Micropaleontology,* **10,** 40–62.

Huxley, T.H. (1868a). On some organisms which live at the bottom of the North Atlantic, in depths of 6000 to 15000 feet. *Report of the British Association for the Advancement of Science,* **1868,** 102.

Huxley, T.H. (1868b). On some organisms living at great depth in the North Atlantic Ocean. *Quarterly Journal of Microscopical Science, New Series,* **8,** 203–12.

Jenkins, D.G., Whittaker, J.E., and Carlton, R. (1986). On the age and correlation of the St. Erth Beds, S.W. England, based on planktonic foraminifera. *Journal of Micropalaeontology,* **5,** 85–92.

Jones, R.W. (1984a). A revised classification of the unilocular hyaline Nodosariida and Buliminida (Foraminifera). *Revista Española de Micropaleontología,* **16,** 91–160.

Jones, R.W. (1984b). On the designation of lectotypes for certain species and subspecies of unilocular hyaline foraminifera; Part 1—those housed in the British Museum (Natural History). *Journal of Micropalaeontology,* **3,** 63–9.

Jones, R.W. (1984c). *Late Quaternary benthonic foraminifera from deep-water sites in the North-East Atlantic and Arctic Oceans.* Unpublished Doctoral Thesis, University College of Wales, Aberystwyth.

Jones, R.W. (1990). Henry Bowman Brady (1835–1891) and the *Challenger* foraminifera. *Bulletin of the British Museum (Natural History), Historical Series,* **18,** 115–43.

Jones, R.W., Bender, H., Charnock, M.A., Kaminski, M.A., and Whittaker, J.E. (1993). Emendation of the foraminiferal genus *Cribrostomoides* Cushman, 1910, and its taxonomic implications. *Journal of Micropalaeontology,* **12,** 181–93.

Joysey, K.A. (1960). Note on the Brady collection of foraminifera. *Micropaleontology,* **6,** 416.

Kaminiski, M.A. and Geroch, S. (1992). *Trochamminoides grzybowskii,* nom. nov., a new name for *Trochammina elegans* Grzybowski, 1898 (Foraminiferida). *Journal of Micropalaeontology,* **11,** 64.

Kaminiski, M.A. and Kuhnt, W. (1991). Depth-related shape variation in *Ammobaculites agglutinans* (d'Orbigny). *Annales Societatis Geologorum Poloniae,* **61,** 221–30.

Kempe, D.R.C. and Buckley, H.A. (1987). Fifty years of oceanography in the Department of Mineralogy, British Museum (Natural History). *Bulletin of the British Museum (Natural History), Historical Series,* **15,** 59–97.

Kennett, J.P. and Srinivasan, M.S. (1983). *Neogene planktonic foraminifera: a phylogenetic atlas.* Hutchinson Ross, Stroudsburg, Pennsylvania.

Kidson, C., Gilbertson, D.D., Haynes, J.R., Heyworth, A., Hughes, C.E., and Whatley, R.C. (1978). Interglacial marine deposits of the Somerset Levels, South West England. *Boreas,* **7,** 215–28.

Knight, R. (1986). Apertural characteristics of certain unilocular foraminifera: methods of study, nomenclature and taxonomic significance. *Journal of Micropalaeontology,* **5,** 37–47.

Lamb, J.L. and Miller, T.H. (1984). Stratigraphic significance of uvigerinid foraminifers in the western hemisphere. *University of Kansas Paleontological Contributions, Article,* **66.**

Langer, M.R. (1992). New recent foraminiferal genera and species from the lagoon at Madang, Papua New Guinea. *Journal of Micropalaeontology,* **11,** 85–93.

Larsen, A.R. (1976). Studies of Recent *Amphistegina*: taxonomy and some ecological aspects. *Israel Journal of Earth-Sciences,* **25,** 1–26.

Larsen, A.R. (1977). A neotype of *Amphistegina lessonii* d'Orbigny, 1826. *Journal of Foraminiferal Research,* **7,** 273–7.

Le Calvez, Y. (1977). Révision des foraminifères de la collection d'Orbigny. II. Foraminifères de l'Île de Cuba—Tome 1. *Cahiers de Micropaléontologie,* **1977,** 1–128.

Lee. S. (Ed.), 1898. *Dictionary of national biography.* Smith, Elder, London.

Leroy, D.O. and Hodgkinson, K.A., 1975. Benthonic foraminifera and some Pteropoda from a deep-water dredge sample, Northern Gulf of Mexico. *Micropaleontology,* **21,** 420–47.

Leroy, L.W. (1944a). Miocene smaller foraminifera of central Sumatra, Netherlands East Indies. *Quarterly of the Colorado School of Mines,* **39,** 9–69.

Leroy, L.W. (1944b). Small foraminifera from the Miocene of west Java, Netherlands East Indies. *Quarterly of the Colorado School of Mines,* **39,** 73–113.

Leroy, L.W. (1964). Smaller foraminifera from the Late Tertiary of southern Okinawa. *United States Geological Survey Professional Paper,* **454–F,** 1–58.

Levy, A., Mathieu, R., Poignant, A., Rosset-Moulinier, M., and Rouvillois, A. (1979). Revision de quelques genres de la famille Discorbidae (Foraminiferida) fondée sur l'observation de leur architecture interne. *Revue de Micropaléontologie,* **22,** 66–8.

Levy, A., Mathieu, R., Poignant, A., Rosset-Moulinier, M., and Rouvillois, A. (1982). Données nouvelles sur *Rotalia trochidiformis* Lamarck (Foraminiferida): émendation du genre *Rotalia* Lamarck, 1804. *Géologie Méditeranéenne,* **9,** 33–41.

Levy, A., Mathieu, R., Poignant, A., Rosset-Moulinier, M., and Rouvillois, A., 1986. Discorbidae and Rotaliidae: a classification to be revised. *Journal of Foraminiferal Research,* **16,** 63–70.

Lewis, K.B. (1979). Foraminifera on the continental shelf and slope off southern Hawke's Bay, New Zealand. *New Zealand Oceanographic Institute Memoirs,* **84,** 1–45.

Lingwood, P.F. (1981). The dispersal of the collecions of H.M.S. *Challenger*: an example of the importance of historical research in tracing a systematically important collection. In *History in the service of systematics* (ed. A. Wheeler and J.H. Price), pp. 71–7. Society for the Bibliography of Natural History (Special Publication), London.

Linklater, E. (1972). *The voyage of the Challenger.* John Murray, London.

Loeblich, A.R., Jr. and Tappan, H. (1955). Revision of some Recent foraminiferal genera. *Smithsonian Miscellaneous Collections,* **128,** 1–37.

Loeblich, A.R., Jr. and Tappan, H. (1961). The status of the foraminiferal genera *Ammodiscus* Reuss and *Involutina* Terquem. *Micropaleontology,* **7,** 189–92.

Loeblich, A.R., Jr. and Tappan, H. (1962). The status and type-species of *Calcarina, Tinoporus* and *Eponides. Contributions from the Cushman Foundation for Foraminiferal Research,* **13,** 33–8.

Loeblich, A.R., Jr. and Tappan, H. (1964). Protista 2: Sarcodina, chiefly 'Thecamoebians' and Foraminiferida. In *Treatise on Invertebrate Palaeontology.* Geological Society of America and University of Kansas Press, Lawrence, Kansas.

Loeblich, A.R., Jr. and Tappan, H. (1974). Recent advances in the classification of the foraminifera. In *Foraminifera,* Volume 1 (ed. R.H. Hedley and C.G. Adams), pp. 1–53. Academic Press, London.

Loeblich, A.R., Jr. and Tappan, H. (1981). Suprageneric revisions of some calcareous foraminifera. *Journal of Foraminiferal Research,* **11,** 159–64.

Loeblich, A.R., Jr. and Tappan, H. (1984a). Suprageneric classification of the Foraminiferida (Protozoa). *Micropaleontology,* **30,** 1–70.

Loeblich, A.R., Jr. and Tappan, H. (1984b). Some new proteinaceous and agglutinated genera of Foraminiferida. *Journal of Paleontology,* **58,** 1158–63.

Loeblich, A.R., Jr. and Tappan, H. (1985a). Some new and redefined genera and families of agglutinated foraminifera I. *Journal of Foraminiferal Research,* **15,** 91–104.

Loeblich, A.R., Jr. and Tappan, H. (1985b). Some new and redefined genera of agglutinated foraminifera II. *Journal of Foraminiferal Research,* **15,** 175–217.

Loeblich, A.R., Jr. and Tappan, H. (1985*c*). Designation of a lectotype for *Cassidulina orientalis* Cushman, 1922, the type-species of *Evolvocassidulina* Eade, 1967. *Journal of Foraminiferal Research*, **15**, 105–7.

Loeblich, A.R., Jr. and Tappan, H. (1986). Some new and redefined genera and families of Textulariina, Fusulinina, Involutinina and Miliolina (Foraminiferida). *Journal of Foraminiferal Research*, **16**, 334–46.

Loeblich, A.R., Jr. and Tappan, H. (1987). *Foraminiferal genera and their classification*. Van Nostrand Reinhold, New York.

Loeblich, A.R., Jr. and Tappan, H. (1989). Implications of wall composition and structure in agglutinated foraminifers. *Journal of Paleontology*, **63**, 769–77.

Łuczkowska, E. (1971). *Inaequalina* n. gen. (Foraminiferida, Miliolina) and its stratigraphic distribution. *Rocznik Polskiego Towarzystwa Geologicznego*, **40**, 439–43.

Łuczkowska, E. (1972). Miliolidae (Foraminiferida) from Miocene of Poland. Part I. Revision of the classification. *Acta Palaeontologica Polonica*, **17**, 341–77.

Łuczkowska, E. (1974). Miliolidae (Foraminiferida) from the Miocene of Poland. Part II. Biostratigraphy, palaeoecology and systematics. *Acta Palaeontologica Polonica*, **19**, 1–161.

Macfadyen, W.A. (1962). *Ammodiscus* Reuss, 1862 (Foraminifera); proposed designation of a type-species under the plenary powers. *Bulletin of Zoological Nomenclature*, **19**, 27–34.

McConnell, A. (1981). *Historical instruments in oceanography*. HMSO, London.

McConnell, A. (Co-Ordinator) (1990). Directory of source materials for the history of oceanography. *UNESCO Technical Papers in Marine Science*, **58**, 40pp.

McCulloch, I. (1977). *Qualitative observations on Recent foraminiferal tests with emphasis on the Eastern Pacific*. University of Southern California, Los Angeles.

McCulloch, I. (1981). *Qualitative observations on Recent foraminiferal tests. Pt. IV: with emphasis on the Allan Hancock Collection, North Atlantic*. University of Southern California, Los Angeles.

Medioli, F.S. and Scott, D.B. (1978). Emendation of the genus *Discoanomalina* Asano and its implications on the taxonomy of some of the attached foraminiferal forms. *Micropaleontology*, **24**, 291–302.

Melville, R.V. (1978). Opinion 1114, *Cornuspira* Schulze, 1854 (Foraminifera): conserved under the plenary powers. *Bulletin of Zoological Nomenclature*, **35**, 108–10.

Mendelson, V.C. (1982). Surface texture and wall structure of some Recent species of agglutinated foraminifera (Textulariina). *Journal of Paleontology*, **56**, 295–307.

Mills, E.L. (1978). Edward Forbes, John Gwyn Jeffreys, and British dredging before the *Challenger* expedition. *Journal of the Society for the Bibliography of Natural History*, **8**, 507–36.

Mills, E.L. (1984). A View of Edward Forbes, Naturalist. *Archives of Natural History*, **11**, 365–93.

Moncharmont-Zei, M. and Sgarrella, F. (1977). Nuove osservazioni sulla struttura del guscio di *Lagena benevistita* Buchner (Foraminiferida). *Bollettino della Società dei Naturalisti in Napoli*, **86**, 103–9.

Moncharmont-Zei, M. and Sgarrella, F. (1978). *Pytine parthenopeia* n. gen. et n. sp. (Nodosariidae, Foraminiferida) del Golfo de Napoli. *Bollettino della Società dei Naturalisti in Napoli*, **87**, 37–50.

Moncharmont-Zei, M. and Sgarrella, F., 1980. *Sipholagena benevistata*, nuovo nome per *Buchneria benevistata* (Buchner). *Bollettino della Società dei Naturalisti in Napoli*, **89**, 95.

Moseley, H.N. (1879). *Notes by a naturalist. An account of observations made during the voyage of H.M.S. "Challenger"*. Macmillan, London.

Murray, J.W. (1971*a*). The W.B. Carpenter collection. *Micropaleontology*, **17**, 105–6.

Murray, J.W., (1971*b*). *An atlas of British Recent Foraminiferids*. Heinemann Educational Books, London.

Murray, J.W. (1981). Some early students of foraminifera in Britain. In *Stratigraphical atlas of fossil foraminifera* (ed. D.G. Jenkins and J.W. Murray), pp. 9–12. Ellis Horwood, Chichester.

Murray, J.W. (1989). Some early students of foraminifera in Britain. In *Stratigraphical atlas of fossil foraminifera*, (2nd edn.) (ed. D.G. Jenkins and J.W. Murray), pp. 13–19. Ellis Horwood, Chichester.

Murray, J.W. and Taplin, C.M. (1984*a*). The W.B. Carpenter collection of foraminifera: a catalogue. *Journal of Micropalaeontology*, **3**, 55–8.

Murray, J.W. and Taplin, C.M. (1984*b*). Larger agglutinated foraminifera collected from the Faeroe Channel and Rockall Trough by W.B. Carpenter. *Journal of Micropalaeontology*, **3**, 59–62.

Nomura, R. (1984). Scanning electron microscopy of *Favocassidulina favus* (Brady). *Journal of Foraminiferal Research*, **14**, 93–100.

Norvang, A. (1968). Interior characteristics of *Bulimina* (Foraminifera). *Proceedings of the 23rd International Geological Congress*: 415–22.

Nuttall, W.L.F. (1927). The localities whence the foraminifera figured in the report of H.M.S. *Challenger* by Brady were derived. *Annals and Magazine of Natural History*, Series 9, **19**, 209–41.

Nuttall, W.L.F. (1931). Additional localities of the "*Challenger*" Foraminifera. *Contributions from the Cushman Laboratory for Foraminiferal Research*, **7**, 46–7.

Nyholm, K.-G. (1961). Morphogenesis and biology of *Cibicides lobatulus*. *Zoologiska Bidrag från Uppsala*, **33**, 157–96.

Nyholm, K.-G. (1973). The ultrastructure of the test in the foraminiferan *Glandulina*. *Zoon, Uppsala*, **1**, 11–15.

O'Herne, L., 1974. A reconsideration of *Amphistegina lessonii* d'Orbigny, 1826 *sensu* Brady, 1884. *Scripta Geologica*, **26**, 1–53.

d'Orbigny, A.D. (1826). Tableau méthodique de la classe des cephalopodes. *Annales des Sciences Naturelles*, Series 1, **7**, 245–314.

Papp, A. and Schmid, M.E. (1985). *Die fossilen Foraminiferen des Tertiaren Beckens von Wien. Revision der Monographie von Alcide d'Orbigny (1846)*. Geologischen Bundesanstalt, Wien.

Parker, F.L. (1962). Planktonic foraminiferal species in Pacific sediments. *Micropaleontology*, **8**, 219–54.

Parker, W.K., Jones, T.R., and Brady, H.B. (1865). On the nomenclature of the foraminifera, Pt. 12: the species enumerated by d'Orbigny in the "Annales des Sciences Naturelles", Vol. 7, 1826. *Annals and Magazine of Natural History*, Series 3, **16**, 15–41.

Parr, W.J. (1941). A new genus, *Planulinoides*, and some species of foraminifera from Southern Australia. *Mining and Geological Journal*, (N.S.), **44**, 218–34.

Patterson, R.T. (1986). *Globofissurella* and *Cerebrina*, two new foraminiferal genera in the family Lagenidae. *Journal of Micropalaeontology*, **5**, 65–9.

Patterson, R.T. and Pettis, R.H. (1986). *Galwayella*, a new foraminiferal genus and new names for two foraminiferal homonyms. *Journal of Foraminiferal Research*, **16**, 74–5.

Patterson, R.T. and Richardson, R.H. (1987). A taxonomic revision of the unilocular foraminifera. *Journal of Foraminiferal Research*, **17**, 212–26.

Patterson, R.T. and Richardson, R.H. (1988). Eight new genera of unilocular foraminifera (Lagenidae). *Transactions of the American Microscopical Society*, **107**, 240–58.

Phleger, F.B. and Parker, F.L. (1951). Ecology of foraminifera, northwest Gulf of Mexico, Pt. II. Foraminiferal species. *Memoirs of the Geological Society of America*, **46**, 1–64.

Phleger, F.B., Parker, F.L., and Peirson, J. (1953). North Atlantic foraminifera. *Reports of the Swedish Deep-Sea Expedition, 1947–1948*, **7** (1).

Poag, C.W. and Skinner, H.C. (1968). Correction of the type species of *Globulina* d'Orbigny. *Tulane Studies in Geology and Paleontology*, **6**, 127–8.

Poignant, A. (1981). Révision de quelques genres de la superfamille Miliolacea. Mise au point bibliographique. I—Les genres . . . quinque- ou tri-loculin, non-planispiralée, sans ouverture trematophorée. *Travaux du Laboratoire de Micropaléontologie de l'Universite Pierre-et-Marie Curie*, **9**, 69–107.

Poignant, A. (1984). La morphologie externe et interne des Oolininae-quelques aspects du tube entosolénien. In *Benthos '83* (ed. H. Oertli), pp. 501–9. Elf Aquitaine, Esso REP & Total CFP, Pau & Bordeaux.

Ponder, R.W. (1972). *Pseudohauerina*: a new genus of the Miliolidae and notes on three of its species. *Journal of Foraminiferal Research*, **2**, 145–56.

Ponder, R.W. (1974*a*). A rapid method of discriminating between quinqueloculine and triloculine tests with examples of its use in foraminiferal taxonomy. *Micropaleontology*, **20**, 194–6.

Ponder, R.W. (1974*b*). The foraminiferal genus *Miliolinella* and its synonyms. *Micropaleontology*, **20**, 236–43.

Popescu, G. (1983). Marine Middle Miocene monothalamous foraminifera from Romania. *Mémoires, Institut de Géologie et de Géophysique, Bucarest*, **31**, 261–80.

Rehbock, P.F. (1979). The early dredgers: "naturalising" in British Seas, 1830–1850. *Journal of the History of Biology*, **12**, 293–368.

Revets, S. (1989). *The Buliminacea (Protista, Foraminiferida): morphology, systematics and evolution*. Unpublished dissertation, Vrije Universitat, Brussel.

Revets, S. (1990*a*). The revision of *Buliminella* Cushman, 1911. *Journal of Foraminiferal Research*, **20**, 336–48.

Revets, S. (1990*b*). The revision of *Buliminoides* Cushman, 1911. *Journal of Foraminiferal Research*, **20**, 50–5.

Revets, S. (1991). The generic revision of the reussellids (Foraminiferida). *Journal of Micropalaeontology*, **10**, 1–15.

Revets, S. (1992). The structure and taxonomic position of *Millettia* Schubert, 1922 (Foraminiferida). *Journal of Micropalaeontology*, **11**, 37–46.

Revets, S. (1993*a*). The revision of the genus *Buliminellita* Cushman and Stainforth, 1947, and *Eubuliminella* gen. nov. *Journal of Foraminiferal Research*, **23**, 141–51.

Revets, S. (1993*b*). The revision of the genus *Elongobula* Finlay, 1939. *Journal of Foraminiferal Research*, **23**, 254–66.

Revets, S. and Whittaker, J.E. (1991). The taxonomic position of *Orthoplecta* Brady, 1884 (Foraminiferida). *Journal of Micropalaeontology*, **9**, 167–72.

Rice, A.L. (1983). Thomas Henry Huxley and the strange case of *Bathybius haeckelii*; a possible alternative explanation. *Archives of Natural History*, **2**, 169–80.

Rice, A.L. (1986). *British oceanographic vessels 1800–1950*. Ray Society, London.

Rice, A.L., Burstyn, H.L., and Jones, A.G.E. (1976). G.C. Wallich M.D.—megalomaniac or mis-used oceanographic genius. *Journal of the Society for the Bibliography of Natural History*, **7**, 423–50.

Ride, W.D.L., Sabrosky, C.W., Bernardi, G., and Melville, R.V. (ed.) (1985). *International Code of Zoological Nomenclature, Third Edition*. ICZN, London.

Rodriguez, C.G., Hooper, K., and Jones, P.C. (1980). The apertural structures of *Islandiella* and *Cassidulina*. *Journal of Foraminiferal Research*, **10**, 48–60.

Rögl, F. and Bolli, H.M. (1973). Holocene to Pleistocene planktonic foraminifera of Leg 15, Site 147 (Carioca Basin (Trench), Caribbean Sea) and their climatic significance. In *Initial Reports of the Deep Sea Drilling Project* (ed. N.T. Edgar *et al.*), **15**, 553–616.

Rögl, F. and Hansen, H.J. (1984). Foraminifera described by Fichtel and Moll in 1798. A revision of Testacea Microscopica. *Neue Denkschriften des Natur-Historischen Museums in Wien*, **3**.

Rouvillois, A. (1974). Un foraminifère meconnu du plateau continental du Golfe de Gascogne: *Pseudoeponides falsobeccarii* n. sp. *Cahiers de Micropaléontologie*, **1974** (3): 3–7.

Saidova, Kh. M. (1970). Benthonic foraminifera in the Kurile-Kamchatka Region based on the data of the 39th Cruise of the R/V "Vityaz". *Trudy Instituta Okeanologii*, **86**, 134–61 (in Russian).

Saidova, Kh. M. (1975). *Benthonic foraminifera of the Pacific Ocean.* Institut Okeanologii P.P. Shishova, Akademiya Nauk SSSR, Moscow (in Russian).

Saidova, Kh. M. (1981). *On an up-to-date system of supraspecific taxonomy of Cenozoic benthonic foraminifera.* Institut Okeanologii P.P. Shishova, Akademiya Nauk SSSR, Moscow (in Russian).

Saito, T., Thompson, P.R., and Breger, D. (1981). *Systematic index of Recent and Pleistocene planktonic foraminifera.* University of Tokyo Press.

Schnitker, D. (1970). Upper Miocene foraminifera from near Grimesland, Pitt County, North Carolina. *North Carolina Department of Conservation and Development, Division of Mineral Resources, Special Publication,* 3.

Schroder, C.J. (1986). Deep-water arenaceous foraminifera in the Northwest Atlantic Ocean. *Canadian Technical Report of Hydrography and Ocean Sciences,* **71**.

Schroder, C.J., Medioli, F.S., and Scott, D.B. (1989). Fragile abyssal foraminifera (including new Komokiacea) from the Nares Abyssal Plain. *Micropaleontology,* **35**, 10–48.

Seiglie, G.A. (1965). Un genero nuevo y dos especies nuevas de foraminiferos de los Testigos, Venezuela. *Boletin del Instituo Oceanografico, Universidad de Oriente,* **4**, 51–9.

Seiglie, G.A. (1969). Observaciones sobre el genero de foraminiferos *Buliminoides* Cushman. *Revista Española de Micropaleontología,* **1**, 327–32.

Seiglie, G.Λ. (1970). Additional observations on the foraminiferal genus *Buliminoides* Cushman. *Contributions from the Cushman Foundation for Foraminiferal Research,* **21**, 112–15.

Seiglie, G.A. and Baker, M.B. (1987). Duquepsamiidae, a new family, and *Duquepsammia,* a new genus of agglutinated foraminifers. *Micropaleontology,* **33**, 263–6.

Seiglie, G.A. Grove, K., and Rivera, J.A. (1977). Revision of some Caribbean Archaisinae, new genera, species and subspecies. *Ecologae Geologicae Helvetiae,* **70**, 855–83.

Sellier de Civrieux, J.M. (1976). Enmiendas a los generos *Rosalina* d'Orbigny, 1826 y *Tretomphalus* Moebius, 1880 (Familia Cymbaloporidae, Orden Foraminiferida). *Boletin del Instituto Oceanografico, Universidad de Oriente,* **15**, 177–97.

Sellier de Civrieux, J.M. (1979). Estudio sistematico y ecologio de los Bolivinitidae Recientes de Venezuela. *Cuadernos Oceanograficos, Instituto Oceanografico, Universidad de Oriente,* **5**, 1–41.

Sen Gupta, B.K. and Schafer, C.T. (1973). Holocene benthonic foraminifera in Leeward Bays of St. Lucia, West Indies. *Micropaleontology,* **19**, 342–65.

Shchedrina, Z.G. (1964). On some changes in the systematics of the order Rotaliida (Foraminifera). *Voprosy Mikropaleontologii,* **8**, 91–101 (in Russian).

Shchedrina, Z.G. (1969). On some changes in the systematics of the families Astrorhizidae and Reophacidae (Foraminifera). *Voprosy Mikropaleontologii,* **11**, 15 (in Russian).

Skinner, H.C. (1961). Revision of '*Proteonina difflugiformis*'. *Journal of Paleontology,* **35**, 1238–40.

Smith, R.K. and Isham, L.B. (1974). Reinstatement of *Mychostomina* Berthelin, 1881, and emendation of *Spirillina* Ehrenberg, 1843, Spirillininae, Spirillinidae and Spirillinacea, all Reuss, 1862. *Journal of Foraminiferal Research,* **4**, 61–8.

Smout, A.H. (1963). The Genus *Pseudedomia* and its phyletic relationships, with remarks on *Orbitolites* and other complex foraminifera. In *Evolutionary trends in foraminifera* (ed. G.H.R. von Koenigswald *et al.*), pp. 224–81. Elsevier, Amsterdam.

Smout, A.H. and Eames, F.E. (1960). The distinction between *Operculina* and *Operculinella*. *Contributions from the Cushman Foundation for Foraminiferal Research,* **11**, 109–14.

Soldani, A. (1789). *Testaceographiae ac Zoophyographiae parvae et microscopicae, Tomus Primus.* Rossi, Siena.

Soldani, A. (1798). *Testaceographiae ac Zoophyographiae parvae et microscopicae, Tomus Primi pars Tertia.* Rossi, Siena.

Srinivasan, M.S. (1966). Descriptions of new species and notes on taxonomy of foraminifera from the upper Eocene and Lower Oligocene of New Zealand. *Transactions of the Royal Society of New Zealand,* **3**, 240–1.

Srinivasan, M.S. and Sharma, V. (1969). The status of the Late Tertiary foraminifera of Kar [Car] Nicobar described by Schwager in 1866. *Micropaleontology,* **15**, 107–10.

Srinivasan, M.S. and Sharma, V. (1980). *Schwager's Car Nicobar Foraminifera in the reports of the Novara expedition—a revision.* Today & Tomorrow's Printers and Publishers, New Delhi.

Stainforth, R.M., Lamb, J.L., Luterbacher, H., Beard, J.H., and Jeffords, R.M. (1975). Cenozoic planktonic foraminiferal zonation and characteristics of index forms. *University of Kansas Paleontological Contributions,* Article, **62**.

Stainforth, R.M., Lamb, J.L., and Jeffords, R.M. (1978). *Rotalia menardii* Parker, Jones & Brady, 1865 (Foraminiferida): proposed suppression of lectotype and neotype Z.N. (S.) 2145. *Bulletin of Zoological Nomenclature,* **34**, 252–62.

Steel, J.W. (1899). *A historical sketch of the Society of Friends 'in scorn called Quakers' in Newcastle and Gateshead, 1653–1898.* G. Robinson, Newcastle-upon-Tyne.

Sverdrup, H.U., Johnson, N.W., and Fleming, R.H. (1942). *The oceans.* Prentice-Hall, Eaglewood Cliffs, New Jersey.

Taylor, S.H., Patterson, R.T., and Choi, Hyo-Won (1985). Occurrence and reliability of internal morphologic features in some Glandulinidae (Foraminiferida). *Journal of Foraminiferal Research,* **15**, 18–23.

Tendal, O.S. and Hessler, R.R. (1977). An introduction to the biology and systematics of the Komokiacea. *Galathea Report,* **14**, 165–94.

Thalmann, H.E. (1932). Nomenclator (Um- und Neubennungen) zu den Tafeln 1 bis 115 in H.B. Brady's Werk über die Foraminiferen der *Challenger*-Expedition, London 1884. *Eclogae Geologicae Helvetiae,* **25**, 293–312.

Thalmann, H.E. (1933a). Nachtrag zum Nomenclator zu Brady's Tafelband der Foraminiferen der "*Challenger*"-Expedition. *Eclogae Geologicae Helvetiae,* **26**, 251–5.

Thalmann, H.E. (1933b). Zwei Neuer Vertreter der Foraminiferen-Gattung *Rotalia* Lamarck, 1804: *R. cubana* nom. nov. und *R. trispinosa* nom. nov. *Eclogae Geologicae Helvetiae*, **26**, 248–51.

Thalmann, H.E. (1937). Mitteilungen über Foraminiferen III. Weitere Nomina Mutata in Brady's Werk über die Foraminiferen der 'Challenger'-Expedition (1884). *Eclogae Geologicae Helvetiae*, **30**, 340–2.

Thalmann, H.E. (1942). Nomina Bradyana mutata. *American Midland Naturalist*, **28**, 463–4.

Thalmann, H.E. (1950). New names and homonyms in foraminifera. *Contributions from the Cushman Foundation for Foraminiferal Research*, **1**, 41–5.

Thalmann, H.E. (1951). Mitteilungen über Foraminifera, Pt. 9. *Eclogae Geologicae Helvetiae*, **43**, 221–5.

Thalmann, H.E. (1955). New names for foraminiferal homonyms, III. *Contributions from the Cushman Foundation for Foraminiferal Research*, **6**, 82.

Thomas, F.C. (1988). Taxonomy and stratigraphy of selected Cenozoic benthic foraminifera, Canadian Atlantic margin. *Micropaleontology*, **34**, 67–82.

Tubbs, P.K. (1990). Note on *Borelis* de Montfort, 1808 (Foraminiferida) and the neotype of its type species (Case 2225/6: see BZN, 45: 116–117, 217–219). *Bulletin of Zoological Nomenclature*, **47**, 45.

van der Zwaan, G.J., Jorissen, F.J., Verhallen, P.J.J.M., and von Daniel, C.H. (ed.) (1986). Atlantic-European Oligocene to Recent *Uvigerina*. *Utrecht Micropaleontological Bulletin*, **35**.

van Marle L.J. (1991). Eastern Indonesian Late Cenozoic smaller benthic foraminifera. *Verhandelingen der Koninklijke Nederlandse Akademie van Wetenschappen, Afd. Natuurkunde, Eerste Reeks*, **34**.

van Morkhoven, F.P.C.M., Berggren, W.A., and Edwards, Anne, S. (1986). *Cenozoic cosmopolitan deep-water benthic foraminifera*. Centres de Recherches Exploration-Production Elf-Aquitaine, Pau, France.

Vella, P. (1961). Upper Oligocene and Miocene uvigerinid foraminifera from Raukumara Peninsula, New Zealand. *Micropaleontology*, **7**, 467–483.

Verhallen, P.J.J.M. (1986). Morphology and function of the internal structures of non-costate *Bulimina*. *Proceedings, Koninklijke Nederlandse Akademie van Wetenschappen*, **89**, 367–85.

Wallich, G.C. (1862). *The North-Atlantic Sea-Bed: comprising a diary of the voyage on board H.M.S. Bulldog, in 1860; and observations on the presence of animal life, and the formation and nature of organic at great depths in the ocean*. John Van Voorst, London.

Wang, P., Zhang, J., and Min, Q. (1985). Distribution of foraminifera in surface sediments of the East China Sea. In *Marine Micropaleontology of China* (ed. P. Wang *et al.*), pp. 34–69. China Ocean Press, Beijing.

Weston J.F. (1984). Wall structure of the agglutinated foraminifera *Eggerella bradyi* (Cushman) and *Karreriella bradyi* (Cushman). *Journal of Micropalaeontology*, **3**, 29–32.

Whittaker, J.E. and Hodgkinson, R.L. (1979). Foraminifera of the Togopi Formation, Eastern Sabah, Malaysia. *Bulletin of the British Museum (Natural History), Geology*, **31**, 1–120.

Wild, J.J. (1877). *Thalassa: an essay on the depth, temperature and currents of the ocean*. Marcus Ward, London.

Woodward, W.B. (1972). Sources of information on the natural history of County Durham. *The Vasculum*, **57** (supplement): 1–40.

Wyville Thomson, C. (1873). *The depths of the sea*. Macmillan, London.

Wyville Thomson, C. (1877). *The voyage of the Challenger. The Atlantic. A preliminary account of the general results of the exploring voyage of H.M.S. Challenger during the year 1873 and the early part of the year 1876*. London.

Yonge, Sir. M. (1972). The inception and significance of the *Challenger* expedition. *Proceedings of the Second International Congress on the History of Oceanography, Edinburgh, 1972 (Challenger Expedition Centenary Volume)*: 1–14.

Zheng, S.Y. (1979). The Recent foraminifera of the Xishe Islands, Guangdong Province, China, II. *Studia Marina Sinica*, **15**, 101–232.

Zheng, S.Y. (1988). *The agglutinated and porcelaneous foraminifera of the East China Sea*. China Ocean Press, Beijing.

Zweig-Strykowski, M. and Reiss, Z. (1975). Bolivinitidae from the Gulf of Elat. *Israel Journal of Earth-Science*, **24**, 97–111.

Taxonomic index

Recognized taxa indicated in **bold**. Err. cit. indicates erroneous citation. Numbers refer to plate numbers.

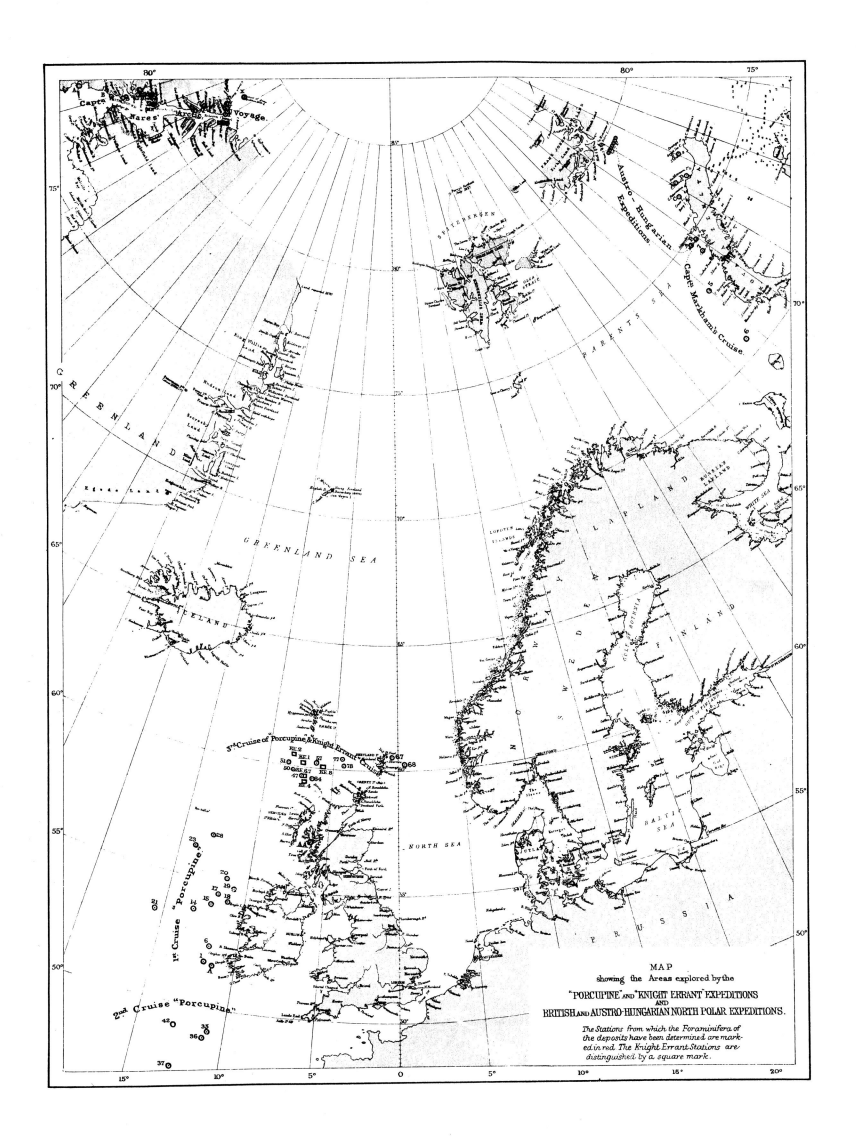

Captⁿ Nares' Arctic Voyage.

Austro-Hungarian Expeditions

Captⁿ Markham's Cruise.

GREENLAND

GREENLAND SEA

ICELAND

SPITZBERGEN

BARENTS SEA

RUSSIAN LAPLAND

WHITE SEA

N O R W A Y   S W E D E N   FINLAND

LOFOTEN ISLANDS

GULF OF BOTHNIA

BALTIC SEA

GULF OF FINLAND

3rd Cruise of Porcupine & Knight Errant.

NORTH SEA

1st Cruise "Porcupine."

2nd Cruise "Porcupine."

PRUSSIA

MAP
showing the Areas explored by the
"PORCUPINE" AND "KNIGHT ERRANT" EXPEDITIONS
AND
BRITISH AND AUSTRO-HUNGARIAN NORTH POLAR EXPEDITIONS.

The Stations from which the Foraminifera of
the deposits have been determined are marked in red. The Knight Errant Stations are
distinguished by a square mark.

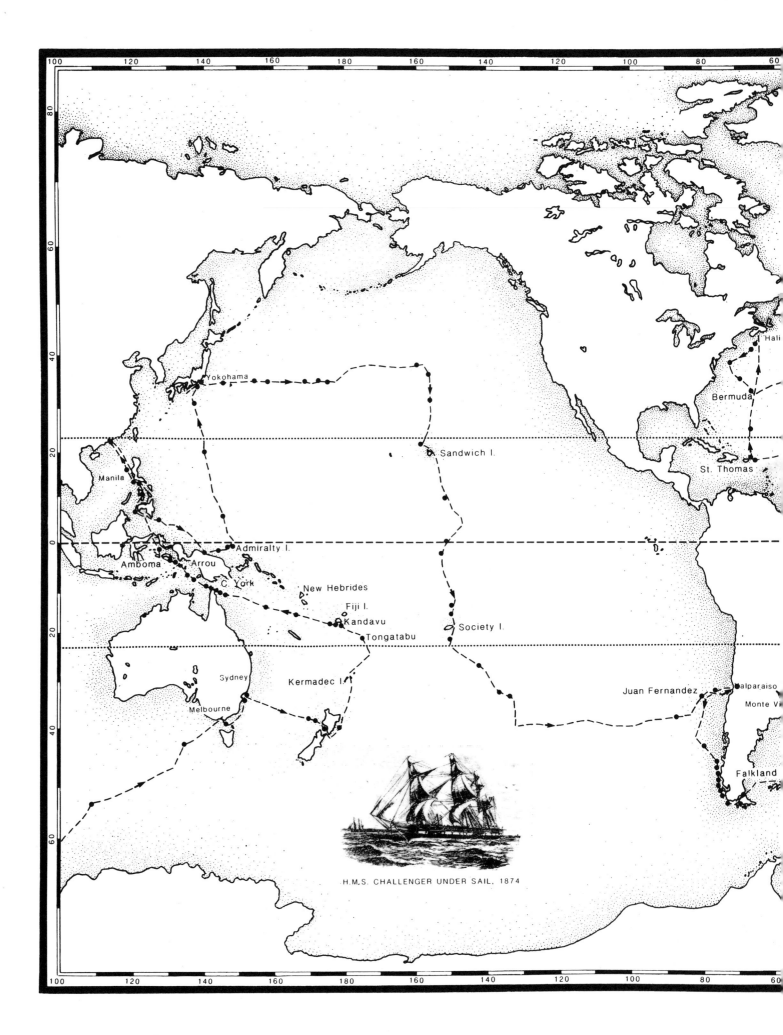

H.M.S. CHALLENGER UNDER SAIL, 1874

Yokohama

Sandwich I.

Manila

Admiralty I.

Amboma   Arrou

C. York

New Hebrides

Fiji I.

Kandavu

Tongatabu

Society I.

Sydney

Kermadec I.

Melbourne

Juan Fernandez   alparaiso

Monte V

Falkland

Hali

Bermuda

St. Thomas